CURRENT ALGEBRA AND ANOMALIES

Sam B. Treiman
Roman Jackiw
Bruno Zumino
Edward Witten

World Scientific

Published by

World Scientific Publishing Co. Pte. Ltd.
P. O. Box 128, Farrer Road, Singapore 9128

Thanks are due to the authors and North-Holland Publishing Co. for permission to reproduce the reprinted articles appearing in this volume.

Library of Congress Cataloging in Publication Data
Main entry under title:

Current algebra and anomalies.

Includes bibliographies.
1. Algebra of currents. 2. Field theory (Physics)
3. Gauge fields (Physics) I. Treiman, Sam B.
II. Jackiw, Roman III. Zumino, Bruno
IV. Witten, Edward
QC793.3.A4C87 1985 539.7'2 85-20328
ISBN 9971-966-96-4
ISBN 9971-966-97-2 (pbk.)

CURRENT ALGEBRA AND ANOMALIES

Printed in Singapore by Kim Hup Lee Printing Co. Pte. Ltd.

PREFACE

This collection comprises surveys on current algebra and anomalies. The first two articles, by Treiman and Jackiw, were lectures delivered at the 1970 Brookhaven Summer School; the next two, by Jackiw and Zumino, were lectures at the 1983 Les Houches Summer School; and the remainder consists of research papers by Bardeen and Zumino, by Witten as well as by Alvarez-Gaumé and Witten.

These days, current algebra is virtually a closed subject. Nevertheless it remains our most successful analysis of fundamental particle interactions. Since the theory is to a large extent model independent, the results have retained their validity, while the development of Quantum Chromodynamics and the electro-weak unification supplies a dynamical basis. Thus the 1970 article, written when research in this area was slowing, gives an account which is still satisfactory today.

Quantum mechanical modification of classical symmetries and algebraic relations — the so-called anomaly phenomenon — began to be appreciated in the last years of current algebra activity and has grown in importance, touching various branches of physics: fundamental particles, gravity and condensed matter. The 1970 articles emphasize the perturbative calculations which form the basis of the subject, while the two articles from 1983 trace subsequent developments, with particular attention to topological and cohomological properties of gauge fields, which provide one way of understanding the effect. Finally, in the research articles, further topological results are presented and ideas developed in connection with chiral anomalies are applied to the older subject of current algebra and to gravity theory, indicating a new direction for further investigation.

The 1983 lectures have been updated to include recent developments. There are no revisions of the 1970 material, save corrections of typographical errors, partly because the later papers provide the necessary new information. However, the reader must be informed that some subjects are incompletely discussed. "Current Algebra and PCAC" does not contain an exposition of the effective Lagrangian approach to that subject. The treatment of scale invariance in "Field Theoretic Investigations in Current Algebra" was written before the full import of the renormalization group and asymptotic freedom was appreciated. While this makes the physical discussion of scaling in deep inelastic scattering out-of-date, the formal results on anomalies in scale and conformal symmetries retain their validity.

Princeton, N. J. S. Treiman
Cambridge, MA. R. Jackiw
Berkeley, CA. B. Zumino
Princeton, N. J. E. Witten

October, 1985

CONTENTS

CURRENT ALGEBRA AND PCAC†

Sam B. Treiman
Joseph Henry Laboratories
Princeton University
Princeton, New Jersey, 08544, USA

†These lecture notes were prepared in 1970 and first appeared in *Lectures on Current Algebra and Its Applications*, by S. B. Treiman, R. Jackiw and D. J. Gross (Princeton University Press, Princeton, N. J., 1972). Much has of course transpired since then, even in the 'classic' applications of PCAC and current algebra. For example, it's good to report that the prediction embodied in Eq. (4.11) is now confirmed experimentally to about 5% rather than the 10% of 1970; the experimental results on $K_{\ell 3}$ decay are no longer in agonizing disagreement with theory (see Eq. (8.5)); weak neutral current effects *do*, after all, enter into the weak interactions (see p. 12), etc. Still, the 1970 notes are reproduced here unchanged — in the hope that the pedagogy and techniques may have some lasting usefulness.

CURRENT ALGEBRA AND PCAC†

Sam B. Treiman*
Joseph Henry Laboratories
Princeton University
Princeton, New Jersey, 08544, USA

† These lecture notes were prepared in 1971 and first appeared in *Lectures on Current Algebra and Its Applications*, by S. B. Treiman, R. Jackiw and D. J. Gross (Princeton University Press, Princeton, N.J., 1972). Much has of course transpired since then, [illegible] As [illegible] current algebra, for example, [illegible] that the prediction embodied in Eq. [illegible] is now confirmed [illegible] the experimental result [illegible] no longer [illegible] disagreement with theory [illegible] But [illegible] these notes have been left unchanged — in the hope that the pedagogy and techniques may have some lasting usefulness.

1. Introduction

The basic ideas for the subject of current algebra were introduced by Gell-Mann [1] as long ago as 1961. But the development proceeded slowly for several years, until 1965, when Fubini and Furlan [2] suggested the appropriate techniques for practical applications and Adler [3] and Weisberger [4] derived their remarkable formula relating β decay parameters to pion-nucleon scattering quantities. This inaugurated the golden age and the literature soon reflected what always happens when a good idea is perceived. In 1967 Renner [5] counted about 500 papers, and the number may well have doubled by now. Of course the number of really distinct advances in understanding is somewhat smaller than may be suggested by the counting of publications. Indeed, the major theoretical outlines have not changed much since 1967, by which time most of the "classical" applications had been worked out. During the last few years numerous variations on the earlier themes have inevitably appeared. But there is also developing an increasingly sophisticated concern about the validity of some of the formal manipulations that people have indulged in to get results. These interesting matters will be discussed by Professor Jackiw in his lectures at the Summer School.

On the experimental side the situation is still unsettled for a number of processes that may well be decisive for the subjects under discussion here. This is the case with $K_{\ell 3}$ decay, where there are lots of data but not all of it consistent; and with $K_{\ell 4}$ decay, where the data is still limited. As for the interesting and perhaps crucial applications to $\pi - \pi$ scattering and to high energy neutrino reactions, relevant experimental information is altogether lacking at present. The neutrino reactions in particular have been attracting a great deal of attention in recent times, and not only from the point of view of what current algebra has to say about them. This subject will be reviewed in the lectures to be given by Professor Gross.

The primary ingredients of current algebra are a set of equal-time commutation relations conjectured by Gell-Mann for the currents that arise in the electromagnetic and weak interactions of hadrons. As will be described and qualified more fully below, the Gell-Mann scheme may be taken to refer to commutators involving the time components of the currents. Various conjectures for the space-space commutators have also been suggested subsequently

It is especially in connection with the latter that possibly dangerous manipulations come into play in the applications. The dangers and the applications will however be left to the other lectures. In any event, commutation relations imply sum rules on the matrix elements of the operators which are involved. But the matrix elements which arise are usually not physically accessible in a practical way. It therefore takes some inventiveness to extract physically useful results; and inevitably this requires approximations or extra assumptions. Tastes can differ here! But one might well wonder what is being tested when, say, an infinite sum is truncated arbitrarily at one or two terms chosen more or less for pure convenience. On the other hand one is likely to be more charitable to the rather discrete demand that a certain dispersion integral shall merely converge, where this doesn't contradict known facts or principles. Even at the very best, however, a high order of rigor is not to be expected. Still, some topics are cleaner than others; and, even though this encroaches on Professor Gross' subject, I shall want later on to discuss the Adler sum rule for neutrino reactions, as exemplifying an especially model insensitive test of the Gell-Mann conjectures.

It will be noticed in any listing of current algebra applications that processes involving pions make a disproportionate appearance. This comes about because the commutation relation hypotheses are nicely matched to another and independent set of ideas about the weak interaction currents. These form the so-called PCAC notion of pion pole dominance for the divergence of the strangeness conserving axial vector current [6]. This too has its independent tests, i.e., independent of the ideas of current algebra, and I will want to take these up. But it is in combination that the two sets of ideas, current algebra and PCAC, display their most impressive predictive powers. A contrast must however be noted. The Gell-Mann conjectures seem to be clearly and consistently posed, so that for them the question is whether they happen to be true for the real world. The PCAC notion, on the other hand, has so far not been sharply stated; so for it the question is not only whether it's true, but – *what is it?*

These lectures will be focused mainly on concrete applications of current algebra and PCAC, with examples taken from the "classical" portion of these subjects. The question, what is PCAC, will no doubt be unduly belabored. But in general, the fare will be standard. It is addressed to people who have not yet had occasion to make themselves experts, and it is intended to serve also as introduction for the more up-to-date material which will appear in the other lecture series here. Luckily, an outstanding published review is provided in the book by Adler and Dashen: *Current Algebras* (W. A. Benjamin Publ., New York, 1968). For more general matters concerning the weak interactions, reference may be made to the book by Marshak, Riazuddin, and Ryan: *Theory of Weak Interactions In Particle Physics* (Wiley-Interscience, New York, 1969). For SU(3) matters, see Gell-Mann and Ne'eman *The Eightfold Way* (W. A. Benjamin Publ., New York, 1964).

Owing to the existence of these excellent works, the present notes will be sparing in references.

2. The Physical Currents

The Adler-Weisberger formula makes an improbable connection between strong interaction quantities and a weak interaction parameter. By generalizing the formula to cover pion scattering on various hadron targets, one can eliminate the weak interaction parameter and obtain connections between purely strong interaction quantities. This remarkable circumstance, and others like it, arises from the PCAC hypothesis concerning a certain physical weak interaction current. In retrospect it would be possible to formulate matters in such a way that no reference is made to the weak interactions: the current in question could be introduced as a purely mathematical object. However, that's not how things happened; and anyway the physical currents that constitute our subject are interesting in their own right. So we begin with a brief review of the way in which these currents arise in the present day description of the weak and electromagnetic interactions of hadrons. The Heisenberg picture is employed throughout for quantum mechanical states and operators. Our metric corresponds to

$$a \cdot b = \mathbf{a} \cdot \mathbf{b} - a_0 b_0 \qquad .$$

2.1 Electromagnetic Hadron Currents

Of all the currents to be dealt with, the electromagnetic is no doubt the most familiar. The coupling of hadrons to the electromagnetic field operator a_λ is described, to lowest order in the unit of electric charge e, by an interaction Hamiltonian density

$$\mathcal{H}^{\text{em}} = e j_\lambda^{\text{em}} a_\lambda \qquad , \tag{2.1}$$

where j_λ^{em} is the hadron electromagnetic current. It is a vector operator formed out of hadron fields in a way whose details must await a decision about fundamental matters concerning the nature of hadrons. It is in line with the trend of contemporary hadron physics to put off such a decision and instead concentrate on symmetry and other forms of characterization which are supposed to transcend dynamical details. Thus, reflecting conservation of electric charge, we assert that the charge operator

$$Q^{\mathrm{em}} = \int \mathrm{d}^3 x j_0 (\mathbf{x}, x_0)$$

is independent of time x_0 and that the current j_λ^{em} is conserved:

$$\partial j_\lambda^{\mathrm{em}} / \partial x_\lambda = 0 \quad .$$

It is customarily assumed that j_λ^{em} is odd under the charge conjugation symmetry transformation defined by the strong interactions. In connection with the discovery of CP violation it has however been suggested that the electromagnetic current may also have a piece which is *even* under charge conjugation; but for present purposes we shall overlook this still unconfirmed possibility. Electromagnetic interactions of course conserve baryon number (N) and strangeness (S) — or equivalently, hypercharge $Y = N + S$; and they conserve the third component I_3 of isotopic spin. According to the familiar formula

$$Q^{\mathrm{em}} = I_3 + Y/2$$

it is generally supposed that j_λ^{em} contains a part which transforms like a scalar $(I = 0)$ under isotopic spin rotations and a part which transforms like the third component of an isovector $(I = 1,\ \Delta I_3 = 0)$:

$$j_\lambda^{\mathrm{em}} = j_\lambda^{\mathrm{em}} (I = 0) + j_\lambda^{\mathrm{em}} (I = 1,\ \Delta I_3 = 0) \quad . \tag{2.2}$$

The two pieces are separately conserved, reflecting conservation of hypercharge and third component of isotopic spin; and the corresponding "charges" measure these quantities:

$$\frac{Y}{2} = \int j_0^{\mathrm{em}} (I = 0)\, \mathrm{d}^3 x$$

$$I_3 = \int j_0^{\mathrm{em}} (I = 1)\, \mathrm{d}^3 x \quad . \tag{2.3}$$

Consider now an electromagnetic process of the sort $\alpha \to \beta + \gamma$, where γ is a real photon of momentum k and α and β are systems of one or more hadrons. To lowest electromagnetic order the transition amplitude is given by

$$e\langle \beta | j_\lambda^{\text{em}} | \alpha \rangle \epsilon_\lambda \quad , \tag{2.4}$$

where ϵ_λ is the photon polarization vector and the states $|\alpha\rangle$ and $|\beta\rangle$ are determined purely by the strong interactions, with electromagnetism switched off. The photon process probes the structure of these states via the current operator j_λ^{em}. Here of course $k = P_\beta - P_\alpha$, $k^2 = 0$, $k \cdot \epsilon = 0$. Off mass shell ($k^2 \neq 0$) electromagnetic probes are provided in processes involving the interactions of electrons or muons with hadrons. For example, to lowest relevant order the process $\text{e} + \alpha \to \text{e} + \beta$ is described by the Feynman diagram shown below, where P_e and P'_e are the initial and final electron momenta and the virtual photon has momentum $k = P_\text{e} - P'_\text{e}$. The amplitude is given by

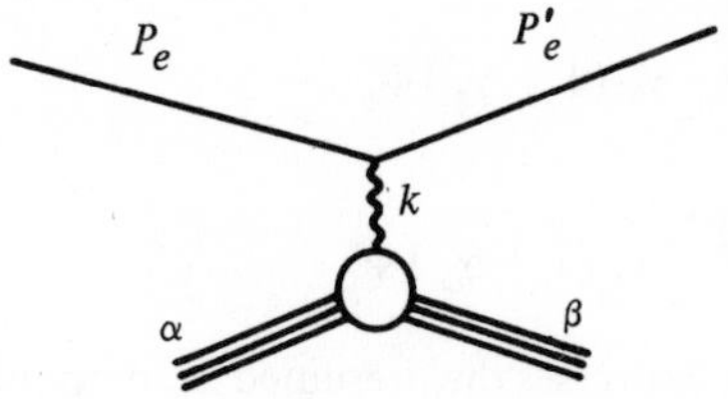

$$e^2 \langle \beta | j_\lambda^{\text{em}} | \alpha \rangle \frac{1}{k^2} \bar{u}(P'_\text{e}) \gamma_\lambda u(P_\text{e}) \quad . \tag{2.5}$$

The matrix element of the electron part of the overall electromagnetic current comes out here as a trivially known factor. All the complexities of the strong interactions are again isolated in the matrix element of the hadron current j_λ^{em}; but now, in general, $k^2 = (P_\beta - P_\alpha)^2$ is not zero. Other related variations have to do with processes such as $\alpha \to \beta + \text{e}^+ + \text{e}^-$ and, for colliding beam experiments, $\text{e}^+ + \text{e}^- \to$ hadrons. In all cases it is the hadronic matrix element that is of interest; and current conservation implies the relation

$$k_\lambda \langle \beta | j_\lambda^{\text{em}} | \alpha \rangle = 0, \qquad k = P_\beta - P_\alpha \quad .$$

2.2 *Weak Lepton Currents*

So much for electromagnetism. Concerning the weak interactions, recall that they group themselves phenomenologically into three classes. (1) Purely leptonic processes, of which muon decay $\mu^- \to e^- + \bar{\nu}_e + \nu_\mu$ and the conjugate μ^+ decay are the sole observed examples. (2) Semi-leptonic processes, i.e., those which involve hadrons and a lepton pair $(e\nu_e)$ or $(\mu\nu_\mu)$. There are many observed examples, e.g., $n \to p + e^- + \bar{\nu}_e$, $\bar{\nu}_\mu + p \to n + \mu^+$, $K \to e + \nu + \pi$, etc. (3) Non-leptonic weak processes, e.g., $\Lambda \to p + \pi^-$, $K^+ \to \pi^+ + \pi^0$, etc.

The purely leptonic muon decay reaction is well described phenomenologically to lowest order in an interaction Hamiltonian which couples an $(e\nu_e)$ "current" with a $(\mu\nu_\mu)$ "current":

$$\mathcal{H}^{\text{leptonic}} = \frac{G}{(2)^{1/2}} \mathcal{L}_\lambda(e\nu)\, \mathcal{L}_\lambda^\dagger(\mu\nu) + h.c. \tag{2.6}$$

where the current operators are given by

$$\mathcal{L}_\lambda(e\nu) = i\bar{\Psi}_{\nu_e}\gamma_\lambda(1+\gamma_5)\Psi_e$$

$$\mathcal{L}_\lambda(\mu\nu) = i\bar{\Psi}_{\nu_\mu}\gamma_\lambda(1+\gamma_5)\Psi_\mu$$

The factor $(1+\gamma_5)$ expresses the presumed 2-component nature of neutrinos and gives to each current both a vector and axial vector part. The coupling constant G has the dimension $(\text{mass})^{-2}$ and can be taken generally as a characteristic (but dimensional) measure of the strength of all classes of weak interactions. From the fit to the muon decay rate, one finds that $Gm^2 \simeq 10^{-5}$ where m is the nucleon mass.

2.3 *Weak Hadron Currents*

The observed semi-leptonic reactions seem to be well described to lowest order in an interaction Hamiltonian which effectively couples the lepton currents to a weak hadron current j_λ^{weak} which, in fact, contains both vector and axial vector parts:

$$\mathcal{H}^{\text{semi leptonic}} = \frac{G}{(2)^{1/2}} j_\lambda^{\text{weak}}\, \mathcal{L}_\lambda^\dagger + h.c. \tag{2.7}$$

where

$$\mathscr{L}_\lambda = \mathscr{L}_\lambda(e\nu) + \mathscr{L}_\lambda(\mu\nu) \quad .$$

Notice that $\mathscr{L}_\lambda(e\nu)$ and $\mathscr{L}_\lambda(\mu\nu)$ couple equally to a common hadron current, reflecting the present belief in e$-\mu$ universality. In the subsequent discussions we will let the symbol ℓ denote either e or μ; and the symbol ν will stand for the appropriate neutrino ν_e or ν_μ.

Consider now a semi-leptonic process such as $\nu + \alpha \to \beta + \ell$, where α and β are hadron systems. To lowest order the transition amplitude is given by

$$\frac{G}{(2)^{1/2}} \langle\beta|j_\lambda^{\text{weak}}|\alpha\rangle\, i\bar{u}(\ell)\gamma_\lambda(1+\gamma_5)u(\nu) \quad , \tag{2.8}$$

where the states $|\alpha\rangle$ and $|\beta\rangle$ are determined solely by the strong interactions. The structure here is similar to that of Eq. (2.5), which describes the process

$$e + \alpha \to e + \beta \quad .$$

The weak lepton current matrix element again appears as a simple and known factor, all the complexities of the strong interactions appearing in the matrix element of the hadron current. The semi-leptonic reactions probe the hadron states via the weak current operator j_λ^{weak}. For a decay process such as

$$\alpha \to \beta + \ell + \bar{\nu}$$

the structure is the same as above, except for the obvious change that the neutrino spinor $u(\nu)$ is replaced by the corresponding antineutrino spinor. Notice that on our present conventions the current j_λ^{weak} *raises* the hadron electric charge by one unit:

$$Q_\beta - Q_\alpha = +1 \quad .$$

For processes in which the hadron charge is lowered by one unit it is the matrix element of the adjoint operator $(j_\lambda^{\text{weak}})^\dagger$ that is encountered; and of course the lepton current matrix elements undergo a corresponding and obvious change.

As with the electromagnetic current, the way in which the weak current is constructed out of fundamental hadron fields cannot presently be attacked in a reliable way. The idea instead is to characterize the current in more general terms, whose implications can be extracted independently of these details. Of course detailed models can well serve here as a source of ideas, and have in fact done so in connection with the ideas of current algebra. As already said, with respect to Lorentz transformation properties, it is known that j_λ^{weak} has both a vector and an axial vector part:

$$j_\lambda^{\text{weak}} = V_\lambda + A_\lambda \ .$$

For some physical processes, e.g.,

$$K^- \to K^0 + e + \bar{\nu}$$

only the vector part contributes; for others, e.g. $\pi \to \mu + \nu$, only A_λ contributes; and for still others, e.g.

$$n \to p + e + \bar{\nu} \quad ,$$

both contribute. The currents preserve baryon number. With respect to strangeness both V_λ and A_λ have pieces which conserve strangeness $(\Delta S = 0)$ and pieces which change it by one unit, ΔS and ΔQ being correlated according to $\Delta S/\Delta Q = +1$. It may be that there are additional pieces, say with $\Delta S/\Delta Q = -1$, or $|\Delta S| > 1$; but in the absence of experimental confirmation we ignore these possibilities for the present. With respect to isospin it is certain that the $\Delta S = 0$ currents contain pieces which transform like the charge-raising part of an isovector $(I = 1)$. For the $\Delta S/\Delta Q = 1$ currents there are pieces which transform like the charge raising member of an isotopic doublet $(I = 1/2)$. Other terms, with different behavior under isotopic spin, do not at present seem to be required. For the various well established pieces we may now summarize the situation by writing

$$V_\lambda = \cos\theta_c \, V_\lambda \, (Y = 0, \ I = 1) + \sin\theta_c \, V_\lambda \, (Y = 1, \ I = 1/2)$$

$$A_\lambda = \cos\theta_c A_\lambda (Y = 0, I = 1) + \sin\theta_c A_\lambda (Y = 1, \ I = 1/2) \tag{2.9}$$

The characterization with respect to the third component of isotopic spin can be left unsymbolized, following as it does from $Q = I_3 + Y/2$. The factors $\cos\,\theta_c$ and $\sin\,\theta_c$ which appear in the above expressions are at this stage gratuitous, as is the factor $G/(2)^{1/2}$ displayed in Eq. (2.7). The point is that no "scale" has as yet been set for the various currents and these factors could all have been absorbed into the currents. Some ideas which serve to set the scales will however be discussed shortly and these will give to the angle θ_c an objective significance. For the present, let us merely assign a name to θ_c: it is the Cabibbo angle.

In connection with strangeness-conserving currents, vector and axial vector, there is one further respect in which they can be classified under operations involving isospin. This has to do with G parity, which involves a familiar isospin rotation that changes the sign of I_3, followed by the operation of charge conjugation that again reverses the sign of I_3. The $Y=0$ currents $V_\lambda(Y=0,\ I=1)$ and $A_\lambda(Y=0,\ I=1)$ are the charge raising members of isotopic triplets. The adjoints $V_\lambda^\dagger$ and $A_\lambda^\dagger$ are therefore the charge lowering members of isotopic triplets. In general, however, it need not be that V_λ and $V_\lambda^\dagger$ belong to the *same* triplet, and similarly for A_λ vs. $A_\lambda^\dagger$. But if a current and its adjoint do in fact belong to a common isotopic multiplet, then a definite G parity quantum number, $G=+1$ or -1, can be assigned to the multiplet. It is in fact known that the vector $Y=0$ current has a part which is *even* under G parity, the axial vector current a part which is *odd* under G parity. There is presently no evidence for the alternative possibilities. Notice that the *isovector* part of the electromagnetic current has $G=+1$, the *isoscalar* part $G=-1$.

The isovector part of the electromagnetic current is thus the neutral member of a $G=1$ isotopic triplet; whereas $V_\lambda(I=1)$ and $V_\lambda^\dagger(I=1)$ are the charge raising and lowering members of what in general might be a different isotopic triplet. It is clearly a tempting proposition that these triplets are in fact one and the same. This is the substance of the celebrated CVC hypothesis of Zeldovitch and Feynman and Gell-Mann. It has received some experimental confirmation and represents, insofar as it is true, a considerable simplication. The electromagnetic currents, isovector and isoscalar, are of course separately conserved. The CVC hypothesis is so-called because it entails also the conservation (C) of the strangeness-conserving weak vector current (VC): $\partial V_\lambda(I=1)/\partial x_\lambda = 0$. Moreover, corresponding matrix elements for $V_\lambda(I=1)$ and j^{em} $(I=1)$ now become related through isospin symmetry considerations. For example, for pions at rest

$$\langle \pi^+ |\, j_0^{\mathrm{em}}\,(I=1)\,| \pi^+ \rangle = 1 \quad .$$

It then follows, as is relevant for pion β-decay, that

$$\langle \pi^+ |\, V_0\,(I=1)\,| \pi^0 \rangle = (2)^{\frac{1}{2}} \quad .$$

The "scale" of the weak vector current $V_\lambda\,(I=1)$ has now acquired a sharp definition.

From the point of view of behavior under the hypercharge and isospin symmetry operations of the strong interactions, the weak and electromagnetic currents have now been characterized in as much detail as is possible. The various currents do not all preserve these symmetries, but they transform in very simple ways. The more complicated options do not seem (at present) to be taken up by nature. Although we're now done with hypercharge and isospin, we're still not finished with strong interaction symmetries. There's SU(3)! It is certainly not an outstandingly exact symmetry, but for the moment let's pretend that it is. It would then be natural to ask how the various currents behave under SU(3). For each of the currents that have been discussed the simplest possibility, consistent with the isospin and hypercharge quantum numbers, is that it transforms like a member of an SU(3) octet. Cabibbo proposed something even stronger, however [7], namely, that all of the vector currents (weak and electromagnetic) belong to a *common* octet. On this view the isoscalar electromagnetic current $(Y=0,\ I=0)$, the triplet of electromagnetic and weak strangeness conserving currents, $(Y=0,\ I=1)$, and the weak strangeness changing currents $(Y=\pm 1,\ I=1/2)$ all stand in the same relation to each other under SU(3) as η^0, $(\pi^\pm,\ \pi^0)$, $(K^+,\ K^-)$. Cabibbo similarly proposed that the various weak axial vector currents all belong to a *common* octet of axial vector operators. These ides, if true, represent an enormous simplification, which leaves only two independent and unrelated sets of objects to be contemplated, the vector and axial vector octets. Within each octet the various pieces are related by symmetry operations, which means also that the relative scales within an octet are fixed. Moreover, for the vector octet the absolute scale has already been fixed, in that this octet provides operators which measure charge and hypercharge. It would seem that there only remains to fix the scale of the axial octet relative to the vector one, something which however seems difficult to imagine for objects which have different Lorentz transformation properties. On the other hand there is already the problem that SU(3) is after all not such a good strong interaction symmetry. The question arises whether anything of these SU(3) symmetry characterizations can be given a useful meaning that survives under the breakdown of SU(3). On both of these issues Gell-Mann's current algebra conjectures have suggested solutions. We will begin our review of these matters very shortly. In the meantime it should be remarked that various applications of the Cabibbo model have been carried out on the assumption of exact SU(3) symmetry for the strong interactions; and treated in this hopeful way the model has in fact been fairly successful in correlating various baryon and meson β decays, with Cabibbo angle $\theta_c \simeq 0.23$ radians. You will no doubt hear more of this from Professor Willis.

2.4 Weak Non-Leptonic Interactions

The weak leptonic and semi-leptonic Hamiltonians each have the form of a current $\times$ current interaction. What about the non-leptonic interactions? Since reactions of this class involve particles which are all subject to strong interaction effects, there is no reliable way to read off from experiment anything of the inner structure of the non-leptonic Hamiltonian — in connection with the possibility, say, that it factors into interesting pieces. This is in contrast with the situation, say, for semi-leptonic reactions, where it can be inferred that the interaction involves the local coupling of a lepton current with a hadron current. On the other hand we can in the usual way attempt to characterize the non-leptonic Hamiltonian with respect to various strong interaction symmetries. It violates C and P invariance (and perhaps also CP symmetry); and it certainly contains pieces which violate strangeness and isotopic spin conservation. Indeed, all of the most familiar weak non-leptonic reactions involve a change of strangeness of one unit. However, recent evidence has been produced for a small degree of parity violation in certain $\Delta S = 0$ nuclear radiative transitions. The effect occurs at a strength level which roughly corresponds to the strength of the usual strangeness-changing reactions. This suggests that the weak non-leptonic interaction contains a (parity violating) $\Delta S = 0$ piece, as well as the usual $|\Delta S| = 1$ pieces. But there is no evidence at present for $|\Delta S| > 1$. With respect to isotopic spin the strangeness-changing pieces appear to transform like the members of an isotopic doublet. This is the famous $\Delta I = 1/2$ rule and it continues to hold up well experimentally, with discrepancies that are perhaps attributable to electromagnetic effects — although the effects are a bit uncomfortably large on this interpretation. In the main, however, things again seem to be simple with respect to isospin and hyperchange. The SU(3) properties of the weak non-leptonic interactions are much less clear. The simplest possibility would be that the $\Delta S = 0$, $\Delta S = 1$, and $\Delta S = -1$ pieces all have octet transformation properties.

Although there is no convincing way to draw from experiment any inferences about the "inner" structure of the non-leptonic Hamiltonian, there does exist a very attractive *theoretical* possibility. Namely, one can imagine that it is built up out of local products of the hadron currents already encountered in the semi-leptonic interactions. An even more compact possibility is that *all* the classes of weak interactions are described by a master current × current coupling. That is, let $J_\lambda = j_\lambda^{\text{weak}} + \mathscr{L}_\lambda$ be a master charge raising current composed of the previously discussed hadron and lepton pieces; and consider the Feynman, Gell-Mann [8] inclusive weak interaction Hamiltonian

$$\mathcal{H}^{\text{weak}} = \frac{G}{(2)^{1/2}} J_\lambda J_\lambda^\dagger \quad . \tag{2.10}$$

This clearly contains the purely leptonic interaction responsible for muon decay, as well as the hadron-lepton couplings responsible for weak semi-leptonic processes. But there are additional pieces: (1) *Self* coupling of the lepton currents,

$$\mathscr{L}_\lambda(e\nu)\,\mathscr{L}_\lambda^\dagger(e\nu) \text{ and } \mathscr{L}_\lambda(\mu\nu)\,\mathscr{L}_\lambda^\dagger(\mu\nu) \quad .$$

This leads to the prediction of processes such as $\nu_e + e \to \nu_e + e$, with conventional strength and structure. (2) Self coupling of the hadron currents, $j_\lambda j_\lambda^\dagger$. This describes non-leptonic interactions and includes $\Delta S = 0$ and $|\Delta S| = 1$ pieces, as wanted. There is the difficulty however that the $|\Delta S| = 1$ couplings contain not only an $I = 1/2$ part but also an $I = 3/2$ part. And with respect to SU(3), in addition to an octet part there is also a term belonging to the representation **27**.

The SU(3) troubles need not be decisive, since the experimental evidence for an octet structure is scanty; and anyhow, SU(3) is not all that good a strong interaction symmetry. But the violation of the $\Delta I = 1/2$ rule on this model could well be serious, unless one wishes to invoke dynamical accidents to quantitatively suppress the violation (or, as it is sometimes said, to enhance the $\Delta I = 1/2$ contribution). It is possible to restore the $\Delta I = 1/2$ rule and, further achieve a pure octet structure, by bringing in appropriate self couplings of the *neutral* members of the hadron current octets. These neutral currents, which have so far tagged along as mathematical objects, do not seem to figure in semi-leptonic interactions with neutral lepton currents, such as $(\nu\bar{\nu})$. To this extent the introduction of self couplings of the neutral hadron currents seems to be artificial. On the other hand, it may be a good idea to find an actual physical significance for these neutral hadron currents. Later on we will discuss some current algebra tests of this scheme.

3. Current Commutation Relations

We have by now assembled a collection of various hadron currents, vector and axial vector, which nature is supposed to employ in electromagnetic and weak semi-leptonic interactions. Each current can be labeled by an index which specifies the isospin and hypercharge quantum numbers (I, I_3, Y). Between the electromagnetic and semi-leptonic interactions, six different vector currents have been encountered; and four axial vector currents arise in the semi-leptonic interactions. Anticipating SU(3) considerations to be taken up shortly, let us complete each set to a full octet, introducing further currents which may or may not have a physical significance but which in any case will be useful as mathematical objects. Moreover it will be convenient to introduce, in place of the quantum numbers, I, I_3, Y, a single SU(3) index α which is appropriate to a real tensor basis. The vector and axial octets are denoted by the symbols V_λ^α, A_λ^α, $\alpha = 1, 2, \ldots, 8$.

In this notation the electromagnetic current is

$$j_\lambda^{\text{em}} = V_\lambda^3 + \frac{1}{(3)^{1/2}} V_\lambda^8 \quad , \tag{3.1}$$

where the first term is the isovector part of the current, the second term the isoscalar part. Similarly, the semi-leptonic hadron current is now written

$$j_\lambda^{\text{weak}} = \cos\theta_c\, j_\lambda^{1+i2} + \sin\theta_c\, j_\lambda^{4+i5} \tag{3.2}$$

and, for the charge-lowering adjoint,

$$(j_\lambda^{\text{weak}})^\dagger = \cos\theta_c\, j_\lambda^{1-i2} + \sin\theta_c\, j_\lambda^{4-i5} \quad . \tag{3.2$'$}$$

Here

$$j_\lambda^\alpha = V_\lambda^\alpha + A_\lambda^\alpha \quad . \tag{3.3}$$

and

$$j_\lambda^{1+i2} \equiv j_\lambda^1 + i j_\lambda^2 \quad , \quad \text{etc.} \tag{3.3$'$}$$

For each vector current V_λ^α let us introduce a generalized "charge" operator Q^α, according to

$$Q^\alpha(x_0) \equiv \int d^3x\, V_0^\alpha(\mathbf{x}, x_0) \quad . \tag{3.4}$$

Conservation of hypercharge in the strong interactions implies that V_λ^8 is a conserved current, hence that Q^8 is in fact a time independent operator.

Up to a factor, it is just the hypercharge operator:

$$Q^8 = \frac{(3)^{1/2}}{2} Y \quad . \tag{3.4'}$$

Similarly, on the CVC hypothesis the vector currents V_λ^1, V_λ^2, V_λ^3 are all conserved and all belong to a common isotopic triplet. The associated charges Q^i, $i = 1, 2, 3$, are therefore time independent and are nothing other than the three components of the isotopic spin operator **I**:

$$Q^i = I^i, \qquad i = 1, 2, 3 \ . \tag{3.5}$$

The Q^i are the generators of isospin symmetry (SU(2)) transformations and satisfy the familiar commutation relations

$$[Q^i, Q^j] = i\epsilon_{ijk} Q^k, \quad i, j, k = 1, 2, 3 \quad . \tag{3.6}$$

Finally, let us for the moment suppose that SU(3) symmetry breaking effects are switched off in the strong interactions. There would then be a full octet of *conserved* vector currents; and on the Cabibbo hypothesis we would identify this with the octet V_λ^α, $\alpha = 1, 2, \ldots, 8$. The members V_λ^6 and V_λ^7 do not enter physically in electromagnetic and semi-leptonic interactions, but we are entitled to carry them along mathematically. In a world with exact SU(3) symmetry for the strong interactions, all eight charges Q^α would be time independent. They would be identified as the generators of SU(3) symmetry transformations and would satisfy the corresponding commutation relations

$$[Q^\alpha, Q^\beta] = i f^{\alpha\beta\gamma} Q^\gamma \quad . \tag{3.7}$$

The SU(3) structure constants $f^{\alpha\beta\gamma}$ are totally antisymmetric in the indices. The non-vanishing $f^{\alpha\beta\gamma}$ (apart from permutations in the indices) are listed below:

$$f^{123} = 1; \ f^{147} = f^{246} = f^{257} = f^{345} = -f^{156} = -f^{367} = 1/2;$$

$$f^{458} = f^{678} = (3)^{1/2}/2 \quad . \tag{3.8}$$

In the mythical world with exact SU(3) symmetry for the strong interactions we can go still farther on Cabibbo's model. For it asserts that not only the charges Q^α but also the current *densities* $V_\lambda^\alpha(x)$ form an SU(3) octet; similarly that the axial vector current densities $A_\lambda^\alpha(x)$ form an octet. These transformation properties are reflected in the more inclusive commutation relations

$$[Q^\alpha, V_\lambda^\beta(x)] = if^{\alpha\beta\gamma} V_\lambda^\gamma(x)$$

$$[Q^\alpha, A_\lambda^\beta(x)] = if^{\alpha\beta\gamma} A_\lambda^\gamma(x) \tag{3.9}$$

In the mythical world under discussion the hadrons would have well-defined SU(3) transformation properties and symmetry relations would connect various matrix elements of the different currents. The Clebsch-Gordan coefficients appearing in these relations of course reflect the structure of the SU(3) groups, hence the structure of the commutation relations of Eq. (3.7). But apart from their technical usefulness in the derivation of the Clebsch-Gordan coefficients, the commutation relations would not be terribly interesting. We all pay passing tribute to the isospin commutation relations of Eq. (3.6) and thereafter look to the Condon-Shortley tables for practical isospin facts. The important thing, rather, is that SU(3) is not really a good strong interaction symmetry. With symmetry breaking terms switched on the hadrons cease having well-defined SU(3) transformation properties, the charges Q^α (apart from $\alpha = 1, 2, 3, 8$) now become time dependent, and the Cabibbo model loses a direct significance so far as symmetry considerations are concerned. It was Gell-Mann's idea that something of the model could nevertheless survive under SU(3) symmetry breakdown. In the first place, he proposed what at the time was a novel way of looking at things: namely, he proposed to characterize the currents in terms of their equal time commutation relations. The equal time commutation relations, formally at least, depend only on the structure of the currents when regarded as functions of the canonical field variables, the q's and p's, so to speak. If the symmetry breaking terms in the Lagrangian are reasonably decent (no derivative couplings, say), then independent of further details the currents will retain the original structure in terms of the canonical fields, and the commutation relations will correspondingly remain unchanged. It is safest to imagine all these things for the space integrated charge densities, i.e., for the charge operators. In short, Gell-Mann suggested that the *equal-time* version of Eq. (3.7) may be exact for the real world. Somewhat stronger is the conjecture that the equal-time version of Eq. (3.9) is exact. Even under the breakdown of SU(3) symmetry for the strong interactions, these conjectures suggest a precise, if abstract, meaning for the notion that the vector and axial vector currents have octet transformation properties. It also sets a relative scale for the various vector currents and similarly for the various axial vector currents. But at this stage the scale of the axial vector current octet is unrelated to that of the vector octet. The connection that is wanted here was again provided by Gell-Mann, who introduced conjectures concerning the equal-time commutators of two axial vector currents. In many ways these represent the most striking part of his scheme. To state these new hypotheses forthwith, let us introduce a set of eight "axial vector charges" $\bar{Q}^\alpha$ defined in analogy with Eq. (3.5)

$$\bar{Q}^\alpha(x_0) = \int d^3 x A_0^\alpha(\mathbf{x}, x_0) \quad . \tag{3.10}$$

Gell-Mann then postulates the equal-time commutation relations

$$[\bar{Q}^\alpha(x_0), \bar{Q}^\beta(x_0)] = i f^{\alpha\beta\gamma} \bar{Q}^\gamma(x_0) \quad . \tag{3.11}$$

Since this is bilinear in axial vector charges, and since the right-hand side contains vector charges, the relative scale becomes fixed as between the axial vector and vector quantities.

Now the equal time version of the commutation relations of Eqs. (3.7) and (3.9) could be motivated by what is at least a roughly visible symmetry of the real world and by the attractive features of the Cabbibo model based on this symmetry. It is less easy to motivate Eq. (3.11). One could do so by contemplating a larger symmetry group, SU(3) x SU(3), with generators $Q^\alpha + \bar{Q}^\alpha$ and $Q^\alpha - \bar{Q}^\alpha$. But this larger symmetry can hardly be said to be visible to the naked eye. Alternatively we might seek movitation in a model. Following Gell-Mann, suppose that the only fundamental fields are those associated with quarks. Then, just as we group, say, the proton and neutron fields in a 2 component vector in isospin space, $\psi = \begin{pmatrix} \psi_p \\ \psi_n \end{pmatrix}$, so for SU(3) let $q = \begin{pmatrix} q_p \\ q_n \\ q_\lambda \end{pmatrix}$ represent the trio of quark fields. And in analogy with the three traceless 2×2 matrices, τ^i, $i = 1, 2, 3$, which operate in the isospin space of the nucleon doublet, here introduce 8 traceless 3×3 matrices λ^α, which operate in the SU(3) space of the quarks. A basis can be chosen such that the λ^α matrices satisfy the commutation relations

$$\left[\frac{\lambda^\alpha}{2}, \frac{\lambda^\beta}{2}\right] = i f^{\alpha\beta\gamma} \frac{\lambda^\gamma}{2} \quad . \tag{3.12}$$

Let us now form vector and axial vector currents in the simplest way possible, as bilinear expressions in the quark fields:

$$V_\lambda^\alpha = i\bar{q}\gamma_\lambda \lambda^\alpha q$$

$$A_\lambda^\alpha = i\bar{q}\gamma_\lambda\gamma_5 \lambda^\alpha q \quad . \tag{3.13}$$

Of course we would be free to multiply the above expression by numerical factors, perhaps different ones for the vector and axial vector currents. But let's not! Using the canonical equal time anticommutation relations for spinor fields, together with the relations of Eq. (3.12), we can work out formally the equal-time commutation relations among the various components of the currents. I will not record the space-space commutators, but for the rest one finds the formal results

$$[V_0^\alpha(x), V_\lambda^\beta(y)]_{x_0=y_0} = if^{\alpha\beta\gamma} V_\lambda^\gamma(x)\delta(\mathbf{x}-\mathbf{y})$$

$$[V_0^\alpha(x), A_\lambda^\beta(y)]_{x_0=y_0} = if^{\alpha\beta\gamma} A_\lambda^\gamma(x)\delta(\mathbf{x}-\mathbf{y})$$

$$[A_0^\alpha(x), A_\lambda^\beta(y)]_{x_0=y_0} = if^{\alpha\beta\gamma} V_\lambda^\gamma(x)\delta(\mathbf{x}-\mathbf{y})$$

$$[A_0^\alpha(x), V_\lambda^\beta(y)]_{x_0=y_0} = if^{\alpha\beta\gamma} A_\lambda^\gamma(x)\delta(\mathbf{x}-\mathbf{y}) \quad . \tag{3.14}$$

These equations evidently contain the results of Eq. (3.11) and of the equal-time version of Eq. (3.9). But these local commutation relations are of course stronger and more far reaching. So far as I am aware, the local *time-time* relations of Eq. (3.14) are at least theoretically tenable. For the *space-time* commutators the local relations are too formal — they run afoul of the so-called Schwinger diseases, which presumably reflect the dangers of proceeding in a purely formal way with objects so singular as quantum fields. I will not enter into these delicacies, since all the applications to be discussed here can be based on the local time-time commutators, or on the presumably safe integrated versions of Eq. (3.14) (charge, current density commutators). So the Eqs. (3.14) will be allowed to stand, formally. As for the space-space commutators, I leave these to later speakers.

With the qualifications stated above, the idea now is to put the commutation relations of Eq. (3.14) forward as definite conjectures for the real world, independent of the quark model and other considerations that motivated them. The question is whether they happen to be true for the real world. So the problem is, how are the physical consequences of these abstract propositions to be extracted, ideally with a minimum of approximation or extra assumptions? In the most direct approach, one can convert the commutation relations into sum rules. In this connection it is worthwhile to recall the famous oscillator strength sum rule for electric dipole radiation in a non-relativistic quantum mechanical system. For simplicity consider a single electron moving one-dimensionally in a potential field. Let the energy eigenvalues be E_i, the corresponding eigenstates $|i\rangle$. The probability for electric dipole radiation between the levels i and j is proportional to the so-called oscillator strength, defined by

$$f_{ij} = \frac{2m}{\hbar^2}(E_j - E_i)|\langle j|x|i\rangle|^2 \quad .$$

The energy levels, eigenstates, and hence the oscillator strengths all depend on the details of the potential field in which the electron moves. Nevertheless there is a constraint on the oscillator strengths that transcends these details, and which follows from the most primitive commutation relation of all, namely

$$[x, p] = i\hbar \quad .$$

It is an elementary exercise, left to the reader, to convert this commutation relation into the sum rule

$$\sum_j f_{ji} = 1 \quad .$$

On the other hand, for the current algebra commutators the practical implications do not spring forward quite so easily. The art of extracting the physics from the current algebra conjectures will be one of our major occupations here. However, since some part of this art makes use of the notions of PCAC, let us drop current algebra for a while and turn to PCAC.

4. PCAC

The pion decay reaction $\pi \to \ell + \bar{\nu}$ plays a central role for the notion of PCAC and we must turn to this first. The process is induced solely by the strangeness-conserving axial vector current and has a very simple structure. For π^- decay, say, the amplitude is given by the expression

$$\frac{G}{(2)^{1/2}} \cos\theta_c \langle 0|A_\lambda^{1+i2}|\pi^-\rangle i\bar{u}(\ell)\gamma_\lambda(1+\gamma_5)u(\bar{\nu}) \quad . \tag{4.1}$$

With p denoting the pion 4-momentum, the hadronic matrix element must clearly have the form

$$\langle 0 | A_\lambda^{1+i2} | \pi^- \rangle = i(2)^{1/2} f_\pi p_\lambda \quad ,$$

where f_π, the so-called pion decay constant, is a parameter with dimensions of a mass. More generally, if we characterize the isospin state of the pion in a real basis, with index $i (i = 1, 2, 3)$, and recall that $\pi^- = (\pi^1 - i\pi^2)/\sqrt{2}$, then the above expression generalizes to

$$\langle 0 | A_\lambda^i | \pi^j \rangle = i f_\pi p_\lambda \delta_{ij} \quad . \tag{4.2}$$

With the symbol μ henceforth denoting the mass of the pion and m_ℓ the mass of the charged lepton, the pion decay rate works out to be

$$\text{Rate}(\pi \to \ell + \bar{\nu}) = \frac{G^2 f_\pi^2}{4\pi} \mu m_\ell^2 \left(1 - \frac{m_\ell^2}{\mu^2}\right)^2 \cos^2\theta_c \quad .$$

The experimental rate corresponds to

$$f_\pi \cos\theta_c \approx 96 \text{ MeV} \quad . \tag{4.3}$$

Since the PCAC idea will focus especially on the matrix element of the divergence of the axial vctor current, let us observe here that

$$\langle 0 | \frac{\partial A_\lambda^i}{\partial x_\lambda} | \pi^j \rangle = i p_\lambda \langle 0 | A_\lambda^i | \pi^j \rangle = \mu^2 f_\pi \delta_{ij} \quad . \tag{4.4}$$

The divergence $\partial A_\lambda^i / \partial x_\lambda$ is a pseudoscalar operator, with odd G parity, unit isospin, zero hypercharge. It has, that is, all the quantum numbers of the ith component of the pion triplet. The essential idea of PCAC is going to be that this operator is a "good" one for use in describing the creation and destruction of pions. What will be involved here, for processes involving a pion, is the notion of continuing the amplitude off mass shell in the pion mass variable. The presumed goodness of the axial vector divergence is supposed to mean that the matrix elements of this operator are very slowly varying in the mass variable as one goes a little way off shell.

These vague statements have, of course, to be elaborated. As a beginning, let us consider the reaction a + b → c + d, parameterized by the usual Mandelstam variables $s = -(p_a + p_b)^2$, $t = -(p_a - p_c)^2$. These quantities can vary continuously over some physical range. It is only over this range that the amplitude is of interest and is initially defined. However, it is known that the amplitude is in fact the boundary value of an analytic function, defined in some larger domain in the complex planes of s and t. That is, the physical amplitude admits of an analytic continuation into "unphysical" regions; and thanks to Cauchy, the continuation is unique. At the end, of course, one is interested in statements that can be made about the *physical* amplitude. But the process of discovering such statements is facilitated by contemplating the amplitude with respect to its full analytic properties. In fact, this has become a major industry.

On the other hand, the mass of a given particle cannot be varied experimentally — over any range whatsoever! So for this kind of parameter the notion of analytic continuation has no objective meaning. Cauchy is not available here. This is not to say that off mass shell amplitudes cannot be *defined* by the theoretical structures that people invent — for the ultimate purpose of describing physical processes. But it does mean that different procedures, dealing with the *same* physics, may validly give different off mass shell results. The familiar LSZ reduction formulas, which we shall be using, would seem to provide a natural basis for defining off mass shell amplitudes. But an arbitrariness is contained in the question: which operator is to be used for describing the creation and destruction of a given kind of particle. Any local operator formed from the fields of the underlying theory will do, provided that it has the appropriate quantum numbers and that it is properly normalized (see below). No doubt the purists have their additional requirements; but the point is that different choices will give the same results on mass shell, but, in general, will differ off mass shell. The "canonical" pion field, for example, even if it appears as a fundamental field of the underlying theory, need not take precedence over other choices. In fact, the PCAC enthusiasts advocate that the role of "pion" field be taken by the axial vector divergence. It can happen, as in the σ-model of Gell-Mann and Levy [9], that the underlying theory contains a canonical pion field ϕ^i and that an axial vector currents A^i_λ can be formed in such a way that the relation $\partial A^i_\lambda / \partial x_\lambda \sim \phi^i$ follows from the equations of motion. In this case the axial vector divergence *is* the canonical pion field, up to a proportionally constant. But this special circumstance does not seem to be crucial to the idea of PCAC. The operational hypothesis of PCAC is that $\partial A^i_\lambda / \partial x_\lambda$ is a gentle operator for the purpose of going off shell in the pion mass variable.

Why one *wants* to go off shell will appear in the subsequent discussions. Briefly, it is because with the PCAC mode of continuation one can make interesting statements when the off shell pion mass variable goes to zero. The hope is that these statements remain true, more or less, back on the physical mass shell.

For use in standard reduction formulas, any choice of pion field operator ϕ^i is supposed to be normalized according to

$$\langle 0|\phi^i|\pi^j\rangle = \delta_{ij} \quad .$$

If we are to make the PCAC choice, the identification must then be

$$\frac{1}{\mu^2 f_\pi} \frac{\partial A_\lambda^i}{\partial x_\lambda} \equiv \phi^i \quad . \tag{4.5}$$

To see in a more concrete way that one is free to make this identification, consider the process $\alpha \to \beta +$ (lepton pair) and focus on the axial vector contribution, as described by the matrix element $\langle\beta|A_\lambda^i|\alpha\rangle$. Let $q = p_\alpha - p_\beta$ be the momentum carried by the lepton pair. The quantity $iq_\lambda\langle\beta|A_\lambda^i|\alpha\rangle$ is the matrix element of the current divergence and let us, in fact, concentrate on this object, i.e., on

$$\langle\beta|\frac{\partial A_\lambda^i}{\partial x_\lambda}|\alpha\rangle \quad .$$

Among the Feynman graphs which contribute to this, single out those in which a single virtual pion connects between the current and the hadrons, as shown. These graphs have a pole at $q^2 = -\mu^2$, the factor $(q^2 + \mu^2)^{-1}$ arising from the pion propagator. The residue of the pole is a product of the factor $\langle 0|\partial A_\lambda^i/\partial x_\lambda|\pi^i\rangle = \mu^2 f_\pi$ and a factor

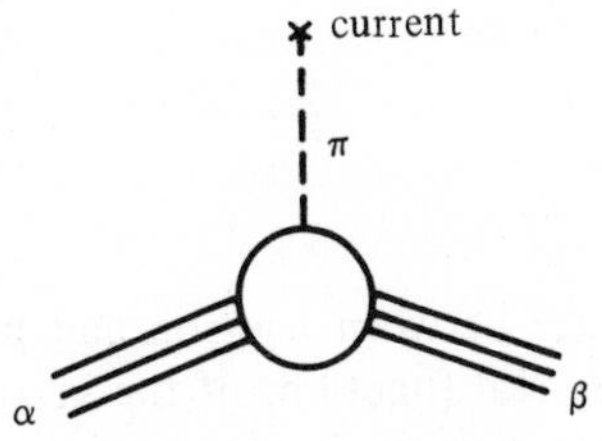

corresponding to the pion mass shell amplitude for the process $\alpha \to \beta + \pi^i$. Thus we can write

$$\langle\beta|\frac{\partial A_\lambda^i}{\partial x_\lambda}|\alpha\rangle = \frac{f_\pi \mu^2}{q^2 + \mu^2} \times \mathrm{Amp}(\alpha \to \beta + \pi^i)$$

$$+ \text{ terms regular at } (q^2 = -\mu^2) \quad .$$

In the limit $q^2 \to -\mu^2$, the pole term dominates, hence

$$\mathrm{Amp}(\alpha \to \beta + \pi^i) = \lim_{q^2 \to -\mu^2} \frac{\mu^2 + q^2}{\mu^2 f_\pi} \langle \beta | \frac{\partial A_\lambda^i}{\partial x_\lambda} | \alpha \rangle \quad . \tag{4.6}$$

But compare this with the standard LSZ formula for the same amplitude, as expressed in terms of "the pion field" ϕ^i:

$$\mathrm{Amp}(\alpha \to \beta + \pi^i) = \lim_{q^2 \to -\mu^2} (\mu^2 + q^2) \langle \beta | \phi^i | \alpha \rangle \quad .$$

4.1 Neutron β Decay

So long as we remain on mass shell, Eq. (4.6) is evidently a mere identity. Upon removing the instruction $q^2 \to -\mu^2$ we *define* an off mass shell amplitude. PCAC acquires content when we assert that the so-defined off shell amplitude is gently varying in q^2, typically over the range at least from $q^2 = -\mu^2$ to $q^2 = 0$. The notion of gentleness is not a very precise one, but let's turn to a simple application where the hypothesis takes on a rather unambiguous meaning operationally [10]. Namely, consider the β decay reaction $\mathrm{n} \to \mathrm{p} + \mathrm{e}^- + \bar{\nu}_\mathrm{e}$ and focus in particular on the axial vector matrix element. The most general structure is given by

$$\langle \mathrm{p} | A_\lambda^{1+i2} | \mathrm{n} \rangle = -i\bar{u}_\mathrm{p} \{ g_\mathrm{A}(q^2) \gamma_\lambda \gamma_5 + i G_\mathrm{A}(q^2) q_\lambda \gamma_5 \} u_\mathrm{n} \quad , \tag{4.7}$$

where

$$q = n - p \quad ,$$

is the momentum transfer between neutron and proton. The form factors $g_\mathrm{A}(q^2)$ and $G_\mathrm{A}(q^2)$ are real functions of the invariant momentum transfer variable q^2. In physical β decay it varies over a small range near $q^2 \approx 0$. For the matrix element of the divergence of the axial vector current we have, using the Dirac equation,

$$\langle \mathrm{p} | \frac{\partial A_\lambda^{1+i2}}{\partial x_\lambda} | \mathrm{n} \rangle = i q_\lambda \langle \mathrm{p} | A_\lambda^{1+i2} | \mathrm{n} \rangle$$

$$= i [2m g_\mathrm{A}(q^2) - q^2 G_\mathrm{A}(q^2)] \, \bar{u}_\mathrm{p} \gamma_5 u_\mathrm{n} \quad . \tag{4.8}$$

With the PCAC identification of pion field, however, we can write

$$\langle \mathrm{p} | \frac{\partial A_\lambda^{1+i2}}{\partial x_\lambda} | \mathrm{n} \rangle = (2)^{1/2} \mu^2 f_\pi \langle \mathrm{p} | \frac{\phi^{1+i2}}{(2)^{1/2}} | \mathrm{n} \rangle \quad . \tag{4.9}$$

The matrix element on the right defines the pion nucleon vertex function $g_r(q^2)$, according to

$$\langle \mathrm{p} | \frac{\phi^{1+i2}}{(2)^{1/2}} | \mathrm{n} \rangle = (2)^{1/2} \frac{g_r(q^2)}{\mu^2 + q^2} \, i\bar{u}_\mathrm{p} \gamma_5 u_\mathrm{n} \quad . \tag{4.10}$$

The vertex function $g_r(q^2)$ has an objective meaning at $q^2 = -\mu^2$, where it becomes the pion-nucleon coupling constant. Off shell, i.e., for $q^2 \neq -\mu^2$, the vertex function does not have an objective significance. It depends on the choice of pion operator used to effect the off shell continuation. Here we have made the PCAC choice. In fact, so far we have done nothing with content. Physical content enters when we conjecture that with the PCAC way of going off mass-shell, $g_r(q^2)$ is a gently varying function. In particular, let us choose $q^2 = 0$ and suppose that $g_r(0) \approx g_r(-\mu^2) \equiv g_r$. From Eqs. (4.8) to (4.10) we then find

$$f_\pi = \frac{m g_\mathrm{A}(0)}{g_r} \quad . \tag{4.11}$$

With $g_\mathrm{A}(0) \simeq 1.22$, $g_r^2/4\pi \approx 14.6$ we find $f_\pi \simeq 87$ MeV, in reasonably good agreement with the empirical value ≈ 96 MeV. More generally, Eqs. (4.8) to (4.10) lead to the relation

$$2m g_\mathrm{A}(q^2) - q^2 G_\mathrm{A}(q^2) = \frac{2g_r(q^2)}{\mu^2 + q^2} \mu^2 f_\pi$$

$$= 2g_r(q^2) f_\pi - \frac{2q^2 g_r(q^2) f_\pi}{\mu^2 + q^2} \quad .$$

The pole on the right-hand side is associated with the form factor $G_\mathrm{A}(q^2)$ on the left side. If $g_r(q^2)$ is slowly varying, as we assume say for $-\mu^2 \leqslant q^2 \leqslant 0$, and if also $g_\mathrm{A}(q^2)$ is slowly varying, then

$$G_\mathrm{A}(0) \simeq \frac{2m g_\mathrm{A}(0)}{\mu^2} \quad . \tag{4.12}$$

We have noted this result because the term proportional to G_A in Eq. (4.7) is responsible for the so-called pseudoscalar coupling in β decay and in muon capture. This has been detected for muon capture in hydrogen and the sign and magnitude of G_A seem to be roughly right.

4.2 PCAC for Neutrino Reactions

Another remarkable application of the PCAC hypothesis was first noticed by Adler in connection with high energy neutrino reactions [11]. These processes are of enormous interest from many points of view and we will return to some of these other aspects later on. Here it is PCAC that is in view. Consider a neutrino-induced reaction of the sort

$$\nu + \mathrm{p} \rightarrow \ell + \beta \qquad ,$$

where β is some multihadron system with zero strangeness and where the target particle for definiteness is taken to be a proton. The amplitude is given by

$$M = \frac{G}{(2)^{1/2}} \langle \beta | j_\lambda^{\text{weak}} | \mathrm{p} \rangle i\bar{u}_\ell \gamma_\lambda (1 + \gamma_5) u_\nu \qquad . \tag{4.13}$$

Let $q = p_\nu - p_\ell$, where p_ν is the neutrino momentum, p_ℓ the outgoing lepton momentum. Suppose that the lepton mass m_ℓ can be ignored, as seems reasonable once the energy is well above threshold; and specialize now to the configurations where the neutrino and lepton have parallel 3-momenta $\mathbf{p}_\ell \parallel \mathbf{p}_\nu$ (we have the laboratory frame in mind, though for massless particles this characterization is an invariant one). For these configurations it is evident that $q^2 = 0$. Moreover a short calculation shows that

$$\bar{u}_\ell \gamma_\nu (1 + \gamma_5) u_\nu = 2 \left(\frac{\epsilon_\nu \epsilon_\ell}{(\epsilon_\nu - \epsilon_\ell)^2} \right)^{1/2} q_\lambda \tag{4.14}$$

where ϵ_ν and ϵ_ℓ are the neutrino and lepton energies. But

$$-iq_\lambda \langle \beta | j_\lambda^{\text{weak}} | \mathrm{p} \rangle = \langle \beta | \partial j_\lambda^{\text{weak}} / \partial x_\lambda | \mathrm{p} \rangle \qquad . \tag{4.15}$$

Thus, for the parallel configurations the amplitude M involves the divergence of the weak current. We are dealing, however, only with the strangeness conserving current and can therefore invoke the CVC hypothesis to eliminate the divergence of the *vector* part of j_λ^{weak}. The matrix element M can then be written, for $q^2 \to 0$ (i.e., for the parallel configurations),

$$M \xrightarrow[q^2 \to 0]{} -(2)^{1/2} G \left[\frac{\epsilon_\nu \epsilon_\ell}{(\epsilon_\nu - \epsilon_\ell)} \right]^{1/2} \langle \beta | \partial A_\lambda^{1+i2} / \partial x_\lambda | \alpha \rangle \cos\theta_c \quad . \tag{4.16}$$

Set this aside for the moment and consider now the *pion* reaction $\pi^+ + p \to \beta$, letting q denote the pion momentum. According to the standard reduction formula, the amplitude for the pionic process is

$$M_\pi = \lim_{q^2 \to -\mu^2} \frac{(q^2 + \mu^2)}{(2)^{1/2}} \langle \beta | \phi^{1+i2} | p \rangle \tag{4.17}$$

where $\phi^{1+i2}/(2)^{1/2}$ is the field that destroys the π^+ (or creates the π^-). Let us now make the PCAC identification for the pion field, according to Eq. (4.5). Moreover, in Eq. (4.17) let us remove the restriction $q^2 \to -\mu^2$ and continue M_π off mass shell in q^2, to the point $q^2 = 0$. At this point the physical neutrino reaction amplitude M becomes related to the continued pion amplitude M_π, and we find

$$|M|^2 \xrightarrow[q^2 \to 0]{} 4G^2 \frac{\epsilon_\nu \epsilon_\ell}{(\epsilon_\nu - \epsilon_\ell)^2} f_\pi^2 \cos^2\theta_c |M_\pi|^2 \quad . \tag{4.18}$$

Again, the physics enters when we invoke the PCAC hypothesis that the pion amplitude continued to zero pion mass is not very different from the physical (i.e., on mass shell) pion amplitude. Hereby, through PCAC a weak process becomes related to a strong one! The more detailed expression of this relation, in terms of cross sections, is a matter now of routine kinematics. To illustrate, let $\sigma_\pi(W)$ be the total cross section for the reaction $\pi^+ + p \to \beta$, where W is the center-of-mass energy, or equivalently, the invariant mass of the hadron system β. For the neutrino reaction, let $\partial\sigma/\partial\Omega\partial W$ be the lab frame differential cross section for forward lepton production, with W the invariant mass of the system β. Then

$$\frac{\partial\sigma}{\partial\Omega\partial W} = \frac{G^2}{\pi^3} f_\pi^2 \cos^2\theta_c \left(\frac{W^2}{m^2}\right) \times \left(\frac{2m\epsilon_\nu - W^2 + m^2}{W^2 - m^2}\right)^2 \left(\frac{W^2 - m^2}{2W}\right) \sigma_\pi(W) \, , \tag{4.19}$$

where m is the nucleon mass, ϵ_ν the laboratory energy of the neutrino. Reliable tests of this relation are unfortunately not yet available.

4.3 The Adler Consistency Condition

A third application of PCAC, again devised by Adler [12] may be illustrated on the example of pion-nucleon scattering. Let p_1 and p_2 be the initial and final nucleon momenta, k and q the initial and final meson momenta. Let the indices α and β denote the isotopic spin states of initial and final mesons. To symbolize all this, write: $k(\alpha) + p_1 \to q(\beta) + p_2$. With the PCAC choice of pion field, the amplitude is given by

$$\begin{aligned} M^{\beta\alpha} &= \lim_{q^2+\mu^2 \to 0} \left(\frac{q^2+\mu^2}{\mu^2 f_\pi} \right) \langle p_2 | \frac{\partial A_\lambda^\beta}{\partial x_\lambda} | p_1, k(\alpha) \rangle \\ &= \lim_{q^2+\mu^2 \to 0} \left(\frac{q^2+\mu^2}{\mu^2 f_\pi} \right) i q_\lambda \langle p_2 | A_\lambda^\beta | p_1, k(\alpha) \rangle \ . \end{aligned} \tag{4.20}$$

The general structure of the amplitude is expressed by

$$M^{\beta\alpha} = \bar{u}(p_2) \left[A^{\beta\alpha} - \frac{i\gamma \cdot (k+q)}{2} B^{\beta\alpha} \right] u(p_1) \tag{4.21}$$

where $A^{\beta\alpha}$ and $B^{\beta\alpha}$ are scalar functions of the two kinematic variables of the problem, say the Mandelstam variables s and t; or more conveniently for present purposes, choose the variables

$$\begin{aligned} \nu &= - \frac{(p_1+p_2)\cdot q}{2m} \\ \nu_B &= \frac{q \cdot k}{2m} \end{aligned} \tag{4.22}$$

Physically, of course, the pion masses are fixed: $q^2 = k^2 = -\mu^2$. The function $B^{\beta\alpha}$ has nucleon pole terms arising from the Feynman graphs shown in the next page.

$$B^{\beta\alpha}\Big|_{\text{pole term}} = \frac{g_r^2}{2m} \left\{ \frac{\tau_\beta \tau_\alpha}{\nu_B - \nu} - \frac{\tau_\alpha \tau_\beta}{\nu_B + \nu} \right\} . \tag{4.23}$$

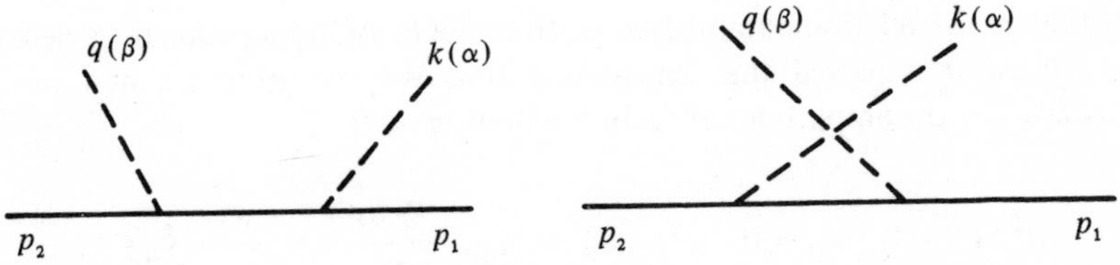

Let us now return to Eq. (4.20), remove the instruction $q^2 \to -\mu^2$, and let the 4-vector q go to zero (so, incidently, $q^2 \to 0$). That is, we are defining an off mass-shell amplitude and are continuing it to $q^2 = 0$ in the mass variable. But as $q \to 0$ we are also taking the "true" variables ν and ν_{B} to unphysical points: $\nu \to 0$, $\nu_{\text{B}} \to 0$. Because of the factor q_λ which appears on the right-hand side of Eq. (4.20), it would appear that $M^{\beta\alpha}$ must vanish as $q \to 0$. This is, of course, the case, except for terms in the matrix element $\langle p_2 | A_\lambda^{\text{B}} | p_1, k(\alpha)\rangle$ which are singular in the limit $q \to 0$. It is easy to see that such terms arise only from the one-nucleon pole diagrams for this matrix element.

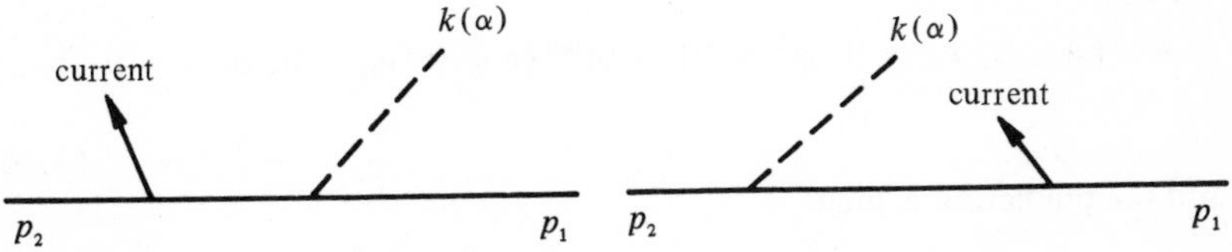

Evaluating these contributions, we find for the limit $q \to 0$

$$M^{\beta\alpha} \xrightarrow[q \to 0]{} \frac{g_{\text{A}} g_r}{f_\pi} \bar{u}(p_2)$$

$$\times \left\{ \delta_{\beta\alpha} - \frac{i\gamma \cdot (k+q)}{2} \frac{1}{2} \left[\frac{\tau_\beta \tau_\alpha}{\nu_{\text{B}} - \nu} - \frac{\tau_\alpha \tau_\beta}{\nu_{\text{B}} + \nu} \right] \right\} u(p_1) \quad . \tag{4.24}$$

Comparing with Eqs. (4.21) and (4.23), we see that the continued function $B^{\beta\alpha}$ has the poles of the mass-shell function $B^{\beta\alpha}$, provided that we set

$$\frac{g_{\text{A}} g_r}{2 f_\pi} = \frac{g_r^2}{2m} \qquad \text{or} \qquad f_\pi = \frac{m g_{\text{A}}}{g_r} \quad .$$

But this is the relation obtained earlier from the PCAC application to β-decay, i.e., PCAC has passed this consistency test. We also obtain a new result however for the amplitude $A^{\beta\alpha}$. In the limit $q \to 0$

$$A^{\beta\alpha}(\nu = 0,\ \nu_B = 0,\ q^2 = 0) = \delta_{\alpha\beta} \frac{g_A g_r}{f_\pi} = \frac{g_r^2}{m} \delta_{\alpha\beta} . \tag{4.25}$$

The amplitude which appears here is evaluated at an unphysical point with respect to all three variables, ν, ν_B, q^2. Physics corresponds to $q^2 = -\mu^2$. The quantities ν and ν_B are continuous variables, but the point $\nu = \nu_B = 0$ lies outside the physical region. In principle this is no problem, however. The physical amplitude admits of a unique analytic continuation to $\nu = \nu_B = 0$; so we may regard $A^{\beta\alpha}(\nu = 0,\ \nu_B = 0,\ q^2 = -\mu^2)$ as knowable, and indeed, Adler has carried out the extrapolation. In relating this quantity to the amplitude on the left side of Eq. (4.25) we now invoke the PCAC hypothesis that the amplitude varies gently as q^2 varies from $-\mu^2$ to zero. Thus we take PCAC to mean

$$A^{\beta\alpha}(\nu = 0,\ \nu_B = 0,\ q^2 = 0) \approx A^{\beta\alpha}(\nu = 0,\ \nu_B = 0,\ q^2 = -\mu^2) . \tag{4.26}$$

and the prediction is then

$$A^{\beta\alpha}(\nu = 0,\ \nu_B = 0,\ q^2 = -\mu^2) \approx \frac{g_r^2}{m} \delta_{\beta\alpha} . \tag{4.27}$$

The experimental fit is quite good. The essential features can be grasped more readily if we consider the pi-nucleon amplitude at physical threshold, $\nu = \mu$, $\nu_B = -\mu^2/m$, which is about as close as we can come physically to $\nu = \nu_B = 0$. Decompose the amplitude $M^{\beta\alpha}$ into parts which are respectively even and odd in the isotopic indices, according to

$$M^{\beta\alpha} = M^{(e)} \delta_{\beta\alpha} + \frac{1}{2} \left[\tau^\beta, \tau^\alpha\right] M^{(0)} , \tag{4.28}$$

and similarly decompose $A^{\beta\alpha}$ and $B^{\beta\alpha}$. Focus on the even amplitude $M^{(e)}$. At *threshold* it is given by

$$M^{(e)} = A^{(e)} + \mu B^{(e)} .$$

For $A^{(e)}$ let us take the Adler prediction of Eq. (4.27), ignoring the small variation between the point $\nu = \nu_B = 0$ and the threshold point $\nu = \mu$, $\nu_B = -\mu^2/m$. For $B^{(e)}$ take the pole approximation of Eq. (4.23), since threshold is fairly close to the pole. Moreover drop terms which are of order μ/m. At threshold one then finds the remarkable prediction

$$M^{(e)} \approx 0 \, . \tag{4.29}$$

In terms of the s-wave scattering lengths $a_{3/2}$ and $a_{1/2}$ (the indices refer to isotopic spin) the prediction is

$$\frac{1}{3}(2a_{3/2} + a_{1/2}) \approx 0 \, .$$

This is well satisfied by the experimental values $a_{1/2} = 0.17\mu^{-1}$, $a_{3/2} = -0.088\mu^{-1}$.

4.4 Discussion

It is important to comment here (as we will do again, from time to time) on some of the ambiguities that arise for the notion of PCAC. In the discussion leading to Eq. (4.11) the PCAC choice of pion field was used to define an off mass shell pion-nucleon vertex function $g_r(q^2)$. Only at the point $q^2 = -\mu^2$ does this quantity have an independent and objective meaning. The relation $f_\pi = mg_A(0)/g_r(0)$ is therefore in itself without content. The physics entered when the PCAC assumption was made that $g_r(q^2)$, in the manner defined, is slowly varying, so that $g_r(0) \approx g_r(-\mu^2) \equiv g_r$. In the simple situation encountered here, where the continued quantity depends only on the mass variable q^2, this seems to be the only reasonable way to interpret the PCAC idea. In other applications, however, the off-shell amplitude depends not only on the mass variable q^2 but on other variables as well. Thus, in the pi-nucleon example, $A^{(e)}$ depends on two physically continuous variables. These were taken to the quantities ν and ν_B. Our procedures then led to a statement concerning $A^{(e)}$ $(\nu = 0,\ \nu_B = 0,\ q^2 = 0)$. As usual, this is free of content until some hypothesis is made about its connection to the mass-shell amplitude $A^{(e)}$ $(\nu = 0, \nu_B = 0, q^2 = -\mu^2)$. To be sure, this latter quantity is itself not physical with respect to the variables ν and ν_B, but that's not the issue — the point $\nu = 0$, $\nu_B = 0$ can, in principle at least, be reached by extrapolation from the physical region. The physics entered when the assumption was made that the variation with q^2 is gentle, for ν and ν_B held fixed, i.e. the operational assumption was that $A^{(e)}(\nu = 0, \nu_B = 0, q^2 = 0) \approx A^{(e)}(\nu = 0, \nu_B = 0, q^2 = -\mu^2)$. But instead of variables ν and ν_B we could originally have chosen a different set of variables, e.g., the Mandelstam quantities s and t. On mass shell the choice is merely a matter of convenience. But off mass shell, the statement that the amplitude is slowly varying in q^2, for fixed ν and ν_B, is in general inequivalent to the same statement made for fixed s and t: the relations connecting one set with the other involve the off shell mass variable q^2. It is only when the *physical* amplitude is in fact slowly varying in the region of interest, i.e., slowly varying in the physically continuous variables, that the specification of these variables becomes irrelevant for the purposes of PCAC. This is presumably the situation for the amplitude $A^{(e)}$ in the region of interest. Here it wouldn't matter much if we had applied the PCAC hypothesis to this amplitude regarded as a function of, say, s and t rather than ν and ν_B. Strictly speaking, however, if the physical amplitude has any variation whatsoever, however slight, a sufficiently extreme alteration in choice of variables could aways be arranged to accomplish any desired change in PCAC predictions. Crudely speaking, one is guided in choice of variables by the requirement that the pion pole singularity at $q^2 = -\mu^2$ shall dominate over other singularities, i.e., that the other singularities be as far away from the point $q^2 = 0$ as possible. But the distance of a singularity and the strength of its contribution are not obviously correlated in a simple way in general.

5. Current Algebra: Methods

A number of techniques that have proved useful for extracting the implications of current algebra are assembled in this section.

Consider the matrix element

$$M_\nu^\beta = i\int \mathrm{d}x\, e^{-iq\cdot x} \langle b | \theta(x_0) [j_\nu^\beta(x), F(0)] | a \rangle \,, \tag{5.1}$$

where j_ν^β is some current operator, specified by the index β; and $F(0)$ is some unspecified local operator. In our subsequent discussions we will see how such matrix elements, with appropriate choices for the operators and states, arise in various physical problems. For the moment we leave interpretations open. Form the contraction

$$iq_\nu M_\nu^\beta = -i\int \mathrm{d}x\, \frac{\partial}{\partial x_\nu}(e^{-iq\cdot x}) \langle b | \theta(x_0) [j_\nu^\beta(x), F(0)] | a \rangle \,,$$

and integrate by parts, ignoring surface terms (along with the anguished cries of purists). Since $\partial\theta(x_0)/\partial x_0 = \delta(x_0)$, this leads to the equation

$$iq_\nu M_\nu^\beta = i\int dx\, e^{-iq\cdot x} \langle b|\theta(x_0) \left[\frac{\partial j_\nu^\beta(x)}{\partial x_\nu}, F(0)\right] |a\rangle + i\int dx\, e^{-iq\cdot x}\, \delta(x_0) \langle b|[j_0^\beta(x), F(0)]|a\rangle \,. \tag{5.2}$$

The first term on the right involves the divergence of the current, which is a useful thing for PCAC purposes; the second term on the right involves an equal time commutator, the sort of thing one likes to see displayed for the subject of current algebra.

For the case where the operator F is itself a current, we consider, in analogy with Eq. (5.1), the matrix element

$$T_{\nu\mu}^{\beta\alpha} = i\int dx\, e^{-iq\cdot x} \langle b|\theta(x_0)[j_\nu^\beta(x), j_\mu^\alpha(0)|a\rangle \,, \tag{5.3}$$

and the analog of Eq. (5.2)

$$iq_\nu T_{\nu\mu}^{\beta\alpha} = i\int dx\, e^{-iq\cdot x} \langle b|\theta(x_0) \left[\frac{\partial j_\nu^\beta(x)}{\partial x_\nu}, j_\mu^\alpha(0)\right] |a\rangle + i\int dx\, e^{-iq\cdot x}\, \delta(x_0)\langle b|[j_0^\beta(x), j_\mu^\alpha(0)]|a\rangle \,. \tag{5.4}$$

We may think of $T_{\mu\nu}^{\beta\alpha}$ as amplitude for the "process": current $(\alpha) + a \to$ current $(\beta) + b$, where q is the momentum carried by the outgoing current and

$$k \equiv p_b - p_a + q \tag{5.5}$$

is the momentum carried by the incoming current. For some later purposes it will be useful to consider a twice contracted equation

$$q_\nu T_{\nu\mu}^{\beta\alpha} k_\mu = k_\mu \int dx\, e^{-iq\cdot x} \langle b|\theta(x_0) \left[\frac{\partial j_\nu^\beta(x)}{\partial x_\nu}, j_\mu^\alpha(0)\right] |a\rangle + k_\mu \int dx\, e^{-iq\cdot x}\, \delta(x_0)\langle b|[j_0^\beta(x), j_\mu^\alpha(0)]|a\rangle \,.$$

Using translation invariance one can express the first term on the right in the form

$$k_\mu \int dx\, e^{ik\cdot x} \langle b|\theta(-x_0)\left[\frac{\partial j_\nu^\beta(0)}{\partial x_\nu}, j_\mu^\alpha(x)\right]|a\rangle .$$

This can be integrated by parts, as before. Then again using translation invariance one finds

$$\begin{aligned} q_\nu T_{\nu\mu}^{\beta\alpha} k_\mu &= i\int dx\, e^{-iq\cdot x} \langle b|\theta(x_0)\left[\frac{\partial j_\nu^\beta(x)}{\partial x_\nu}, \frac{\partial j_\mu^\alpha(0)}{\partial x_\mu}\right]|a\rangle \\ &+ k_\mu \int dx\, e^{-iq\cdot x}\,\delta(x_0)\langle b|[j_0^\beta(x), j_\mu^\alpha(0)]|a\rangle \\ &+ i\int dx\, e^{-iq\cdot x}\,\delta(x_0)\langle b|\left[j_0^\alpha(0), \frac{\partial j_\nu^\beta(x)}{\partial x_\nu}\right]|a\rangle . \end{aligned} \tag{5.6}$$

Finally, consider an amplitude of the form

$$F_\mu^\alpha = i\int dx\, dy\, e^{-iq\cdot x - ik\cdot y}\langle b|T(j_\mu^\alpha(x)B(y)C(0))|a\rangle , \tag{5.7}$$

where $T(\ \ )$ denotes time-ordering. Following the procedures used above one finds

$$\begin{aligned} iq_\mu F_\mu^\alpha &= i\int dx\, dy\, e^{-iq\cdot x - ik\cdot y}\langle b|T\left(\frac{\partial j_\mu^\alpha(x)}{\partial x_\mu} B(y)C(0)\right)|a\rangle \\ &+ i\int dx\, dy\, e^{-iq\cdot x - ik\cdot y}\,\delta(x_0 - y_0)\langle b|T([j_0^\alpha(x), B(y)]\, C(0))|a\rangle \\ &+ i\int dx\, dy\, e^{-iq\cdot x - ik\cdot y}\,\delta(x_0)\langle b|T([j_0^\alpha(x), C(0)]\, B(y))|a\rangle . \end{aligned} \tag{5.8}$$

Several important applications of current algebra ideas (the Cabibbo-Radicati sum rule for Compton scattering, the Adler sum rule for neutrino induced reactions, the Adler-Weisberger relation for pi-nucleon scattering) can be treated on the same footing, up to a point. For these cases we focus on the amplitude $T_{\nu\mu}^{\beta\alpha}$ defined by Eq. (5.3) and specialize to the situation where the states $|a\rangle$ and $|b\rangle$ are identical states corresponding to a single hadron (proton, say) of momentum p. To bring this out, write $|a\rangle = |b\rangle = |p\rangle$. From Eq. (5.4): $k = q$. Moreover, if the hadron has spin, let it be understood that $T^{\beta\alpha}_{\nu\mu}$ is averaged over spin. The different applications will be distinguished by the choice of currents and by the interpretation of certain general results which are now to be discussed. For a while longer we continue with formalism, leaving these distinguishing features till later on. However, in all the cases under present discussion we take j^α to be a charge-raising current and j^β to be its adjoint, a charge-lowering current. For brevity, write $j^\alpha = j$, $j^\beta = j^\dagger$. Then in the present situation we are studying

$$T_{\mu\nu} = i\int dx\, e^{-iq\cdot x} \langle p|\theta(x_0)[j_\nu^\dagger(x), j_\mu(0)]|p\rangle \,. \tag{5.9}$$

Since $T_{\mu\nu}$ is understood to be averaged over spin of the hadron state $|p\rangle$, it is a tensor that can depend only on the momenta p and q. It can therefore be decomposed as follows:

$$T_{\mu\nu} = A p_\mu p_\nu + B_1 \delta_{\mu\nu} + B_2 p_\mu q_\nu + B_3 q_\mu p_\nu + B_4 q_\mu q_\nu + B_5 \epsilon_{\mu\nu\alpha\beta} q_\alpha p_\beta \,, \tag{5.10}$$

where the coefficients $A, B_1, \ldots B_5$ are scalar functions of two covariant variables — the "mass" variable q^2 and the laboratory energy of the current

$$\nu = -p\cdot q/m, \qquad p^2 = -m^2 \,.$$

Notice that the function B_5 arises only in a parity violating situation, from interference of vector and axial vector components of j and $j^\dagger$.

Refer now to Eq. (5.4) and decompose the first term on the right according to

$$\int dx\, e^{-iq\cdot x} \langle p|\theta(x_0)\left[\frac{\partial j_\nu^\dagger(x)}{\partial x_\nu}, j_\mu(0)\right]|p\rangle = D p_\mu + E q_\mu \,, \tag{5.11}$$

where D and E are scalar functions of q^2 and ν. Equation (5.4) also contains an equal time commutator term and for this we invoke the hypotheses of current algebra. For the current j_μ we will be dealing, variously with the choices V_μ^{1+i2}, A_μ^{1+i2}, and $j^{\text{weak}} = \cos\theta_c(V_\mu^{1+i2} + A_\mu^{1+i2}) + \sin\theta_c (V_\mu^{4+i5} + A_\mu^{4+i5})$. Let us recall then some of the relevant commutation relations of current algebra:

$$[V_0^{1-i2}(x), V_\mu^{1+i2}(0)]_{x_0=0} = [A_0^{1-i2}(x), A_\mu^{1+i2}(0)]_{x_0=0}$$
$$= -2V_\mu^3(x)\delta(\mathbf{x}) , \qquad (5.12)$$

$$[V_0^{4-i5}(x), V_\mu^{4+i5}(0)]_{x_0=0} = [A_0^{4-i5}(x), A_\mu^{4+i5}]_{x_0=0}$$
$$= -\left\{(3)^{1/2} V_\mu^8(x) + V_\mu^3\right\}\delta(\mathbf{x}) . \qquad (5.12')$$

The following may also be noted

$$[V_0^{1-i2}(x), A_\mu^{1+i2}(0)]_{x_0=0} = [A_0^{1-i2}(x), V_\mu^{1+i2}(0)]_{x_0=0}$$
$$= -2A_\mu^3(x)\delta(\mathbf{x}) , \qquad (5.13)$$

$$[V_0^{4-i5}(x), A_\mu^{4+i5}(0)]_{x_0=0} = [A_0^{4-i5}(x), V_\mu^{4+i5}(0)]_{x_0=0}$$
$$= -\left\{(3)^{1/2} A_\mu^8(x) + A_\mu^3(x)\right\}\delta(\mathbf{x}) ; \qquad (5.13')$$

but these won't figure into the present considerations since we are averaging over hadron spins and therefore $\langle p|A_\mu^\alpha|p\rangle = 0$. Finally, observe that

$$\langle p|V_\mu^8|p\rangle = \frac{p_\mu}{m} I_3 , \qquad (5.14)$$

$$\langle p|V_\mu^8|p\rangle = \frac{p_\mu}{m}\frac{(3)^{1/2}}{2} Y , \qquad (5.15)$$

where Y and I_3 are the hypercharge and isospin quantum numbers of the hadron. In all the cases considered here, therefore, the equal time commutator term in Eq. (5.4) can be written

$$\int d\mathbf{x}\, e^{-iq\cdot x}\,\delta(x_0)\langle p|[j_0^\dagger(x), j_\mu(0)]|p\rangle = Cp_\mu , \qquad (5.16)$$

where the commutator constant C will depend on the particular case under discussion. Now insert Eqs. (5.10), (5.11), and (5.16) into Eq. (5.4), which really constitutes two equations, one for the coefficients of p_μ, another for the coefficients of q_μ. The former yields the relation

$$(q \cdot p)A + q^2 B_2 = D + C \,. \tag{5.17}$$

Recall that A, B_2, and D all depend on the variables ν and q^2, whereas C is a constant determined by the commutation relations.

Now comes the major technical assumption. The function A, B_2, D all satisfy dispersion relations in the variable ν, for q^2 fixed (we are always concerned here only with space-like q). Let us suppose that all three in fact satisfy unsubtracted dispersion relations. For example

$$A(\nu, q^2) = \frac{1}{\pi}\int_{-\infty}^{\infty} \mathrm{d}\nu' \; \frac{\operatorname{Im} A(\nu', q^2)}{\nu' - \nu} \quad ,$$

and similarly for B_2 and D. But in Eq. (5.17) the function A appears multiplied by the factor $q \cdot p = -m\nu$. In the limit $\nu \to \infty$, the functions B_2 and D make no contribution in Eq. (5.17) and therefore

$$\lim_{\nu \to \infty} \, (q \cdot p)A = \frac{m}{\pi}\int_{-\infty}^{\infty} \mathrm{d}\nu' \operatorname{Im} A(\nu', q^2) = C \,. \tag{5.18}$$

This is the basic result.

The absorptive part of $T_{\mu\nu}$ is given by

$$\operatorname{Abs} T_{\mu\nu} \equiv t_{\mu\nu} = \frac{1}{2}\int \mathrm{d}x \, e^{-iq\cdot x} \langle p \,|\, [\, j_\nu^\dagger(x), j_\mu(0)] \,|\, p \rangle \,, \tag{5.19}$$

and its tensor decomposition is related to that of Eq. (5.10) in a simple way:

$$t_{\mu\nu} = \operatorname{Im} A \, p_\mu p_\mu + \operatorname{Im} B_1 \, \delta_{\mu\nu} + \ldots \tag{5.20}$$

Expanding $t_{\mu\nu}$ in contributions from a complete set of states $|s\rangle$ one has

$$t_{\mu\nu} = \frac{(2\pi)^4}{2} \sum_s \Big\{ \langle p \,|\, j_\nu^\dagger \,|\, s\rangle\langle s \,|\, j_\mu \,|\, p\rangle \delta(s - p - q)$$

$$- \langle p \,|\, j_\mu \,|\, s\rangle \langle s \,|\, j_\mu^\dagger \,|\, p\rangle \delta(s + q - p) \Big\} \,. \tag{5.21}$$

For $\nu > 0$ only the first term makes a contribution, whereas for $\nu < 0$ only the second term contributes. Let us now introduce a new matrix element $\widetilde{T}_{\nu\mu}$, which differs from $T_{\nu\mu}$ only in that the charge-raising and charge-lowering currents are interchanged:

$$\widetilde{T}_{\nu\mu} = i\int dx\, e^{-iq\cdot x} \langle p|\theta(x_0)[j_\nu(x), j_\mu^\dagger(0)]|p\rangle , \tag{5.22}$$

with decomposition as Eq. (5.10) but $A \to \widetilde{A}$, $B_1 \to \widetilde{B}_1$, etc. The absorptive part of $\widetilde{T}_{\nu\mu}$ is

$$\widetilde{t}_{\nu\mu} = \frac{1}{2}\int dx\, e^{-iq\cdot x} \langle p|[j_\nu(x), j_\mu^\dagger(0)]|p\rangle ; \tag{5.23}$$

expanding this in the manner of Eq. (5.21) one finds that

$$t_{\mu\nu}(p, -q) = -\widetilde{t}_{\mu\nu}(p, q) , \tag{5.24}$$

hence in particular

$$\mathrm{Im}\, A(-\nu, q^2) = \mathrm{Im}\, \widetilde{A}(\nu, q^2) . \tag{5.25}$$

The sum rule of Eq. (5.18) can thus be written

$$\frac{1}{\pi}\int_0^\infty d\nu\, [\mathrm{Im}\, A(\nu, q^2) - \mathrm{Im}\, \widetilde{A}(\nu, q^2)] = \frac{C}{m} . \tag{5.26}$$

This result would have been obtained if we had carried out the whole discussion for the isotopically "odd" amplitude $\frac{1}{2}(T_{\mu\nu} - \widetilde{T}_{\nu\mu})$. It is only for this amplitude that the no-subtraction assumption had to be made. It may be said that standard Regge model ideas support the assumption.

The Infinite Momentum Frame Method

The result of Eq. (5.18), or its equivalent Eq. (5.26), is sufficiently important to merit discussion from an alternative viewpoint. Start with the absorptive amplitude $t_{\nu\mu}$ defined in Eq. (5.19) and integrate over q_0, holding the three-vectors $\mathbf{q}$ and $\mathbf{p}$ fixed. Using

$$\int dq_0\, e^{iq_0 x_0} = 2\pi\,\delta(x_0) ,$$

we find

$$\frac{1}{\pi p_0}\int dq_0\, t_{\nu\mu}(\mathbf{q}, \mathbf{p}, q_0) = \frac{1}{p_0}\int dx\, e^{-iq\cdot x}\,\delta(x_0)$$
$$\langle p\,|[j_\nu^\dagger(x), j_\mu(0)]\,|p\rangle\,,$$

where, for convenience, we have divided through by a factor p_0. Now choose time components for both of the indices ν and μ and refer to Eq. (5.16):

$$\frac{1}{\pi p_0}\int dq_0\, t_{00}(\mathbf{q}, \mathbf{p}, q_0) = C\,, \tag{5.27}$$

where

$$t_{00} = \operatorname{Im} A\, p_0^2 - \operatorname{Im} B_2 + \ldots\ . \tag{5.28}$$

The scalar functions $\operatorname{Im} A$, $\operatorname{Im} B_1, \ldots$ depend on the variables $q^2 = |\mathbf{q}|^2 - q_0^2$ and $\nu = -q\cdot p/m = (q_0 p_0 - \mathbf{q}\cdot\mathbf{p})/m$. Choose a frame where $\mathbf{q}\cdot\mathbf{p} = 0$, so that

$$q_0 = \frac{m\nu}{p_0}\,,\quad q^2 = \mathbf{q}^2 - \frac{m^2\nu^2}{p_0^2}\,. \tag{5.29}$$

Converting (5.27) to an integral over ν, we have

$$\frac{m}{\pi p_0^2}\int d\nu\left\{\operatorname{Im} A\, p_0^2 - \operatorname{Im} B_1 + \ldots\right\} = C\,, \tag{5.30}$$

where

$$\operatorname{Im} A = \operatorname{Im} A\left(\nu, q^2 = \mathbf{q}^2 - \frac{m^2\nu^2}{p_0^2}\right), \text{ etc.}$$

In the integration over ν, the quantity q^2 itself varies, in such a way that $q^2 \to \infty$ as $|\nu| \to \infty$. However, there is still freedom left in the choice of Lorentz frame. Suppose we adopt the "infinite momentum" frame for the target state $|p\rangle$, where $p_0 \to \infty$. Make the *assumption* now that the limiting operation can be taken under the integral in Eq. (5.30). That is, suppose it is legitimate, for each fixed ν, to first pass to the limit $p_0 \to \infty$, the integration over ν following thereafter. If this is allowed then only the $\operatorname{Im} A$ term survives in Eq. (5.30) and moreover the argument $q^2 \to \mathbf{q}^2$

becomes independent of ν. We then find the earlier result

$$\frac{1}{\pi}\int_{-\infty}^{\infty} d\nu \, \mathrm{Im}\, A(\nu, q^2) = \frac{C}{m} \, ,$$

where the integration over ν is carried out for fixed q^2.

6. Some Applications of Current Algebra

The basic result of Eq. (5.26) is exploited in this section for Compton scattering, high energy neutrino reactions, and pi-nucleon scattering. The first two applications are independent of the notions of PCAC and thus, apart from the additional technical assumptions that have gone into Eq. (5.26), they constitute rather clear tests of current algebra. The pi-nucleon discussion will culminate in the Adler-Weisberger formula and will bring in also the hypotheses of PCAC.

6.1 Cabibbo-Radicati Sum Rule

Consider the forward elastic scattering of a photon of momentum q on a hadron target of momentum p, specializing to the case of no helicity flip for photon or hadron [13]. For definiteness take the hadron to be a proton. Let ϵ be the photon polarization vector, so that $\epsilon^2 = 1$, $q^2 = 0$, $q \cdot \epsilon = 0$. The scattering amplitude is expressed by $e^2 \epsilon_\nu T_{\nu\mu} \epsilon_\mu$, where $T_{\nu\mu}$ is given by Eq. (5.3), with $j^\alpha = j^\beta = j^{\mathrm{em}}$. Recall that the electromagnetic current is composed of an isoscalar part and an isovector part. The absorptive part of the amplitude $T_{\nu\mu}$ is related, through the optical theorem (unitarity), to the total cross section for production of hadrons in the collision of a photon with the hadron target. In principle, by use of both proton and neutron targets and by an isotopic analysis of the final hadron states, one can separately determine the contributions to the cross section coming from the isoscalar and isovector parts of the current, and the latter in turn can be decomposed into contributions from final states with $I = \frac{1}{2}$ and $I = \frac{3}{2}$. Insofar as we have something to say about the absorptive part of the Compton amplitude, it is not unphysical then to consider Compton scattering induced solely by the isovector part of the electromagnetic current: $j_\lambda^{\mathrm{em}} \to V_\lambda^3$. But in fact, let us consider the scattering of "charged" isovector photons, where we employ the charge-raising and charge-lowering members of the triplet of which V_λ^3 is the neutral member.

At the end our results can be related through isospin symmetry considerations to quantities of direct physical interest. The amplitude to be considered is therefore

$$T_{\nu\mu} = i\int dx\, e^{-iq\cdot x} \langle p|\theta(x_0)[V_\nu^{1-i2}(x),\ V_\mu^{1+i2}(0)]|p\rangle\ . \quad (6.1)$$

Its absorptive part is

$$t_{\nu\mu} = \frac{1}{2}\int dx\, e^{-iq\cdot x} \langle p|[V_\nu^{1-i2}(x),\ V_\mu^{1+i2}(0)]|p\rangle\ . \quad (6.2)$$

These have the tensor decompositions given respectively by Eqs. (5.10) and (5.20), of course with $B_5 = 0$.

Because we are dealing here with conserved currents it follows that

$$q_\nu t_{\nu\mu} = t_{\nu\mu} q_\mu = 0;\ q_0^2\ t_{00} = q_i q_j t_{ij},\ i,j = 1,2,3\ . \quad (6.3)$$

For a while let us retain q^2 at a non-zero value, although of course at the end $q^2 \to 0$. The other scalar variable of the problem is $\nu = -q\cdot p/m$, the laboratory energy of the photon. From Eq. (6.3) it is easily established that

$$\mathrm{Im}\, A(\nu, q^2 = 0) = 0\ ; \quad (6.4)$$

but

$$\frac{\partial}{\partial q^2}\ \mathrm{Im}\, A(\nu, q^2|_{q^0=0} = \frac{\mathrm{Im}\, B_1(\nu, 0)}{m^2\nu^2}\ , \quad (6.5)$$

a result which will shortly be required.

The amplitude $T_{\nu\mu}$, hence the scalar quantities $A, B_1, \ldots$ refer to the process $\gamma^+ + p \to \gamma^+ + p$ for $\nu \geqslant 0$. But these scalar quantities can be analytically continued to $\nu < 0$ and refer there to the crossed reaction $\gamma^- + p \to \gamma^- + p$. That is, let $\tilde{A}(\nu, q^2)$, $\tilde{B}$, (ν, q^2), etc., refer to the crossed reaction, with physical energy $\nu > 0$. Then $\mathrm{Im}\, A(-\nu, q^2) = \mathrm{Im}\, \tilde{A}(\nu, q^2)$, etc. Refer now to Eqs. (5.26), (5.16), (5.14), (5.12). The commutator coefficient C is given by

$$C = \frac{-2}{m} I_3\ , \quad (6.6)$$

where I_3 is the isospin quantum number of target ($I_3 = \frac{1}{2}$ for proton);

and Eq. (5.26) now reads

$$\frac{m^2}{\pi}\int_0^{\infty} d\nu\,[\mathrm{Im}\,\tilde{A}\,(\nu, q^2) \;-\; \mathrm{Im}\,A\,(\nu, q^2)] \;=\; 1\;. \tag{6.7}$$

For $\mathrm{Im}\,A$, which corresponds to $\gamma^+ + \mathrm{p} \to \gamma^+ + \mathrm{p}$, the integral begins at the continuum threshold: $-(q+p)^2 = (m+\mu)^2$. For $\mathrm{Im}\,\tilde{A}$, which corresponds to $\gamma^- + \mathrm{p} \to \gamma^- + \mathrm{p}$ the continuum begins at the same threshold, but there is a "pole" contribution from the one-neutron intermediate state, i.e., $\tilde{A}$ has a pole at $-(q+p)^2 = m^2$ coming from $\gamma^- + \mathrm{p} \to \text{neutron} \to \gamma^- + \mathrm{p}$. The residue depends on the isovector electromagnetic form factors of the nucleon, evaluated at momentum transfer q^2. Separating off this term explicitly, we find

$$1 \;=\; [F_1^{\mathrm{V}}(q^2)]^2 \;-\; \frac{q^2}{4m^2}\,[F_2^{\mathrm{V}}(q^2)]^2$$
$$+\;\frac{m^2}{\pi}\int_{\text{thresh.}}^{\infty} d\nu\,[\mathrm{Im}\,\tilde{A}\,(\nu, q^2) - \mathrm{Im}\,A\,(\nu, q^2)]\;, \tag{6.8}$$

where $F_1^{\mathrm{V}}(q^2)$ and $F_2^{\mathrm{V}}(q^2)$ are respectively the isovector charge and anomolous magnetic moment form factors of the nucleon: $F_1^{\mathrm{V}}(0) = 1$, $F_2^{\mathrm{V}}(0) = \mu_p^{(a)} - \mu_n^{(a)}$, the difference of anomalous magnetic moments of protons and neutrons. Since $\mathrm{Im}\,\tilde{A} = \mathrm{Im}\,A = 0$ for $q^2 = 0$, the above equation is a trivial identity for the limit $q^2 = 0$. However, differentiate Eq. (6.8) with respect to q^2 and then set $q^2 = 0$, using Eq. (6.5). In this way one finds

$$2\left(\frac{dF_1^{\mathrm{V}}}{dq^2}\right)_{q^2=0} \;-\; \frac{1}{4m^2}\,[F_2^{\mathrm{V}}(0)]^2 \;+\; \frac{1}{\pi}\int_{\text{thresh.}}^{\infty}\frac{d\nu}{\nu^2}\,[\mathrm{Im}\,\tilde{B}_1(\nu, 0)$$
$$-\;\mathrm{Im}\,B_1(\nu, 0)] \;=\; 0\;. \tag{6.9}$$

To interpret the integrand in terms of physical quantities, recall that the amplitude for, say, $\gamma^+ + \mathrm{p} \to \gamma^+ + \mathrm{p}$ scattering is given by $\frac{1}{2}e^2\,\epsilon_\nu\,T_{\nu\mu}\,\epsilon_\mu$, where the factor $\frac{1}{2}$ arises because it is $V_\mu^{1+i2}/(2)^{1/2}$ that creates γ^+, destroys γ^-, etc. In Eq. (6.9) we have passed to the limit $q^2 = 0$, so that $\epsilon\cdot q = 0$, $\epsilon^2 = 1$. Let us work in the laboratory frame ($\mathbf{p} = 0$), where ν represents the photon energy, and let us choose the gauge where $\epsilon\cdot p = 0$.

The absorptive amplitude then reduces to

$$\operatorname{Im} F \equiv \frac{1}{2} e^2 \epsilon_\nu t_{\nu\mu} \epsilon_\mu = \frac{1}{2} e^2 \operatorname{Im} B_1(\nu, 0) \quad . \tag{6.10}$$

For $\gamma^- + \mathrm{p} \to \gamma^- + \mathrm{p}$ scattering $B_1 \to \widetilde{B}_1$. The expression in Eq. (6.10), and its analog for $\gamma^- + \mathrm{p}$ scattering, is the imaginary part of the Feynman amplitude F for forward scattering. The ordinary scattering amplitude f (the thing whose absolute square gives $\mathrm{d}\sigma/\mathrm{d}\Omega$) differs from this by a factor of 4π. By the optical theorem, however,

$$\operatorname{Im} f = \frac{\operatorname{Im} F}{4\pi} = \frac{\nu\sigma(\nu)}{4\pi} \, .$$

Thus, with $\sigma^+(\nu)$ and $\sigma^-(\nu)$ the total cross sections for hadron production in $\gamma^+ + \mathrm{p}$ and $\gamma^- + \mathrm{p}$ collisions, the integral in Eq. (6.9) can be written

$$\frac{2}{e^2\pi} \int_{\text{thresh.}}^{\infty} \frac{\mathrm{d}\nu}{\nu} [\sigma^-(\nu) - \sigma^+(\nu)] \, .$$

It remains only to relate the quantities σ^- and σ^+ to cross sections for processes induced by neutral isovector photons corresponding to the isovector current V_λ^3. For the latter one can in principle isolate the cross sections for hadron production in states with $I = \frac{1}{2}$ and $I = \frac{3}{2}$. Let $\sigma_{1/2}^{\mathrm{V}}$ and $\sigma_{3/2}^{\mathrm{V}}$ be corresponding cross sections. These can be related to σ^- and σ^+ by a simple isotopic analysis. In this way one finds the Cabibbo-Radicati formula

$$\frac{1}{2\pi^2\alpha} \int_{\text{thresh.}}^{\infty} \frac{\mathrm{d}\nu}{\nu} [\sigma_{1/2}^{\mathrm{V}}(\nu) - 2\sigma_{1/2}^{\mathrm{V}}(\nu)]$$

$$= \left(\frac{F_2^{\mathrm{V}}(0)}{2m}\right)^2 - 2\left(\frac{\mathrm{d}F_1^{\mathrm{V}}}{\mathrm{d}q^2}\right)_{q^2=0} \, . \tag{6.11}$$

Although the cross sections which appear here are in principle accessible to experiment, the practical demands are clearly very great. Gilman and Schnitzer [14] have attempted to test the formula by saturating the integral with contributions from low lying baryon resonances. The trends are reasonably satisfactory.

6.2 *The Adler Sum Rule for Neutrino Reactions*

The procedures employed above can be adapted with little change to a discussion of high energy neutrino reactions. In Eq. (5.9) j_μ is now taken to be the total weak semi-leptonic current

$$j_\mu = j_\mu^{\text{weak}} = \cos\theta_c\,(V_\mu^{1+i2} + A_\mu^{1-i2}) + \sin\theta_c\,(V_\mu^{4+i5} + A_\mu^{4+i5}), \tag{6.12}$$

and $j_\mu^\dagger$ is the charge-lowering adjoint. With this choice of currents $T_{\nu\mu}$ describes the improbable reaction lepton pair + target → lepton pair + target. The absorptive part $t_{\nu\mu}$ is more physically practical: it enters into the description of the reactions ν + target → ℓ + hadrons. Again, in Eq. (5.9) an average over target spin is understood. Otherwise, in this equation we leave open the specification of the state $|p\rangle$. Again, the tensor decomposition of $T_{\nu\mu}$ is given in Eq. (5.10). But for present purposes it is conventional to rearrange the terms and to introduce a special notation. We write out the decomposition as follows:

$$T_{\nu\mu} = T_1\left(\delta_{\nu\mu} - \frac{q_\nu q_\mu}{q^2}\right) + \frac{T_2}{m^2}\left(p_\nu - \frac{p\cdot q}{q^2}\,q_\nu\right)\left(p_\mu - \frac{p\cdot q}{q^2}\right)$$
$$- \frac{1}{2}\,\frac{T_3}{m^2}\,\epsilon_{\nu\mu\alpha}\,q_\alpha\,p_\beta + \frac{T_4}{m^2}\,q_\nu q_\mu$$
$$+ \frac{T_5}{m^2}\,q_\nu p_\mu + \frac{T_6}{m^2}\,p_\nu q_\mu\ . \tag{6.13}$$

The absorptive part $t_{\nu\mu}$ has the same structure, with $T_i \to \operatorname{Im} T_i$. But it is now customary to write

$$W_i(q^2,\nu) = \frac{1}{\pi}\,\operatorname{Im} T_i(q^2,\nu)\ . \tag{6.14}$$

It is only the "structure functions" W_1, W_2, W_3 that will enter into the present discussion. The W_i are scalar functions of the variables q^2 and $\nu = -q\cdot p/m$, where m is the mass of the hadron target. Notice that in taking over results from our earlier discussion, we must make the identification

$$\operatorname{Im} A \to \frac{\pi W_2}{m^2}\ . \tag{6.15}$$

In the present notation, and with $\widetilde{W}_2(\nu, q^2) = -W_2(-\nu, q^2)$, Eq. (5.26) can be written

$$\int_0^\infty d\nu \left[W_2(\nu, q^2) - \widetilde{W}_2(\nu, q^2) \right] \equiv mC \,, \tag{6.16}$$

where the commutator coefficient C is defined by Eq. (5.16). With the present choice of currents [see Eq. (6.12)], and from Eqs. (5.12), (5.13), (5.14), (5.15), we obtain the coefficient C and hence the Adler sum rule

$$\int_0^\infty d\nu \left[\widetilde{W}_2(\nu, q^2) - W_2(\nu, q^2) \right] = 2 \left[2I_3 \cos^2 \theta_c + \left(I_3 + \frac{3}{2} Y \right) \sin^2 \theta_c \right] . \tag{6.17}$$

Here Y and I_3 are the hypercharge and isospin quantum numbers of the hadron target. In the above integral we have not bothered to separate off the contributions arising from the one-baryon intermediate states. Notice on the right-hand side of Eq. (6.17) that the term proportional to $\cos^2 \theta_c$ arises from the strangeness-conserving currents; the other term from the strangeness changing currents. Notice also that the right-hand side of this equation is independent of the variable q^2.

Having derived the Adler sum rule of Eq. (6.17), let us now interpret the structure functions which appear in it. For this purpose consider the reaction

$$\nu + \text{hadron target} \to \ell + \text{hadrons} \,. \tag{6.18}$$

For the moment suppose that some definite system, s, of final hadrons is being contemplated, where the symbol s names the system and also stands for its total 4-momentum. Let $q(\nu)$, $q(\ell)$, and p be respectively the momenta of neutrino, outgoing lepton, and hadron target, so that $q(\nu) + p = q(\ell) + s$. Let ϵ be the laboratory energy of the neutrino, ϵ' the energy of the lepton, and θ the laboratory angle between lepton and neutrino directions of motion. In the following discussion the lepton mass is systematically neglected. Now define

$$q = q(\nu) - q(\ell), \quad \nu = -q \cdot p/m \,, \tag{6.19}$$

and observe that

$$q^2 = 4\epsilon\epsilon' \sin^2 \frac{\theta}{2} \geqslant 0 \,, \qquad \nu = \epsilon - \epsilon' \geqslant 0 \,. \tag{6.20}$$

The invariant mass M of the hadron system s is related to these quantities by

$$M^2 = -(q+p)^2 = 2m\nu - q^2 + m^2 \ . \tag{6.21}$$

So much for kinematics!

The amplitude for the reaction (6.18) is given by

$$\text{Amp} = \frac{G}{(2)^{1/2}} \langle s | j_\mu^{\text{weak}} | p \rangle i \bar{u}_\ell \gamma_\mu (1+\gamma_5) u_\nu \ . \tag{6.22}$$

Let us now compute the differential cross section in its dependence on energy ϵ' and angle θ of the outgoing lepton, but summed over *all* final hadron states which are kinematically accessible. An average over target spin is also taken (unpolarized target). In computing this differential cross section we encounter a quantity

$$\frac{(2\pi)^4}{2} \sum_s \langle p | j_\nu^+ | s \rangle \langle s | j_\mu | p \rangle \delta(s-p-q) \ .$$

Recall that the variable ν satisfies $\nu \geqslant 0$; and observe that the above quantity is just the absorptive part $t_{\nu\mu}$ of the amplitude $T_{\nu\mu}$ (for $\nu > 0$ only the first sum in Eq. (5.21) contributes to $t_{\nu\mu}$). But from Eqs. (6.13) and (6.14)

$$\frac{1}{\pi} t_{\nu\mu} = W_1 \left(\delta_{\mu\nu} - \frac{q_\nu q_\mu}{q^2} \right) + \frac{W_2}{m^2} \left(p_\nu - \frac{p \cdot q}{q^2} q_\nu \right) \times \left(p_\mu - \frac{p \cdot q}{q^2} q_\mu \right) - \frac{1}{2} \frac{W_3}{m^2} \epsilon_{\nu\mu\alpha\beta} q_\alpha p_\beta + \dots \ . \tag{6.23}$$

In the approximation where we neglect the lepton mass, $q_\mu \bar{u}_\ell \gamma_\mu (1+\gamma_5) u_\nu = 0$; hence the remaining structure functions W_4, W_5, W_6 make no contribution to the cross section. Carrying out the remaining arithmetic, we find

$$\frac{\partial \sigma^{(\nu)}}{\partial \Omega \, \partial \epsilon'} = \frac{G^2}{2\pi^2} \epsilon'^2 \left\{ 2 W_1 \sin^2 \frac{\theta}{2} + W_2 \cos^2 \frac{\theta}{2} - \left(\frac{\epsilon + \epsilon'}{m} \right) W_3 \sin^2 \frac{\theta}{2} \right\} . \tag{6.24}$$

The superscript (ν) on $d\sigma^{(\nu)}$ reminds us that we are dealing with the neutrino reaction (6.18). The structure functions obey the following positivity conditions

$$\left(1+\frac{\nu^2}{q^2}\right) W_2 \geqslant W_1 \geqslant \frac{1}{2m}(\nu^2+q^2)^{1/2}\,|W_3| \geqslant 0\ . \qquad (6.25)$$

For antineutrino reactions

$$\bar{\nu} + \text{hadrons} \to \ell^+ + \text{hadrons}$$

the whole kinematic analysis is as above. But here one encounters the structure functions $\widetilde{W}_i(\nu, q^2)$. The other change is in the overall sign of the last term in Eq. (6.24). The cross section formula for the antineutrino process is

$$\frac{\partial\sigma^{(\bar{\nu})}}{\partial\Omega\,\partial\epsilon'} = \frac{G^2}{2\pi^2}\,\epsilon'^2 \left\{2\widetilde{W}_1 \sin^2\frac{\theta}{2} + \widetilde{W}_2 \cos^2\frac{\theta}{2} + \left(\frac{\epsilon+\epsilon'}{m}\right)\widetilde{W}_3 \sin^2\frac{\theta}{2}\right\}. \qquad (6.26)$$

Whether for the neutrino or antineutrino case, the differential cross section depends on three variables, $\epsilon, \epsilon', \theta$, or equivalently ϵ, ν, q^2. The structure functions depend on only two variables ν and q^2. In principle the three structure functions can thus be disentangled; and Eq. (6.17) represents an important sum rule relating to one type of structure function.

As yet there is very little experimental evidence that bears on this Adler sum rule. The importance of the sum rule should be evident. It brings in parts of current algebra involving commutators of strangeness-changing as well as strangeness concerning currents, vector and axial; and for $q^2 \neq 0$ one is testing the *local* commutation relations.

6.3 The Adler-Weisberger Formula

In the next sections we discuss a number of problems in which current algebra and PCAC are used jointly. The first and still the most striking application culminates in the Adler-Weisberger formula. It merits discussion from different angles. One approach can be based closely on the methods of the present section and it is natural to take this up here. In the next section we return to the Adler-Weisberger formula from another point of view.

Return again to amplitude $T_{\mu\nu}$ of Eq. (5.9), defined this time for axial vector currents

$$j_\mu = A_\mu^{1+i2}, \quad j_\mu^\dagger = A_\mu^{1-i2} .$$

The tensor decomposition of $T_{\nu\mu}$, the absorptive part $t_{\nu\mu}$, etc., are as discussed before; and the basic result of the current algebra hypothesis is still expressed in Eq. (5.26). For definiteness take the hadron target to be a proton. Then the commutator coefficient which appears in Eq. (5.26) has, according to Eq. (6.6), the value $C = -1/m$. Now in Eq. (6.26) the continuum part of the integral starts at $\nu = \mu + (\mu^2 + q^2)/2m$. But there is an isolated contribution to $\mathrm{Im}\,\tilde{A}$ arising from the one-neutron intermediate state. In the remaining discussion we specialize to $q^2 = 0$; and we separate off the one-nucleon contribution explicitly. This evidently brings in the axial vector form factor $g_A(q^2 = 0) \equiv g_A$ of β-decay. One finds

$$g_A^2 + \frac{m^2}{\pi} \int_{\text{thresh.}}^{\infty} d\nu \,[\mathrm{Im}\,\tilde{A}(\nu, 0) - \mathrm{Im}\,A(\nu, 0)] = 1 . \tag{6.27}$$

From the tensor decomposition of the absorptive amplitude $t_{\nu\mu}$ (see Eq. 5.20)) it is easily verified that

$$q_\nu t_{\nu\mu} q_\mu \xrightarrow[q^2 \to 0]{} m^2 \nu^2 \,\mathrm{Im}\,A(\nu, 0) . \tag{6.28}$$

But from Eq. (5.21) for $\nu > \mu$ (when only the first sum contributes), and $q^2 \to 0$, one has

$$m^2 \nu^2 \,\mathrm{Im}\,A(\nu, 0) = q_\nu q_\mu \frac{(2\pi)^4}{2} \sum_s \langle p | A_\nu^{1-i2} | s \rangle$$

$$\times \langle s | A_\mu^{1+i2} | p \rangle \delta(s - p - q) . \tag{6.29}$$

Notice however that

$$q_\nu \langle p | A_\nu | s \rangle_{q=s-p} = -i \langle p | \frac{\partial A_\nu}{\partial x_\nu} | s \rangle$$

$$q_\mu \langle s | A_\mu | p \rangle_{q=s-p} = i \langle s | \frac{\partial A_\mu}{\partial x_\mu} | p \rangle .$$

With the PCAC definition of pion field, we can write

$$\frac{\partial A_\mu^{1+i2}}{\partial x_\mu} = (2)^{\frac{1}{2}} \mu^2 f_\pi \phi \, ; \quad \frac{\partial A_\mu^{1-i2}}{\partial x_\mu} = (2)^{\frac{1}{2}} \mu^2 f_\pi \phi^\dagger \; ,$$

where ϕ creates π^+, destroys π^-, and conversely for $\phi^\dagger$. Thus, for $\nu > \mu$ and $q^2 = 0$

$$m^2 \nu^2 \, \mathrm{Im}\, A(\nu, 0) = 2\mu^4 f_\pi^2 \frac{(2\pi)^4}{2} \sum_s \langle p | \phi^\dagger | s \rangle \times \langle s | \phi | p \rangle \delta(s - p - q) \; . \tag{6.30}$$

But apart from the factor $2f_\pi^2$, the quantity on the right can now be recognized to be the imaginary (absorptive) part of the forward amplitude for $\pi^+ + \mathrm{p}$ scattering, except of course that $q^2 = 0$, hence the pion has zero mass. That is, for $\pi^+ + \mathrm{p} \to \pi^+ + \mathrm{p}$ forward scattering the amplitude is

$$M^+(\nu) = \lim_{q^2 \to -\mu^2} (q^2 + \mu^2)^2 \, i \int dx \, e^{-iq \cdot x} \times \langle p | \theta(x_0) [\phi^\dagger(x), \phi(0)] \, p \rangle \; . \tag{6.31}$$

The absorptive part is

$$\mathrm{Im}\; M^{(+)} = \frac{1}{2} \lim_{q^2 \to -\mu^2} (q^2 + \mu^2)^2 \int dx \, e^{-iq \cdot x} \langle p | [\phi^\dagger(x), \phi(0)] | p \rangle \; . \tag{6.32}$$

When expanded in a sum of contributions over a complete set of states, $\mathrm{Im}\, M^{(+)}$ for $\nu > \mu$ and $q^2 = 0$ is given by

$$\mathrm{Im} M^{(+)} = \mu^4 \frac{(2\pi)^4}{2} \sum_s \langle p | \phi^\dagger | s \rangle \langle s | \phi | p \rangle \delta(s - p - q) \; ;$$

so indeed, for $\nu > \mu$ and $q^2 = 0$,

$$\mathrm{Im}\, A(\nu, 0) = \frac{2}{m^2 \nu^2} f_\pi^2 \; \mathrm{Im}\, M^{(+)}(\nu) \; . \tag{6.33}$$

In the same way one finds that $\mathrm{Im}\, \widetilde{A}\; (\nu, 0)$, for $\nu > \mu$, is related to the imaginary part of the forward amplitude for $\pi^- + p$ scattering (with pions of zero mass):

$$\mathrm{Im}\,\widetilde{A}\,(\nu, 0) = \frac{2}{m^2 \nu^2} f_\pi^2 \;\mathrm{Im}\, M^{(+)}(\nu) \quad . \tag{6.34}$$

The sum rule therefore can be written in the form

$$g_A^2 + \frac{2}{\pi} f_\pi^2 \int_{\text{thresh.}}^{\infty} \frac{d\nu}{\nu^2} \left[\mathrm{Im}\, M^{(-)}(\nu) - \mathrm{Im}\, M^{(+)}(\nu)\right] = 1 \; . \tag{6.35}$$

As always with PCAC affairs this result for off mass shell amplitudes is without content until the assumption is made that the amplitude for $q^2 = 0$ are not so different from the physical ones, $q^2 = -\mu^2$. Let's make this PCAC assumption and then relate the imaginary part of the amplitude to the total pi-nucleon cross section, via the optical theorem

$$\mathrm{Im}\, M^{(\pm)}(\nu) = |\mathbf{q}|\, \sigma^{\pm}(\nu) \; ,$$

where

$$|\mathbf{q}| = (\nu^2 - \mu^2)^{1/2} \; .$$

Then

$$g_A^2 + \frac{2}{\pi} f_\pi^2 \int_\mu^\infty d\nu \, \frac{|\mathbf{q}|}{\nu^2} \left[\sigma^{(-)}(\nu) - \sigma^{(+)}(\nu)\right] = 1 \; . \tag{6.36}$$

This is the Adler-Weisberger formula. It can be regarded as an equation which determines g_A in terms of f_π and cross section data. This yields $|g_A| = 1.21$. Or one can employ the earlier PCAC result to express f_π itself in terms of g_A : $f_\pi = m g_A / g_r$. Then g_A is determined solely by strong interaction quantities. This yields $|g_A| = 1.15$. On either view the results are remarkable when one recalls that $g_A^{\exp} \approx 1.23$.

Concerning the current algebra input to the Adler-Weisberger formula, note that the pion momentum q has been set to zero. So what is involved here is the equal time commutator of axial vector *charges:* $[\bar{Q}^i, \bar{Q}^j] = i\epsilon^{ijk} Q^k$.

7. The A-W Formula and Generalizations

The Adler-Weisberger formula, as it is written in Eq. (6.35), has the appearance of a sum rule. The content can however be displayed in the form of a low energy theorem on pi-nucleon scattering, and it is this way of looking at things that we wish to elaborate now. Consider the isotopically antisymmetric amplitude

$$M^{(0)}(\nu) = \frac{M^{(-)}(\nu) - M^{(+)}(\nu)}{2} , \tag{7.1}$$

where $M^{(-)}$ and $M^{(+)}$ are the amplitudes for forward $\pi^- \mathrm{p}$ and $\pi^+ \mathrm{p}$ scattering. We suppose that $M^{(0)}(\nu)$ has been continued off pion mass shell in the PCAC manner already described. In the end interesting results will be obtained for the limit $q^2 = 0$, where q^2 is the pion mass variable. For the moment, let us only require that q^2 be real and $\geqslant -\mu^2$. For fixed q^2, $M^{(0)}(\nu)$ is an analytic function of the complex variable ν, with nucleon poles at $\nu = \pm q^2/2m$ and with a cut running along the real ν axis except for the gap

$$-\mu - \left(\frac{q^2 + \mu^2}{2m}\right) < \nu < \mu + \left(\frac{q^2 + \mu^2}{2m}\right) .$$

Moreover $M^{(0)}(\nu)$ is an odd function of ν, $M^0(\nu) = -M^{(0)}(-\nu)$; and $M^{(0)}(\nu)/\nu$ presumably satisfies an unsubtracted dispersion relation. In general, because it is an odd function, $M^{(0)}$ must vanish at the origin $\nu = 0$. But for the special case that concerns us, $q^2 \to 0$, there is the delicacy that the poles move to the origin. These poles arise from the one-neutron intermediate state and the delicacy can be circumvented by restoring the so-far neglected mass difference $\Delta m = m_{\mathrm{n}} - m_{\mathrm{p}}$ between neutron and proton. Extracting the pole terms explicitly, we then have the dispersion relation for $q^2 = 0$,

$$\frac{M^{(0)}(\nu)}{\nu} = \frac{g_r^2}{2m^2} \frac{(\Delta m)^2}{(\Delta m)^2 - \nu^2} + \frac{2}{\pi} \int_{\text{thresh.}}^{\infty} \mathrm{d}\nu' \frac{\mathrm{Im}\, M^{(0)}(\nu')}{\nu'^2 - \nu^2} , \tag{7.2}$$

where the continuum threshold is at $\nu' = \mu + \mu^2/2m$. It now follows that

$$\lim_{\nu \to 0} \frac{M^{(0)}(\nu)}{\nu} = \frac{g_r^2}{2m^2}$$

$$+ \frac{1}{\pi} \int_{\text{thresh.}}^{\infty} \frac{d\nu'}{\nu'^2} [\operatorname{Im} M^{(-)}(\nu') - \operatorname{Im} M^{(+)}(\nu')] \; . \tag{7.3}$$

Compare this with Eq. (6.35), employing for f_π the PCAC result $f_\pi = m g_A / g_r$. Then the sum rule of Eq. (6.35) becomes the low energy theorem

$$\frac{M^{(0)}(\nu)}{\nu} \xrightarrow[\nu \to 0]{} \frac{g_r^2}{2m^2 g_A} . \tag{7.4}$$

Let us now see how to get this result more directly. For this purpose turn to Eq. (5.6), with $|a\rangle = |b\rangle = |p\rangle$ and $k = q$. Take for the currents $j_\mu^\alpha = A_\mu^\alpha$, $j_\nu^\beta = A_\nu^\beta$, $\alpha, \beta = 1, 2, 3$. For practice we now work with isotopic indices, in a real basis. Via the PCAC identification of the pion field, the first term on the right side of Eq. (5.6) becomes related to the amplitude for forward pi-nucleon scattering, with α and β as the isotopic indices of the initial and final pions, q the pion momentum:

$$M^{\beta\alpha}(\nu) = \lim_{q^2 \to -\mu^2} \frac{i(q^2 + \mu^2)^2}{\mu^4 f_\pi^2} \int dx \, e^{-iq \cdot x}$$

$$\langle p | \theta(x_0) \left[\frac{\partial A_\nu^\beta(x)}{\partial x_\nu} , \frac{\partial A_\mu^\alpha(0)}{\partial x_\mu} \right] | p \rangle \, . \tag{7.5}$$

We go off mass shell by dropping the instruction $q^2 \to -\mu^2$, and we pass to the limit $q \to 0$, retaining terms only up to first order in the four vector q (hence $q^2 = 0$, and ν is retained up to first order). Thus the first term on the right side of Eq. (5.6) is replaced by $f_\pi^2 M^{\beta\alpha}(\nu)$. The left-hand side of Eq. (5.6) is second order in q, $(k = q)$ on our artifice (?) of retaining the finite neutron-proton mass difference, so to our order the left-hand term is set equal to zero. The second term on the right is a familiar equal time commutator

$$q_\mu \int dx\, \delta(x_0) \langle p | [A_0^\beta(x), A_\mu^\alpha(0)] | p \rangle = i\epsilon^{\beta\alpha\gamma} q_\mu$$

$$\times \langle p | V_\mu^\gamma | p \rangle = -i\epsilon_{\beta\alpha\gamma} \frac{\tau^\gamma}{2} \nu \,. \tag{7.6}$$

Finally, the third term on the right side of Eq. (5.6) involves the equal time commutator of a current with the divergence of a current. It is not part of the standard apparatus of current algebra and must be regarded in its details, as model dependent. Nevertheless some useful things can be said about this so-called "σ-term". It is evidently even in the vector $\mathbf{q}$ and so, to first order at least, it is independent of $\mathbf{q}$:

$$\sigma^{\beta\alpha} \equiv i \int d^3x \langle p | [A_0^\beta(\mathbf{x}, x_0) \frac{\partial A_\mu^\alpha}{\partial x_\mu} (0, x_0)] | p \rangle$$

$$= i \int d^3x \langle p | [A_0^\beta(0, x_0), \frac{\partial A_\mu^\alpha}{\partial x_\mu} (\mathbf{x}, x_0)] | p \rangle$$

$$= i \int d^3x \langle p | [A_0^\beta(\mathbf{x}, x_0), \frac{\partial A_0^\alpha}{\partial x_0} (0, x_0)] | p \rangle \,. \tag{7.7}$$

The first equality employs translation invariance and a change of integration variable: $\mathbf{x} \to -\mathbf{x}$. The second is Gauss' theorem. The third again employs translation invariance. But

$$\sigma^{\beta\alpha} = i \frac{\partial}{\partial x_0} \int d^3x \langle p | [A_0^\beta(\mathbf{x}, x_0), A_0^\alpha(0, x_0)] | p \rangle$$

$$-i \int d^3x \langle p | \left[\frac{\partial A_0^\beta(\mathbf{x}, x_0)}{\partial x_0}, A^\circ(0, x_0) \right] | p \rangle \,.$$

The second term on the right is however equal to $\sigma^{\alpha\beta}$; hence

$$\sigma^{\beta\alpha} - \sigma^{\alpha\beta} = i \frac{\partial}{\partial x_0} \int d^3x \langle p | [A_0^\beta(\mathbf{x}, x_0), A_0^\alpha(0, x_0)] | p \rangle \,.$$

But the right-hand matrix element is evidently time independent. Hence

$$\sigma^{\beta\alpha} = \sigma^{\alpha\beta} \ . \tag{7.8}$$

Let us now write for the pi-nucleon amplitude as it enters Eq. (5.6)

$$M^{\beta\alpha} = M^{(e)}\,\delta_{\beta\alpha} + i\epsilon^{\beta\alpha\gamma}\,\frac{\tau^\gamma}{2}\,M^{(0)} \ . \tag{7.9}$$

Evidently the isotopically antisymmetric part $M^{(0)}$ receives no contribution from the $\sigma^{\beta\alpha}$ term. One then finds

$$\frac{M^{(0)}(\nu)}{\nu} \xrightarrow[\nu\to 0]{} \frac{1}{2f_\pi^2} ,$$

which agrees with Eq. (7.4) when the relation $f_\pi = mg_A/g_r$ is employed. The point $\nu = 0$ is of course unphysical for real pions. So in applying the low energy theorem in the real world one continues to $\nu = 0$ precisely by using the dispersion relation that converts the low energy theorem back into the A-W sum rule.

But there's another approach. Take the $\nu = 0$ prediction for $M^{(0)}(\nu)/\nu$ and believe it at threshold for physical pi-nucleon scattering, i.e., at $\nu = \mu$:

$$\frac{M^{(0)}(\mu)}{\mu} \approx \frac{g_r^2}{2m^2 g^2} \ . \tag{7.10}$$

After all, the exact mass shell dispersion relation reads

$$\frac{M^{(0)}(\nu)}{\nu} = \frac{g_r^2}{2m^2}\,\frac{\mu^2}{\nu^2 - (\mu^2/2m)^2} + \frac{1}{\pi}\int_\mu^\infty \frac{d\nu'}{\nu'^2 - \nu^2}\,[M^{(-)}(\nu') - M^{(+)}(\nu')] \ .$$

Set $\nu = \mu$, drop corrections of order μ^2/m^2, and, in the dispersion denominator $(\nu'^2 - \mu^2)$, drop the meson mass. Then

$$\frac{M^{(0)}(\mu)}{\mu} \approx \frac{g_r^2}{2m^2} + \frac{1}{\pi}\int_\mu^\infty \frac{d\nu'}{\nu'^2}\,[M^{(-)}(\nu') - M^{(+)}(\nu')] \ .$$

Compare with Eq. (6.35), setting $f_\pi = mg_A/g_r$, and observe that the result agrees with Eq. (7.10).

The threshold formula of Eq. (7.10) can be related to a statement about the s-wave scattering length, according to

$$M^{(0)}(\mu) = 4\pi\left(1+\frac{\mu}{m}\right)a^{(0)} , \qquad a^{(0)} = \frac{1}{2}\left[a^{(-)} - a^{(+)}\right]$$

$$= \frac{1}{3}\left(a_{1/2} - a_{3/2}\right) ; \tag{7.11}$$

where the subscripts in the last equality refer to isotopic spin. On the present interpretation of the Adler-Weisberger formula we have for the pi-nucleon scattering length

$$a^{(0)} = \frac{L}{1+(\mu/m)} , \quad L \equiv \left(\frac{\mu}{2m^2 g_A^2}\right)\left(\frac{g_r^2}{4\pi}\right) \approx 0.11\mu^{-1} . \tag{7.12}$$

Experimentally $a^{(0)} \approx 0.086\mu^{-1}$. The fit is quite satisfactory.

On the present procedure current algebra and PCAC have furnished a prediction for the isotopically antisymmetric amplitude. Concerning the isotopically even amplitude, $M^{(e)}$, we run up against the so-called σ term of Eq. (7.7). On the other hand, recall Adler's PCAC consistency result, Eq. (4.27). We in fact supposed this result could be applied at physical threshold and were led to the prediction that the isotopically even amplitude vanishes there [see Eq. (4.29)]. All these small shifts in venue presumably involve only small errors, of order, say $(\mu/m)^2$, where m is the target (proton) mass. The success which was encountered suggests that the σ term must be similarly small. We are then tempted to generalize these results to threshold behavior for pions scattering on an arbitary massive hadron target [16]. Systematically neglecting the σ terms, one finds that Eqs. (7.9), (7.10), and (4.29) generalize to

$$M^{\beta\alpha} = \frac{-\mu}{f_\pi^2}\left(I_\pi^\gamma\right)_{\beta\alpha} \langle t|I^\gamma|t\rangle . \tag{7.13}$$

Here I_π^γ is the isotopic spin operator in the SU(2) space of pions; I^γ is the same in the isospin space of the hadron target t. Let I_t be the isotopic spin of the target. Then the threshold amplitude M^I for scattering in a state of total isotopic spin I is

$$M^I(\mu) = \frac{\mu}{f_\pi^2}\left\{\frac{I_t(I_t+1)+2-I(I+1)}{2}\right\} . \tag{7.14}$$

Equivalently, for scattering lengths:

$$a_I = \frac{L}{1+(\mu/m_t)}\left\{I_t(I_t+1)-2-I(I+1)\right\} . \tag{7.14'}$$

Weinberg's Treatment of π-π Scattering

In our interpretation of the Adler-Weisberger formula describe above, we simply took over the prediction for $M^{(0)}(\nu)/\nu$ at $\nu=0$ and decided to believe it at threshold, $\nu=\mu$, which is as near to $\nu=0$ as one can get physically. On this view it can be said that we expanded $M^{(0)}$ in powers of the momentum q, fixed the linear term on the basis of current algebra and PCAC, and then accepted that the expansion to first order is sufficiently good at threshold. Weinberg has argued that in general this is a proper way to operate with PCAC [17]. Namely, interesting results are obtained typically only in the limit where a pion momentum goes to zero. This not only puts the mass into an unphysical region but other quantities also become unphysical (e.g. $\nu=-q\cdot p/m$). A systematic way to cope with the fact that all components of q have gone to zero is then to imagine expanding an amplitude in powers of q, using current algebra and PCAC to fix the coefficients of the low order terms as far as possible, then supposing that the first few terms in fact provide an adequate approximation near "threshold" (where the various invariants involving q are "as small' as physically possible). In the pi-nucleon problem the expansion was carried out to first order in q.

An especially interesting application of these ideas was made by Weinberg for the problem of π-π scattering. Consider this reaction, symbolized by

$$k(a)+p(c)\to q(b)+\ell(d) ,$$

where k, p, q, ℓ represent the momenta of the pions and a, b, c, d are their isotopic labels. Here if all the pions are to be treated on an equal footing the amplitude has to be expanded in powers of all the momenta; and it is clear that only even powers can occur. Suppose that we are interested in the behavior of the amplitude near threshold; how far must the expansion be carried? The operational answer of course is: only so far that all the coefficients are determined, and no farther! Here this means an expansion to second order in the momenta.

Introduce the usual Mandelstam variables

$$s = -(p+k)^2, \quad t = -(k-q)^2, \quad u = -(p-q)^2$$

$$s+t+u = -(p^2+k^2+q^2+\ell^2) \ . \tag{7.15}$$

Bose statistics, crossing symmetry, and isospin invariance dictate the following structure for the amplitude, to second order in momenta,

$$\begin{aligned} M^{ba}_{cd} &= \delta_{ab}\,\delta_{cd}\,[A + B(s+u) + Ct] \\ &\quad + \delta_{ad}\,\delta_{bc}\,[A + B(s+t) + Cu] \\ &\quad + \delta_{ac}\,\delta_{bd}\,[A + B(u+t) + Cs] \ , \end{aligned} \tag{7.16}$$

where A, B, C are constants, independent of the pion momenta. To check the Bose requirement observe that the amplitude is indeed even, e.g., under $a \leftrightarrow c$ and $k \leftrightarrow p$ (hence $s \to s$, $t \leftrightarrow u$). Similarly, crossing symmetry requires that the amplitude be even under $c \leftrightarrow b$, $p \leftrightarrow -q$ (hence $s \leftrightarrow t$, $u \leftrightarrow u$); and we see that this is satisfied too. And so on. What is remarkable about Eq. (7.16) is that the amplitude does not depend explicitly on the "mass" variables k^2, p^2, q^2, ℓ^2, except for their appearance in the relation $s+t+u = -(k^2+p^2+q^2+\ell^2)$. Expressed in terms of the parameters A, B, C, the physical s-wave scattering lengths a_I (I is the total isotopic spin) are given by

$$a_0 = \frac{1}{32\pi\mu}\,[5A + 8\mu^2 B + 12\mu^2 C]$$

$$a_2 = \frac{1}{32\pi\mu}\,[2A + 8\mu^2 B] \ . \tag{7.17}$$

Let's now proceed to determine the coefficients A, B, C. Introduce the amplitude

$$T_{\nu\mu} = i\int dx\, e^{-iq\cdot x}\,\langle \ell(d)|\theta(x_0)\,[A^b_\nu(x), A^a_\mu(0)]\,|p(c)\rangle$$

and refer to Eq. (5.6). The first term on the right is, up to factors, the π-π

amplitude M_{cd}^{ba}. The third term on the right, the "σ-term" σ_{cd}^{ba}, is symmetric in the indices b and a, as we have already discussed, when $k = q \to 0$. Let $q = k \to 0$, but keep $p = \ell$ on mass shell. To first order in q and k the left side of Eq. (5.6) can be neglected — there are no pole-term delicacies in the π-π problem. The second term on the right is a familiar commutator

$$k_\mu \int dx\, e^{-iq\cdot x}\, \delta(x_0) \langle \ell(d) | [A_0^b(x), A_\mu^a(0)] | p(c) \rangle$$

$$= i\epsilon^{ba\gamma} \langle \ell(d) | V_\mu^\gamma | p(c) \rangle k_\mu$$

$$\approx (i\epsilon^{ba\gamma})(i\epsilon^{d\gamma c}) 2p\cdot k \quad .$$

Thus, for $k = q \to 0$, to first order in $k = q$, we have

$$M_{cd}^{ba} = \frac{1}{f_\pi^2} [\delta_{bc}\, \delta_{da} - \delta_{ba}\, \delta_{dc}]\, 2p\cdot k + \sigma_{dc}^{ba} \quad . \tag{7.18}$$

But to this order

$$2p\cdot k \to \frac{u - s}{2} \quad . \tag{7.19}$$

The first term on the right side of Eq. (7.18) is antisymmetric in the indices a and b, whereas σ_{dc}^{ba} is symmetric. Comparing the antisymmetric parts of Eqs. (7.16) and (7.18), and using Eq. (7.19), one finds

$$C - B = \frac{1}{f_\pi^2} \simeq \frac{g_r^2}{m^2 g_A^2} \quad .$$

From Eq. (7.17) it then follows that

$$2a_0 - 5a_2 = 6L, \qquad L = \left(\frac{\mu}{2m^2 g_A^2}\right)\left(\frac{g_r^2}{4\pi}\right) \simeq 0.11\mu^{-1} \quad . \tag{7.20}$$

Next, consider the implications of the Adler PCAC consistency condition. It asserts that the amplitude must vanish when the momentum of any one pion goes to zero, all the other pions being held on mass shell. Kinematically this corresponds to the point $s = t = u = \mu^2$; and the PCAC condition yields the result

$$A + \mu^2 (2B + C) = 0 \quad . \tag{7.21}$$

To complete the analysis one more relation is needed, and here Weinberg introduces an extra physical assumption concerning the σ term in Eq. (7.18). With $p = \ell$ on mass shell, and to zeroeth order in $k = q$, the σ term is symmetric in the indices a and b. Weinberg now supposes that it is in fact proportional to $\delta_{ab}\,\delta_{cd}$. This property is at any rate true in the σ model of Gell-Mann and Levy. To zeroeth order in $k = q$ we then have

$$M_{cd}^{ba} \propto \delta_{ab}\,\delta_{cd}, \qquad s = u = \mu^2,\ t = 0 \ .$$

From Eq. (7.16) it then follows that

$$A + \mu^2 (B + C) = 0 \ . \tag{7.22}$$

Altogether one finds

$$B = 0,\ A = -\mu^2 C,\ C \approx \frac{g_r^2}{m^2 g_A^2} \ , \tag{7.23}$$

and for the scattering lengths

$$a_0 = \frac{7}{4} L \ , \quad a_2 = -\frac{1}{2} L \ . \tag{7.24}$$

The definition of the length L, as given in Eq. (7.20), is based on the relation $f_\pi = m g_A / g_r$. In fact $(f_\pi)_{\text{exp}}$ is larger than $m g_A / g_r$ by about 10%, so that if $(f_\pi)_{\text{exp}}$ were used L would decrease by $\sim$20%. However it doesn't seem profitable to worry about such fine points, PCAC being the rough notion that it is. The important thing about Weinberg's treatment of π-π scattering is that it leads to what intuitively seem to be surprisingly small scattering lengths. Unfortunately, direct tests still lie in the remote future. Various modes of indirect access to π-π scattering have been widely discussed, based for example on Chew-Low extrapolation techniques for π + nucleon $\rightarrow 2\pi$ + nucleon. But these methods seem least reliable for lower energy π-π scattering, which is what concerns us here. In principle a rather clean attack can be based on analysis of the spectrum structure for $K_{\ell 4}$ decay; and the difficult experiments are now underway in several laboratories.

8. Semi-Leptonic Decay Processes

In the application of current algebra to Compton scattering and to neutrino reactions, the currents in question arose in their direct role as interaction currents. In the derivation of the Adler-Weisberger formula, the currents entered the analysis solely through PCAC. In various other situations one encounters a mixture of these elements: one or more currents enter the analysis as "proper" interaction currents. Others enter via PCAC to represent soft pions. The photoproduction reaction $\gamma + N \to N + \pi$, and the closely related electroproduction reaction $e + N \to e + N + \pi$, are obvious examples when the pion is to be treated as soft. These processes have in fact been much discussed, but they do not lend themselves to a brief review. Instead we shall concentrate on two applications to weak decay processes: $K \to \pi + \ell + \nu$ and $K \to \pi + \pi + \ell + \nu$.

8.1 $K_{\ell 3}$ Decay

For definiteness consider the reaction

$$K^- \to \pi^0 + \ell + \bar{\nu} \ .$$

Let the symbols K and q represent the kaon and pion momenta. Observe that this reaction is induced solely by the vector part of the strangeness changing weak current, V_μ^{4+i5}; so the amplitude is given by

$$\text{Amp} = \frac{G}{(2)^{1/2}} \sin\theta_c \langle q | V_\mu^{4+i5} | K \rangle \, i\bar{u}_\ell \gamma_\mu (1 + \gamma_5) u_{\bar{\nu}} \ .$$

The hadronic matrix element has the structure

$$M_\mu \equiv \langle q | V_\mu^{4+i5} | K \rangle = \frac{1}{(2)^{1/2}} \left\{ f_+ (K+q)_\mu + f_- (K-q)_\mu \right\} \ , \qquad (8.1)$$

where the form factors $f_\pm$ are scalar functions of the momentum transfer variable $Q^2 \equiv -(K-q)^2$. When we define an off mass shell continuation, as we will shortly do, the form factors will come to depend also on the pion mass variable q^2. Indeed, using the PCAC *definition* of pion field and the standard LSZ reduction procedure one writes

$$M_\mu = \lim_{q^2 \to -\mu^2} \frac{i(q^2+\mu^2)}{\mu^2 f_\pi} \int dx\, e^{-iq\cdot x} \times \langle 0|\theta(x_0)\left[\frac{\partial A_\nu^3(x)}{\partial x_\nu},\ V_\mu^{4+i5}(0)\right]|K\rangle\ . \tag{8.2}$$

Let us now drop the instruction $q^2 \to -\mu^2$ and use the above expression to define an off shell amplitude. In the usual way an interesting result will be found for the unphysical limit $q \to 0$. Namely, define

$$T_{\nu\mu} = i\int dx\, e^{-iq\cdot x}\langle 0|\theta(x_0)[A_\nu^3(x), V_\mu^{4+i5}(0)]|K\rangle\ ,$$

as in Eq. (5.3); and from Eq. (5.4) observe that

$$iq_\nu T_{\nu\mu} = \frac{\mu^2 f_\pi}{q^2+\mu^2} M_\mu + i\int dx\, e^{-iq\cdot x}\,\delta(x_0) \times \langle 0|[A_0^3(x), V_\mu^{4+i5}(0)]|K\rangle\ .$$

In the limit $q \to 0$ the left side vanishes (there are no pole term delicacies here) and

$$M_\mu(K, q\to 0) = \frac{-i}{f_\pi}\int dx\,\delta(x_0)\langle 0|[A_0^3(x),\ V_\mu^{4+i5}(0)]|K\rangle$$

$$= \frac{-i}{2f_\pi}\langle 0|A_\mu^{4+i5}|K\rangle\ . \tag{8.3}$$

But

$$\langle 0|A_\mu^{4+i5}|K\rangle = i(2)^{1/2} f_K K_\mu\ , \tag{8.4}$$

is evidently the hadron matrix element that arises for $K \to \ell + \nu$ decay; the parameter f_K plays the same role here as does the paramter f_π for $\pi \to \ell + \nu$ decay. Thus, regarding the $K_{\ell 3}$ form factors as (scalar) functions of the kaon and pion momenta, K and q, we have

$$f_+(K, q=0) + f_-(K, q=0) = f_K/f_\pi\ . \tag{8.5}$$

This is the basic result of a standard PCAC approach to $K_{\ell 3}$ decay. For the rest, the question is how to interpret it. The $K_{\ell 3}$ problem has received a great deal of attention, both experimental and theoretical. From the present point of view the $K_{\ell 3}$ problem is perhaps a decisive one, because the steps leading to Eq. (8.5) are so straightforward and seemingly in the spirit of current algebra and PCAC. What is especially interesting then is that, on its most natural interpretation, the result expressed in Eq. (8.5) seems to disagree seriously with the present experimental trend.

The form factors are scalar functions of K and q. But physically, $K^2 = -m_K^2$ and $q^2 = -\mu^2$ are fixed; so there is only one physically variable quantity, customarily chosen to be the invariant momentum transfer

$$Q^2 \equiv -(K-q)^2 \; .$$

The form factors have however been continued off mass shell and have come to depend also on the mass variable q^2. Regarded as functions of Q^2 and q^2, the form factors $f_\pm$ are to be evaluated at $Q^2 = m_K^2$, $q^2 = 0$ for purposes of Eq. (8.5). This point is unphysical with respect to both variables, since physically $q^2 = -\mu^2$ and $m_\ell^2 < Q^2 < (m_K - \mu)^2$. However, the physical form factors are analytic in Q^2 and they can in principle be extrapolated to $Q^2 = m_K^2$; i.e., we may regard $f_\pm(Q^2 = m_K^2, q^2 = -\mu^2)$ as knowable. For Eq. (8.5), however, we need $f_\pm(Q^2 = m_K^2, q^2 = 0)$. As usual the equation has no content until it is said how the off mass shell quantities are related to on shell quantities. For this we invoke the PCAC hypothesis, which presumably says that the form factors are gently varying in q^2; so we are tempted to set $f_\pm(Q^2, q^2 = 0) \approx f_\pm(Q^2, q^2 = -\mu^2)$. But is it indeed so that the q^2 dependence is gentle when Q^2 is held fixed; or does gentleness obtain when some other quantity is held fixed, e.g., $2K \cdot q = Q^2 + q^2 - m_K^2$? If it were to happen that the physical form factors are slowly varying in Q^2 (over the region of interest) they would of course also be slowly varying in $K \cdot q$, or any other variable. In this situation the choice of variables would make no difference for the purposes of PCAC. If the form factors were rapidly varying, the choice of variables would matter and PCAC would lose meaning (in the absence of further instructions, not yet available).

Regarding the physical form factors as functions of Q^2, one customarily employs a linear parametrization of the data, according to

$$f_\pm(Q^2, q^2 = -\mu^2) = f_\pm(0, -\mu^2) \left\{ 1 + \lambda_\pm \frac{Q^2}{\mu^2} \right\} \; . \tag{8.6}$$

There is no compelling evidence to suggest that a linear fit won't do, although the last word may well not yet be in on this question. For f_+ the slope parameter seems to be in the vicinity of $\lambda_+ \approx 0.03$. This is small; nevertheless it implies a variation of about 40% in f_+, between $Q^2 = 0$ and $Q^2 = m_K^2$. For f_- the slope parameter is less well determined; the evidence seems to be compatible with $|\lambda_-| << \lambda_+$.

For the moment let us accept the linear parameterization of Eq. (8.6) — this is an experimentally resolvable matter. More dubious, let us for the moment guess that the form factors vary gently in q^2 when Q^2 is held fixed: let's take PCAC to apply when the variables for the off shell form factors are taken to be q^2 and Q^2. Then Eq. (8.5) becomes

$$\left(1 + \lambda_+ \frac{m_K^2}{\mu^2}\right) + \xi(0, -\mu^2)\left(1 + \lambda_- \frac{m_K^2}{\mu^2}\right) = \frac{f_K}{f_\pi f_+(0, -\mu^2)} \quad , \tag{8.7}$$

where

$$\xi(0, -\mu^2) = \frac{f_-(0, -\mu^2)}{f_+(0, -\mu^2)} \quad . \tag{8.8}$$

We may recall that the overall $K_{\ell 3}$ decay amplitude contains a Cabibbo factor $\sin\theta_c$. But the same factor occurs for $K_{\ell 2}$ decay. Hence the ratio $f_K/f_+(0, -\mu^2)$ is experimentally measurable, up to algebraic sign, independently of the Cabibbo factor. On the other hand, from $\pi_{\ell 2}$ decay one measures $f_\pi \cos\theta_c$. The Cabibbo angle is rather small; and it is well enough known so that $\cos\theta_c$ is not sensitive to the residual uncertainties. Thus the right-hand side of Eq. (8.7) is reasonably well known, up to algebraic sign. Now if SU(3) were an exact strong interaction symmetry, the Cabibbo model would predict $f_- = 0$, hence $\xi = 0$; also $f_+(0, -\mu^2) = 1$. Moreover, it would follow that $f_K = f_\pi$. So the right side of Eq. (8.7) would be equal to unity. Empirically, the *magnitude* is about 1.28; and we may believe enough of SU(3) to accept that the algebraic sign is positive. Thus experimentally

$$\frac{f_K}{f_\pi f_+(0, -\mu^2)} \approx 1.28 \ .$$

If now we also accept that $\lambda_+ \approx 0.03$ and that $\lambda_- \approx 0$, then Eq. (8.7) leads to the prediction $\xi(0, -\mu^2) \approx -0.1$. The experiments [18] however, seem to be converging on a much more negative, $\xi(0, -\mu^2) \approx -1$!

It may happen that this large discrepancy will go away: the ξ parameter as determined from Eq. (8.7) is fairly sensitive to the slope parameters and, for that matter, to the assumption of linearity in Eq. (8.6). Should a substantial discrepancy nevertheless survive, one could of course call current algebra into question — at least that part of the doctrine that figures in $K_{\ell 3}$ decay. But it is PCAC that would most likely be suspect. As repeatedly emphasized, PCAC is not a notion that can be stated with any precision in universal terms. Indeed, with no change in its basic spirit we could have gotten a different prediction for $K_{\ell 3}$ decay. Suppose that the presumed gentleness in off shell extrapolation applies when some quantity other than Q^2 is held fixed, e.g., a quantity $X = C_1(Q^2 + \rho q^2 + C_2)$, where C_1 and C_2 are irrelevant constants, ρ a relevant one. For example, with $K \cdot q$ chosen as the "natural" variable X, one has $\rho = 1$. A little arithmetic shows that the left-hand side of Eq. (8.7) would acquire a correction term $\rho[\lambda_+ + \xi\lambda_-]$. Once given that λ_+ and/or λ_- are non-vanishing, by a suitable choice of "natural" variable one can get any answer he wishes. To be sure, some reasonably guidelines have to be followed. In order to favor pion pole dominance one wants to choose variables in such a way that other singularities in the q^2 plane are as unimportant — "far away" — as possible. But this is hard to formulate in a general way. Empirically successful procedures discovered for one problem do not, *a priori,* indicate what procedures are to be followed in another problem.

Choice of variables is one ambiguity. But there are other and perhaps deeper ambiguities. Let us return for a moment to the case of forward pi-nucleon scattering: $q(\alpha) + p \to q(\beta) + p$. Using the PCAC choice of pion field, we had earlier represented the amplitude according to

$$M^{\beta\alpha} = \frac{(q^2 + \mu)^2}{\mu^4 f_\pi^2} \; i \int dx\, e^{-iq\cdot x} \langle p | \theta(x_0)$$
$$\times \left[\frac{\partial A_\nu^\beta(x)}{\partial x_\nu} , \frac{\partial A_\mu^\alpha(0)}{\partial x_\mu} \right] | p \rangle \, . \tag{8.9}$$

This is of course exact on mass shell, $q^2 \to -\mu^2$, but we used this expression to define the amplitude off mass shell, supposing that the amplitude changes slowly between $q^2 = -\mu^2$ and $q^2 = 0$. Alternatively, we could have proceeded in two steps, as follows. Define

$$\overline{M}^{\beta\alpha} = \frac{(q^2+\mu^2)}{\mu^2 f_\pi} \langle p, q(\beta) | \frac{\partial A^\alpha_\mu}{\partial x_\mu} | p \rangle$$

$$= -i \frac{(q^2+\mu^2)}{\mu^2 f_\pi} q_\mu \langle p, q(\beta) | A^\alpha_\mu | p \rangle$$

$$= \frac{(q^2+\mu^2)^2}{\mu^4 f_\pi^2} q_\mu \int \mathrm{d}x e^{-iq\cdot x}$$

$$\times \langle p | \theta(x_0) \left[\frac{\partial A^\beta_\nu(x)}{\partial x_\nu}, A^\alpha_\mu(0) \right] | p \rangle . \qquad (8.10)$$

On mass shell this is again exact, and $\overline{M}^{\beta\alpha} = M^{\beta\alpha}$ there. But off mass shell $M^{\beta\alpha}$ and $\overline{M}^{\beta\alpha}$ represent two different kinds of continuations, although each seems to be within the spirit of PCAC. At $q^2 = 0$ the two expressions differ precisely by the σ-term:

$$\overline{M}^{\beta\alpha} - M^{\beta\alpha} \xrightarrow[q^2 \to 0]{} \frac{i}{f_\pi^2} \int \mathrm{d}x\, e^{-iq\cdot x}\, \delta(x_0)$$

$$\times \langle p | A^\alpha_0(x), \frac{\partial A^\beta_\nu(0)}{\partial x_\nu} | p \rangle .$$

(8.11)

To be sure this difference, empirically, did not seem to amount to much for pion-nucleon scattering (i.e., the σ term seems small). But if the difference *had* been significant the problem would have arisen as to which PCAC path has to be followed. Theorists of course can always be relied on to find arguments that favor the empirically successful choices when choices have to be made. Notice that in Weinberg's treatment of π-π scattering a non-vanishing σ term was essential.

Similar ambiguities show up for $K_{\ell 3}$ decay. We defined the off mass shell amplitude on the basis of Eq. (8.2) and were then led to an interesting result for the combination $f_+ + f_-$ in the limit $q \to 0$. Suppose instead that we now focus on the scalar amplitude

$$D \equiv i\langle q | \frac{\partial V_\mu^{4+i5}}{\partial x_\mu} | K\rangle = -(K-q)_\mu \langle q | V_\mu^{4+i5} | K\rangle$$

$$= \frac{1}{(2)^{1/2}} [m_K^2 - \mu^2) f_+ + Q^2 f_-] \ . \tag{8.12}$$

According to the standard reduction formula

$$D(K, q) = -\frac{(q^2 + \mu^2)}{\mu^2 f_\pi} \int dx\, e^{-iq\cdot x} \langle 0 | \theta(x_0)$$

$$\times \left[\frac{\partial A_\nu^3(x)}{\partial x_\nu} , \frac{\partial V_\mu^{4+i5}(0)}{\partial x_\mu} \right] | K\rangle \ , \tag{8.13}$$

where the limit $q^2 \to -\mu^2$ is understood for the physical amplitude. However, let us drop this instruction and base an off shell continuation on this expression. From Eq. (5.2), for the limit $q \to 0$ (where the left-hand side vanishes) we then find

$$D(K, q = 0) = \frac{1}{f_\pi} \int dx\, \delta(x_0)$$

$$\times \langle 0 | \left[A_0^3(x), \frac{\partial V_\mu^{4+i5}(0)}{\partial x_\mu} \right] | K\rangle \ . \tag{8.14}$$

In contrast to the earlier procedure, we are now continuing a different combination of f_+ and f_- off mass shell and in a different way. Here we obtain a result for the quantity

$$f_+^D(K, q = 0) + \frac{m_K^2}{m_K^2 - \mu^2} f_-^D(K, q = 0) \ ,$$

where the superscript D is to remind us that off shell form factors defined here need not be the same as the off shell quantities defined by the earlier procedure — although *on* mass shell they must agree. Since $m_K^2 >> \mu^2$ the above combination is very nearly $f_+^D + f_-^D$ and we have

$$f_+^D (K, q=0) + f_-^D (K, q=0) \simeq \frac{(2)^{1/2}}{m_K^2 f_\pi} \int dx\, \delta(x_0)$$

$$\times \langle 0 | \left[A_0^3 (x), \frac{\partial V_\mu^{4+i5}(0)}{\partial x_\mu} \right] | K \rangle \,. \qquad (8.15)$$

Quite apart from the question of choice of variables, which shall we believe extrapolates gently off mass shell, $f_+ + f_-$ or $f_+^D + f_-^D$?

The procedure which leads to Eq. (8.5) has the advantage that the right-hand side is known, from the standard hypotheses of current algebra. The right-hand side of Eq. (8.15) on the other hand is not specified by current algebra or any other general doctrine. This of course does not preclude new hypotheses being advanced. Still, wishing to have a definite prediction one is tempted to shape the meaning of PCAC in that way which leads to Eq. (8.5). Since Eq. (8.15) nevertheless seems also to be in the spirit of PCAC, one hopes that the right hand side has the "right" value f_K/f_π. If there were some independent way to determine the right side, and if it gave the "wrong" value, one would have to choose between the alternatives, or let experiment choose, or devise other alternatives — or, what could well happen, discover that experiment doesn't accord with any reasonable alternative.

As said, the commutator term in Eq. (8.15) is not part of the usual current algebra apparatus. But some ideas which bear on this commutator are much under discussion these days. It would carry us too far afield to discuss these matters in detail. But briefly, recall that the algebra of the vector and axial vector charges, $Q^\alpha = \int d^3x\, V_0^\alpha$ and $\bar{Q}^\alpha = \int d^3x A_0^\alpha$ is that associated with the group SU(3) × SU(3): the quantities $Q_R^\alpha \equiv Q^\alpha - \bar{Q}^\alpha$ commute among themselves like generators of SU(3); similarly the $Q_L^\alpha \equiv Q^\alpha + \bar{Q}^\alpha$ commute among themselves like SU(3) generators; but the Q_R^α and Q_L^α have vanishing equal time commutators. If the strong interactions were invariant under SU(3) × SU(3), the charges would be conserved, and the symmetry would reflect itself either through the occurrence of parity doublings of all particles or through the existence of an octet of massless (Goldstone) bosons. Although the real world is surely far from being SU(3) × SU(3) invariant, perhaps some remnants can be perceived. The charges, at any rate, are supposed to have the SU(3) × SU(3) algebra; and maybe the pseudoscalar octet particles ($\pi, K, \bar{K}, \eta$) are the Goldstone bosons.

It has been suggested by Gell-Mann [20] that the strong interaction Hamiltonian has a part which is SU(3) × SU(3) invariant and a part, $\mathcal{H}'$, which breaks the symmetry, badly perhaps, but in a simple way. He conjectured, namely, that

$$\mathcal{H}' = \alpha_0 u_0 + \alpha_8 u_8 \quad , \tag{8.16}$$

where α_0 and α_8 are strength parameters and u_0 and u_8 belong to a *common* multiplet of scalar operators which transform under SU(3) × SU(3) according to the representation $(\mathbf{3}, \bar{\mathbf{3}}) + (\bar{\mathbf{3}}, \mathbf{3})$. The operator u_0 preserves ordinary SU(3) symmetry, whereas u_8 transforms like the eighth component of an ordinary SU(3) octet. The multiplet $(\mathbf{3}, \bar{\mathbf{3}}) + (\bar{\mathbf{3}}, \mathbf{3})$ contains 18 members in all: nine scalar operator u_α ($\alpha = 0, 1, 2 \ldots 8$), and nine pseudo-scalar operators v_α. The stated transformation properties specify the equal time commutators of the charges with the u's and v's; e.g.,

$$[Q^q(x_0), u_b(\mathbf{x}, x_0)] = i f_{abc}\, u_c\, \delta(\mathbf{x}) \;.$$

We do not write out all the commutation relations here. The important thing is that, in the presumed absence of derivative couplings, the current divergences can now be specified:

$$\frac{\partial V_\mu^\alpha(x)}{\partial x_\mu} = i[\mathcal{H}'(x), Q^\alpha(x_0)] \;. \tag{8.17}$$

$$\frac{\partial A_\mu^\alpha(x)}{\partial x_\mu} = i\,]\,\mathcal{H}'(x), \bar{Q}^\alpha(x_0)] \;. \tag{8.18}$$

From the commutation relations one finds, in particular,

$$\frac{\partial V_\mu^{4+i5}}{\partial x_\mu} = i\,\frac{(3)^{1/2}}{2}\,\alpha_8\, u^{4+i5} \quad , \tag{8.19}$$

and

$$[A_0^3(\mathbf{x}, 0), u^{4+i5}(0)] = \frac{i}{2}\, v^{4+i5}(0)\,\delta(\mathbf{x}) \;. \tag{8.20}$$

But also

$$\frac{\partial A_\mu^{4+i5}}{\partial x_\mu} = \left[\left(\frac{2}{3}\right)^{1/2} \alpha_0 - \frac{1}{2(3)^{1/2}} \alpha_8\right] v^{4+i5}(x) \ . \tag{8.21}$$

Putting these results together, and recalling that

$$\langle 0 | \frac{\partial A_\mu^{4+i5}}{\partial x_\mu} | K \rangle = (2)^{1/2} m_K^2 f_K \quad ,$$

we find for the right-hand side of Eq. (8.15) the expression

$$\left(\frac{f_K}{f_\pi}\right) \cdot \left(\frac{3\alpha_8}{\alpha_8 - 2(2)^{1/2} \alpha_0}\right) \quad . \tag{8.22}$$

To achieve agreement with Eq. (8.5) one requires

$$\alpha_8 \approx -(2)^{1/2} \alpha_0 \ . \tag{8.23}$$

But with this choice of parameters Eq. (8.18) and the basic commutation relations lead to

$$\frac{\partial A_\mu^i}{\partial x_\mu} = 0, \quad i = 1, 2, 3 \ .$$

That is, the ratio α_8 / α_0 specified by Eq. (8.23) corresponds to invariance under the subgroup SU(2) × SU(2), the strangeness-conserving axial vector currents being divergenceless. The real world is of course not quite like this. But consider the familiar relations

$$\langle 0 | \frac{\partial A_\mu^i}{\partial x_\mu} | \pi^j \rangle = \mu^2 f_\pi \, \delta_{ij} \ , \qquad i, j = 1, 2, 3 \ ,$$

where the square of the pion mass appears on the right-hand side. The pion is by a substantial margin the lightest of all hadrons; and in this limited sense it may be said that the axial vector divergence is indeed small, if not quite vanishing. So the relation of Eq. (8.23) is generally regarded as being a reasonably good approximation — approximate SU(2) × SU(2) symmetry (or PCAC — partially conserved axial vector current). On the other hand, suppose all of this is wrong and that, at the opposite extreme SU(2) × SU(2) is much more badly broken than SU(3): $|\alpha_8| << |\alpha_0|$. In this extreme the right hand side of Eq. (8.15) would approach zero! And the prediction based on Eq. (8.15) would be $\xi \approx -1$! That is, with SU(2) × SU(2) badly broken (relative to SU(3) the two approaches to $K_{\ell 3}$ decay described here lead to very different results. Of course the result expressed in Eq. (8.22) is built on a particular model of SU(3) × SU(3) breaking. And within the model the *quantitative* prediction depends on a parameter α_8/α_0 which can, *a priori*, be freely adjusted. Independent evidence on both matters is clearly needed. It is not our purpose here to select between Eqs. (8.5) and (8.15), or to fix the parameters in Eq. (8.15). What is brought out, however, is again the lack of precision of the PCAC notion.

8.2 $K_{\ell 4}$ *Decay*

Consider the $K_{\ell 4}$ decay reaction

$$K^- \to \pi^+ + \pi^- + \ell + \bar{\nu} \quad .$$

Here contributions arise both from the axial vector and vector parts of the strangeness-changing current. The structure of the corresponding hadron matrix elements is

$$\langle q^{(+)}, q^{(-)} | A_\mu^{4+i5} | K \rangle = \frac{i}{m_K} \{ F_i (q^{(-)} + q^{(+)})_\mu$$

$$+ F_2 (q^{(-)} - q^{(+)})_\mu + F_3 (k - q^{(-)} - q^{(+)})_\mu \} \quad , \qquad (8.24)$$

$$\langle q^{(+)}, q^{(-)} | V_\mu^{4+i5} | K \rangle = \frac{i}{m_K} F_4 \, \epsilon_{\mu\nu\rho\sigma} K_\nu q_\rho^{(+)} q_\sigma^{(-)} \quad . \qquad (8.25)$$

The form factors F_i are scalar functions of the momenta $K, q^{(+)}, q^{(-)}$, so they depend on three physical variables, say $K \cdot q^{(+)}, K \cdot q^{(-)}, (q^{(+)} + q^{(-)})^2$. However we will soon be going off shell in the pion masses, so the F_i will come to depend on the mass variables $q^{(+)2}, q^{(-)2}$.

All of the PCAC ambiguities which have been belabored in connection with $K_{\ell 3}$ decay will arise also for $K_{\ell 4}$ decay — with respect to choice of variables (there are more of them here) and mode of off mass shell continuation.

All the warnings and branchings need not again be stated. But because two pions are involved in $K_{\ell 4}$ decay a new kind of feature arises. We have the option to treat one pion or the other, or both, as soft; and it is this aspect of the problem that we want to especially focus on. The vector current matrix element of Eq. (8.25) vanishes when either pion has zero momentum and we shall not learn anything about it on present methods. So consider the axial vector matrix element. As before, the question arises whether PCAC is to be applied to this matrix element, or to the one involving the divergence of the current. Here let us follow the traditional path in order to get definite results [17], but this is without prejudice as to the relative merit of the alternative procedure.

Define the amplitude

$$M_\mu^{(-)} = i\,\frac{q^{(-)2}+\mu^2}{(2)^{1/2}\mu^2 f_\pi}\int \mathrm{d}x\, e^{-iq^{(-)}\cdot x}\,\langle q^{(+)}|\,\theta(x_0)$$

$$\times\left[\frac{\partial A_\nu^{1+i2}(x)}{\partial x_\nu}\,,\,A_\mu^{4+i5}(0)\right]|K\rangle\,, \qquad (8.26)$$

and observe that *on* mass shell, $q^{(-)2}\to-\mu^2, M_\mu^{(-)}=\langle q^{(+)}\,q^{(-)}|A_\mu^{4+i5}|K\rangle$. We use Eq. (8.26) to define a continuation off mass shell for the π^- meson, with π^+ remaining on mass shell. In the standard way, one finds for the limit $q^{(-)}\to 0$

$$M_\mu^{(-)}\xrightarrow[q^{(-)}\to 0]{}\frac{-i}{(2)^{1/4} f_\pi}\int \mathrm{d}x\,\delta(x_0)$$

$$\times\ \langle q^{(+)}|[A_0^{1+i2}(x),A_\mu^{4+i5}(0)]\,|K\rangle\ . \quad (8.27)$$

But the commutator on the right side of the equation is zero! So we immediately learn that

$$F_1(K,q^{(+)},q^{(-)}=0) = F_2(K,q^{(+)},q^{(-)}=0)\ . \qquad (8.28)$$

$$F_3(K,q^{(+)},q^{(-)}=0) = 0\ . \qquad (8.29)$$

Next define

$$M_\mu^{(+)} = i\,\frac{q^{(+)2}+\mu^2}{(2)^{1/2}\mu^2 f_\pi}\int \mathrm{d}x\, e^{-iq^{(+)}\cdot x}$$

$$\times\ \langle q^{(-)}|\theta(x_0)\left[\frac{\partial A_\nu^{1+i2}(x)}{\partial x_\nu}\,,\,A_\mu^{4+i5}(0)\right]|K\rangle\,, \tag{8.30}$$

which provides a basis for continuing off shell in the π^+ mass. In the limit $q^{(+)}\to 0$ one finds

$$M_\mu^{(+)} \xrightarrow[q^{(+)}\to 0]{} \frac{-i}{(2)^{1/2} f_\pi}\int \mathrm{d}x\,\delta(x_0)$$

$$\times\ \langle q^{(-)}|[A_0^{1-i2}(x), A_\mu^{4+i5}(0)]|K\rangle$$

$$= \frac{-i}{(2)^{1/2} f_\pi}\langle q^{(-)}|V_\mu^{6+i7}|K\rangle$$

$$= \frac{i}{f_\pi}\langle q^{(-)}|V_\mu^{4+i5}|K\rangle\,. \tag{8.31}$$

The last step follows from isospin invariance and produces the matrix element for $K_{\ell 3}^-$ decay, with $q^{(-)}$ the π^0-momentum there. Thus

$$F_3(K, q^{(+)}=0, q^{(-)}) = \frac{m_K}{(2)^{1/2} f_\pi}[f_+(K, q^{(-)}) + f_-(K, q^{(-)})]\,, \tag{8.32}$$

$$F_1(K, q^{(+)}=0, q^{(-)}) + F_2(K, q^{(+)}=0, q^{(-)})$$

$$= (2)^{1/2}\frac{m_K}{f_\pi} f_+(K, q^{(-)})\,. \tag{8.33}$$

What is striking here is that, unless $f_+ + f_- \approx 0$, F_3 has very different values for the two limits $q^{(-)} \to 0$ and $q^{(+)} \to 0$. On the other hand the present results are at least compatible with F_1 and F_2 being slowly varying functions of their arguments. On contracting the hadron matrix element with the lepton current matrix element, one finds that the form factor F_3 gets to be multiplied by the lepton mass. So F_3 is essentially undetectable in $K_{\ell 4}$ decay and nothing is known about it experimentally. For F_1 and F_2, in crudest approximation we are at present entitled to treat them as essentially constant (this neglects the rather small variation of f_+ in the variable $(K - q^{(-)})^2$). Hence

$$F_1 = F_2 \approx \frac{m_K}{(2)^{1/2} f_\pi} f_+ \, . \tag{8.34}$$

The present available experimental evidence is in rough accord with these predictions, at the 50% level, say. Although nothing is known about the form factor F_3, the very different values obtained theoretically for the two limits discussed above calls for some explanation, even though no paradox is involved. Such an explanation has been provided by Weinberg, again within the framework of current algebra and PCAC. He considered the additional information that becomes available when one allows both pions to become simultaneously soft. Rather than develop the rather lengthy techniques needed for dealing with such an analysis, let us see the essence of Weinberg's argument in a more intuitive way. In effect the rapid dependence on the momenta $q^{(+)}$ and $q^{(-)}$ which is displayed (theoretically) by F_3 arises from the Feynman diagram shown. At the vertex which couples the virtual meson to the current A_μ^{4+i5} there occurs the factor $(2)^{1/2} f_K (K - q^{(-)} - q^{(+)})_\mu$, where $K - q^{(-)} - q^{(+)}$ is the momentum of the virtual K-meson. This already shows that the above diagram contributes exclusively to the form factor F_3. The K-meson propagator contributes a factor $\{(K - q^{(-)} - q^{(+)})^2 + m_K^2\}^{-1}$, and in the limit where both $q^{(+)}$ and $q^{(-)}$ are small this is

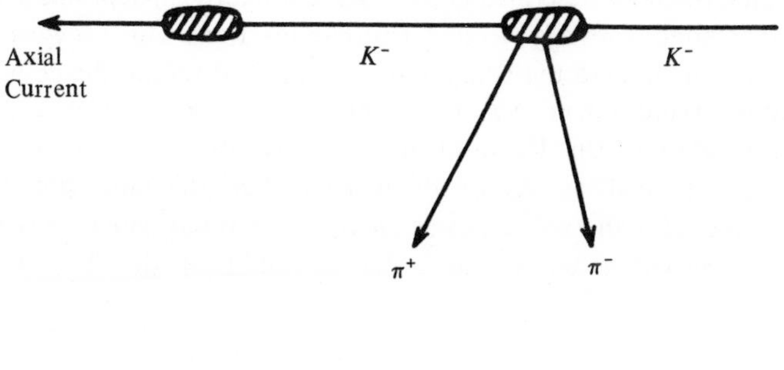

is proportional to $[2K\cdot(q^{(-)}+q^{(+)})]^{-1}$. The $(KK\pi\pi)$ vertex can itself be discussed by the methods of current algebra and PCAC. For present purposes, however, it is enough to notice, on the basis of the Adler and PCAC consistency condition, that in the soft pion limit the pions come out in a relative p-wave state. Hence this vertex contributes a factor proportional to $K\cdot(q^{(-)}-q^{(+)})$.

Altogether then, in the limit $q^{(-)}\to 0$, $q^{(+)}\to 0$, the diagram makes a contribution to F_3 which is of the form

$$C_1\,\frac{K\cdot(q^{(-)}-q^{(+)})}{K\cdot(q^{(-)}+q^{(+)})}\,,$$

where C_1 is some constant. Notice that this function is formally of zeroeth order in the momenta and yet is momentum dependent. All the other contributions to F_3, from non-pole diagrams, can be lumped into an additive constant, so long as we stick to zeroeth order in the momenta. Thus

$$F_3 = C_1\,\frac{K\cdot(q^{(-)}-q^{(+)})}{K\cdot(q^{(-)}+q^{(+)})} + C_2$$

The constants C_1 and C_2 can now be adjusted by matching this expression to the results obtained earlier for the two limits $q^{(-)}\to 0$ and $q^{(+)}\to 0$. In this way we find an expression for F_3 which "interpolates" between these limits, namely, the result [21]

$$F_3 = \frac{m_K}{(2)^{1/2}f_\pi}\,[f_+ + f_-]\,\frac{K\cdot q^{(-)}}{K\cdot(q^{(-)}+q^{(+)})}\,. \qquad (8.35)$$

9. Non-Leptonic Decays

The soft pion methods described in the last section for semi-leptonic processes can be transferred, formally, to non-leptonic reactions which involve pions. But one encounters here the commutator of an axial vector charge with the non-leptonic Hamiltonian. So it becomes necessary to take a position as to the structure of this Hamiltonian, or at any rate as to the commutators which enter the analysis. As we discussed earlier, the most popular and attractive model for the non-leptonic Hamiltonian is one which ascribes to it a current $\times$ current structure. In order to build in the $\Delta I = 1/2$ rule for the strangeness-changing interactions, it is necessary however to invoke the coupling of neutral as well as charged currents. And if the Hamiltonian is to have an octet character, so far as SU(3) is concerned, the structure becomes fully specified. Namely, the strangeness-changing Hamiltonian becomes the sixth component $\mathcal{H}^6$ of an octet of operators $\mathcal{H}^\gamma$, $\gamma = 1, 2 \ldots 8$, where

$$\mathcal{H}^\gamma = \text{const.}\, d_{\gamma\alpha\beta}\, j_\mu^\alpha\, j_\mu^\beta$$

$$j_\mu^\alpha = V_\mu^\alpha + A_\mu^\alpha \quad . \tag{9.1}$$

The coefficients $d_{\gamma\alpha\beta}$ are totally symmetric in their indices and are defined by the anticommutation relations among the SU(3) matrices λ^α:

$$\left\{ \frac{\lambda^\alpha}{2}, \frac{\lambda^\beta}{2} \right\} = \frac{1}{3}\, \delta_{\alpha\beta} + d_{\gamma\alpha\beta}\, \frac{\lambda^\gamma}{2} \quad . \tag{9.2}$$

The weak non-leptonic Hamiltonian $\mathcal{H}^6$ is of course charge conserving and hermitian; and it contains both a $\Delta S = 1$ and a $\Delta S = -1$ piece.

Consider now a process

$$A \to B + \pi^\alpha \quad ,$$

where the index α describes the isospin state of the pion. Let q denote the momentum of the pion. To lowest order in the weak interactions, the amplitude for this process is given by

$$\text{Amp}\,(A \to B + \pi^\alpha) = i\, \frac{q^2 + \mu^2}{\mu^2 f_\pi} \int dx\, e^{-iq\cdot x}\, \langle B\,|\,\theta\,(x_0)$$

$$\times \left[\frac{\partial A_\nu^\alpha(x)}{\partial x_\nu}\,,\, \mathcal{H}^6 \right] |\,A\rangle \,, \tag{9.3}$$

where the instruction $q^2 \to -\mu^2$ is understood. In the usual way we remove this instruction and employ the above expression to effect an off mass shell continuation; and as usual we find an interesting result for the limit $q \to 0$.

Namely

$$\text{Amp}(\text{A} \to \text{B} + \pi^{\alpha}) \xrightarrow[q \to 0]{} \frac{-i}{f_\pi} \int dx\, \delta(x_0) \langle \text{B} | [A_0^{\alpha}(x), \mathcal{H}^6(0)] | \text{A} \rangle . \tag{9.4}$$

On the model which has been adopted here for the weak Hamiltonian the above commutator can be worked out explicitly. One finds

$$\int d^3x \, [A_0^{\alpha}(\mathbf{x}, 0), \mathcal{H}^6(0)] = i f_{\alpha 6 \beta} \mathcal{H}^{\beta} ; \tag{9.5}$$

and therefore

$$\text{Amp}(\text{A} \to \text{B} + \pi^{\alpha}) \xrightarrow[q \to 0]{} \frac{1}{f_\pi} f_{\alpha 6 \beta} \langle \text{B} | \mathcal{H}^{\beta} | \text{A} \rangle . \tag{9.6}$$

This is the central result, and a useful one whenever the matrix element on the right is itself relateable to a physical process or is otherwise discussable.

9.1 K → 3π Decays

Let us begin by reviewing the standard phenomenology of these reactions. To a reasonably good approximation the $\text{K} \to 3\pi$ decays appear to fit the $\Delta I = 1/2$ rule; and to an even better approximation they are in accord with CP invariance. Our model Hamiltonian of course has these features built into it from the start. Accordingly, if we ignore the slight CP impurities in the states K_S^0 and K_L^0, the reaction $\text{K}_\text{S}^0 \to 3\pi^0$ is supposed to be forbidden; and for $\text{K}_\text{S}^0 \to \pi^+\pi^-\pi^0$ the pions emerge in an $I = 0$ state, which is totally antisymmetric and therefore centrifugally inhibited. Neither of these processes has as yet been seen. For the remaining processes $\text{K}^+ \to \pi^+ + \pi^+ + \pi^-$, $\text{K}^+ \to \pi^+\pi^0\pi^0$, $\text{K}_\text{L}^0 \to \pi^+\pi^-\pi^0$, $\text{K}_\text{L}^0 \to 3\pi^0$, the final pions are supposed to emerge in states of unit isotopic spin. Kinematically, the reaction $\text{K} \to 3\pi$ is fully specified by any two of the pion energies $\omega_1, \omega_2, \omega_3$, as measured in the K meson rest frame. Of course, $m_K = \omega_1 + \omega_2 + \omega_3$. To a good approximation the maximum possible energy of any one pion is

$$\omega_{\text{max}} \approx \frac{2}{3} Q + \mu ; \quad Q \equiv m_\text{K} - 3\mu .$$

One usually supposes that the amplitude for a particular $\text{K} \to 3\pi$ reaction can be well represented by the first few terms in a power series expansion in the energies ω_i; and indeed, phenomenologically an expansion up to linear terms only seems to work remarkably well. To this order the $\Delta I = 1/2$ rule and Bose statistics impose a very simple structure, the amplitudes for all four processes under discussion being expressible in terms of two parameters. Namely, one finds

$$\mathrm{Amp}\,(\mathrm{K}^+ \to 2\pi^+ + \pi^-) = 2\rho\left(1 + \frac{1}{2}\lambda_{++-}\, t_-\right)$$

$$\mathrm{Amp}\,(\mathrm{K}^+ \to 2\pi^0 + \pi^+) = \rho\left(1 + \frac{1}{2}\lambda_{00+}\, t_+\right)$$

$$\mathrm{Amp}\,(\mathrm{K}^0_{\mathrm{L}} \to \pi^+\pi^-\pi^0) = -\rho\left(1 + \frac{1}{2}\lambda_{+-0}\, t_0\right)$$

$$\mathrm{Amp}\,(\mathrm{K}^0_{\mathrm{L}} \to 3\pi^0) = 3\rho \quad , \tag{9.7}$$

where

$$t_i \equiv \frac{3}{Q}(\omega_i - \mu) - 1 \; , \tag{9.8}$$

and

$$\lambda_{00+} = \lambda_{+-0} = -2\lambda_{++-} \; . \tag{9.9}$$

Insofar as our theoretical approximations enforce this phenomenology, it is enough then to consider the single process $\mathrm{K}^+ \to \pi^+\pi^+\pi^-$ in order to get at the two independent parameters ρ and λ_{++-}.

From Eq. (9.6) we get two pieces of information by considering in turn the limit where the π^--momentum $q^{(-)}$ goes to zero or a π^+-momentum $q^{(+)}$ goes to zero. In either case it is only one pion at a time that is taken off mass shell.

Taking $q^{(-)} \to 0$, we have

$$\mathrm{Amp}\,(\mathrm{K}^+ \to \pi^+ + \pi^+ + \pi^-) \xrightarrow[q^{(-)}\to 0]{} - \frac{i}{2(2)^{1/2} f_\pi}\langle \pi^+\pi^+ | \mathcal{H}^{4+i5} | \mathrm{K}^+\rangle = 0 \; . \tag{9.10}$$

The matrix element on the right vanishes, simply from consideration of strangeness conservation. With $q^{(+)} \to 0$, we have

$$\mathrm{Amp}\,(K^+ \to \pi^+ + \pi^+ + \pi^-) \xrightarrow[q^{(+)} \to 0]{} \frac{i}{2(2)^{1/2} f_\pi} \langle \pi^+ \pi^- | \mathcal{H}^{4-i5} | K^+ \rangle$$

$$= \frac{1}{2f_\pi} \mathrm{Amp}\,(K_S^0 \to \pi^+ \pi^-) \; . \qquad (9.11)$$

The second step follows purely from isospin considerations.

As usual with PCAC, these results for off mass shell amplitudes are purely formal until we say how they are to be related to the on shell $K \to 3\pi$ amplitude. Which variables are to be held fixed in order to produce a "gentle" dependence on off shell pion mass? Because it will lead to happy results, let's suppose that the mass dependence is gentle when the pion energies ω_i are held fixed. In effect, suppose that the parameters ρ and λ_{++-} do not depend on the pion mass variable. From Eq. (9.10) we then find

$$\mathrm{Amp}\,(K^+ \to 2\pi^+ + \pi^-) = 0 \qquad \text{for} \quad \omega_- = 0 \; .$$

Hence

$$\lambda_{++-} = \frac{2Q}{m_K} \; . \qquad (9.12)$$

In connection with Eq. (9.11), notice that $q^{(+)} = 0$ corresponds to $\omega_- = m_K/2$. Hence

$$\rho = \frac{1}{6f_\pi} \mathrm{Amp}\,(K_S^0 \to \pi^+ + \pi^-) \; . \qquad (9.13)$$

The result expressed by Eqs. (9.12) and (9.13) are in remarkably good agreement with experiment. Of course, our choice of variables in the interpretation of PCAC was somewhat arbitrary, but let us not be belabor this point!

9.2 *Hyperon Decay*

Consider the reaction

$$\Sigma^+ \to n + \pi^+ \; ,$$

and in particular its s-wave part. Take the pion off mass shell in the usual way and pass to the limit $q = 0$. In this limit we deal with the s-wave amplitude A_S and find

$$A_S(\Sigma^+ \to n + \pi^+) \xrightarrow[q \to 0]{} \frac{i}{2(2)^{1/2} f_\pi} \langle n | \mathcal{H}^{4-i5} | \Sigma^+ \rangle = 0, \tag{9.14}$$

where the vanishing of the matrix element on the right side follows purely from the consideration of strangeness conservation. Making the usual PCAC assumption that the off shell variation is gentle, we are hereby led to the prediction that $\Sigma^+ \to n + \pi^+$ is essentially pure p-wave. This is in remarkably good agreement with experiment.

Current algebra and PCAC are not always so successful as in the above applications. But let us not dwell on troubles . . .

References

[1] M. Gell-Mann, Physics 1 (1964) 63 and references therein.
[2] S. Fubini and G. Furlan, Physics 1 (1965) 229.
[3] S. Adler, Phys. Rev. 140 (1965) B736.
[4] W. Weisberger, Phys. Rev. 143 (1966) 1302.
[5] B. Renner, Current Algebras and Their Applications (Pergamon Press, Oxford 1968).
[6] Y. Nambu, Phys. Rev. Lett 4 (1966) 380; J. Bernstein, S. Fubini, M. Gell-Mann and W. Thirring, Nuovo Cimento 17 (1969) 757; M. Gell-Mann, Phys. Rev. 125 (1962) 1067.
[7] N. Cabibbo, Phys. Rev. Lett. 10 (1963) 531.
[8] R. P. Feynman and M. Gell-Mann, Phys. Rev. 109 (1958) 193.
[9] M. Gell-Mann and M. Lévy, Nuovo Cimento 16 (1960) 705.
[10] M. L Goldberger and S. B. Treiman, Phys. Rev. 110 (1958) 1178.
[11] S. L. Adler, Phys. Rev. 135 (1964) B963.
[12] S. L. Adler, Phys. Rev. 137 (1965) B1022.
[13] N. Cabibbo and L. A. Radicati, Phys. Lett. 19 (1966) 697.
[14] F. Gilman and H. Schnitzer, Phys. Rev. 150 (1966) 1562.
[15] S. L. Adler, Phys. Rev. 143 (1966) 1144.
[16] S. Weinberg, Phys. Rev. Lett. 17 (1966) 616.
[17] C. G. Callan and S. B. Treiman, Phys. Rev. Lett. 16 (1966) 153.
[18] M. K. Gaillard and L. M. Chounet, CERN Report 70-74 (1970). These authors present a comprehensive survey of the $K_{\ell 3}$ situation.
[19] M. K. Gaillard, Nuovo Cimento 61 (1969) 499; R. A. Brandt and G. Preparata, to be published.
[20] M. Gell-Mann, Phys. Rev. 125 (1962) 1067; see also Ref. 1.

[21] S. Weinberg, Phys. Rev. Lett. 17 (1966) 336.
[22] Ref. 17. Also see Y. Hara and Y. Nambu, Phys. Rev. Lett. 16 (1966) 875; D. K. Elias and J. C. Taylor, Nuovo Cimento 44A (1966) 518. H. D. I. Abarbanel, Phys. Rev. 153 (1967) 1547.
[23] M. Suzuki, Phys. Rev. Lett. 15 (1965) 986.

FIELD THEORETIC INVESTIGATIONS IN CURRENT ALGEBRA†

Roman Jackiw
Center for Theoretical Physics
Laboratory for Nuclear Science and Department of Physics
Massachusetts Institute of Technology
Cambridge, MA 02139, USA

†This contribution first appeared in *Lectures on Current Algebra and Its Applications*, by S. B. Treiman, R. Jackiw and D. J. Gross (Princeton University Press, Princeton, N. J., 1972).

1. Introduction

The techniques of current algebra have been developed to circumvent two difficulties which hamper progress in particle physics. These are (1) a lack of knowledge of the precise laws which govern elementary processes, other than electromagnetism; (2) an inability of solving any of the realistic models which have been proposed to explain dynamics. It was in this context that Gell-Mann [1], in a brilliant induction from non-relativistic quantum mechanics, proposed his now famous charge algebra, which subsequently has been extended to the local algebra of charge and current densities. Just as the canonical, non-relativistic Heisenberg commutator between the momentum $p \equiv \delta L/\delta \dot{q}$ and position q [2], $i[p,q] = 1$, is independent of the specific form of the Lagrangian L, and can be used to derive useful, interaction independent results such as the Thomas-Reiche-Kuhn sum rule [3]; so also it should be possible to exhibit interaction-independent relations in relativistic dynamics, which can be exploited to yield physical information without solving the theory. This program has led to the algebra of current commutation relations. It has had successes in two broad categories of application: low energy theorems and high energy sum rules.

It is my purpose in these lectures to discuss research of the last two years which has elucidated the form of commutators in those model field theories which had served to derive and motivate the algebra of currents. The remarkable results that have been obtained indicate that current commutators are *not* independent of interactions, and that some of the current algebra predictions can be circumvented. It is of course possible to *postulate* the relevant current commutators in their minimal, Gell-Mann, form and to dispense with the theoretical structure, which in the first instance led to their derivation. I shall not be taking this point of view, since such a postulational approach provides no basis for the validity of these relations. Moreover, and more importantly, the minimal current algebra has led to certain predictions which are in conflict with experiment; while the anomalies that have been found offer the possibility of eliminating some difficulties.

Examples of results which we shall be discussing in great detail are the following two: (1) The Sutherland-Veltman theorem [4] predicts that the effective coupling constant for $\pi^0 \to 2\gamma$ should vanish for zero pion mass in any field theory which exhibits current algebra, PCAC and electromagnetism, minimally and gauge invariantly coupled. However, in the σ model, which possesses all these properties, this object does not vanish [5]. (2) The Callan-Gross sum rule [6] predicts that in a large class of quark models the longitudinal cross section, for total electroproduction off protons, vanishes in a certain high energy domain, the so-called deep inelastic limit. Explicit calculation of this cross section in the relevant models yields a non-vanishing result [7].

It must be emphasized that the above calculations, which provide evidence for conflict betweeen the formal, canonical reasoning of current algebra and explicit computation, are in no way ambiguous. The calculations, performed of course by perturbative techniques, are well defined consequences of the dynamics of the theory. Evidently the formal, canonical properties of the theory which lead to the erroneous predictions, are not maintained.

Our program is the following. We shall need to make frequent use of some technical results concerning Green's functions, commutators and Ward-Takahashi identities. Thus first we shall explore the canonical and the space-time constraints which limit the possible structure of these objects, and we shall discuss the Bjorken-Johnson-Low definition of the commutator. Next we shall perform the calculations relevant to the two examples quoted above. It will then be shown how the minimal current algebra must be modifed to avoid "false" theorems like those of Sutherland-Veltman and Callan-Gross. The theoretical and experimental consequences of these modifications will be examined. Finally we shall study in the context of our discoveries the recently popularized topic of broken scale invariance [8].

References

[1] An excellent survey of current algebra is to be found in the book by S. L. Adler and R. Dashen, Current Algebra, Benjamin/Cummings, Reading MA (1968).

[2] Throughout these lectures, we set $\hbar$ and c to unity. The metric we use is $g^{\mu\nu} = g_{\mu\nu} = \mathrm{diag}\{1, -1, -1, -1\}$. The anti-symmetric epsilon tensor $\epsilon^{\mu\alpha\beta\gamma}$ is normalized by $\epsilon^{0123} = 1$. The Dirac γ_5 matrix is defined to be Dirac self-adjoint, $\gamma_5 \equiv \gamma^0\gamma^1\gamma^2\gamma^3$, hence it is anti-hermitian: $\gamma_5^\dagger = -\gamma_5$, $\bar{\gamma}_5 \equiv \gamma_0\gamma_5^\dagger\gamma_0 = \gamma_5$. Greek indices are space-time while Latin indices are space.

[3] A modern treatment of these "classical" sum rules is given by H. A. Bethe and R. Jackiw, Intermediate Quantum Mechanics, 3rd edition, Benjamin/Cummings, Reading MA (1986).

[4] D. G. Sutherland, Nucl. Phys. B2 (1967) 433; M. Veltman, Proc. Roy. Soc. A301 (1967) 107.

[5] J. S. Bell and R. Jackiw, Nuovo Cimento 60 (1969) 47.
[6] C. G. Callan, Jr. and D. J. Gross, Phys. Rev. Lett. 22 (1969) 156.
[7] R. Jackiw and G. Preparata, Phys. Rev. Lett. 22 (1969) 975, (E) 22 (1969) 1162; S. L. Adler and Wu-Ki Tung, Phys. Rev. Lett. 22 (1969) 978.
[8] A brief summary of the material covered here appears in R. Jackiw, Non-Canonical Behaviour in Canonical Theories, CERN preprint, TH-1065.

2. Canonical and Space-Time Constraints in Current Algebra

2.1 Canonical Theory of Currents

An arbitrary field theory is described by a Lagrange density $\mathscr{L}$ which we take to depend on a set of independent fields ϕ and on their derivatives $\partial^\mu \phi \equiv \phi^\mu$. The canonical formalism rests on the following equal time commutators (ETC) [1].

$$i[\pi^0(t,\mathbf{x}), \phi(t,\mathbf{y})] = \delta(\mathbf{x}-\mathbf{y}) ,$$

$$i[\pi^0(t,\mathbf{x}), \pi^0(t,\mathbf{y})] = i[\phi(t,\mathbf{x}), \phi(t,\mathbf{y})] = 0 . \tag{2.1}$$

Here π^0 is the time component of the canonical 4-momentum.

$$\pi^\mu = \frac{\delta \mathscr{L}}{\delta \phi_\mu} . \tag{2.2}$$

The Euler-Langrange equation of the theory is

$$\partial_\mu \pi^\mu = \frac{\delta \mathscr{L}}{\delta \phi} . \tag{2.3}$$

Consider now an infinitesmal transformation which changes $\phi(x)$ to $\phi(x) + \delta\phi(x)$. The explicit form for $\delta\phi(x)$ is assumed known; we have in mind a definite, though unspecified transformation. It is interesting to inquire what conditions on $\mathscr{L}$ insure this transformation to be a *symmetry* operation for the theory. This can be decided by examining what happens to $\mathscr{L}$ under the transformation.

$$\delta\mathscr{L} = \frac{\delta\mathscr{L}}{\delta\phi}\delta\phi + \frac{\delta\mathscr{L}}{\delta\phi^\mu}\delta\phi^\mu = \frac{\delta\mathscr{L}}{\delta\phi}\delta\phi + \pi_\mu \partial^\mu \delta\phi . \tag{2.4}$$

If *without the use of equations of motion* we can show that $\delta\mathcal{L}$ is a total divergence of some object Λ^μ,

$$\delta\mathcal{L} = \frac{\delta\mathcal{L}}{\delta\phi}\,\delta\phi + \pi_\mu \partial^\mu \delta\phi = \partial_\mu \Lambda^\mu \ , \tag{2.5}$$

then the action, $I = \int d^4x\,\mathcal{L}$, is not affected by the transformation, and the transformation is a symmetry operation of the theory. The conserved current can now be constructed in the following fashion. *With the help of the equations of motion* (2.3), an alternate formula for $\delta\mathcal{L}$ can be given which is always true, regardless whether or not we are dealing with a symmetry operation. We have from (2.3) and (2.4),

$$\delta\mathcal{L} = \partial_\mu \pi^\mu \delta\phi + \pi^\mu \partial_\mu \delta\phi = \partial_\mu(\pi^\mu \delta\phi) \ . \tag{2.6a}$$

Equating this with (2.5) yields

$$0 = \partial_\mu[\pi^\mu \delta\phi - \Lambda^\mu] \ . \tag{2.6b}$$

Hence the conserved current is

$$J_\mu = \pi_\mu \delta\phi - \Lambda_\mu \ . \tag{2.7}$$

Two situations are now distinguished. If $\Lambda^\mu = 0$, we say that we are dealing with an *internal symmetry*; otherwise we speak of a *space-time* symmetry [2]. Examples of the former are the SU(3) × SU(3) currents of Gell-Mann.

$$\delta^a \phi = T^a \phi \ . \tag{2.8}$$

T^a is a representation matrix of the group; it is assumed that the fields transform under a definite representation. The internal group index a labels the different matrices. The internal symmetry current is

$$J_\mu^{\ a} = \pi_\mu T^a \phi \ . \tag{2.9}$$

Space-time symmetries are exemplified by translations.

$$\delta^\alpha \phi = \partial^\alpha \phi \ ,$$

$$\delta^\alpha \mathcal{L} = \partial^\alpha \mathcal{L} \quad , \tag{2.10a}$$

$$\Lambda_\mu^\alpha = g_\mu^\alpha \mathscr{L} \quad . \tag{2.10b}$$

Now the transformations are labeled by the space-time index α. The conserved quantity is the canonical energy-momentum tensor $\theta_c{}^{\mu\alpha}$.

$$\theta_c{}^{\mu\alpha} = \pi^\mu \phi^\alpha - g^{\mu\alpha}\mathscr{L} \quad . \tag{2.11}$$

In the subsequent we shall reserve the symbol J_μ and the term "current" for *internal* symmetries.

It is clear that when the internal transformation (2.8) is not a symmetry operation, i.e. $\delta\mathscr{L} \neq 0$, it is still possible to define the current (2.9), which is not conserved. By virtue of the canonical formalism, the charge density satisfies a model-independent ETC, regardless whether or not the current is conserved [3].

$$\begin{aligned}
&[J_0{}^a(t,\mathbf{x}),\ J_0{}^b(t,\mathbf{y})] \\
&\quad = [\pi_0(t,\mathbf{x})\, T^a \phi(t,\mathbf{x}),\ \pi_0(t,\mathbf{y})\, T^b \phi(t,\mathbf{y})] \\
&\quad = i\,\pi_0(t,\mathbf{x})\, [T^a, T^b]\, \phi(t,\mathbf{x})\, \delta(\mathbf{x}-\mathbf{y}) \\
&\quad = -f_{abc}\, J_0{}^c(t,\mathbf{x})\, \delta(\mathbf{x}-\mathbf{y}) \ .
\end{aligned} \tag{2.12}$$

We have used the group property of the representation matrices

$$[T^a, T^b] = i f_{abc}\, T^c \ . \tag{2.13}$$

Similarly the charge,

$$Q^a(t) = \int d^3x\, J_0{}^a(t,\mathbf{x}) \ , \tag{2.14}$$

which for conserved currents is a time independent Lorentz scalar, generates the proper transformation on the fields, even in the non-conserved case.

$$\begin{aligned}
&i\,[Q^a(t),\ \phi(t,\mathbf{x})] \\
&\quad = i\int d^3y\, [\pi_0(t,\mathbf{y})\, T^a \phi(t,\mathbf{y}),\ \phi(t,\mathbf{x})] \\
&\quad = T^a \phi(x) = \delta^a \phi(x) \ .
\end{aligned} \tag{2.15}$$

It should be remarked here that although conserved and non-conserved internal symmetry currents and charges satisfy the ETC (2.12) and (2.15), the space-time currents do not, in general, satisfy commutation relations which are insensitive to conservation, or lack thereof of the appropriate quantity; see Exercise 2.5.

The importance of relations (2.12) and (2.15) is that they have been derived without reference to the specific form of $\mathscr{L}$; i.e., without any commitment of dynamics. Thus it appears that they are *always* valid, and that any consequence that can be derived from (2.12) and (2.15) will necessarily be true. But it must be remembered that the Eqs. (2.12) and (2.15) have been obtained in a very formal way; all the difficulties of local quantum theory have been ignored. Thus we have not worried about multiplying together two operators at the same space-time point, as in (2.9); nor have we inquired whether or not the equal time limit of an unequal time commutator really exists as in (2.1), (2.12) and (2.15). It will eventually be seen that the failure of current algebra predictions can be traced to precisely these problems.

A word about non-conserved currents. It turns out that in applications of the algebra of non-conserved currents, it is necessary to make assumptions about the divergence of the current. The assumption that is most frequently made is that $\partial^\mu J_\mu$ is a "gentle" operator, though the precise definition of "gentle" depends on the context. We shall spell out in detail what we mean by "gentle"; however for the moment the following concept of "gentleness" will delimit the non-conserved currents which we shall consider. The dimension of a current, in mass units, is 3. This follows from the fact that the charge, which is dimensionless, is a space integral of a current component. Therefore $\partial^\mu J_\mu$ has dimension 4. However if the dynamics of the theory is such that all *operators* which occur in $\partial^\mu J_\mu$ carry dimension less than 4, then we say that $\partial^\mu J_\mu$ is "partially conserved".

As an explicit example, consider the axial current constructed from fermion fields, $J_5{}^\mu = i\bar{\psi}\gamma^\mu\gamma^5\psi$; and assume that the fermions satisfy the equation of motion

$$i\gamma^\mu\partial_\mu\psi = -m\psi + e\gamma^\mu A_\mu + g\phi\gamma^5\psi\ .$$

Here A^μ and ϕ are vector and pseudo-scalar boson fields respectively. Recall that the dimension of a fermion field is $\frac{3}{2}$ while that of a boson field is 1. (This is seen from the Lagrangian, which necessarily has dimension 4, so that the action, $I = \int d^4x\mathscr{L}$, is dimensionless. The fermion Lagrangian contains $i\bar{\psi}\gamma^\mu\partial_\mu\psi$; the derivative carries one unit of dimension, this leaves

3 for $\bar{\psi}\psi$, hence ψ has dimension $\frac{3}{2}$. The boson Lagrangian contains $\partial^\mu\phi\partial_\mu\phi$; the two derivatives use up 2 units of dimension; hence ϕ has dimension 1.) Evidently $J_5{}^\mu$ possesses in this model the divergence $\partial_\mu J_5{}^\mu = 2m\bar{\psi}\gamma^5\psi + 2g\bar{\psi}\psi\phi$. The operator $\bar{\psi}\gamma^5\psi$ has dimension 3, while $\bar{\psi}\psi\phi$ has dimension 4. Hence we say that $J_5{}^\mu$ is partially conserved only in the absence of the pseudo-scalar coupling.

Although model independent commutators for current components have been derived from canonical transformation theory, the use of these results for physical predictions requires a tacit dynamical assumption which we must expose here. The point is that in the context of transformation theory it is always possible to add to the canonical current a divergence of an anti-symmetric tensor.

$$J^\mu \to J^\mu + \partial_\lambda X^{\lambda\mu} \ ,$$

$$X^{\lambda\mu} = -X^{\mu\lambda} \ . \tag{2.16}$$

Such additions, called "super-potentials", do not change the charges nor the divergence properties of the current. (Conservation of $\partial_\lambda X^{\lambda\mu}$ is assured by the anti-symmetry of the super-potentials. The fact that the super-potential does not contribute to the charges is seen as follows: $\int d^3x\,\partial_\lambda X^{\lambda 0} = \int d^3x\,\partial_i X^{i0} = 0$.) It may be that the modified current possesses a physical significance, greater than that of the canonical expression. Indeed this state of affairs occurs with the energy momentum tensor. For reasons which I shall discuss presently, the canonical expression (2.11), is usually replaced in physical discussions by the symmetric, Belinfante form; see Exercise 2.3.

$$\theta_B{}^{\mu\alpha} = \theta_c{}^{\mu\alpha} + \frac{1}{2}\partial_\lambda X^{\lambda\mu\alpha} \ , \tag{2.17a}$$

$$X^{\lambda\mu\alpha} = -X^{\mu\lambda\alpha} = \pi^\lambda\Sigma^{\mu\alpha}\phi - \pi^\mu\Sigma^{\lambda\alpha}\phi - \pi^\alpha\Sigma^{\lambda\mu}\phi \ . \tag{2.17b}$$

Here $\Sigma^{\alpha\beta}$ is the spin matrix appropriate to the field ϕ.

The modified expressions for currents will in general possess commutators which differ from the canonical ones given above (2.12). Thus our insistence on the canonical commutators, rather than some others, requires an assumption that the canonical currents have a unique physical significance. This significance can be derived from the seemingly well established fact that the electromagnetic and weak interactions are governed by the *canonical* electromagnetic and SU(3) × SU(3) currents respectively. The physical significance of the Belinfante tensor follows from the belief that gravitational interactions are described by Einstein's general relativity. In that theory gravitons couple to $\theta_B{}^{\mu\nu}$, and not to $\theta_c{}^{\mu\nu}$. (In our discussion of scale transformations, Section 7, we shall argue that a new improved energy-momentum tensor should be introduced; and correspondingly gravity theory should be modified.) It is possible to develop a general formalism based directly on the dynamical role of currents. In this context one can derive current commutators without reference to canonical transformation theory. The results are of course the same, and we shall not discuss this approach here [4].

We conclude this section by recording another commutator which can be established by canonical reasoning; see Exercise 2.4.

$$i[\theta^{00}(t,\mathbf{x}), J_0{}^a(t,\mathbf{y})] = \partial^\mu J_\mu{}^a(x)\,\delta(\mathbf{x}-\mathbf{y}) + J_i^a(x)\,\partial^i\,\delta(\mathbf{x}-\mathbf{y}) \;. \tag{2.18}$$

This has the important consequence that the divergence of a current can be expressed as a commutator.

$$i[\theta^{00}(t,\mathbf{x}), Q^a(t)] = \partial^\mu J_\mu{}^a(x) \;. \tag{2.19}$$

Formulas (2.18) and (2.19) are insensitive to the choice of θ^{00}; both the canonical and the Belinfante tensor lead to the same result. Other ETC between selected componets of $\theta^{\alpha\beta}$ and $J_\mu{}^a$ can also be derived, in a model independent fashion. We do not pursue this topic here; one can read about it in the literature [5].

2.2 Space-Time Constraints on Commutators

Although interesting physical results can be obtained from the charge-density algebra, (2.12), the applications that we shall study require commutation relations between other components of the currents. These cannot be derived canonically in a model independent form. For example, $J_k{}^a$ involves π_k, see (2.9); but the dependence of π_k on the canonical variables π^0 and ϕ is not known in general, and one cannot compute commutators involving π_k in an abstract fashion.

It is possible to determine the $[J_a{}^0, J_b{}^k]$ ETC by investigating the space-time constraints which follow from the fact (2.12) is supposed to hold in all Lorentz frames. As an example, consider the once integrated version of (2.12) for the case of conserved currents.

$$[Q^a, J_0{}^b(0)] = -f_{abc}\, J_0{}^c(0) \;. \tag{2.20}$$

An infinitesimal Lorentz transformation can be effected on (2.20) by commuting both sides with M^{0i}, the generator of these transformations.

$$[M^{0i}, [Q^a, J_0{}^b(0)]]$$

$$= [[M^{0i}, Q^a], J_0{}^b(0)] + [Q^a, [M^{0i}, J_0{}^b(0)]] \; ,$$

$$= -f_{abc} [M^{0i}, J_0{}^c(0)] \; . \qquad (2.21a)$$

The second equality in (2.21a) follows from the first by use of the Jacobi identity. All the commutators with M^{0i} may be evaluated, since the commutator of $M^{\alpha\beta}$ with $J_\mu{}^a$ is known from the fact that the current transforms as a vector.

$$i[M^{\alpha\beta}, J_\mu{}^a(x)]$$

$$= (x^\alpha \partial^\beta - x^\beta \partial^\alpha) J_\mu{}^a(x) + (g_\mu^\alpha g^{\beta\nu} - g^{\alpha\nu} g_\mu^\beta) J_\nu{}^a(x) \; . \qquad (2.21b)$$

It now follows that (remember the current is assumed conserved)

$$[Q^a, J_i^b(0)] = -f_{abc} J_i^c(0) \; . \qquad (2.22a)$$

The local version of (2.22a) is

$$[J_0{}^a(t, \mathbf{x}), J_i^b(t, \mathbf{y})]$$

$$= -f_{abc} J_i^c(x)\, \delta(\mathbf{x}-\mathbf{y}) + S_{ij}^{ab}(y)\, \partial^j \delta(\mathbf{x}-\mathbf{y}) + \ldots \; . \qquad (2.22b)$$

In (2.22b) we have inserted a gradient of a δ function; the dots indicate the possible higher derivatives of δ functions which may be present. Of course, all these gradients must disappear upon integration over $\mathbf{x}$, so that (2.22a) is regained. Such gradient terms in the ETC are called Schwinger terms (ST) [6].

Further constraints can be obtained by commuting the *local* commutator (2.12) with P^0 and M^{0i}. However, the strongest results are arrived at by commuting (2.12) with θ^{00}, rather than with once integrated moments of θ^{00} which is what P^0 and M^{0i} are. ($P^0 = \int d^3x\, \theta^{00}(0, \mathbf{x})$; $M^{0i} = -\int d^3x\, x^i\, \theta^{00}(0, \mathbf{x})$.) Thus we are led to consider

$$i\,[\theta^{00}(0, \mathbf{z}), [J_0{}^a(0, \mathbf{x}), J_0{}^b(0, \mathbf{y})]]$$

$$= -f_{abc}\, \delta(\mathbf{x}-\mathbf{y})\, i\,[\theta^{00}(0, \mathbf{z}), J_0{}^c(0, \mathbf{x})] \; . \tag{2.23a}$$

The left-hand side is rewritten in terms of the Jacobi identity; then (2.18) is used to evaluate the $[\theta^{00}, J_0{}^a]$ ETC. The result, for conserved currents, is

$$[J_0{}^b(0, \mathbf{y}), J_k^a(0, \mathbf{z})]\, \partial^k\, \delta(\mathbf{x}-\mathbf{z}) + [J_0{}^a(0, \mathbf{x}), J_k^b(0, \mathbf{z})]\, \partial^k\, \delta(\mathbf{z}-\mathbf{y})$$

$$= -f_{abc}\, \delta(\mathbf{x}-\mathbf{y})\, J_k^c(0, \mathbf{z})\, \partial^k\, \delta(\mathbf{z}-\mathbf{y}) \; . \tag{2.23b}$$

The most general form for the $[J_0{}^a, J_i^b]$ ETC consistent with the constraint (2.23b) is (see Exercise 2.6)

$$[J_0{}^a(0, \mathbf{x}), J_i^b(0, \mathbf{y})]$$

$$= -f_{abc}\, J_i^c(0, \mathbf{x})\, \delta(\mathbf{x}-\mathbf{y}) + S_{ij}^{ab}(0, \mathbf{y})\, \partial^j\, \delta(\mathbf{x}-\mathbf{y}) \; , \tag{2.24a}$$

$$S_{ij}^{ab}(0, \mathbf{y}) = S_{ji}^{ba}(0, \mathbf{y}) \; . \tag{2.24b}$$

Thus we have determined the $[J_0{}^a, J_i^b]$ ETC up to *one* derivative of the δ function; all higher derivatives should vanish. The surviving ST possess the symmetry (2.24b). It will be shown later that the ST cannot vanish. The same conclusions can be obtained when the current is partially conserved, as long as the divergence of the current is sufficiently gentle so that no ST is produced when it is commuted with $J_a{}^0$.

The above methods can be used to obtain additional constraints on various current commutators. One exploits the Jacobi identity, and the model independent commutators between selected components of $\theta^{\alpha\beta}$ and J^μ. We do not present these results here, since they are only of limited interest. However one result is sufficiently elegant to deserve explicit mention. If $S_{ij}^{ab} = \delta_{ij} S^{ab}$, where S_{ab} is a Lorentz scalar, then the $[J_i^a, J_j^b]$ ETC does not have any derivatives of δ functions [7].

2.3 Space-Time Constraints on Green's Functions

We must also discuss the space-time structure of Green's functions and Ward identities. The reason for emphasizing this topic here is that the theorems of current algebra concern themselves with Green's functions: scattering amplitudes, decay amplitudes and the like; while the most felicitous way of obtaining these results is by the use of Ward identities.

Consider the T product of two operators A and B.

$$\begin{aligned} \mathrm{T}(x) &= \mathrm{T}A(x)B(0) , \\ &= \theta(x_0)A(x)B(0) + \theta(-x_0)B(0)A(x) . \end{aligned} \tag{2.25}$$

Matrix elements of $\mathrm{T}(x)$ are related to Green's functions. However, a Green's function must be Lorentz covariant, while $\mathrm{T}(x)$ need not have this property because of the time ordering. It is necessary, in the general case, to add to $\mathrm{T}(x)$ another non-covariant term, called a seagull, $\tau(x)$, so that the sum is covariant. The sum of a time ordered product with the covariantizing seagull is called a T^* product.

$$\mathrm{T}^*(x) = \mathrm{T}(x) + \tau(x) . \tag{2.26}$$

It is required that $\mathrm{T}^*(x)$ and $\mathrm{T}(x)$ coincide for $x_0 \neq 0$; hence $\tau(x)$ has support only at $x_0 = 0$; i.e., $\tau(x)$ will involve δ functions of x_0 and derivatives thereof.

We now investigate under what conditions $\mathrm{T}(x)$ is not covariant. We also show how to construct the covariantizing seagull. Finally we examine under what conditions Feynman's conjecture concerning the cancellation of Schwinger terms against divergences of seagulls is valid. (Feynman's conjecture will be explained, when we come to it.) To effect this analysis it is necessary to assume that the $[A,B]$ ETC is known.

$$[A(0,\mathbf{x}), B(0)] = C(0)\delta(\mathbf{x}) + S^i(0)\,\partial_i\,\delta(\mathbf{x}) . \tag{2.27}$$

In offering (2.27) we have assumed, for simplicity, one ST; higher derivatives can easily be accommodated by the present technique.

Our analysis [8] makes use of the device of writing non-covariant expressions in a manifestly covariant, but frame dependent notation. A unit time-like vector n^μ, and a space-like projection $P^{\mu\nu}$ are introduced.

$$n^0 > 0\,, \quad n^2 = 1\,,$$

$$P^{\mu\nu} = g^{\mu\nu} - n^\mu n^\nu\,. \tag{2.28}$$

In terms of n, the n-dependent T product has the form

$$\mathrm{T}(x;n) = \theta(x\cdot n)A(x)B(0) + \theta(-x\cdot n)B(0)A(x)\,, \tag{2.29}$$

while the ETC is

$$[A(x), B(0)]\;\delta(x\cdot n) = C(n)\,\delta^4(x) + S^\alpha(n)P_{\alpha\beta}\,\partial^\beta\,\delta^4(x)\,. \tag{2.30}$$

The T* product is n-independent.

$$\mathrm{T}^*(x) = \mathrm{T}(x;n) + \tau(x;n)\,. \tag{2.31}$$

We now vary n. However, since n is constrained to be time-like, only space-like variations of n are permitted. Hence we operate on (2.31) by $P^{\alpha\beta}(\delta/\delta n^\beta)$.

$$0 = P^{\alpha\beta}\frac{\delta}{\delta n^\beta}\mathrm{T}(x;n) + P^{\alpha\beta}\frac{\delta}{\delta n^\beta}\tau(x;n)\,. \tag{2.32}$$

The first term in (2.32) is evaluated from (2.29).

$$P^{\alpha\beta}\frac{\delta}{\delta n^\beta}\mathrm{T}(x;n) = P^{\alpha\beta}x_\beta\,\delta(x\cdot n)\,[A(x), B(0)]\,. \tag{2.33a}$$

Inserting the commutator from (2.30), we find

$$P^{\alpha\beta}\frac{\delta}{\delta n^\beta}\mathrm{T}(x;n) = -P^{\alpha\beta}\,S_\beta(n)\,\delta^4(x)\,. \tag{2.33b}$$

The above equation shows that the T product is not covariant (n-independent) whenever the ETC of the relevant operators contains a ST.

To proceed with the construction of the seagull, (2.33b) is inserted in (2.32), and a differential equation for τ is obtained.

$$P^{\alpha\beta}\,S_\beta(n)\,\delta^4(x) = P^{\alpha\beta}\frac{\delta}{\delta n^\beta}\tau(x;n)\,. \tag{2.34a}$$

The space-like projection $P^{\alpha\beta}$ may be cancelled from (2.34a), if an arbitrary expression proportional to n^β is introduced. However, since $S_\beta(n)$ is defined by (2.30) only up to terms proportional to n^β, such arbitrary contributions can be absorbed into the definition of $S_\beta(n)$. Hence we have

$$S_\beta(n)\,\delta^4(x) = \frac{\delta}{\delta n^\beta}\tau(x;n) . \tag{2.34b}$$

The solution of this differential equation is

$$\tau(x;n) = \int^n dn'_\beta\, S^\beta(n')\,\delta^4(x) + \tau_0(x) . \tag{2.35}$$

We have thus constructed the covariantizing seagull from the Schwinger term. Of course the seagull is not uniquely determined; an arbitrary Lorentz covariant term $\tau_0(x)$ may be added. We shall see below how $\tau_0(x)$ is specified further by Ward identities. (It can be shown that Lorentz covariance insures that the line integral in (2.35) is line independent [8].)

2.4 Space-Time Constraints on Ward Identities

Let us now consider the case that A and B are vector operators; for example currents. We are interested in the T and the T* products of these quantities.

$$\begin{aligned} \mathrm{T}^{\mu\nu}(x,y;n) = {} & \theta([x-y]\cdot n)A^\mu(x)B^\nu(y) \\ & + \theta([y-x]\cdot n)B^\nu(y)A^\mu(x) , \end{aligned} \tag{2.36a}$$

$$\mathrm{T}^{*\mu\nu}(x,y) = \mathrm{T}^{\mu\nu}(x,y;n) + \tau^{\mu\nu}(x,y;n) . \tag{2.36b}$$

(In contrast to our previous notation, we indicate here explicitly the coordinate dependence of B^ν, which earlier was set of zero.) In applications of current algebra one frequently desires to obtain the Ward identities which are satisfied by $T^{*\mu\nu}(x,y)$; i.e., one wants to know the formula for

$$\begin{aligned}\frac{\partial}{\partial x^\mu} T^{*\mu\nu}(x,y) = {} & \theta([x-y]\cdot n)\,\partial_\mu A^\mu(x)\,B^\nu(y) \\ & + \theta([y-x]\cdot n)\,B^\nu(y)\,\partial_\mu A^\mu(x) \\ & + \delta([x-y]\cdot n)\,[n_\mu A^\mu(x), B^\nu(y)] \\ & + \frac{\partial}{\partial x^\mu}\tau^{\mu\nu}(x,y;n)\ ,\end{aligned} \tag{2.37a}$$

$$\begin{aligned}\frac{\partial}{\partial y^\nu} T^{*\mu\nu}(x,y) = {} & \theta([x-y]\cdot n)\,A^\mu(x)\,\partial_\nu B^\nu(y) \\ & + \theta([y-x]\cdot n)\,\partial_\nu B^\nu(y)\,A^\mu(x) \\ & - \delta([x-y]\cdot n)\,[A^\mu(x), n_\nu B^\nu(y)] \\ & + \frac{\partial}{\partial y^\nu}\tau^{\mu\nu}(x,y;n)\ .\end{aligned} \tag{2.37b}$$

To evaluate such expressions, it is necessary to know the $[A^0, B^\nu]$ and the $[B^0, A^\nu]$ ETC's completely, since these objects occur in the divergence of the T product. Also one needs to know the divergence of the seagull, since that object is also present in the T* product. In the specific case of SU(3) × SU(3) currents: $A^\mu = J_a{}^\mu$, $B^\nu = J_b{}^\nu$; we have shown above that the term proportional to the δ function in the $[J_a{}^0, J_b{}^\nu]$ ETC is known explicitly; however we have not determined the ST proportional to the gradient of the δ function; see (2.12) and (2.24). Consequently the co-variantizing seagull also cannot be determined. An obstacle thus has arisen in our program of calculating the Ward identity.

This obstacle is in practice overcome with the help of Feynman's conjecture [9]. Observe that if, for the moment, we pretend that there are no gradient terms in the ETC, then we may also set the seagull to zero. In this fictitious case, the Ward identity *can* be derived, since all we need to use is the known coefficient of the δ function. Feynman's conjecture is the statement that this "naive" procedure gives the correct answer. That is, one hopes that whatever the form of the ST, the associated seagull has the property that its divergence always cancels against the ST in the ETC, and one is left only with the term proportional to the δ function.

It is clear that Feynman's conjecture is crucial for progress in current algebraic analysis. Hence we now present an analysis [8] of it. Our conclusion will be a criterion for the general conditions which assure the validity of the conjecture. Later when we study anomalies, we shall see that one of the reasons for the conflict between current algebraic predictions and explicit calculations is the failure of Feynman's conjecture.

We return now to the general discussion of the T* product of the two vector operators A^μ and B^ν. We record here their commutator in abstract form.

$$[A^\mu(x), B^\nu(y)]\, \delta([x-y]\cdot n) = C^{\mu\nu}(y;n)\,\delta^4(x-y) + S^{\mu\nu\,\alpha}(y;n)\,P_{\alpha\beta}\,\partial^\beta\,\delta^\mu(x-y)\,. \quad (2.38)$$

It is clear that the previous formulas for the seagull still hold.

$$\tau^{\mu\nu}(x,y;n) = \tau^{\mu\nu}(y;n)\,\delta^4(x-y) + \tau_0{}^{\mu\nu}(x,y)$$

$$\tau^{\mu\nu}(y;n) = \int^n \mathrm{d}n'_\beta\, S^{\mu\nu\,\beta}(y;n')\,, \quad (2.39)$$

$$\frac{\delta}{\delta n^\beta}\tau^{\mu\nu}(y;n) = S^{\mu\nu\,\beta}(y;n)\,. \quad (2.40)$$

Now let us insert (2.38) into (2.37). We find

$$\begin{aligned}\frac{\partial}{\partial x^\mu}\mathrm{T}^{*\mu\nu}(x,y) = {} & \theta([x-y]\cdot n)\,\partial_\mu A^\mu(x)\,B^\nu(y) \\ & + \theta([y-x]\cdot n)\,B^\nu(y)\,\partial_\mu A^\mu(x) \\ & + n_\mu\, C^{\mu\nu}(y;n)\,\delta^4(x-y) \\ & + n_\mu\, S^{\mu\nu\,\alpha}(y;n)\,P_{\alpha\beta}\,\partial^\beta\,\delta^4(x-y) \\ & + \tau^{\mu\nu}(y;n)\frac{\partial}{\partial x^\mu}\delta^4(x-y) + \frac{\partial}{\partial x^\mu}\tau_0{}^{\mu\nu}(x,y)\,.\end{aligned}$$

(2.41)

For conserved currents, the T product on the right-hand side of (2.41) is absent. The remaining terms must be covariant, since the T* product on the left-hand side possesses this property. Therefore, it must be true that

$$n_\mu C^{\mu\nu}(y;n) = I_1{}^\nu(y) , \tag{2.42}$$

$$\tau^{\mu\nu}(y;n) + n_\beta S^{\beta\nu\,\alpha}(y;n) P^\mu_\alpha = I_1{}^{\mu\nu}(y) . \tag{2.43}$$

In the above $I_1{}^\nu(y)$ and $I_1{}^{\mu\nu}(y)$ are Lorentz covariant, n-independent quantities. For partially conserved currents, we again obtain the same result when we interpret "partial conservation" to mean that $\partial_\mu A^\mu$ is a sufficiently gentle operator, so that it does not give rise to a ST when commuted with B^ν. In that case the T product in (2.41) is already covariant, and the remaining terms again satisfy (2.42) and (2.43). Similar results are obtained by diverging with $\partial/\partial y^\mu$.

$$n_\nu C^{\mu\nu}(y;n) - \frac{\partial}{\partial y^\nu} \tau^{\mu\nu}(y;n) = I_2{}^\nu(y) , \tag{2.44}$$

$$\tau^{\mu\nu}(y;n) + n_\beta S^{\mu\beta\,\alpha}(y;n) P^\nu_\alpha = I_2{}^{\mu\nu}(y) . \tag{2.45}$$

$I_2{}^\nu$ and $I_2{}^{\mu\nu}$ are also covariant, not necessarily equal to $I_1{}^\nu$ and $I_1{}^{\mu\nu}$. The solution to (2.42) is

$$C^{\mu\nu}(y;n) = n^\mu I_1{}^\nu + P^\mu_\alpha C^{\alpha\nu}(y;n) , \tag{2.46}$$

while (2.43) and (2.45) require

$$n_\mu \tau^{\mu\nu}(y;n) = n_\mu I_1{}^{\mu\nu}(y) , \tag{2.47a}$$

$$n_\nu \tau^{\mu\nu}(y;n) = n_\nu I_2{}^{\mu\nu}(y) . \tag{2.47b}$$

Equations (2.46) and (2.47) represent constraints which must be satisfied by $C^{\mu\nu}(y;n)$ and $\tau^{\mu\nu}(y;n)$ which follow from Lorentz covariance.

Next we inquire under what conditions on the ETC can Feynman's conjecture be established; i.e., we require that divergences of seagulls cancel against ST, so that all gradient terms are absent from the Ward identity. Evidently for this to be true in the μ Ward identity, we must have, according to (2.41) and (2.43), that the following combination be free of gradients of δ functions.

$$I_1{}^{\mu\nu}(y)\, \partial_\mu \delta^4(x-y) + \frac{\partial}{\partial x^\mu} \tau_0{}^{\mu\nu}(x,y) . \tag{2.48a}$$

Similarly, the ν Ward identity sets the requirement that no gradients of δ functions occur in

$$-I_2^{\mu\nu}(y)\,\partial_\nu\delta^4(x-y) + \frac{\partial}{\partial y^\nu}\tau_0^{\mu\nu}(x,y) . \qquad (2.48b)$$

Equations (2.48) show that *it is always possible to satisfy Feynman's conjecture in one of the two Ward identities.* For example, to satisfy (2.48a) we may set

$$\tau_0^{\mu\nu}(x,y) = -I_1^{\mu\nu}(y)\,\delta^4(x-y) . \qquad (2.49a)$$

Alternatively, to satisfy (2.48b) $\tau_0^{\mu\nu}(x,y)$ can be chosen to be

$$\tau_0^{\mu\nu}(x,y) = -I_2^{\mu\nu}(y)\,\delta^4(x-y) . \qquad (2.49b)$$

However, in the general case, it need not be possible to satisfy *both* requirements since (2.48) may overdetermine the equation for $\tau_0^{\mu\nu}$, and no solution need exist.

A sufficient condition for the solution to both Eqs. (2.48) is that there be no ST in the time-time component of the commutators. The reason for this can be seen from (2.43) and (2.47). We have

$$\begin{aligned} n_\nu n_\beta S^{\beta\nu\alpha}(y;n)P^\mu_\alpha &= n_\nu I_1^{\mu\nu}(y) - n_\nu \tau^{\mu\nu}(y;n) \\ &= n_\nu[I_1^{\mu\nu}(y) - I_2^{\mu\nu}(y)] . \end{aligned} \qquad (2.50)$$

If there are no ST in the time-time ETC, the left-hand side vanishes. Consequently (2.50) implies that $I_1^{\mu\nu}(y) = I_2^{\mu\nu}(y)$, and the two solutions (2.49) become identical; thus *both* Ward identities can satisfy Feynman's conjecture. (It is also possible to derive a necessary and sufficient condition for the existence of a solution to (2.48); see Ref. 8.)

We conclude that if the *time*-component algebra, which we derived canonically, and which does not possess a ST, survives in the complete theory, then Feynman's conjecture can be satisfied. On the other hand, if the time-component algebra acquires a ST, it may not be possible to effect Feynman's conjecture. It will be seen that this indeed happens for the $\pi^0 \to 2\gamma$ problem.

2.5 *Schwinger Terms*

As a final example of space-time constraints on current commutators, we show that the ST which we allowed for the $[J_a{}^0, J_b^i]$ ETC cannot be zero. For simplicity we study only the conserved, electromagnetic currents; hence the group indices a and b are suppressed.

Consider the vacuum expectation value of the $[J^0, J^i]$ ETC.

$$\langle 0|[J^0(0,\mathbf{x}), J^i(0,\mathbf{y})]|0\rangle = \langle 0|C^i(\mathbf{x},\mathbf{y})|0\rangle\,, \tag{2.51}$$

$C^i(\mathbf{x}, \mathbf{y})$ is, by definition, the $[J^0, J^i]$ ETC. It has support only at $\mathbf{x} = \mathbf{y}$; thus it is composed of the δ function and derivatives thereof. However, from (2.22) we see that the δ function contribution is absent, since $a = b$. Therefore $C^i(\mathbf{x}, \mathbf{y})$ is non-vanishing only to the extent that ST are non-vanishing. We now prove that $C^i(\mathbf{x}, \mathbf{y})$ has non-zero vacuum expectation value; hence the ST is present. (In order to keep the discussion as general as possible, we do not use the specific form for $C^i(\mathbf{x}, \mathbf{y})$, derived in (2.24).)

To effect the analysis, we begin by differentiating (2.51) with respect to $\mathbf{y}$, and use current conservation.

$$\langle 0|[J^0(0,\mathbf{x}), -\partial_0 J^0(0,\mathbf{y})]|0\rangle = \langle 0|\frac{\partial}{\partial y^i}C^i(\mathbf{x},\mathbf{y})|0\rangle\,. \tag{2.52a}$$

The time derivative may be expressed as a commutator with the Hamiltonian.

$$\begin{aligned}\langle 0|[J^0(0,\mathbf{x}), \partial_0 J^0(0,\mathbf{y})]|0\rangle &= \langle 0|[J^0(0,\mathbf{x}), i[H, J^0(0,\mathbf{y})]]|0\rangle \\ &= i\langle 0|J^0(0,\mathbf{x})HJ^0(0,\mathbf{y}) \\ &\quad + J^0(0,\mathbf{y})HJ^0(0,\mathbf{x})|0\rangle\,.\end{aligned} \tag{2.52b}$$

In presenting (2.52b) we have used the fact that the vacuum has zero energy. Finally we multiply the above by $f(\mathbf{x})f(\mathbf{y})$, where $f(\mathbf{x})$ is an arbitrary real function; and integrate over $\mathbf{x}$ and $\mathbf{y}$. One is left with

$$i\int d^3x\, d^3y\, \langle 0|C^i(\mathbf{x},\mathbf{y})|0\rangle\, f(\mathbf{x})\frac{\partial}{\partial y^i}f(\mathbf{y}) = 2\langle 0|FHF|0\rangle$$

$$F = \int d^3x\, f(\mathbf{x})J^0(0,\mathbf{x})\,. \tag{2.53}$$

The right-hand side is non-zero. The reason for this is that the operator F possesses in general non-vanishing matrix elements between the vacuum and other states which necessarily carry positive energy, i.e.,

$$\begin{aligned}\langle 0 | FHF | 0 \rangle &= \sum_{nm} \langle 0 | F | m \rangle \langle m | H | n \rangle \langle n | F | 0 \rangle \\ &= \sum_{n} E_n \, | \langle 0 | F | n \rangle |^2 > 0 \, . \end{aligned} \tag{2.54}$$

(The vacuum cannot be an eigenstate of F, since J^0 has zero vacuum expectation value.) It therefore follows that $C^i(\mathbf{x}, \mathbf{y})$ is non-zero, and the ST cannot vanish [10]. The proof can be extended to non-conserved currents; hence one knows that there must be a ST in the $[J_a{}^0, J_a^i]$ ETC as well, where a is any internal group index.

When attention was first drawn to the existence of this ST, great interest was aroused because canonical computation in many instances leads to a vanishing result which, as we have seen, is inconsistent with other properties of local quantum theory: positivity and Lorentz covariance. In spinor electrodynamics for example one believes the formula for J^μ to be $\bar{\psi} \gamma^\mu \psi$; one further believes that the ETC $[J^0, \psi]$ is given by

$$[J^0(0, \mathbf{x}), \psi(0)] = -\psi(0) \, \delta(\mathbf{x}) \, . \tag{2.55}$$

From this it then follows that the $[J^0, J^i]$ ETC vanishes. (In scalar electrodynamics, on the other hand, a non-vanishing result is obtained.) Similar comments apply to SU(3) x SU(3) quark currents; there too canonical evaluation leads to a vanishing result.

This then is an example of a commutator anomaly; historically it predates the current emphasis of this topic. The reasons which were advanced at one time for ignoring these anomalies, were based on the belief that physical predictions are insensitive to the ST; essentially by virtue of the Feynman conjecture discussed earlier. We now know that this optimism was unjustified.

The present argument for the existence of the ST does not indicate whether this object is a c number or an operator — only the vacuum expectation value has been shown to be non-zero. Model calculations are also inconclusive. In scalar electrodynamics and in the σ model it comes out canonically to be a q number; in the algebra of fields it is a c number. In theories where canonical evaluation gives zero, more careful computation yields a quadratically divergent c-number result in spinor electrodynamics, a q number in quark models with spin-zero gluons and a c number in quark models with vector gluons [11].

To settle the question of the nature of this object one must turn to experiment. Unfortunately only very limited experimental probes have been discovered so far. The vacuum expectation value of the ST in the electromagnetic current commutator can be expressed in terms of an integral over the total cross section for lepton annihilation [12]. Recently a sum rule has been derived relating the single proton connected matrix element of the same object with the total electroproduction cross section [13]. The latter result tests the q-number nature of the electromagnetic ST. Preliminary conclusions indicate that this ST is a c number, although very likely it is a quadratically divergent object.

The experimental evidence for the nature of the SU(3) × SU(3) ST is even more scanty. The only thing one can say, at the present time, is that the apparent validity of the SU(2) × SU(2) Weinberg first sum rule [14] strongly indicates that S_{ab}^{ij} possesses no $I = 1$ part; i.e., $S_{ab}^{ij} = S_{ba}^{ij}$, $a, b = 1, 2, 3$, for vector and axial vector currents.

The vacuum expectation value sum rule is given in the literature; we shall not examine it here; see Exercise 3.2. We shall discuss the proton matrix element relation in connection with the Callan-Gross sum rule in Section 5. Finally, we mention that the existence of non-canonical structures analogous to the ST has also been demonstrated for ETC between selected components of $\theta^{\mu\nu}$; see the first two papers in Ref. 4.

2.6 *Discussion*

The constraints which we have obtained in this section follow from very general model independent considerations. However, a cautionary reminder must be inserted here concerning the validity of our results. Whenever very *detailed* and formal properties of field theory are used to arrive at a conclusion, we must remember that the troubles of local quantum field theory, previously alluded to, may invalidate the argument. On the other hand, when *general* properties of the theory are exploited, greater confidence may be placed in the result. Subsections 2.1 and 2.2 above are examples of the former; thus we must not be surprised when the canonical constraints are evaded. Subsections 2.3, 2.4 and 2.5 rely merely on Lorentz covariance, positivity and the existence of equal time commutators. So far no one has found counter examples to the conclusions given there.

References

[1] We ignore the complications which arise with Fermion operators: *anti*-commutation relations, etc. One can verify that these complications do not modify our final conclusions.

[2] Our distinction between *internal* and *space-time* symmetries must be refined to account for the possibility of adding total divergences to a Lagrangian without affecting dynamics. Thus even when

$$\delta \mathscr{L} = \partial_\mu \Lambda^\mu \neq 0 ,$$

we would still call this an *internal symmetry* if it is possible to find a dynamically equivalent Lagrangian $\mathscr{L}'$ such that

$$\mathscr{L}' = \mathscr{L} + \partial_\mu X^\mu ,$$

$$\delta \mathscr{L}' = \delta \mathscr{L} + \partial_\mu \delta X^\mu = \partial_\mu [\Lambda^\mu + \delta X^\mu] = 0 .$$

When it is possible to remove $\partial_\mu \Lambda^\mu$ by this method, we are dealing with an internal symmetry.

[3] A more conventional form for the current commutator is

$$[J_0^{\,a}(t,\mathbf{x}), J_0^{\,b}(t,\mathbf{y})] = i f_{abc} J_0^{\,c}(t,\mathbf{x})\, \delta(\mathbf{x}-\mathbf{y}) .$$

This form is equivalent to (2.12), once one replaces $J_0^{\,a}$ by $iJ_0^{\,a}$.

[4] This formalism has been given by J. Schwinger, Phys. Rev. 130 (1963) 800. Various applications of this point of view are found in the work of Schwinger, and D. G. Boulware and S. Deser, J. Math. Phys. 8 (1967) 1468; D. J. Gross and R. Jackiw, Phys. Rev. 163 (1967) 1688.

[5] D. J. Gross and R. Jackiw, Phys. Rev. 163 (1967) 1688; R. Jackiw, Phys. Rev. 175 (1968) 2058; S. Deser and L. K. Morrison, J. Math. Phys. 11 (1970) 596.

[6] J. Schwinger, Phys. Rev. Lett. 3 (1959) 296. These gradient terms were first discovered by T. Goto and I. Imamura, Prog. Theor. Phys. 14 (1955) 196.

[7] This argument is detailed in the papers by D. J. Gross and R. Jackiw, Ref. 5.

[8] This analysis is presented by D. J. Gross and R. Jackiw, Nucl. Phys. B14 (1969) 269.

[9] R. P. Feynman, unpublished; see also M. A. Bég, Phys. Rev. Lett. 17 (1966) 333.

[10] The derivation presented here is due to Schwinger, Ref. 6; Goto and Imamura, Ref. 6, previously had presented a different argument toward the same end.

[11] D. G. Boulware and R. Jackiw, Phys. Rev. 186 (1969) 1442; D. G. Boulware and J. Herbert, Phys. Rev. D2 (1970) 1055; see also Section 6.

[12] J. D. Bjorken, Phys. Rev. 148 (1966) 1467; V. N. Girbov, B. L. Ioffe and I. Ya. Pomeranchuk, Phys. Lett. 24B (1967) 554; R. Jackiw and G. Preparata, Phys. Rev. Lett. 22 (1969) 975; (E) 22 (1969) 1162.

[13] R. Jackiw, R. Van Rogen and G. B. West, Phys. Rev. D2 (1970) 2473. See also L. S. Brown in *Lectures in Theoretical Physics*, XII B, eds. K. Mahanthappa and W. E. Brittin, Gordon Breach, New York (1971); J. M. Cornwall, D. Corrigan and R. E. Norton, Phys. Rev. Lett. 24 (1970) 1141.

[14] S. Weinberg, Phys. Rev. Lett. 18 (1967) 507.

3. The Bjorken-Johnson-Low Limit

In order to understand the failure of formal, current algebraic predictions, we shall need to calculate the ETC in order to ascertain whether or not the canonical value is maintained. The first difficulty that is encountered in this program is that the ETC frequently is ambiguous, and the result one obtains may depend on the rules one adopts towards the handling of ambiguous expressions. For example, how one proceeds to equal times from an unequal time commutator can affect the answer. A related, further problem is that in calculating commutators of operators, which themselves are products of other operators, ambiguities and infinities arise in forming these products. I have in mind, for example, the construction of currents which are bilinear in fermion fields. There have appeared in the literature many evaluations of commutators which yield different results; this variety is traceable to the various ways one can choose to handle the attendant ambiguities.

Fortunately, for our purposes, we can prescribe a unique method for calculating commutators. The reason for this is that we are comparing the current algebraic predictions with explicit dynamical predictions of the theory. For a consistent check, the ETC must be evaluated by the same techniques as the solutions of the theory. The only known tool for calculating physical consequences of a field theory is renormalized perturbation theory, which provides one with finite, well-defined Green's functions. Thus the ETC must be computed from the (known) Green's function. This is achieved by the method of Bjorken, Johnson and Low (BJL) [1, 2], which we now discuss.

Consider a matrix element of the T product of two operators A and B.

$$\mathrm{T}(q) \equiv \int d^4x\, e^{iq\cdot x} \langle\alpha\,|\,\mathrm{T}A(x)B(0)\,|\,\beta\rangle\,. \tag{3.1}$$

The BJL definition of the matrix elements of the $[A, B]$ ETC is

$$\lim_{q_0\to\infty} q_0\mathrm{T}(q) = i\int d^3x\, e^{-i\mathbf{q}\cdot\mathbf{x}} \langle\alpha\,|\,[A(0,\mathbf{x}), B(0)]\,|\,\beta\rangle\,. \tag{3.2}$$

Alternatively we can say that the $1/q_0$ term in $\mathrm{T}(q)$, at large q_0, determines the commutator. In all our subsequent calculations, we shall determine the commutator from the (known) T products by Eq. (3.2). If the above limit diverges, this is interpreted as the statement that that particular matrix element of the ETC is divergent.

As a justification of the BJL formula, one can present "derivations" of it, which are valid if various mathematical manipulations are permitted.

These derivations serve merely to *motivate* the result, and to insure that in non-singular situations the BJL definition corresponds to the usual ones for the ETC. Three derivations of increasing amount of rigor are offered.

The first method [3] begins by rewriting (3.1) as

$$\mathrm{T}(q) = -i\frac{1}{q_0}\int \mathrm{d}^4x \left(\frac{\partial}{\partial x_0} e^{iq\cdot x}\right) \langle \alpha | \mathrm{T}A(x)B(0) | \beta \rangle \,. \tag{3.3a}$$

The time integration is performed by parts, the surface terms are dropped, and one has

$$q_0 \mathrm{T}(q) = i\int \mathrm{d}^4x\, e^{iq\cdot x} \langle \alpha | \mathrm{T}\dot{A}(x)B(0) | \beta \rangle$$

$$+\, i\int \mathrm{d}^3x\, e^{-i\mathbf{q}\cdot\mathbf{x}} \langle \alpha | [A(0,\mathbf{x}), B(0)] | \beta \rangle \,. \tag{3.3b}$$

We now observe that the first term on the right-hand side of (3.3b) is a Fourier transform of an object which in the x_0 variable possesses singularities that are no worse than discontinuities. Hence as $q_0 \to \infty$, this term should vanish, according to the Riemann-Lesbegue lemma. Therefore, in the limit as $q_0 \to \infty$, (3.3b) reproduces (3.2). This derivation also indicates how anomalous, non-canonical results may be obtained from the BJL definition. It may turn out that a *canonical* computation of $\dot{A}$, by use of the operator equations of motion of the theory, may give an expression which leads to a singular behavior of $\int \mathrm{d}^4x\, e^{iq\cdot x} \langle \alpha | \mathrm{T}\dot{A}(x)B(0) | \beta \rangle$. Thus, in spite of the formalism, this term may survive in the large q_0 limit, and add a non-vanishing contribution of the *canonical* value of $i\int \mathrm{d}^3x\, e^{-i\mathbf{q}\cdot\mathbf{x}} \langle \alpha | [A(0,\mathbf{x}) B(0)] | \beta \rangle$. This results in a non-canonical value for $\lim_{q_0\to\infty} q_0 \mathrm{T}(q)$, and is interpreted as a non-canonical value for the (by definition) commutator.

The second derivation is Bjorken's original presentation [1]. Define

$$\rho(q_0, \mathbf{q}) = \int \mathrm{d}^4x\, e^{iq\cdot x} \langle \alpha | A(x) B(0) | \beta \rangle \,, \tag{3.4a}$$

$$\bar{\rho}(q_0, \mathbf{q}) = \int \mathrm{d}^4x\, e^{iq\cdot x} \langle \alpha | B(0) A(x) | \beta \rangle \,. \tag{3.4b}$$

By use of the integral representation for the step function,

$$\theta(x_0) = \frac{i}{2\pi}\int d\alpha\, e^{-ix_0\alpha}\,\frac{1}{\alpha+i\epsilon}\,, \tag{3.5}$$

the T product (3.1) may be written in the form

$$\mathrm{T}(q) = \frac{i}{2\pi}\int_{-\infty}^{\infty}\frac{d\alpha}{\alpha+i\epsilon}[\rho(q_0-\alpha,\mathbf{q})+\bar{\rho}(q_0+\alpha,\mathbf{q})]$$

$$= \frac{i}{2\pi}\int_{-\infty}^{\infty} dq_0'\left[\frac{\rho(q_0',\mathbf{q})}{q_0-q_0'+i\epsilon}-\frac{\bar{\rho}(q_0',\mathbf{q})}{q_0-q_0'-i\epsilon}\right]. \tag{3.6a}$$

This representation for the T product is called the Low representation. Multiplying by q_0 and passing to the $q_0\to\infty$ limit, leaves

$$\lim_{q_0\to\infty} q_0\,\mathrm{T}(q) = \frac{i}{2\pi}\int_{-\infty}^{\infty} dq_0'\,[\rho(q_0',\mathbf{q})-\bar{\rho}(q_0',\mathbf{q})]\,. \tag{3.6b}$$

According to the definitions (3.4), (3.6b) is just (3.2).

$$\lim_{q_0\to\infty} q_0\,\mathrm{T}(q) = i\int d^4x\, e^{-iq\cdot x}\,\delta(x_0)(\langle\alpha|A(x)B(0)|\beta\rangle$$

$$-\langle\alpha|B(0)A(x)|\beta\rangle)$$

$$= i\int d^3x\, e^{-i\mathbf{q}\cdot\mathbf{x}}\langle\alpha|[A(0,\mathbf{x}),B(0)]|\beta\rangle\,. \tag{3.6c}$$

This derivation is useful in that it shows explicitly that the ETC, as given by BJL definition, is physically interesting. The point is that according to (3.6b) the commutator has been expressed in terms of the spectral functions ρ and $\bar{\rho}$. These in turn are related to directly measurable matrix elements of A and B.

$$\rho(q_0,\mathbf{q}) = \sum_n (2\pi)^4\,\delta^4(q+p_\alpha-p_n)\langle\alpha|A(0)|n\rangle\langle n|B(0)|\beta\rangle\,,$$

$$\bar{\rho}(q_0,\mathbf{q}) = \sum_n (2\pi)^4\,\delta^4(q+p_n-p_\beta)\langle\alpha|B(0)|n\rangle\langle n|A(0)|\alpha\rangle\,. \tag{3.7}$$

The third derivation is due to Johnson and Low [2]. Consider the position space T product

$$t(x) = \int \frac{d^4 q}{(2\pi)^4} e^{-iq\cdot x}\, \mathrm{T}(q) \ . \tag{3.8}$$

The ETC may be defined as

$$C(0, \mathbf{x}) = \lim_{x_0 \to 0^+} t(x) - \lim_{x_0 \to 0^-} t(x) \ . \tag{3.9}$$

As $x_0 \to 0^+$, the q_0 integral in (3.8) may be extended into the complex q_0 plane, by closing the contour in the lower half plane; similarly as $x_0 \to 0^-$, the integral can be closed in the upper half plane. In the limit $x_0 \to 0^\pm$, the integrals along the real q_0-axis cancel between the two terms in (3.9), and one is left with

$$C(0, \mathbf{x}) = \oint \frac{dq_0}{2\pi} \int \frac{d^3 q}{(2\pi)^3}\, e^{i\mathbf{q}\cdot\mathbf{x}}\, \mathrm{T}(q) \ . \tag{3.10a}$$

Assume now that $\mathrm{T}(q)$ has an expansion in inverse powers of q_0. The clockwise contour integral of $(q_0)^{-n}$ is $-2\pi i \delta_{n,1}$; so that only the $1/q_0$ part of $\mathrm{T}(q)$ contributes. Call that part $\mathrm{T}_{-1}(\mathbf{q})$. We are left with

$$C(0, \mathbf{x}) = -i \int \frac{d^3 q}{(2\pi)^3}\, e^{i\mathbf{q}\cdot\mathbf{x}}\, \mathrm{T}_{-1}(\mathbf{q}) \ . \tag{3.10b}$$

The Fourier transform of this is equivalent to (3.2). Johnson and Low further show that if the leading singularity of $\mathrm{T}(q)$ at large q_0 is $(1/q_0)\log q_0$, then the previous derivation remains correct, and the equal time commutator is logarithmically divergent. Similarly if the singularity is of the form $q_0 \log q_0 = (1/q_0)(q_0^2 \log q_0)$; this should be interpreted as a quadratically divergent commutator.

It should be remarked here that when the BJL theorem was first discovered it was applied in many instances to a *derivation* of the high energy behavior of amplitudes. The idea was to write

$$\mathrm{T}(q) \sim \frac{i}{q_0} \int d^3 x\, e^{-i\mathbf{q}\cdot\mathbf{x}} \langle \alpha | [A(0,\mathbf{x}), B(0)] | \beta \rangle \ , \tag{3.11}$$

and to evaluate the right-hand side of (3.11) canonically. However, the use of the canonical commutator in this context is unjustified, as Johnson and Low already demonstrated in their paper [2]. The fact that the high energy behavior of amplitudes *is not*, in general, correctly given by this technique has recently been described as " the breakdown of the BJL theorem".

Note that the BJL theorem defines the commutator from the T product, rather than from the covariant T* product. However, in perturbation theory one calculates only the covariant object. Hence the T product must be separated. This is achieved by remembering that the difference btween T and T* is local in position space, hence it is a polynomial of q_0 in momentum space. Therefore, before applying the BJL technique to the expressions calculated in perturbation theory, all polynomials in q_0 must be dropped.

It is clear that the expansion in inverse powers of q_0 can be extended beyond the first. From (3.3a), it is easy to see that if the $[A, B]$ ETC vanishes, then we have

$$\lim_{q_0 \to \infty} q_0^2 \, \mathrm{T}(q) = -\int \mathrm{d}^3x \, e^{-i\mathbf{q}\cdot\mathbf{x}} \langle \alpha | [\dot{A}(0, \mathbf{x}), B(0)] | \beta \rangle \,. \tag{3.12}$$

Again, if this limit is divergent, then this matrix element of the $[\dot{A}, B]$ ETC is infinite. Eventually commutators with sufficient number of time derivatives probably are infinite, since it is unlikely that the expansion in inverse powers of q_0 can be extended without limit.

References

[1] J. D. Bjorken, Phys. Rev. 148 (1966) 1467.
[2] K. Johnson and F. E. Low, Prog. Theor. Phys. (Kyoto), Supp. 37–38, (1966) 74.
[3] This method was developed in conservations with Prof. I. Gerstein.

4. The $\pi^0 \to 2\gamma$ Problem

4.1 Preliminaries

The neutral pion is observed to decay into 2 photons with a width of the order of 10 eV. This experiment measures the matrix element $M(p, q) = \langle \pi, k | \gamma, p; \gamma', q \rangle$; p and q are the 4-momenta of the photons, $k = p + q$ is the 4-momentum of the pion. $M(p, q)$ has the form

$$\epsilon_\mu(p) \, \epsilon'_\nu(q) \, \mathrm{T}^{\mu\nu}(p, q) \,,$$

i.e., $T^{\mu\nu}(p, q)$ is the previous matrix element with the photon polarization vectors $\epsilon_\mu(p)\epsilon'_\nu(q)$ removed. The tensor $T^{\mu\nu}$ has the following structure.

$$T^{\mu\nu}(p, q) = \epsilon^{\mu\nu\alpha\beta} p_\alpha q_\beta T(k^2) . \tag{4.1}$$

This is dictated by Lorentz covariance and parity conservation (the pion is a pseudoscalar, $T^{\mu\nu}$ must be a pseudo-tensor, hence the factor $\epsilon^{\mu\nu\alpha\beta}$). Gauge invariance $(p_\mu T^{\mu\nu}(p,q) = 0 = q_\nu T^{\mu\nu}(p,q))$ and Bose symmetry $(T^{\mu\nu}(p,q) = T^{\nu\mu}(q,p))$ are seen to hold.

We shall keep q^2 and p^2, the photon variables, on their mass shell $q^2 = p^2 = 0$. The pion variable k^2 is, of course, equal to the pion mass squared μ^2, but for our arguments we allow it to vary away from this point. This continuation off the mass shell may be effected by the usual LSZ method.

$$\begin{aligned} T^{\mu\nu}(p, q) &= \epsilon^{\mu\nu\alpha\beta} p_\alpha q_\beta T(k^2) \\ &= (\mu^2 - k^2)\langle 0 | \phi(0) | \gamma, p ; \gamma', q \rangle . \end{aligned} \tag{4.2}$$

Here ϕ is an interpolating field for the pion. We, of course, do not assert that it is *the* pion field — such an object may not exist. It merely is some local operator which has a non-vanishing matrix element between the vacuum and the single pion state, normalized to unity $\langle 0 | \phi(0) | \pi \rangle = 1$.

4.2 Sutherland-Veltman Theorem

Following Sutherland and Veltman [1], we now prove that if the divergence of the axial current is used as the pion interpolating field, then $T(0) = 0$, as long as the conventional current algebraic ideas are valid. This is a *mathematical* fact, without direct experimental content. However, since μ^2 is small, compared to all other mass parameters relevant to this problem, one may expect that $T(\mu^2) \approx T(0)$. This smoothness hypothesis is based on the supposition that the divergence of the axial current is a "gentle" operator whose matrix elements do not have any dynamically unnecessary rapid variation. This is the content of PCAC, which is very successful notion in other contexts. Unfortunately, in the present application, one cannot understand the *experimental* fact that $T(\mu^2) \neq 0$.

After Sutherland and Veltman pointed out this *experimental* failure of PCAC, the most widely accepted explanation was that $T(k^2)$ *was* rapidly varying, for unknown reasons. This is not impossible, since it has happened before in current algebra-PCAC applications that a source of rapid variation for a particular amplitude was at first overlooked. However, in the present instance, as the years passed by, no reason was forthcoming to explain the putative rapid variation.

The Sutherland-Veltman argument begins by representing the off mass shell pion amplitude (4.2) by

$$T^{\mu\nu}(p,q) = e^2(\mu^2 - k^2)\int d^4x\, d^4y\, e^{-ip\cdot x}\, e^{-iq\cdot y}$$
$$\times \langle 0|T^* J^\mu(x) J^\nu(y)\phi(0)|0\rangle \;. \tag{4.3}$$

Here J^μ is the electromagnetic current. The pion field is replaced by the divergence of the neutral axial current $J_5{}^\alpha$.

$$\phi(0) = \frac{\partial_\alpha J_5{}^\alpha(0)}{F\mu^2} \;. \tag{4.4}$$

$F\mu^2$ is the appropriate factor which assures the proper normalization for the pion field, defined by (4.4).

$$\langle 0|J_5{}^\alpha(0)|\pi\rangle \equiv ip^\alpha F \;,$$
$$\langle 0|\partial_\alpha J_5{}^\alpha(0)|\pi\rangle = \mu^2 F \;. \tag{4.5}$$

Thus

$$T^{\mu\nu}(p,q) = \frac{e^2(\mu^2-k^2)}{F\mu^2}\int d^4x\, d^4y\, e^{-ip\cdot x}\, e^{-iq\cdot y}$$
$$\times \langle 0|T^* J^\mu(x)J^\nu(y)\,\partial_\alpha J_5{}^\alpha(0)|0\rangle \;,$$
$$= \frac{e^2(\mu^2-k^2)}{F\mu^2}\int d^4x\, d^4y\, e^{-ip\cdot x}\, e^{-iq\cdot y}$$
$$\times\, \partial_\alpha \langle 0|T^* J^\mu(x)J^\nu(y)J_5{}^\alpha(0)|0\rangle \;,$$
$$= \frac{(\mu^2-k^2)}{F\mu^2}\, k_\alpha T^{\alpha\mu\nu}(p,q) \;; \tag{4.6a}$$

where $T^{\alpha\mu\nu}(p,q)$ is defined by

$$T^{\alpha\mu\nu}(p,q) = -ie^2\int d^4x\, d^4y\, e^{-ip\cdot x}\, e^{-iq\cdot y}$$

$$\times\ \langle 0|T^* J^\mu(x) J^\nu(y) J_5{}^\alpha(0)|0\rangle\ . \tag{4.6b}$$

The justification for passing from the first to the second term on the right-hand side of (4.6a) is the current algebra satisfied by J_5^0 and J^μ: apart from possible ST the currents commute.

$$[J_5^0(0,\mathbf{x}), J^\mu(0)] = 0 + \mathrm{ST}\ . \tag{4.7}$$

The ST is handled by one of three ways. One may simply assume that it is absent; since the ETC does not involve two identical currents, there is no proof that a ST must be present. Alternatively a weaker assumption is that the ST is a c number. It is easy to see that since the vacuum expectation value of a current vanishes, a c-number ST would not interfere with passing the derivative through the T* product. Finally the weakest assumption that one can make is Feynman's conjecture — without discussing the nature of any possible ST, it is asserted that the naive procedure is the correct one, due to cancellation of ST with divergences of seagulls.

The tensor $T^{\alpha\mu\nu}(p,q)$ must possess odd parity, because $J_5{}^\alpha$ is a pseudovector; it must satisfy the Bose symmetry: $T^{\alpha\mu\nu}(p,q) = T^{\alpha\nu\mu}(q,p)$; finally it must be transverse to p_μ and q_ν: $p_\mu T^{\alpha\mu\nu}(p,q) = 0$, $q_\nu T^{\alpha\mu\nu}(p,q) = 0$. The last condition follows from the conservation of J^μ and the current algebra satisfied by J^0 with J^μ and $J_5{}^\alpha$. Again all these commutators vanish apart from possible ST; the latter being ignored in this calculation.

$$[J^0(0,\mathbf{x}), J^\mu(0)] = 0 + \mathrm{ST}\ , \tag{4.8a}$$

$$[J^0(0,\mathbf{x}), J_5{}^\mu(0)] = 0 + \mathrm{ST}\ . \tag{4.8b}$$

The following form for $T^{\alpha\mu\nu}(p,q)$ is the most general structure, free from kinematical singularities, satisfying the above requirements (remember that $p^2 = q^2 = 0$).

$$\begin{aligned} T^{\alpha\mu\nu}(p,q) = {} & \epsilon^{\mu\nu\omega\phi} p_\omega q_\phi k^\alpha F_1(k^2) \\ & + (\epsilon^{\alpha\mu\omega\phi} q^\nu - \epsilon^{\alpha\nu\omega\phi} p^\mu) p_\omega q_\phi F_2(k^2) \\ & + (\epsilon^{\alpha\mu\omega\phi} p^\nu - \epsilon^{\alpha\nu\omega\phi} q^\mu) p_\omega q_\phi F_3(k^2) \\ & + \epsilon^{\alpha\mu\nu\omega}(p_\omega - q_\omega) k^2 F_3(k^2)/2\ . \end{aligned} \tag{4.9}$$

It now follows that

$$k_\alpha \mathrm{T}^{\alpha\mu\nu}(p,q) = \epsilon^{\mu\nu\omega\phi} p_\omega q_\phi k^2 [F_1(k^2) - F_3(k^2)] \ . \tag{4.10a}$$

Comparison with (4.6a) and (4.1) finally gives

$$\mathrm{T}(k^2) = \frac{(\mu^2 - k^2)}{F\mu^2} k^2 [F_1(k^2) - F_3(k^2)] \ . \tag{4.10b}$$

As we mentioned, the F_i are free from kinematical singularities; since we are working to lowest order in electromagnetism, they do not possess dynamical singularities at $k^2 = 0$. Hence we find, as promised, $\mathrm{T}(0) = 0$. Note that PCAC is not used to obtain the mathematical statement $\mathrm{T}(0) = 0$. This hypothesis becomes necessary only when $\mathrm{T}(0)$ is related to $\mathrm{T}(\mu^2)$. It will now be shown that even the mathematical prediction is invalid in the σ model.

4.3 Model Calculation

We calculate [2] the off mass shell pion decay constant, $\mathrm{T}(k^2)$, in the σ model where all the assumptions of the Sutherland-Veltman theorem seem to be satisfied [3]. The Lagrangian is

$$\begin{aligned} \mathscr{L} = {} & \bar{\psi}(i\gamma^\mu \partial_\mu - m)\psi + \tfrac{1}{2}\partial_\mu\phi\partial^\mu\phi - \tfrac{1}{2}\mu^2\phi^2 + \tfrac{1}{2}\partial_\mu\sigma\partial^\mu\sigma \\ & - \tfrac{1}{2}(\mu^2 + 2\lambda F^2)\sigma^2 + e\bar{\psi}\gamma^\mu\psi A_\mu + g\bar{\psi}(\sigma + \phi\gamma_5)\psi \\ & - \lambda[(\phi^2 + \sigma^2)^2 - 2F\sigma(\sigma^2 + \phi^2)] \ . \end{aligned} \tag{4.11}$$

Here ψ, ϕ and σ are fields for the "proton", "pion" and σ particle, each possessing the respective masses m, μ and $(\mu^2 + 2\lambda F^2)^{1/2}$. The proton interacts with the pion and σ in a chirally symmetric fashion with strength g. The proton also has an electromagnetic interaction; since we work to lowest order in that interaction, it is sufficient to consider the electromagnetic potential A^μ as an external perturbation. There are also meson self-couplings with strength λ, which are necessary for the consistency of the model, but which do not affect the present discussion. The parameter F is equal to $2mg^{-1}$. All isospin effects are ignored, since they are irrelevant to the argument.

The model possesses a neutral axial current $J_5{}^\alpha$ whose divergence according to the equations of motion of the theory is the pion field.

$$J_5{}^\alpha = i\bar{\psi}\gamma^\alpha\gamma^5\psi + 2(\sigma\partial^\alpha\phi - \phi\partial^\alpha\sigma) - F\partial^\alpha\phi\,, \tag{4.12a}$$

$$\partial_\alpha J_5{}^\alpha = \mu^2 F\phi\,. \tag{4.12b}$$

The electromagnetic current $J^\mu = \bar{\psi}\gamma^\mu\psi$ and the axial current satisfy conventional current commutators. Of course, no ST is present canonically, in the time-component algebra so we cannot ascertain whether or not Feynman's conjecture is satisfied.

In this theory the pion can decay into two photons by dissociating first into a virtual proton-antiproton pair, which then emits two photons. The lowest order graphs are those of Fig. 4-1. These have the integral representation

$$\mathrm{T}^{\mu\nu}(p,q) = \Gamma^{\mu\nu}(p,q) + \Gamma^{\nu\mu}(q,p)\,,$$

$$\Gamma^{\mu\nu}(p,q) = ige^2\int\frac{d^4r}{(2\pi)^4}\,\mathrm{Tr}\,\gamma^5\,[\gamma_\alpha r^\alpha + \gamma_\alpha p^\alpha - m]^{-1}$$

$$\times\ \gamma^\mu[\gamma_\alpha r^\alpha - m]^{-1}\gamma^\nu[\gamma_\alpha r^\alpha - \gamma_\alpha q^\alpha - m]^{-1}\,. \tag{4.13a}$$

The integral appears to diverge linearly; however, after the trace is performed one is left with a finite expression.

$$\Gamma^{\mu\nu}(p,q) = 4mige^2\,\epsilon^{\mu\nu\alpha\beta}p_\alpha q_\beta\int\frac{d^4r}{(2\pi)^4}$$

$$\times\ [(r+p)^2 - m^2]^{-1}\,[r^2 - m^2]^{-1}\,[(r-q)^2 - m^2]^{-1}\,. \tag{4.13b}$$

The remaining evaluation is elementary [4]. The answer is

$$\Gamma^{\mu\nu}(p,q) = \frac{mge^2}{4\pi^2}\epsilon^{\mu\nu\alpha\beta}p_\alpha q_\beta\int_0^1 dx\int_0^{1-x} dy\,[m^2 - k^2xy]^{-1}\,. \tag{4.13c}$$

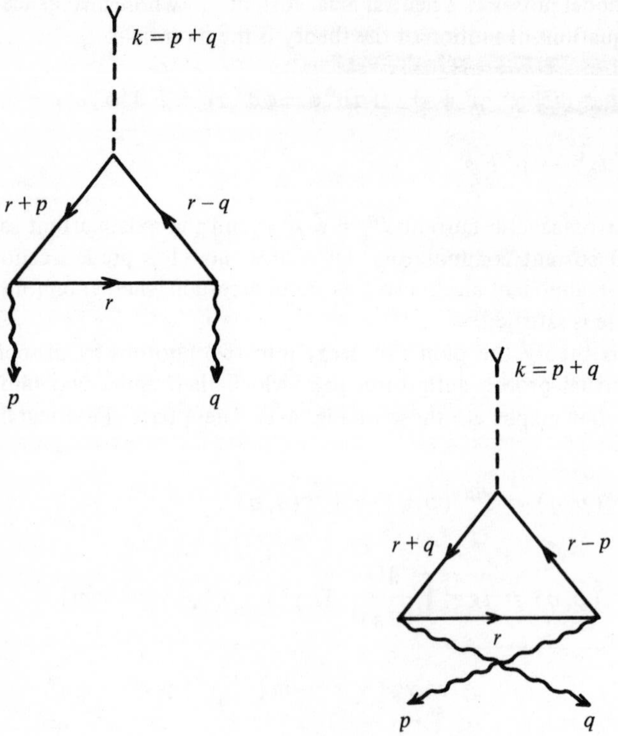

Fig. 4-1. Feynman graphs describing the $\pi^0 \to 2\gamma$ amplitude, in lowest order for the theory given by Eq. (4.11).

In the notation (4.1) we find

$$T(k^2) = \frac{mge^2}{2\pi^2} \int_0^1 dx \int_0^{1-x} dy\, [m^2 - k^2 xy]^{-1} ,$$

$$T(0) = \frac{ge^2}{4\pi^2 m} = \frac{e^2}{2\pi^2 F} \neq 0 . \tag{4.14}$$

For future reference, note that the large m behavior of $T(k^2)$ is $ge^2/4\pi^2 m$.

Our calculation has demonstrated that the *mathematical* portion of the Sutherland-Veltman theorem is false. Since $\mathrm{T}(k^2)$ is perfectly smooth for small k^2, $\mathrm{T}(k^2) \approx \mathrm{T}(0)\,[1 + k^2/(12m^2)]$, we see that the *experimental* part of that theorem is also incorrect in this model. The reason does not lie in any unexpected rapid variation, but rather in the failure of conventional current algebra.

4.4 *Anomalous Ward Identity*

To gain further understanding into the problem, we calculate the function $\mathrm{T}^{\alpha\mu\nu}(p,q)$, (4.6b). The relevant Feynman graphs are given in Fig. 4-2. In the first two graphs, Fig. 4-2a, the axial current attaches directly to the Fermion loop. In the last two, Fig. 4-2b, it passes first through the virtual pion with the coupling $2mg^{-1}$, thus acquiring the necessary pion pole. The integral representation is

$$\mathrm{T}^{\alpha\mu\nu}(p,q) = \mathrm{T}_1{}^{\alpha\mu\nu}(p,q) + \mathrm{T}_2{}^{\alpha\mu\nu}(p,q) \ , \tag{4.15a}$$

$$\mathrm{T}_1{}^{\alpha\mu\nu}(p,q) = \Gamma^{\alpha\mu\nu}(p,q) + \Gamma^{\alpha\nu\mu}(q,p) \ , \tag{4.15b}$$

$$\Gamma^{\alpha\mu\nu}(p,q) = ie^2 \int \frac{d^4 r}{(2\pi)^4} \operatorname{Tr} \gamma^5 \gamma^\alpha \, [\gamma_\beta r^\beta + \gamma_\beta p^\beta - m]^{-1} \times \gamma^\mu \, [\gamma_\beta r^\beta - m]^{-1} \gamma^\nu \, [\gamma_\beta r^\beta - \gamma_\beta q^\beta - m]^{-1} \ , \tag{4.15c}$$

$$\mathrm{T}_2{}^{\alpha\mu\nu}(p,q) = -\frac{2mg^{-1}}{k^2 - \mu^2} k^\alpha \, \mathrm{T}^{\mu\nu}(p,q) \ . \tag{4.15d}$$

Evidently the verification of the axial Ward identity

$$(F\mu^2)^{-1}(\mu^2 - k^2) k_\alpha \, \mathrm{T}^{\alpha\mu\nu}(p,q) = \mathrm{T}^{\mu\nu}(p,q) \ , \tag{4.16a}$$

which was used in the derivation of the Sutherland-Veltman theorem, is equivalent to showing

$$k_\alpha \mathrm{T}_1{}^{\alpha\mu\nu}(p,q) = 2mg^{-1} \, \mathrm{T}^{\mu\nu}(p,q) \ . \tag{4.16b}$$

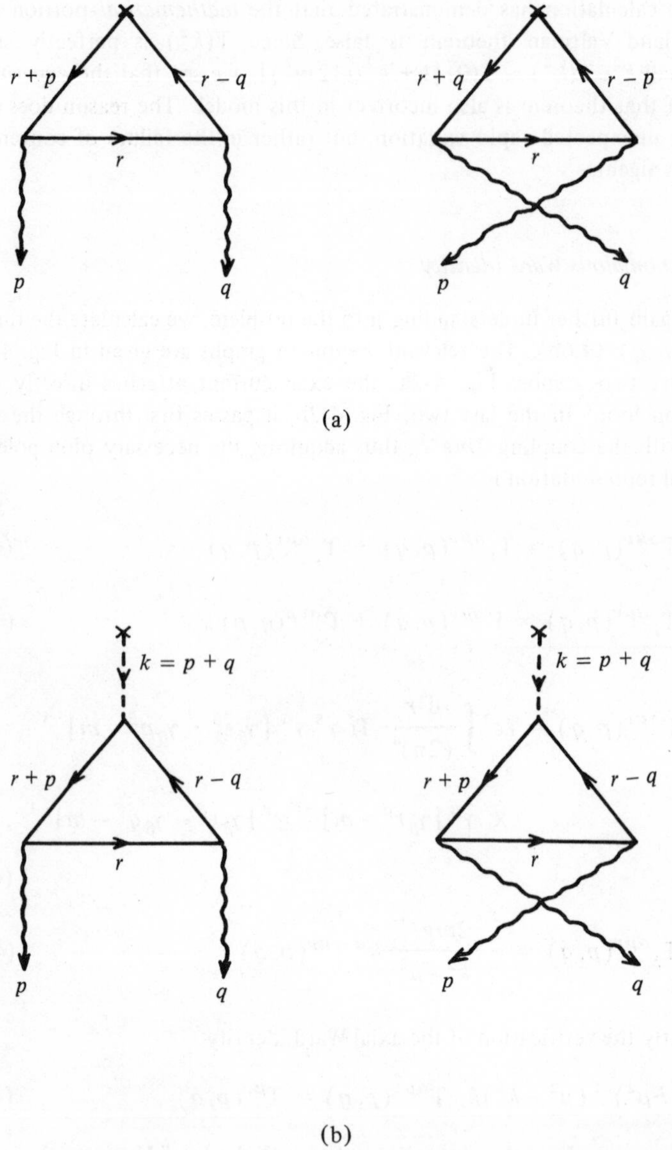

Fig. 4-2. Feynman graphs describing the $J_5^{\alpha} \to 2\gamma$ amplitude, in lowest order for the theory given by Eq. (4.11).

The vector Ward identity, i.e., gauge invariance, is also necessary for the theorem. In the present notation it is equivalent to

$$p_\mu \mathrm{T}_1{}^{\alpha\mu\nu}(p,q) = q_\nu \mathrm{T}_1{}^{\alpha\mu\nu}(p,q) = 0 \; . \tag{4.17}$$

We now demonstrate that both (4.16) and (4.17) cannot be maintained simultaneously for $\mathrm{T}_1{}^{\alpha\mu\nu}$ as given by (4.15).

The important property of the graphs in Fig. 4-2a which is responsible for this anomalous behavior is their linear divergence. Unlike in the evaluation of $\mathrm{T}^{\mu\nu}$, (4.13), performing the trace does not remove this divergence. In a linearly divergent integral it is illegitimate to shift the variable of integration. This is easily seen on one dimension. Consider

$$\Delta(a) = \int_{-\infty}^{\infty} \mathrm{d}x \, [f(x+a) - f(x)] \; . \tag{4.18a}$$

If one can shift the integration variable in the first integral $x+a \to x$; one can conclude that $\Delta(a)=0$. To see that $\Delta(a)$ need not vanish, let us expand the integrand.

$$\Delta(a) = \int_{-\infty}^{\infty} \mathrm{d}x \, [a f'(x) + \frac{a^2}{2} f''(x) + \ldots] \; . \tag{4.18b}$$

Integrating by parts, we find

$$\Delta(a) = a[f(\infty) - f(-\infty)] + \frac{a^2}{2}[f'(\infty) - f'(-\infty)] + \ldots \; . \tag{4.18c}$$

When the integral $\int_{-\infty}^{\infty} \mathrm{d}x \, f(x)$ converges (or at most diverges logarithmically) we have $0 = f(\pm\infty) = f'(\pm\infty) \ldots$, and $\Delta(a)$ vanishes. However, for a linearly divergent integral $0 \neq f(\pm\infty)$, $0 = f'(\pm\infty) \ldots$ and $\Delta(a)$ need not vanish.

$$\Delta(a) = a[f(\infty) - f(-\infty)] \; . \tag{4.18d}$$

Such a contribution is called a "surface term". This state of affairs persists in 4-dimensional, Minkowski space integrals. Consider

$$\Delta^\mu(a) = i \int \frac{\mathrm{d}^4 r}{(2\pi)^4} \left[\frac{r^\mu + a^\mu}{([r+a]^2 - m^2)^2} - \frac{r^\mu}{(r^2 - m^2)^2} \right] \; . \tag{4.19a}$$

Here a is an arbitrary 4-vector. The surface term may be evaluated. The result is non-vanishing; see Exercise 4.1.

$$\Delta^\mu(a) = -\frac{a^\mu}{32\pi^2} \quad . \tag{4.19b}$$

The consequence of this for our problem is that the integral $\Gamma^{\alpha\mu\nu}$ (4.15c), which contributes to $T_1{}^{\alpha\mu\nu}$, is not uniquely defined. The point is that in exhibiting (4.15c) we have routed the integration momentum r in a particular fashion: the fermion leg between the two photons carries r. However any other routing could also be chosen, so that the fermion leg between the photons carries the 4-momentum $r+a$, where a is an arbitrary four vector. If the integral were not linearly divergent, then a shift of integration would return this routing to the previous one; but in the present instance such shifts produce surface terms.

In conventional evaluations of divergent Feynman graphs, such ambiguities are usually ignored. Typically cut-offs are introduced, which eliminate these problems and then the cut-offs are removed by the renormalization procedure. For our purposes we need to keep track of all the possible sources of ambiguity. Thus we replace the expression for $\Gamma^{\alpha\mu\nu}$, (4.15c), by a class of expressions, parametrized by an arbitrary 4-vector a^μ.

$$\Gamma^{\alpha\mu\nu}(p,q\,|\,a) = \Gamma^{\alpha\mu\nu}(p,q) + \Delta^{\alpha\mu\nu}(p,q\,|\,a) \ , \tag{4.20a}$$

$$\begin{aligned} &\Delta^{\alpha\mu\nu}(p,q\,|\,a) \\ &= ie^2 \int \frac{d^4 r}{(2\pi)^4} \operatorname{Tr} \gamma^5 \gamma^\alpha \; [\gamma_\beta r^\beta + \gamma_\beta a^\beta + \gamma_\beta p^\beta - m]^{-1} \\ &\quad \times \gamma^\mu [\gamma_\beta r^\beta + \gamma_\beta a^\beta - m]^{-1} \gamma^\nu [\gamma_\beta r^\beta + \gamma_\beta a^\beta - \gamma_\beta q^\beta - m]^{-1} \\ &\quad - [\gamma_\beta r^\beta + \gamma_\beta p^\beta - m]^{-1} \gamma^\mu [\gamma_\beta r^\beta - m]^{-1} \\ &\quad \times \gamma^\nu [\gamma_\beta r^\beta - \gamma_\beta q^\beta - m]^{-1} \quad . \end{aligned} \tag{4.20b}$$

The surface term is evaluated, see Exercise 4.2.

$$\Delta^{\alpha\mu\nu}(p,q\,|\,a) = -\frac{e^2}{8\pi^2} \epsilon^{\alpha\mu\nu\beta} a_\beta \ . \tag{4.21}$$

The arbitrariness of a_β is limited somewhat by the plausibility requirement that no vectors, other than those already present in the problem at hand, should be introduced. Hence we set $a_\beta = (a+b)p_\beta + bq_\beta$. Corresponding to the class of functions $\Gamma^{\alpha\mu\nu}(p,q\,|\,a)$ we have a class of functions $T_1{}^{\alpha\mu\nu}(p,q\,|\,a)$. Accordingly (4.15b) and (4.21)

$$T_1{}^{\alpha\mu\nu}(p,q\,|\,a) = T_1{}^{\alpha\mu\nu}(p,q) - \frac{e^2}{8\pi^2}\, a\epsilon^{\alpha\mu\nu\beta}(p_\beta - q_\beta)\,. \tag{4.22}$$

Note that Bose symmetry has been maintained.

Any member of the class of functions $T_1{}^{\alpha\mu\nu}(p,q\,|\,a)$ may be considered "correct". The various functions differ among themselves only by a polynomial in p and q, i.e., by a covariant seagull. We now attempt to determine a by imposing the axial and the vector Ward identities. We are hoping that $T_1{}^{\alpha\mu\nu}(p,q\,|\,a)$, for some definite value of a, will satisfy these identities. It will be seen that no such value for a exists.

It is possible to evaluate $T_1{}^{\alpha\mu\nu}(p,q)$ as given by (4.15), and therefore to exhibit an explicit formula for $T_1{}^{\alpha\mu\nu}(p,q\,|\,a)$. The evaluation is effected by conventional methods, except it must be always remembered that shifts of integration variables produce non-vanishing, but well defined terms. A remarkable thing that occurs is that the end result is finite; symmetric integration removes the linear divergences as well as the sub-dominant logarithmic divergence. Thus a finite, unambiguous formula for $T_1{}^{\alpha\mu\nu}(p,q\,|\,a)$ may be arrived at. (The illegitimacy of shifts of integration follows from the *superficial* divergence properties of integral, even if accidentally the result is finite.) The detailed evaluation of $T_1{}^{\alpha\mu\nu}(p,q)$ has been given in the literature [5]; for our present purposes of verifying the Ward identities we do not need this formula, the integral representation (4.15) will suffice.

Consider first the axial Ward identity. We wish to learn the form of $k_\alpha T_1{}^{\alpha\mu\nu}(p,q)$. From (4.15c) it follows that

$$\begin{aligned} &k_\alpha \Gamma^{\alpha\mu\nu}(p,q) \\ &\quad = (p_\alpha + q_\alpha)\,\Gamma^{\alpha\mu\nu} \\ &\quad = ie^2 \int \frac{d^4r}{(2\pi)^4} \operatorname{Tr} \gamma^5 (\gamma_\beta p^\beta + \gamma_\beta q^\beta)\,[\gamma_\beta r^\beta + \gamma_\beta p^\beta - m]^{-1} \\ &\qquad \times \gamma^\mu [\gamma_\beta r^\beta - m]^{-1} \gamma^\nu [\gamma_\beta r^\beta - \gamma_\beta q^\beta - m]^{-1}\,. \end{aligned} \tag{4.23a}$$

After rewriting $\gamma_\beta p^\beta + \gamma_\beta q^\beta$ as $2m + (\gamma_\beta p^\beta + \gamma_\beta r^\beta - m) - (\gamma_\beta r^\beta - \gamma_\beta q^\beta + m)$ we have

$$k_\alpha \Gamma^{\alpha\mu\nu}(p,q)$$

$$= (2mg^{-1}) ige^2 \int \frac{d^4 r}{(2\pi)^4} \gamma^5 [\gamma_\beta r^\beta + \gamma_\beta p^\beta - m]^{-1}$$

$$\times \gamma^\mu [\gamma_\beta r^\beta - m]^{-1} \gamma^\nu [\gamma_\beta r^\beta - \gamma_\beta q^\beta - m]^{-1}$$

$$+ ie^2 \int \frac{d^4 r}{(2\pi)^4} \mathrm{Tr}\, \gamma^5 \gamma^\mu [\gamma_\beta r^\beta - m]^{-1} \gamma^\nu [\gamma_\beta r^\beta - \gamma_\beta q^\beta - m]^{-1}$$

$$- ie^2 \int \frac{d^4 r}{(2\pi)^4} \mathrm{Tr}\, \gamma^5 [\gamma_\beta r^\beta - \gamma_\beta q^\beta + m] [\gamma_\beta p^\beta + \gamma_\beta r^\beta - m]^{-1}$$

$$\times \gamma^\mu [\gamma_\beta r^\beta - m]^{-1} \gamma^\nu [\gamma_\beta r^\beta - \gamma_\beta q^\beta - m]^{-1} \ . \tag{4.23b}$$

The first integral is recognized as $2mg^{-1}$ times $\Gamma^{\mu\nu}(p,q)$; see (4.13). In the third integral $\gamma_\beta r^\beta - \gamma_\beta q^\beta + m$ may be taken through the γ^5, thus changing the overall sign and the sign of m. Then the cyclicity of the trace allows one to transpose that term to the end of that expression, thus cancelling the last propagator. We are now left with

$$k_\alpha \Gamma^{\alpha\mu\nu}(p,q) = 2mg^{-1} \Gamma^{\mu\nu}(p,q)$$

$$+ ie^2 \int \frac{d^4 r}{(2\pi)^4} \mathrm{Tr}\, \gamma^5 \gamma^\mu [\gamma_\beta r^\beta - m]^{-1} \gamma^\nu [\gamma_\beta r^\beta - \gamma_\beta q^\beta - m]^{-1}$$

$$+ ie^2 \int \frac{d^4 r}{(2\pi)^4} \mathrm{Tr}\, \gamma^5 [\gamma_\beta p^\beta + \gamma_\beta r^\beta - m]^{-1} \gamma^\mu [\gamma_\beta r^\beta - m]^{-1} \gamma^\nu \ . \tag{4.23c}$$

Each of the two integrals must vanish since it is impossible to form a two-index pseudotensor which depends on only one vector. We find therefore

$$k_\alpha T_1{}^{\alpha\mu\nu}(p,q) = 2mg^{-1} T^{\mu\nu}(p,q) \ ,$$

$$k_\alpha T_1{}^{\alpha\mu\nu}(p,q\,|\,a) = 2mg^{-1} T^{\mu\nu}(p,q) + \frac{a e^2}{4\pi^2} \epsilon^{\mu\nu\alpha\beta} p_\alpha q_\beta \ . \tag{4.24}$$

The conclusion is that in order to satisfy the axial Ward identity the routing of the integration variable must be as in Fig. 4-2; i.e., a must be set to zero. Note that this verification required no shifts of integration variable. The vector Ward identity, i.e., gauge invariance cannot be established in the same fashion, as we now demonstrate.

We wish to learn the form of $p_\mu \mathrm{T}_1{}^{\alpha\mu\nu}(p,q)$. According to (4.15)

$$
\begin{aligned}
p_\mu \mathrm{T}_1{}^{\alpha\mu\nu}(p,q) &= p_\mu \Gamma^{\alpha\mu\nu}(p,q) + p_\mu \Gamma^{\alpha\mu\nu}(q,p) \\
&= ie^2 \int \frac{\mathrm{d}^4 r}{(2\pi)^4} \operatorname{Tr} \gamma^5 \gamma^\alpha [\gamma_\beta r^\beta + \gamma_\beta p^\beta - m]^{-1} \\
&\quad \times \gamma_\beta p^\beta [\gamma_\beta r^\beta - m]^{-1} \gamma^\nu [\gamma_\beta r^\beta - \gamma_\beta q^\beta - m]^{-1} \\
&\quad + ie^2 \int \frac{\mathrm{d}^4 r}{(2\pi)^4} \operatorname{Tr} \gamma^5 \gamma^\alpha [\gamma_\beta r^\beta + \gamma_\beta q^\beta - m]^{-1} \\
&\quad \times \gamma^\nu [\gamma_\beta r^\beta - m]^{-1} \gamma_\beta p^\beta [\gamma_\beta r^\beta - \gamma_\beta p^\beta - m]^{-1} .
\end{aligned}
\tag{4.25a}
$$

Use of the identities

$$
\begin{aligned}
&[\gamma_\beta r^\beta + \gamma_\beta p^\beta - m]^{-1} \gamma_\beta p^\beta [\gamma_\beta r^\beta - m]^{-1} \\
&\quad = [\gamma_\beta r^\beta - m]^{-1} - [\gamma_\beta r^\beta + \gamma_\beta p^\beta - m]^{-1} ,
\end{aligned}
$$

$$
\begin{aligned}
&[\gamma_\beta r^\beta - m]^{-1} \gamma_\beta p^\beta [\gamma_\beta r^\beta - \gamma_\beta p^\beta - m]^{-1} \\
&\quad = [\gamma_\beta r^\beta - \gamma_\beta p^\beta - m]^{-1} - [\gamma_\beta r^\beta - m]^{-1} ,
\end{aligned}
$$

allows (4.25a) to be written as

$$p_\mu T_1{}^{\alpha\mu\nu}(p,q)$$

$$= ie^2 \int \frac{d^4 r}{(2\pi)^4} \operatorname{Tr} \gamma^5 \gamma^\alpha [\gamma_\beta r^\beta - m]^{-1} \gamma^\nu [\gamma_\beta r^\beta - \gamma_\beta q^\beta - m]^{-1}$$

$$- ie^2 \int \frac{d^4 r}{(2\pi)^4} \operatorname{Tr} \gamma^5 \gamma^\alpha [\gamma_\beta r^\beta + \gamma_\beta p^\beta - m]^{-1} \gamma^\nu [\gamma_\beta r^\beta - \gamma_\beta q^\beta - m]^{-1}$$

$$+ ie^2 \int \frac{d^4 r}{(2\pi)^4} \operatorname{Tr} \gamma^5 \gamma^\alpha [\gamma_\beta r^\beta + \gamma_\beta q^\beta - m]^{-1} \gamma^\nu [\gamma_\beta r^\beta - \gamma_\beta p^\beta - m]^{-1}$$

$$- ie^2 \int \frac{d^4 r}{(2\pi)^4} \operatorname{Tr} \gamma^5 \gamma^\alpha [\gamma_\beta r^\beta + \gamma_\beta q^\beta - m]^{-1} \gamma^\nu [\gamma_\beta r^\beta - m]^{-1} \ . \tag{4.25b}$$

The first and last integrals in (4.25b) vanish because they are two index pseudotensors depending on one vector. The remaining two integrals could be made to cancel against each other if shifts of integration were allowed. Unfortunately such shifts lead to a finite contribution. The value of the surface term evaluated is (see Exercise 4.3)

$$p_\mu T_1{}^{\alpha\mu\nu}(p,q) = \frac{e^2}{4\pi^2} \epsilon^{\alpha\mu\nu\beta} p_\mu q_\beta \ . \tag{4.25c}$$

Therefore

$$p_\mu T_1{}^{\alpha\mu\nu}(p,q\,|\,a) = \frac{e^2}{4\pi^2} \epsilon^{\alpha\mu\nu\beta} p_\mu q_\beta [1 + \frac{a}{2}] \ . \tag{4.26a}$$

Bose symmetry, which has been maintained all along, insures a similar Ward identity in the ν index.

$$q_\nu T_1{}^{\alpha\mu\nu}(p,q\,|\,a) = -\frac{e^2}{4\pi^2} \epsilon^{\alpha\mu\nu\beta} p_\nu q_\beta [1 + \frac{a}{2}] \ . \tag{4.26b}$$

It is seen that the choice for a which insures the vector Ward identity, $a = -2$, is different from the choice that insures the axial Ward identity, $a = 0$. The conclusion is that there is no way of evaluating $T_1^{\alpha\mu\nu}(p,q)$ so that both Ward identities are satisfied. This remarkable result is even more striking when it is remembered that $\Gamma_1^{\alpha\mu\nu}(p,q)$ is not divergent in the explicit evaluation.

One might inquire whether it is possible to add to $T_1^{\alpha\mu\nu}$ a further seagull, which then would restore both Ward identities. If such a seagull were to exist, one would gladly insert it into the definition of $T_1^{\alpha\mu\nu}$ even though it did not arise "naturally" from the integration. It should be clear that no such further additions are possible. Any seagull one adds must be a three-index pseudotensor, and a polynomial in p and q. Bose symmetry limits it to be proportional to $\epsilon^{\alpha\mu\nu\beta}(p_\beta - q_\beta)$. This is precisely the arbitrariness which we have previously allowed for; see (4.22); and it is not sufficient to establish both Ward identities.

Faced with the impossibility of maintaining both Ward identities, we must decide which one we shall accept and which one we shall abandon, i.e., we wish to choose a. It is recognized that the vector Ward identity is a consequence of gauge invariance, while the axial Ward identity is a consequence of an equation of motion, $\partial_\mu J_5^\mu = F\mu^2 \phi$. Clearly the former is a much more important principle, and a should be set equal to -2. If there were a physical principle which assured the conservation of the axial current as well, we would be faced with a much more problematical situation. Thus we should be grateful that massless neutral pions do not, in fact, occur in nature. We conclude, therefore, that the reason for the violation of the Sutherland-Veltman theorem, $T(0) = 0$, is the violation of the axial Ward identity. Once a modified Ward identity is used, the Sutherland-Veltman theorem is modified, and the new conclusion agrees with the explicit evaluation. With the choice for a which assures gauge invariance, the Ward identities are

$$p_\mu T^{\alpha\mu\nu}(p,q) = q_\nu T^{\alpha\mu\nu}(p,q) = 0 \ , \tag{4.27a}$$

and

$$k_\alpha T^{\alpha\mu\nu}(p,q) = \frac{F\mu^2}{\mu^2 - k^2} T^{\mu\nu}(p,q) - \frac{e^2}{2\pi^2}\epsilon^{\mu\nu\alpha\beta} p_\alpha q_\beta \ . \tag{4.27b}$$

The Sutherland-Veltman derivation is now modified at the crucial step (4.6a). Instead of that equation, we have

$$\begin{aligned}\epsilon^{\mu\nu\alpha\beta} p_\alpha q_\beta T(k^2) &= T^{\mu\nu}(p,q) \\ &= \frac{\mu^2 - k^2}{F\mu^2}\left\{k_\alpha T^{\alpha\mu\nu}(p,q) + \frac{e^2}{2\pi^2}\epsilon^{\mu\nu\alpha\beta} p_\alpha q_\beta\right\}.\end{aligned} \tag{4.28}$$

The first term in the brackets is as before; therefore

$$\mathrm{T}(k^2) = \frac{\mu^2 - k^2}{F\mu^2}\left\{k^2\,[F_1(k^2) - F_3(k^2)] + \frac{e^2}{2\pi^2}\right\} ,$$

$$\mathrm{T}(0) = \frac{e^2}{2\pi^2 F} . \tag{4.29}$$

This agrees with the explicit calculations, (4.14).

The phenomenon of the violation of a Ward identity in perturbation theory should be familiar from quantum electrodynamics. For example, the vacuum-polarization tensor and the photon-photon scattering amplitude, as calculated perturbatively in spinor electrodynamics, are not transverse to the photon momenta as they should be. The conventional way of restoring gauge invariance is by the Pauli-Villars regulator technique. It is instructive to demonstrate the workings of that technique in the present context.

Recall that according to the Pauli-Villars regulator method, an amplitude involving a loop integration is considered to be a function of the mass of the particles circulating in the loop. A "regulated" amplitude is defined as the difference between the given amplitude and the same amplitude with the mass evaluated at a "regulator" mass. Finally the physical amplitude is regained by letting the regulator mass pass to infinity. Thus for the pion decay amplitude we have

$$\mathrm{T}^{\mu\nu}_{\mathrm{Reg}}(p,q) = \mathrm{T}^{\mu\nu}(p,q\,|\,m) - \mathrm{T}^{\mu\nu}(p,q\,|\,M) , \tag{4.30a}$$

$$\mathrm{T}^{\mu\nu}_{\mathrm{Physical}}(p,q) = \lim_{M\to\infty} \mathrm{T}^{\mu\nu}_{\mathrm{Reg}}(p,q) . \tag{4.30b}$$

According to (4.14), $\mathrm{T}^{\mu\nu}(p,q\,|\,M)$ vanishes as $\epsilon^{\mu\nu\alpha\beta}p_\alpha q_\beta(ge^2/4\pi^2 M)$ for large M, hence

$$\mathrm{T}^{\mu\nu}_{\mathrm{Physical}}(p,q) = \mathrm{T}^{\mu\nu}(p,q\,|\,m) . \tag{4.30c}$$

This is as it should be, since $\mathrm{T}^{\mu\nu}(p,q)$ was evaluated unambiguously from a finite integral. For the axial current amplitude on the other hand, we have

$$\mathrm{T}^{\alpha\mu\nu}_{1,\mathrm{Reg}}(p,q\,|\,a) = \mathrm{T}_1{}^{\alpha\mu\nu}(p,q\,|\,a\,|\,m) - \mathrm{T}_1{}^{\alpha\mu\nu}(p,q\,|\,a\,|\,M) , \tag{4.31a}$$

$$\mathrm{T}^{\alpha\mu\nu}_{1,\mathrm{Physical}}(p,q) = \lim_{M\to\infty} \mathrm{T}^{\alpha\mu\nu}_{1,\mathrm{Reg}}(p,q\,|\,a) . \tag{4.31b}$$

Consider now the vector Ward identity. According to (4.26a)

$$p_\mu T^{\alpha\mu\nu}_{1,\mathrm{Reg}}(p,q\,|\,a) = \frac{e^2}{4\pi^2}\epsilon^{\alpha\mu\nu\beta}\,p_\mu q_\beta\,[1+\frac{a}{2}] - \frac{e^2}{4\pi^2}\epsilon^{\alpha\mu\nu\beta}\,p_\mu q_\beta\,[1+\frac{a}{2}] = 0\,, \qquad (4.32a)$$

$$p_\mu T^{\alpha\mu\nu}_{1,\mathrm{Physical}}(p,q) = 0\,. \qquad (4.32b)$$

For the axial Ward identity, we have according to (4.24) and (4.30c)

$$\begin{aligned} k_\alpha T^{\alpha\mu\nu}_{1,\mathrm{Reg}}(p,\,q\,|\,a) &= 2mg^{-1}\,T^{\mu\nu}(p,q\,|\,m) + \frac{a\,e^2}{4\pi^2}\epsilon^{\mu\nu\alpha\beta}\,p_\alpha q_\beta \\ &\quad - 2Mg^{-1}\,T^{\mu\nu}(p,q\,|\,M) - \frac{a\,e^2}{4\pi^2}\epsilon^{\mu\nu\alpha\beta}\,p_\alpha q_\beta \\ &= 2mg^{-1}\,T^{\mu\nu}_{\mathrm{Physical}}(p,q) - 2Mg^{-1}\,T^{\mu\nu}(p,q\,|\,M)\,, \end{aligned} \qquad (4.33a)$$

$$\begin{aligned} &k_\alpha T^{\alpha\mu\nu}_{1,\mathrm{Physical}}(p,q) \\ &= 2mg^{-1}\,T^{\mu\nu}_{\mathrm{Physical}}(p,q) - \lim_{M\to\infty} 2Mg^{-1}\,T^{\mu\nu}(p,q\,|\,M)\,. \end{aligned} \qquad (4.33b)$$

Since $\lim_{M\to\infty} 2Mg^{-1}\,T^{\mu\nu}(p,q\,|\,M) = \frac{e^2}{2\pi^2}\epsilon^{\mu\nu\alpha\beta}\,p_\alpha q_\beta$, we are left with

$$\begin{aligned} &k_\alpha T^{\alpha\mu\nu}_{1,\mathrm{Physical}}(p,q) \\ &= 2mg^{-1}\,T^{\mu\nu}_{\mathrm{Physical}}(p,q) - \frac{e^2}{2\pi^2}\epsilon^{\mu\nu\alpha\beta}\,p_\alpha q_\beta\,. \end{aligned} \qquad (4.33c)$$

It is seen that the Pauli-Villars technique automatically evaluates the gauge invariant expression for the amplitude. It selects $a = -2$, which, as we have seen, leads to a violation of the axial Ward identity.

In conclusion we remark that the troubles we found with the matrix element of the axial current are not restricted to the σ model. It is clear that it is the triangle graph which leads to difficulties. Such a graph occurs in quantum electrodynamics, in a quark model, indeed in any model which possesses an axial current which is bilinear in fermion fields. This observation will permit us to generalize the present results beyond the specific σ model [6].

4.5 *Anomalous Commutators*

We have seen that the evaluation of the vector, vector, axial vector triangle graph, Fig. 4-2a, results in a formula for $T_1^{\alpha\mu\nu}(p,q)$ which does not satisfy the Ward identities one would naively expect. Our next task is to understand the breakdown of the Ward identities in terms of the anomalous commutators, which must be responsible for this state of affairs [7].

According to the BJL theorem, the ETC between the various currents may be evaluated from the high energy behavior of the triangle graph. Evidently we now must go off the photon mass shell $p^2 = q^2 = 0$, so that the time component of the 4-momentum can be sent to infinity independently, as is required by the BJL technique. It turns out, for our purposes, to be sufficient to go off mass shell for one photon only. Thus we are led to consider

$$\begin{aligned}\overline{T}_1^{\alpha\mu}(p,q) &= -ie\int d^4x\, e^{-ip\cdot x}\,\langle 0|T J^\mu(x)\bar{J}_5^\alpha(0)|\gamma q\rangle \\ &= \overline{T}_1^{\alpha\mu\nu}(p,q)\,\epsilon_\nu(q)\,,\end{aligned} \tag{4.34a}$$

$$\begin{aligned}\overline{T}_1^{\alpha\mu\nu}(p,q) &= -ie\int d^4x\, d^4y\, e^{-ip\cdot x}\, e^{-iq\cdot y} \\ &\quad\times \langle 0|T J^\mu(x)J^\nu(y)\bar{J}_5^\alpha(0)|0\rangle\Big|_{\substack{q^2=0\\ p^2\neq 0}}\end{aligned} \tag{4.34b}$$

Here the bar over $\overline{T}_1^{\alpha\mu}$ and $\overline{T}_1^{\alpha\mu\nu}$ serves to remind us that one photon is on the mass shell. The bar over $\bar{J}_5^\alpha$ indicates that we are not considering the full axial current of the σ model, (4.12a), but only the part bilinear in the nucleon fields; thus the lowest order matrix element involves only the problematical triangle graph, Fig. 4-2a. Note also that we are interested in the T product, *not* the covariant T* product. It is the former object that determines the ETC by the BJL definition.

According to the discussion of Section 3, the following formula for the ETC will enable us to calculate it.

$$\lim_{p_0 \to \infty} p_0 \bar{\mathrm{T}}_1{}^{\alpha\mu}(p,q)$$

$$= -e \int d^3x\, e^{i\mathbf{p}\cdot\mathbf{x}} \langle 0 | [J^\mu(0,\mathbf{x}), \bar{J}_5{}^\alpha(0)] | \gamma q \rangle \,. \tag{4.35}$$

Our program, therefore, is the following. We evaluate the triangle graph as before, except that the photon with 4-momentum p is not off mass shell. From the explicit formula for that amplitude, which is a covariant T* product, as it must be since it arises from covariant Feynman rules, we extract the non-covariant T product by dropping all seagulls — all polynomials in p_0. This provides us with an explicit formula for $\bar{\mathrm{T}}_1{}^{\alpha\mu}(p,q) = \bar{\mathrm{T}}_1{}^{\alpha\mu\nu}(p,q)\,\epsilon_\nu(q)$. Note that the present evaluation does not suffer from the ambiguities which beset the calculation of $\mathrm{T}_1{}^{\alpha\mu\nu}(p,q)$ in the previous subsection. The reason is that all the previously encountered ambiguities are seagulls, which we are neglecting.

The evaluation of the relevant triangle graph appears in the literature [8]. The integral yields a finite result as before, and the limit indicated in (4.35) is performed. The resulting commutators are summarized by the following formulas.

$$[J^0(t,\mathbf{x}), J_5^0(t,\mathbf{y})] = 2c\, {}^*F^{0j}(y)\, \partial_j \delta(\mathbf{x}-\mathbf{y}) \,, \tag{4.36a}$$

$$[J^0(t,\mathbf{x}), J_5{}^i(t,\mathbf{y})] = c\, {}^*F^{ij}(y)\, \partial_j \delta(\mathbf{x}-\mathbf{y}) \,, \tag{4.36b}$$

$$[J^i(t,\mathbf{x}), J_5^0(t,\mathbf{y})] = -c\,\partial_j\, {}^*F^{ij}(x)\, \delta(\mathbf{x}-\mathbf{y}) + c\, {}^*F^{ij}(y)\, \partial_j \delta(\mathbf{x}-\mathbf{y}) \,. \tag{4.36c}$$

Here $c = ie/4\pi^2$ and ${}^*F^{\mu\nu}$ is the antisymmetric, conserved electromagnetic dual tensor: ${}^*F^{\mu\nu} = \frac{1}{2}\epsilon^{\mu\nu\alpha\beta}(\partial_\alpha A_\beta - \partial_\beta A_\alpha)$. In offering (4.36), we do not imply that we have derived the ETC by any operator technique. All that is meant is that the limit (4.35) is non-vanishing and the non-zero expression may be regained by evaluating the appropriate matrix element of (4.36). For example, explicit computation shows that

$$\lim_{p_0 \to \infty} p_0 \bar{\mathrm{T}}_1{}^{00}(p,q) = -\frac{e^2}{2\pi^2}\epsilon^{0\mu\nu\alpha}\, p_\mu\, q_\nu\, \epsilon_\alpha(q) \,. \tag{4.37}$$

The right-hand side of (4.37) is also obtained when (4.36a) is inserted into the right-hand side of (4.35). Therefore, properly speaking, all we have shown is that the ETC (4.36) has non-canonical contributions whose vacuum-one photon matrix element is equal to the same matrix element of the operators appearing in the right-hand side of (4.36).

We see that the ETC between time components has acquired a non-canonical ST, Eq. (4.36a). Therefore according to the discussion of Section 2, the Feynman conjecture may not be satisfied for both Ward identities. To see that indeed the conjecture is violated [9], the ETC (4.36) is expressed in the formalism of Section 2.

$$[J^\mu(x), J_5{}^\nu(y)]\,\delta([x-y]\cdot n)$$

$$= C^{\mu\nu}(x;n)\,\delta^4(x-y) + S^{\mu\nu|\alpha}(y;n)\,P_{\alpha\beta}\,\partial^\beta\delta^4(x-y)\ . \quad (4.38a)$$

The ST, according to (4.36), is

$$S^{\mu\nu|\alpha}(y;n)$$

$$= c\,{}^*F^{\mu\gamma}(y)\,[g^\alpha_\gamma n^\nu + g^{\alpha\nu} n_\gamma] + c\,{}^*F^{\nu\gamma}(y)\,[g^\alpha_\gamma n^\mu + g^{\alpha\mu} n_\gamma]\ . \quad (4.38b)$$

Equation (4.38b) may be expressed as a total divergence.

$$S^{\mu\nu|\alpha}(y;n)$$

$$= c\,{}^*F^{\mu\gamma}(y)\frac{\delta}{\delta n_\alpha}[n_\gamma n^\mu] + c\,{}^*F^{\nu\gamma}(y)\frac{\delta}{\delta n_\alpha}[n_\gamma n^\mu]\ . \quad (4.38c)$$

Hence the seagull which covariantizes the T product of J^μ and $J_5{}^\alpha$ is given by

$$\tau^{\mu\nu}(x,y;n) = \int^n S^{\mu\nu|\alpha}(y;n')\,dn'_\alpha\,\delta^4(x-y) + \tau_0{}^{\mu\nu}(x,y)$$

$$= \tau^{\mu\nu}(y;n)\,\delta^4(x-y) + \tau_0{}^{\mu\nu}(x,y)\ ,$$

$$\tau^{\mu\nu}(y;n) = c\,{}^*F^{\mu\gamma}(y)n_\gamma n^\nu + c\,{}^*F^{\nu\gamma}(y)n_\gamma n^\mu\ . \quad (4.39)$$

In the above, as before, $\tau_0{}^{\mu\nu}(x,y)$ is a covariant seagull, as yet undermined. The covariant quantities $I_1{}^{\mu\nu}$ and $I_2{}^{\mu\nu}$ defined from $\tau^{\mu\nu}(y;n)$ by (2.47) may now be evaluated.

$$n_\mu \tau^{\mu\nu}(y;n) = n_\mu I_1{}^{\mu\nu}(y) = c\,{}^*F^{\nu\gamma}(y) n_\gamma \; ,$$

$$n_\nu \tau^{\mu\nu}(y;n) = n_\nu I_2{}^{\mu\nu}(y) = c\,{}^*F^{\mu\gamma}(y) n_\gamma \; . \tag{4.40a}$$

Evidently we have

$$I_2{}^{\mu\nu}(y) = -I_1{}^{\mu\nu}(y) = c\,{}^*F^{\mu\nu}(y) \; . \tag{4.40b}$$

Finally to evaluate the covariant seagull we make use of the relations (2.48). These require that the following combinations be free of gradients of the δ function

$$\frac{\partial}{\partial x^\mu}\tau_0{}^{\mu\nu}(x,y) - c\,{}^*F^{\mu\nu}(y)\,\partial_\mu \delta^4(x-y) \; , \tag{4.41a}$$

$$\frac{\partial}{\partial y^\nu}\tau_0{}^{\mu\nu}(x,y) - c\,{}^*F^{\mu\nu}(y)\,\partial_\nu \delta^4(x-y) \; . \tag{4.41b}$$

The first of the above equations assures the validity of Feynman's conjecture in the μ, vector Ward identity; while the second effects this state of affairs in the ν, axial Ward identity.

It may be verified that a solution to both conditions (4.41) is

$$\tau_0{}^{\mu\nu}(x,y)$$
$$= c\,{}^*F^{\mu\nu}(y)\,\delta^4(x-y) + 2c\,\epsilon^{\mu\nu\alpha\beta} A_\alpha(y)\,\partial_\beta \delta^4(x-y) \; . \tag{4.42}$$

Hence it appears that Feynman's conjecture can be satisfied in both indices. However, the seagull (4.42) is unacceptable for the following reason. The explicit dependence of the seagull on the vector potential A_α indicates that gauge invariance has been lost. Recall that all the operators, which we are here considering, are to be sandwiched between the vacuum and one-photon state. If one of these operators is the vector potential, then this matrix element will not be gauge invariant. Therefore we must reject this seagull, and content ourselves with one which allows one or the other of the two

Ward identities to satisfy Feynman's conjecture. Such a seagull may easily be shown to be

$$\tau_0{}^{\mu\nu}(x,y) = -(1+a)c\,{}^*F^{\mu\nu}(y)\,\delta^4(x-y)\ . \tag{4.43}$$

Here a is an arbitrary parameter, which is determined only when it is decided which Ward identity is to be satisfied. The choice $a=-2$ effects cancellation of Schwinger terms and seagulls in the μ, vector identity; while $a=0$ performs this service in ν, axial identity. If the former choice is made, one then finds

$$\frac{\partial}{\partial y^\nu}\mathrm{T}^*J^\mu(x)J_5{}^\nu(y)$$
$$= \mathrm{T}^*J^\mu(x)\,\partial_\nu J_5{}^\nu(y) + 2c\,{}^*F^{\mu\nu}(y)\,\partial_\nu\delta^4(x-y)\ . \tag{4.44}$$

The second term on the right-hand side of (4.44) is the anomaly.

Thus we have understood why the Ward identities are not satisfied; the ETC between J^μ and $J_5{}^\alpha$ departs from its canonical value, acquiring non-canonical contributions. These non-canonical terms are consequences of the intrinsic singularities of local field theory. They have the property that naive current algebraic manipulations become invalid.

4.6 *Anomalous Divergence of Axial Current*

The anomalies of the triangle graph, which we have understood in terms of non-canonical commutators and modified Ward identities, may also be shown to lead to a modified divergence equation of the neutral, gauge invariant axial current [10].

$$\partial_\mu J_5{}^\mu = J_5 + \frac{e^2}{8\pi^2}\,{}^*F^{\mu\nu}F_{\mu\nu}\ . \tag{4.45}$$

Here J_5 is the naive value of the divergence, derived by application of the equations of motion of whatever model we have under consideration.

We consider the fermion part of the axial current.

$$J_5{}^\mu(x) = i\bar\psi(x)\gamma^5\gamma^\mu\psi(x)\ . \tag{4.46}$$

Since it is known that the equal time anti-commutator of ψ and $\bar\psi$ involves a three-dimensional δ function, we must expect that $\lim_{x\to y}\bar\psi(x)\psi(y)$ is singular. Hence the definition (4.46) for $J_5{}^\mu(x)$ is necessarily singular. To regulate this singularity, a small separation is introduced in a preliminary definition for $J_5{}^\mu$.

$$J_5{}^\mu(x|\epsilon) = i\bar{\psi}(x+\epsilon/2)\gamma^5\gamma^\mu\psi(x-\epsilon/2)\ . \tag{4.47a}$$

In the presence of electromagnetism, which we shall always consider to be described by an *external* field (i.e., we work to lowest order in electromagnetism), the definition (4.47a) is not gauge invariant. If the electromagnetic potential A^μ is replaced by $A^\mu + \partial^\mu\Lambda$, where Λ is arbitrary, and the fermion fields are allowed to change correspondingly, $\psi(x)\to e^{ie\Lambda(x)}\psi(x)$, then no changes should occur in quantities of physical interest. The formula (4.47a) does not have this property. A modified expression can be constructed which is gauge invariant.

$$J_5{}^\mu(x|\epsilon|a)$$
$$= i\bar{\psi}(x+\epsilon/2)\gamma^5\gamma^\mu\psi(x-\epsilon/2)\exp\left(iea\int_{x-\epsilon/2}^{x+\epsilon/2} A^\alpha(y)\,dy_\alpha\right)\ . \tag{4.47b}$$

In (4.47b) a should be set equal to 1 for gauge invariance. However, we prefer to leave this constant unspecified for the time being. The local physical current is obtained by choosing ϵ to be small, averaging over the directions of ϵ and letting $\epsilon^2 = \epsilon_\mu\epsilon^\mu \to 0$. The method of defining singular products of operators by introducing a small separation is called the "point splitting technique".

We now wish to calculate the divergence of (4.47b). To do so, we need the equation of motion for ψ. We shall here assume that the only interaction is with the external electromagnetic field. More general, interactions have been discussed in the literature [10].

$$i\gamma_\mu\partial^\mu\psi = m\psi - e\gamma_\mu A^\mu\psi\ . \tag{4.48}$$

By virtue of (4.48), the divergence of $J_5{}^\mu(x|\epsilon|a)$ is

$$\partial_\mu J_5{}^\mu(x|\epsilon|a) = J_5(x|\epsilon|a) - ie\,J_5{}^\mu(x|\epsilon|a)$$
$$\times\left[A_\mu(x+\epsilon/2) - A_\mu(x-\epsilon/2) - a\,\partial_\mu\int_{x-\epsilon/2}^{x+\epsilon/2} A_\nu(y)\,dy^\nu\right],$$
$$= J_5(x|\epsilon|a) - ie\,J_5{}^\mu(x|\epsilon|a)\epsilon^\alpha[\partial_\alpha A_\mu(x) - a\partial_\mu A_\alpha(x) + O(\epsilon)]\ . \tag{4.49}$$

Here $J_5(x|\epsilon|a)$ is the regulated, split point formula for the naive divergence in this model: $2m\bar{\psi}\gamma^5\psi$. The usual naive result $\partial_\mu J_5{}^\mu = J_5$ is regained from (4.49) if ϵ is set to zero, uncritically. Then the last term in (4.49) appears to vanish. This is legitimate when $J_5{}^\mu(x|\epsilon|a)$ is well behaved as $\epsilon\to 0$. On the other hand, if a matrix element of $J_5{}^\mu(x|\epsilon|a)$ diverges as $\epsilon\to 0$, a finite result may remain. Since the dimension of $J_5{}^\mu$ is $(\text{length})^{-3}$, one may expect a cubic divergence. However, the pseudovector character of $J_5{}^\mu$ reduces the divergence by two powers, leaving a possible linear divergence. We now show that such a divergence is indeed present, and modifies the naive formula for $\partial_\mu J_5{}^\mu(x|\epsilon|a)$.

Consider the vacuum expectation value of $\partial_\mu J_5{}^\mu(x|\epsilon|a)$.

$$\langle 0|\partial_\mu J_5{}^\mu(x|\epsilon|a)|0\rangle = \langle 0|J_5(x|\epsilon|a)|0\rangle$$
$$-ie\,\epsilon^\alpha\langle 0|J_5{}^\mu(x|\epsilon|a)|0\rangle[\partial_\alpha A_\mu(x) - a\,\partial_\mu A_\alpha(x) + O(\epsilon)]\ . \tag{4.50}$$

The vacuum element of $J_5{}^\mu(x|\epsilon|a)$ is non-vanishing because it is computed in the presence of an external electromagnetic field. We have for the last term in (4.50)

$$-i\epsilon^\alpha\langle 0|J_5{}^\mu(x|\epsilon|a)|0\rangle$$
$$= \epsilon^\alpha\langle 0|\bar{\psi}(x+\epsilon/2)\,\gamma^5\gamma^\mu\psi(x-\epsilon/2)|0\rangle\exp\left(iea\int_{x-\epsilon/2}^{x+\epsilon/2} A_\alpha(y)\,dy^\alpha\right)$$
$$= -\mathrm{Tr}\,\gamma^5\gamma^\mu\epsilon^\alpha\langle 0|T\psi(x-\epsilon/2)\,\bar{\psi}(x+\epsilon/2)|0\rangle\exp\left(iea\int_{x-\epsilon/2}^{x+\epsilon/2} A^\alpha(y)\,dy_\alpha\right)$$
$$= -\mathrm{Tr}\,\gamma^5\gamma^\mu\epsilon^\alpha\,G(x-\epsilon/2, x+\epsilon/2)\exp\left(iea\int_{x-\epsilon/2}^{x+\epsilon/2} A^\alpha(y)\,dy_\alpha\right)\ . \tag{4.51}$$

The fermion propagator function, $G(x,y)$, in the external field A^μ, has been introduced. In offering (4.51), ϵ^0 is taken to be positive.

$G(x,y)$ possesses an expansion in powers of A^μ which may be summarized graphically as in Fig. 4-3. The double line represents G; the single line is the free fermion propagator $S(x)$; while the $\times$ represents an interaction with the external field. $S(x)$ behaves as $1/x^3$ as $x \to 0$. Therefore the successive terms in the series for $G(x,y)$ behave, when $x \to y$, as $(x-y)^{-3}$, $(x-y)^{-2}$, $(x-y)^{-1}$, $\log(x-y)$, etc. For our calculation of $G(x-\frac{1}{2}\epsilon,\ x+\frac{1}{2}\epsilon)$ we need terms which do not vanish, for small ϵ, when multiplied by ϵ. Therefore we set

$$\begin{aligned} G(x-\tfrac{1}{2}\epsilon, x+\tfrac{1}{2}\epsilon) \\ &= S(-\epsilon) + ie\int d^4y\, S(x-\tfrac{1}{2}\epsilon-y)\,\gamma^\alpha\, S(y-x-\tfrac{1}{2}\epsilon)\,A_\alpha(y) \\ &\quad - e^2\int d^4y\, d^4z\, S(x-\tfrac{1}{2}\epsilon-y)\,\gamma^\alpha\, S(y-z)\,\gamma^\beta \\ &\quad \times S(z-x-\tfrac{1}{2}\epsilon)A_\alpha(y)\,A_\beta(z) + O(\log\epsilon)\ . \end{aligned} \tag{4.52a}$$

$$S(x) = i\int \frac{d^4p}{(2\pi)^4}\, e^{-ip\cdot x}\,[\gamma_\alpha p^\alpha - m]^{-1}\ . \tag{4.52b}$$

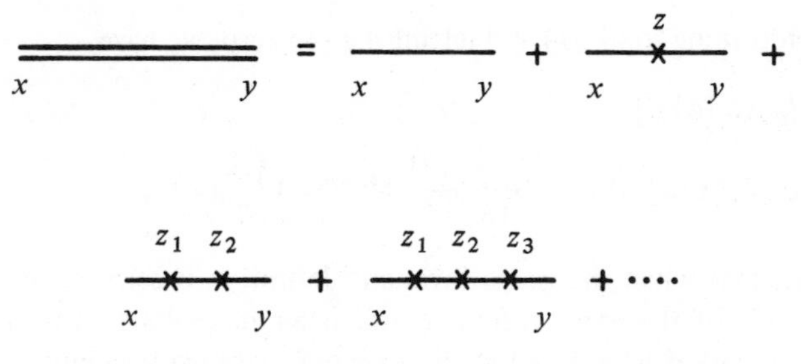

Fig. 4-3. Fermion propagator $G(x,y)$ in an external field.

By C invariance, only the contribution to G which is linear in A^μ is of interest. That term in (4.52a) has the following momentum representation.

$$ie\int \frac{d^4p\, d^4q}{(2\pi)^8}\, e^{i\epsilon\cdot p}\, e^{-ix\cdot q}\, S(p+\tfrac{1}{2}q)\,\gamma^\alpha\, S(p-\tfrac{1}{2}q)\,A_\alpha(q)\ . \tag{4.53}$$

Therefore (4.51) becomes

$$- \mathrm{Tr}\,[\epsilon^\alpha \gamma^5 \gamma^\mu G(x - \tfrac{1}{2}\epsilon, x + \tfrac{1}{2}\epsilon)]$$

$$= - ie\, \mathrm{Tr}\, \gamma^5 \gamma^\mu \int \frac{\mathrm{d}^4 p\, \mathrm{d}^4 q}{(2\pi)^8} \epsilon^\alpha e^{i\epsilon \cdot p}\, e^{-ix \cdot q}$$

$$\times\ S(p + \tfrac{1}{2}q)\gamma^\nu S(p - \tfrac{1}{2}q) A_\nu(q) + O(\epsilon \log \epsilon)$$

$$= e\, \mathrm{Tr}\, \gamma^5 \gamma^\mu \int \frac{\mathrm{d}^4 p\, \mathrm{d}^4 q}{(2\pi)^8} e^{i\epsilon \cdot p}\, e^{-ix \cdot q} A_\nu(q) \frac{\partial}{\partial p_\alpha}$$

$$\times\ S(p + \tfrac{1}{2}q)\gamma^\nu S(p - \tfrac{1}{2}q) + O(\epsilon \log \epsilon)\ . \tag{4.54a}$$

The last equality follows by integration by parts. We now set ϵ to zero. The p-integral is just a surface term; it is easily evaluated by the symmetric methods exemplified in the exercises. The remaining q-integral inverts the Fourier transform of $A_\nu(q)$. The final result for (4.54a) is

$$- \mathrm{Tr}\,[\epsilon^\alpha \gamma^5 \gamma^\mu G(x - \tfrac{1}{2}\epsilon, x + \tfrac{1}{2}\epsilon)]_{\epsilon \to 0} = - \frac{e}{8\pi^2} {}^*F^{\mu\alpha}(x)\ . \tag{4.54b}$$

Therefore returning to (4.51) and letting ϵ go to zero, we have

$$\langle 0 | \partial^\mu J_\mu{}^5(x|a)|0\rangle$$

$$= \langle 0 | J_5(x|a)|0\rangle + \frac{e^2(1+a)}{16\pi^2} {}^*F^{\mu\nu}(x) F_{\mu\nu}(x)\ . \tag{4.55}$$

It is seen that when the gauge invariant definition is selected, $a = 1$, then the divergence of the axial current contains an anomalous term. The naive divergence equation is regained at the expense of gauge invariance if $a = -1$. One may give a simple, heuristic argument which illuminates the origin of this anomalous divergence term. Consider the naive axial current in the model discussed in this subsection. In order to assure gauge invariance, a Pauli-Villars regulator field Ψ is introduced, and correspondingly a regulated axial current is defined.

$$J_5{}^\mu \Big|_{\mathrm{Reg}} = J_5{}^\mu - \mathcal{J}_5{}^\mu\ . \tag{4.56a}$$

$\mathcal{J}_5{}^\mu$ is constructed from the regulator fields, in the same fashion as $J_5{}^\mu$ is constructed from the usual fields. The physical axial current is regained by letting the mass M of the regulator field pass to infinity. The divergence of (4.56a) is

$$\partial_\mu J_5{}^\mu \Big|_{\text{Reg}} = 2m\bar{\psi}\gamma^5\psi - 2M\bar{\Psi}\gamma^5\Psi \; . \tag{4.56b}$$

Now when $M \to \infty$, the regulator field contribution to (4.56b) may leave a non-vanishing remainder if the matrix elements of $\bar{\Psi}\gamma^5\Psi$ behave as M^{-1} for large M. Detailed calculation shows that this indeed occurs.

Note that the anomalous divergence does not directly affect our previous derivation of the Sutherland-Veltman theorem for $\pi^0 \to 2\gamma$. The amplitude $\mathrm{T}^{\mu\nu}$, which is considered in (4.3), is already $0(e^2)$, the two photons having been contracted out of the state. Hence to order e^2 we need not inquire into any modification of $\partial_\mu J_5{}^\mu$. The anomaly in that argument came from the commutators and seagull, as was explicitly demonstrated. Nevertheless it is possible to use an anomalous divergence equation to give an alternate derivation of the true Sutherland-Veltman theorem [11]. The photons are not contracted out of their state, and we consider (4.2) in conjunction with (4.45).

$$\begin{aligned}
\mathrm{T}^{\mu\nu}(p,q) &= \epsilon^{\mu\nu\alpha\beta} p_\alpha q_\beta \mathrm{T}(k^2) = (\mu^2 - k^2) \langle 0 | \phi(0) | \gamma, p; \gamma', q \rangle \\
&= \frac{e^2}{8\pi^2 F\mu^2} (k^2 - \mu^2) \langle 0 | {}^*F^{\mu\nu}(0) F_{\mu\nu}(0) | \gamma, p; \gamma', q \rangle \\
&\quad + \frac{1}{F\mu^2} (\mu^2 - k^2)\, \partial_\alpha \langle 0 | J_5{}^\alpha(0) | \gamma, p; \gamma', q \rangle \; .
\end{aligned} \tag{4.57a}$$

The last term in (4.57a) is the divergence of a gauge invariant, three index pseudotensor, hence the original Sutherland-Veltman argument applies. We conclude that it will not contribute to T(0). Note that in this derivation we do not pull the divergence through any T* product, so we need not concern ourselves with commutators. The matrix element of the anomaly may be evaluated to lowest order in electromagnetism. Its value is

$$\begin{aligned}
&\frac{e^2}{8\pi^2 F\mu^2} (k^2 - \mu^2) \langle 0 | {}^*F^{\mu\nu}(0) F_{\mu\nu}(0) | \gamma, p; \gamma', q \rangle \\
&= \frac{e^2}{2\pi^2 F} \frac{(\mu^2 - k^2)}{\mu^2} \epsilon^{\mu\nu\alpha\beta} p_\alpha q_\beta \; .
\end{aligned} \tag{4.57b}$$

Therefore T(0), as before, is $e^2/2\pi^2 F$.

This exercise shows that the anomalies in commutators, which are encountered in the original derivation, and the anomalous divergence are two sides of the same coin. One must be present when the other is.

4.7 Discussion

We conclude this treatment of the anomalies of the neutral axial-vector current with a discussion of various disconnected, but important topics.

(1) Consider massless spinor electrodynamics [12]. The present arguments indicate that it is impossible to define a conserved gauge invariant axial current, in spite of the fact that chirality is a symmetry of the theory. Nevertheless there does exist a conserved, gauge invariant axial charge [11]. This charge is constructed as follows. Define

$$\tilde{J}_5{}^{\mu} = J_5{}^{\mu} - \frac{e^2}{4\pi^2} {}^*F^{\mu\nu} A_\nu \, . \tag{4.58a}$$

$J_5{}^{\mu}$ is gauge invariant, but its conservation is broken by the anomaly. On the other hand $\tilde{J}_5{}^{\mu}$ is conserved, but not gauge invariant. The charge Q_5 constructed from $\tilde{J}_5{}^{\mu}$ is time-independent.

$$Q_5 = \int d^3x \, \tilde{J}_5^0(x) \, . \tag{4.58b}$$

Performing a gauge transformation on Q_5 : $\delta A_\nu = \partial_\nu \Lambda$, we see that Q_5 is gauge invariant, even though $\tilde{J}_5{}^{\mu}$ is not.

$$\begin{aligned}
\delta Q_5 &= \int d^3x \left(-\frac{e^2}{4\pi^2} {}^*F^{0\nu}(x) \right) \partial_\nu \Lambda(x) \\
&= \int d^3x \left(-\frac{e^2}{4\pi^2} {}^*F^{0i}(x) \right) \partial_i \Lambda(x) \\
&= \int d^3x \, \frac{e^2}{4\pi^2} \Lambda(x) \, \partial_i {}^*F^{0i}(x) \\
&= 0 \, .
\end{aligned} \tag{4.58c}$$

The conservation and antisymmetry of ${}^*F^{\mu\nu}$ has been used.

Therefore, in spite of the trouble with the *local* axial current, globally axial symmetry can be implemented in the model. This has the consequence that any property of the model, based on axial symmetry, will be maintained in perturbation theory. For example, the anomaly in the divergence cannot be used to generate a mass for the election [13].

In massless electrodynamics, one may describe the anomaly as a clash between two symmetry principles: gauge invariance and chirality. In perturbation theory it is impossible to maintain both, though either one can be satisfied. Such a clash between the conservation of two symmetry currents has been encountered before in the model field theory of spinor electrodynamics in two dimensions [14].

(2) An important question is whether or not higher order effects modify the anomaly. An argument may be given to the end that in spinor electrodynamics and in the σ model, they do not [11]. The argument is as follows: for definiteness we consider the former theory. To fourth order in e, the axial-vector, vector, vector triangle graph has the insertions represented in Fig. 4-4. If the photon integration is carried out *after* the fermion loop integration, then the fermion loop integral is completely convergent. Therefore all shifts of integration, which are required to verify the Ward identities may be performed with impunity. Thus, it is argued, that no anomalies will be present in this or higher orders.

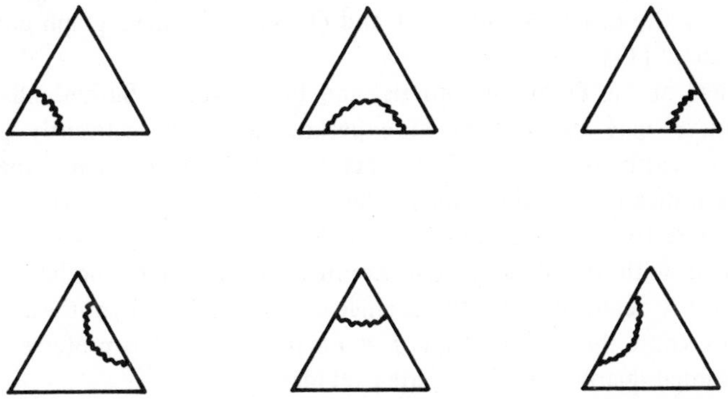

Fig. 4-4. Radiative corrections to the axial-vector, vector, vector triangle graph.

This argument may be criticized because it is somewhat formal [15]. The rules of renormalized perturbation theory require one to perform the photon integrals first, and then the renormalized vertex and propagator corections are to be inserted into the triangle skeleton. In order to resolve this question, explicit calculations for the graphs in Fig. 4-4 have been performed, and they support the formal argument [16]. It should be noticed, however, that the fourth order contribution is characterized by the fact that all indicated corrections are either self-energy or vertex *insertions.* This feature is not true in higher orders, and perhaps something new will be found there [17].

(3) The models in which the axial anomaly has been exposed: spinor electrodynamics and σ model, do not have much dynamical significance for hadron physics. However, as we have stated before, *any* theory with fermion fields out of which the axial vector current is constructed, will possess the anomaly, as long as electromagnetism is coupled minimally. Therefore, it is to be expected that in general PCAC should be modified in the presence of electrodynamics. Thus when we consider the neutral member of the octet of axial currents, $\mathscr{F}_3^{\mu}(x)$, (this current is $\frac{1}{2}$ of the previously defined J_5^{μ}) PCAC should be modified by

$$\partial_\mu \mathscr{F}_3^{\mu} = F\mu^2 \phi_3 + c \frac{e^2}{8\pi^2} {}^*F^{\mu\nu} F_{\mu\nu} \ . \tag{4.59}$$

Here c is constant which, of course, we cannot derive theoretically in a general fashion. The contribution to c from the triangle graph is determined by the coupling of fermion fields to the axial and vector currents. In the σ model this contribution is $\frac{1}{2}$. In a general quark triplet model where the charges of the quarks are Q, $Q-1$ and $Q-1$, the triangle graph contributes $Q-\frac{1}{2}$ to c [11].

By use of the PCAC hypothesis, and the corrected Sutherland-Veltman theorem, c may be determined experimentally. The currently published $\pi^0 \to 2\gamma$ width of 7.37 ± 1.5 eV sets $|c|$ at 0.44. Further experimental analysis indicates that most likely the sign of c is positive. Thus the theoretical value for c as given by the σ model, triangle graph, $c = \frac{1}{2}$, is in good agreement with the data. If one assumes that in quark models the entire value of c is determined by the triangle graph — a bold hypothesis since one does not know the nature of quark dynamics — $Q = 1$ is preferred, and the conventional quarks with $Q = \frac{1}{3}$ are excluded.

Perhaps at the present stage of understanding of hadron physics, one should not expect to be able to calculate c theoretically. Just as the coefficient of the usual term in $\partial_\mu \mathscr{F}_3^{\mu}$ is taken from experiment — $F\mu^2$ has not been calculated — so also we should content ourselves with an experimental determination of the anomaly.

When c is fitted to the pion data, and a model for SU(3) $\times$ SU(3) symmetry breaking is adopted, modified Sutherland-Veltman theorems for $\eta \to 2\gamma$ and

$X \to 2\gamma$ may be derived. Such calculations have been performed in the context of the $(3, \bar{3}) \oplus (\bar{3}, 3)$ symmetry breaking scheme [18]. The results for the η width are consistent with experiment (~ 1 keV), while the X width comes out remarkably enhanced beyond 80 keV. Present experimental data (< 360 keV) provides no check. Such a check would be very interesting, since this large value is very difficult to understand from any different point of view.

(4) One may wonder why we speak of a modification of PCAC; why one cannot continue using the divergence of the axial current as the pion interpolating field. The answer lies in part in our model calculations where we found that the term ${}^*F^{\mu\nu}F_{\mu\nu}$ is manifestly not smooth when its matrix elements vary off the pion mass shell. It is this property which we have abstracted, and which we assume holds in nature, as well as in models. Our assumption is supported by the observation that the dimension of the anomaly is 4, and there is no reason to believe that such an operator is smooth. Finally, by adopting the present philosophy, the experimentally unsatisfactory prediction of Sutherland and Veltman is avoided.

(5) The discovery of anomalous Ward identities in the present context engendered a systematic study of all useful Ward identities in SU(3) × SU(3) models [19]. Although other anomalous Ward identities have been found, they seem to be without interest. The only other triangle graph anomaly is in the triple axial vector vertex, which has not, as yet, been used in physical predictions.

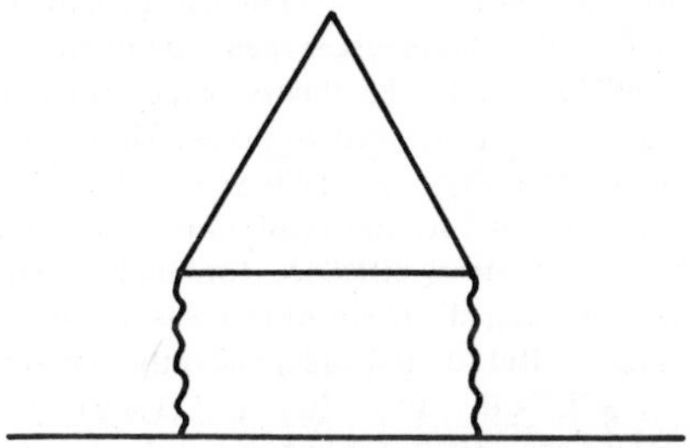

Fig. 4-5. Divergent contribution to axial vector vertex function.

(6) In spinor electrodynamics, the proper electromagnetic vertex function is renormalized by the same infinite constant which effects electron wave function renormalization. This desirable state of affairs is a consequence of the Ward identity satisfied by that function. Before the discovery of anoma-

lous PCAC, it was thought that the proper vertex function of the axial vertex current possesses this property as well, as consequence of the axial Ward identity [20]. The anomaly has destroyed this result; the axial vector vertex function remains infinite after wave function renormalization. The offending graph is the one of Fig. 4-5. One consequence of this is that radiative corrections to neutrino-lepton elastic scattering are infinite [11].

References

[1] D. G. Sutherland, Nucl. Phys. B2 (1967) 433; M. Veltman, Proc. Roy. Soc. A301 (1967) 107.

[2] The argument here follows J. S. Bell and R. Jackiw, Nuovo Cimento 60A (1969) 47.

[3] M. Gell-Mann and M. Levy, Nuovo Cimento 16 (1960) 705.

[4] The evaluation of this graph was first performed by H. Fukuda and Y. Miyamoto, Prog. Theor. Phys. 4 (1949) 347 and J. Steinberger, Phys. Rev. 76 (1949) 1180. Steinberger considered pion decay in the old PS-PS model of π-nucleon interaction. In that theory the pion decays into two photons; the lowest order graphs being given by Fig. 4-1. It presumably is only an accident that this completely implausible calculation gives a result in excellent agreement with experiment.

[5] This careful and unambiguous evaluation is given in Ref. 2.

[6] A historical note is here in order. The first people to calculate the $\pi^0 \to 2\gamma$ process in field theory were H. Fukuda, Y. Miyamoto and J. Steinberger, Ref. 4. In addition to the PS-PS calculation, where the π-N vertex is γ^5, Steinberger also calculated the same amplitude in the PV-PS model, where the π-N vertex is $ik_\alpha\gamma^5\gamma^\alpha$. The second calculation is identical to our Pauli-Villars regulator method evaluation of $T^{\alpha\mu\nu}$. Steinberger then attempted to verify the equivalence theorem between PS-PS and PV-PS theory, which is based on the formal Ward identity (4.16b); and of course failed to do so. He noted this puzzle, and then ceased being a theoretical physicist. Two years later, J. Schwinger, Phys. Rev. 82 (1951) 664, gave an analysis and resolution of the problem. This work was essentially forgotten, and its significance for modern ideas of current algebra and PCAC was not appreciated. The problem was rediscovered in the σ model by J. S. Bell and R. Jackiw, Ref. 2; and independently and simultaneously in spinor electrodynamics by S. L. Adler, Phys. Rev. 177 (1969) 2426. Schwinger's analysis is similar to the one we shall present in subsection 6.

[7] This presentation follows R. Jackiw and K. Johnson, Phys. Rev. 182 (1969) 1459. The same anomalous commutators have also been given by S. L. Adler and D. G. Boulware, Phys. Rev. 184 (1969) 1740.

[8] L. Rosenberg, Phys. Rev. 129 (1963) 2786. Rosenberg's formula appears also in Adler's paper, Ref. 6.

[9] That Feynman's conjecture is violated in $\pi^0 \to 2\gamma$ was first pointed out by R. Jackiw and K. Johnson, Ref. 7. The present analysis follows the paper of D. J. Gross and R. Jackiw, Nucl. Phys. B14 (1969) 269.

[10] Equation (4.45) was first derived by J. Schwinger, Ref. 3. The same form was given by S. L. Adler, Ref. 6, on the basis of an investigation in spinor electrodynamics. The presentation in this section is analogous to Schwinger's and has been given in the contemporary literature by R. Jackiw and K. Johnson, Ref. 7; C. R. Hagen, Phys. Rev. 177 (1969) 2622 and B. Zumino, Proceedings of Topical Conference on Weak Interactions, CERN, Geneva, p. 361 (1969).
[11] This analysis is due to S. L. Adler, Ref. 6.
[12] It is possible that such a theory has physical significance. It has been suggested by K. Johnson, R. Willey and M. Baker, Phys. Rev. 163 (1967) 1699, that the bare mass of the electron is zero and that the physical mass is entirely of dynamical origin.
[13] For example, it might be thought the anomaly allows a mass to be generated in a massless theory without the intervention of Goldstone Bosons. That this is not the case in theories with Abelian vector mesons has been demonstrated by H. Pagels (private communication).
[14] J. Johnson, Phys. Lett. 5 (1963) 253.
[15] R. Jackiw and K. Johnson, Ref. 7.
[16] S. L. Adler and W. Bardeen, Phys. Rev. 182 (1969) 1517.
[17] Ambiguities in Ward identities containing two loop integrations have been found by E. Abers, D. Dicus and V. Teplitz, Phys. Rev. D3 (1971) 485.
[18] S. L. Glashow, R. Jackiw and S. Shei, Phys. Rev. 187 (1969) 1916. An alternate treatment of SU(3) × SU(3) symmetry breaking together with axial current anomalies has been considered by G. Gounaris, Phys. Rev. D1 (1970) 1426 and Phys. Rev. D2 (1970) 2734. Gounaris' theory makes use of a detailed phenomenological Lagrangian. Correspondingly his results are more precise than those of the above reference, though, of course, they are model dependent. Gounaris gives predictions for off mass shell photon processes, as well as for the processes considered by Glashow, Jackiw and Shei. Agreement is obtained for $\pi^0 \to 2\gamma$ and $\eta \to 2\gamma$; but the prediction for $X^0 \to 2\gamma$ is decreased by a factor of 10.
[19] K. Wilson, Phys. Rev. 181 (1969) 1909; I. Gerstein and R. Jackiw, Phys. Rev. 181 (1969) 1955; W. Bardeen, Phys. Rev. 184 (1969) 1849; R. W. Brown, C. C. Shih and B. L. Young, Phys. Rev. 186 (1969) 1491; D. Amati, G. Bouchiat and J. L. Gervais, Nuovo Cimento 65A (1970) 55.
[20] G. Preparata and W. Weisberger, Phys. Rev. 175 (1968) 1965.

5. Electroproduction Sum Rules

5.1 Preliminaries

In the electroproduction experiments, an electron is scattered off a nucleon target, typically a proton. The hadronic final states that are observed, are thought to arise, in the context of lowest order electromagnetism, from the inelastic interaction between an off mass shell photon and the proton. The process is depicted in Fig. 5-1. By measuring total inelastic cross sections, that is by summing over all final states, one obtains a determination of the

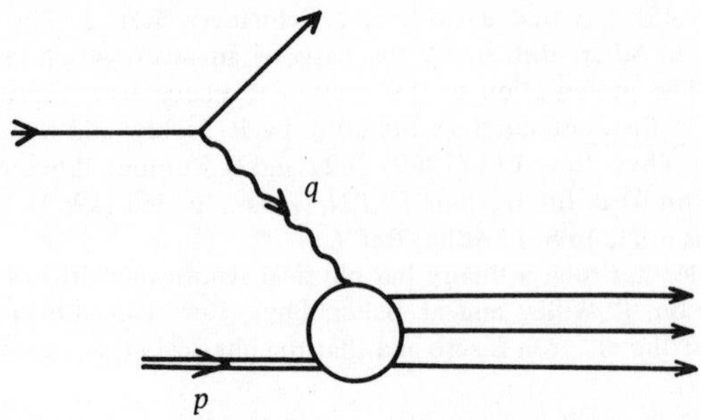

Fig. 5-1. Inelastic electron-proton scattering.

one proton connected matrix element of the commutator function of the electromagnetic current [1]

$$C^{\mu\nu}(p,q) = \int d^4x\, e^{iq\cdot x} \langle p \,|\, [J^\mu(x), J^\nu(0)] \,|\, p \rangle \,. \tag{5.1}$$

Here $|\, p\rangle$ is the target state of 4-momentum p^μ, $p^2 = m^2$. We shall assume that an averaging has been performed over the spin states of the proton. The momentum transferred by the lepton pair to the photon is q which is spacelike, $q^2 < 0$, $q_0 > 0$. Since the proton is the lowest state, we have $q^2 \geqslant -2q\cdot p$. We use the symbol ν for $q\cdot p$ and ω for $-q^2/2q\cdot p$. Because the photon is off its mass shell, it has transverse and longitudinal components. Correspondingly one speaks of two cross sections — the transverse σ_T and the longitudinal σ_L.

Since J^μ is conserved $C^{\mu\nu}(q,p)$ must be transverse in q. A gauge invariant, Lorentz covariant decompositions of $C^{\mu\nu}(q,p)$ is

$$C^{\mu\nu}(q,p) = -\left(g^{\mu\nu} - \frac{q^\mu q^\nu}{q^2}\right)\widetilde{F}_1 + \frac{1}{p\cdot q}\left(p^\mu - q^\mu \frac{q\cdot p}{q^2}\right)\left(p^\nu - q^\nu \frac{q\cdot p}{q^2}\right)\widetilde{F}_2 \,. \tag{5.2}$$

The notation here is slightly unconventional; in the literature $\widetilde{F}_1$ is usually called W_1 and $\widetilde{F}_2$ corresponds to νW_2. The Lorentz invariant functions $\widetilde{F}_i$ depend on q^2 and ν; frequently we shall choose ω and ν or ω and q^2 as the independent variables. The combination $\widetilde{F}_2 - 2\omega \widetilde{F}_1$ is called $\widetilde{F}_L$.

The total transverse and longitudinal cross sections are directly expressible in terms of the $\widetilde{F}_i$. It is easy to show that the functions $\widetilde{F}_i$ are dimensionless. $\widetilde{F}_2$ and $\widetilde{F}_L$ are even in ν, $\widetilde{F}_1$ is odd.

A remarkable fact can be deduced from the experimental data. At large values of ν and q^2, the $\widetilde{F}_i$ cease to depend on ν and q^2 separately; they become functions of only the ratio q^2/ν. (This is even more remarkable when it is observed that the energies at which this phenomenon occurs are not particularly large on the relevant mass scale.) This phenomenon is called the "scaling behavior" and leads to the following hypothesis. It is assumed that the experimental data indicates that as ν gets large, at fixed ω, the limit of $\widetilde{F}_i(\omega,\nu)$ exists [2]. We shall here accept this scaling hypothesis

$$\lim_{\nu\to\infty} \widetilde{F}_i(\omega,\nu) = F_i(\omega) < \infty \ . \tag{5.3}$$

One can show that $F_2(\omega) \geqslant 2\omega F_1(\omega) \geqslant 0$ since the F_i are proportional to limiting terms of non-negative cross sections. The region of large ν and q^2, at fixed ratio has been described as "deep inelastic".

The scaling hypothesis about the existence of scaling functions $F_i(\omega)$ has led to some very interesting applications of current algebra. It has been possible to derive sum rules, which relate properties of the $F_i(\omega)$ to ETC of electromagnetic currents. Our program is to indicate the derivation of the sum rules. Then we shall show that these applications of current algebraic reasoning are not verified by explicit calculations. Thus we shall demonstrate the existence of commutator anomalies in the present context.

5.2 Derivation of Sum Rules, Naive Method

We shall now derive the two sum rules which will be the objects of the present discussion. Consider the $0i$ component of (5.2) in the frame where the three vectors $\mathbf{q}$ and $\mathbf{p}$ are parallel, $\mathbf{p} = |\mathbf{p}|\hat{\mathbf{q}}$. Since the starting point and end results are covariant, no loss of generality occurs by working in a particular frame. Now pass into the deep inelastic region by letting $p_0 \to \infty$, subject to the constraints $p^2 = m^2$, i.e., $p_0 \simeq |\mathbf{p}|$; and ω constant, i.e.,

$$\omega = -\frac{q_0^2 - \mathbf{q}^2}{2(|\mathbf{p}|q_0 - |\mathbf{p}||q|)} = -\frac{1}{2|\mathbf{p}|}(q_0 + |\mathbf{q}|) \ ,$$

$$q_0 = -2p_0\omega - |\mathbf{q}| \ .$$

In this limit ν gets large as $-2p_0^2\,\omega$.

From (5.2) we have

$$C^{0i}(q,p) = -\frac{q^0 q^i}{2\omega\nu}\widetilde{F}_1 + \frac{q^i}{\nu}\left(p^0 + \frac{q^0}{2\omega}\right)\left(\frac{|\mathbf{p}|}{|\mathbf{q}|} + \frac{1}{2\omega}\right)\widetilde{F}_2 \quad . \tag{5.4a}$$

In (5.4a) the vector $\mathbf{p}$ has been replaced by the vector $|\mathbf{p}|\mathbf{q}$. In the second term in the second parentheses $|\mathbf{p}|/|\mathbf{q}|$ dominates $1/2\omega$, since $|\mathbf{p}|$ is getting large and $|\mathbf{q}|$ is fixed. Also

$$p^0 + \frac{q^0}{2\omega} \sim -\frac{|\mathbf{q}|}{2\omega} \quad .$$

Thus the following expression for C^{0i} may be used in the deep inelastic limit.

$$C^{0i}(q,p) \rightarrow -\frac{q^0\,q^i}{2\omega\nu}\widetilde{F}_1 - \frac{q^i\,|\mathbf{p}|}{2\omega\nu}\widetilde{F}_2 \quad . \tag{5.4b}$$

Next replace q^0/ν by $1/p_0$ and $|\mathbf{p}|/\nu$ by $-\tfrac{1}{2}p_0\,\omega$. We therefore find

$$2p_0\,C^{0i}(q,p) \rightarrow \frac{1}{2}q^i\,\frac{F_{\mathrm{L}}(\omega)}{\omega^2} \quad . \tag{5.4c}$$

An alternate formula for $C^{0i}(q,p)$, in the deep inelastic region, may be given from (5.1).

$$2p_0\,C^{0i}(q,p) \rightarrow 2p_0\int \mathrm{d}^4x\, e^{-2ip_0\,\omega x_0}\; e^{-i|\mathbf{q}|x_0}\; e^{-i\mathbf{q}\cdot\mathbf{x}}$$
$$\times\ \langle p|[J^0(x),J^i(0)]|p\rangle \quad . \tag{5.5}$$

Combining (5.5) with (5.4c), and integrating over ω gives

$$\lim_{p_0\to\infty}\int \mathrm{d}\omega\, 2p_0\,C^{0i}(q,p) = \frac{1}{2}q^i\int_{-\infty}^{\infty}\mathrm{d}\omega\,\frac{F_{\mathrm{L}}(\omega)}{\omega^2}$$

$$= \lim_{p_0\to\infty}\; 2\pi\int \mathrm{d}^4x\, 2p_0\,\delta(2p_0 x_0)\, e^{-i|\mathbf{q}|x_0}$$
$$\times\ e^{-i\mathbf{q}\cdot\mathbf{x}}\,\langle p|[J^0(x),J^i(0)]|p\rangle\ ,$$

$$= 2\pi \int d^3x\, e^{-i\mathbf{q}\cdot\mathbf{x}} \lim_{p_0\to\infty} \langle p\,|\,[J^0(0,\mathbf{x}), J^i(0)]\,|\,p\rangle \ . \tag{5.6}$$

This is the desired ST sum rule [3], to which we referred in Section 2. It shows that, for consistency of the result, the ST matrix element must be a Lorentz scalar; otherwise the $p_0 \to\infty$ limit diverges.

$$\langle p\,|\,[J^0(0,\mathbf{x}), J^i(0)]\,|\,p\rangle = i\,\langle p\,|\,S\,|\,p\rangle\, \partial^i \delta(\mathbf{x}) \ , \tag{5.7a}$$

$$\langle p\,|\,S\,|\,p\rangle = \frac{1}{4\pi}\int_{-\infty}^{\infty} d\omega\, \frac{F_L(\omega)}{\omega^2} \ . \tag{4.7b}$$

The integration extends over negative ω which are unphysical for electroproduction. However, if we assume that the symmetry of $\widetilde{F}_L(\omega,\nu) = \widetilde{F}_L(-\omega,-\nu)$ is maintained at infinite ν, as well as finite ν, we conclude that $F_L(\omega)$ is even in ω. Also the integration cuts off at $\omega = 1$, since $q^2 + 2\nu > 0$.

$$\langle p\,|\,S\,|\,p\rangle = \frac{1}{2\pi}\int_{0}^{1} d\omega\, \frac{F_L(\omega)}{\omega^2} \ . \tag{5.8}$$

If valid, this sum rule measures the q-number ST. (Only the q-number part of the ST contributes, since the connected matrix element is under consideration here — the vacuum expectation value of S has been subtracted out.) At the present time F_L is consistent with zero, a strong indication that the ST is a c number. Alternatively, if it is known that the ST is a c number, (5.8) predicts that F_L vanishes since F_L is non-negative. However, if $F(0)$ does not vanish sufficiently rapidly, so that the sum rule diverges it is not certain whether one should interpret that as a consistent statement about the infinite nature of this matrix element of the ST, or whether the sum rule is invalid. We shall return to the question of validity below.

The second sum rule which we consider is the Callan-Gross relation [4]. One begins with the i,j components of $C^{\mu\nu}$ but $\mathbf{q}$ is no longer taken parallel to p; indeed $\mathbf{q}$ is set to zero.

$$C^{ij}(q,p) = \delta^{ij}\,\widetilde{F}_1 - \frac{p^i p^j}{2p_0^2\,\omega}\,\widetilde{F}_2 \ , \tag{5.9a}$$

p_0 is sent to infinity in the deep inelastic region.

$$2\omega C^{ij}(q,p) \to \delta^{ij} 2\omega F_1(\omega) - \hat{p}^i \hat{p}^j F_2(\omega)$$

$$\to -p^i p^j F_{\mathrm{L}}(\omega) + (\delta^{ij} - \hat{p}^i \hat{p}^j) 2\omega F_1(\omega) \ . \qquad (5.9b)$$

The alternate expression for C^{ij} follows from (5.1).

$$2\omega C^{ij}(q,p)$$

$$\to 2\int \mathrm{d}^4 x\, \omega e^{-i2\omega x_0 p_0} \langle p | [J^i(x), J^j(0)] | p \rangle \ . \qquad (5.10)$$

Combining (5.9b) and (5.10) and integrating over ω yields

$$\lim_{p_0 \to \infty} \int_{-\infty}^{\infty} \mathrm{d}\omega\, 2\omega\, C^{ij}(q,p)$$

$$= (\delta^{ij} - \hat{p}^i \hat{p}^j) \int_{-\infty}^{\infty} \mathrm{d}\omega\, 2\omega F_1(\omega) - \hat{p}^i \hat{p}^j \int_{-\infty}^{\infty} \mathrm{d}\omega\, F_{\mathrm{L}}(\omega)$$

$$= i\pi \lim_{p_0 \to \infty} \frac{1}{p_0^2} \int \mathrm{d}^4 x\, \delta'(x_0) \langle p | [J^i(x), J^j(0)] | p \rangle$$

$$= -i\pi \lim_{p_0 \to \infty} \frac{1}{p_0^2} \int \mathrm{d}^3 x \langle p | [\dot{J}^i(0, \mathrm{x}), J^j(0)] | p \rangle \ . \qquad (5.11a)$$

The final sum rule is

$$(\delta^{ij} - \hat{p}^i \hat{p}^j) \int_0^1 \mathrm{d}\omega\, 2\omega F_1(\omega) - \hat{p}^i \hat{p}^j \int_0^1 \mathrm{d}\omega\, F_{\mathrm{L}}(\omega)$$

$$= -i \frac{\pi}{2} \lim_{p_0 \to \infty} \frac{1}{p_0^2} \int \mathrm{d}^3 x \langle p | [\dot{J}^i(0, \mathrm{x}), J^j(0)] | p \rangle \ . \qquad (5.11b)$$

The assumed properties of the F_i have been used to write the ω-integral over physical ω.

Note that the $\mathbf{x}$ integration picks out only the coefficient of the δ function in the ETC; all ST disappear upon integration. Also the limiting procedure $p_0 \to \infty$ selects only those parts of the ETC which are components of a second rank Lorentz tensor. The matrix element of such objects can be bilinear in $\mathbf{p}$, hence will survive in the limit. Matrix elements of Lorentz scalars must be independent of $\mathbf{p}$, hence they will vanish in the $p_0 \to \infty$ limit, when multiplied by p_0^{-2}. For consistency of (5.11b) it must be assumed that the $\langle p | [\dot{J}^i, J^j] | p \rangle$ ETC does not possess any tensors of rank 4 or higher in the coefficient of the δ function. Such tensors behave as $|\mathbf{p}|^4$, and (5.11b) would diverge.

This sum rule may also be used in two ways. It converts experimental data into information about the commutator. Alternatively if the commutator is known, it makes predictions about the results of experiment. It is the latter route which was chosen by Callan and Gross. By computing the $[\dot{J}^i, J^j]$ in a wide variety of quark models, they found that (see Exercise 5.1):

$$\int d^3x \langle p | [\dot{J}^i(0,\mathbf{x}), J^j(0)] | p \rangle = A(\delta^{ij}\mathbf{p}^2 - p^i p^j) + B\,\delta^{ij} \; . \tag{5.12}$$

It therefore follows from (5.11b) that $\int_0^1 d\omega F_L(\omega)$ and therefore $F_L(\omega)$ should identically vanish in these quark models. (The same conclusion is, of course, obtained from the Schwinger term sum rule, if the $[J^0, J^i]$ ETC is evaluated canonically.)

Clearly the Callan-Gross sum rule is more convergent at $\omega = 0$ than the ST sum rule. Unfortunately its predictive power is weaker, since knowledge is required of a commutator whose form can be deduced only after an application of canonical equations of motion for fields as well as canonical commutation relations. Further sum rules may be derived which relate integrals over $F_i(\omega)$, weighted by positive powers of ω, to ETC of currents involving many time derivatives. As the relevant power of ω increases, the convergence in ω is improved. The price paid is that the relevant commutator can be determined only after many applications of the equations of motion.

The present derivations, although very simple and direct, are heuristic in the sense that they require an interchange of limit and integral. Over this mathematical manipulation we have no control, and we cannot assess its validity. Also the technique of passing into the deep inelastic region by requiring p_0 and q_0 to become large at fixed $\mathbf{q}$, forces q^2 to be time-like, which is outside the physical region for this process. We now give an alternate derivation which exposes some of the mathematical underpinnings.

5.3 *Derivation of Sum Rules, Dispersive Method*

To effect the second derivation [5], we consider the forward Compton amplitude, whose absorptive part is the commutator function (5.1).

$$
\begin{aligned}
T^{*\mu\nu}(q,p) &= i\int d^4x\, e^{iq\cdot x}\,\langle p|T^* J^\mu(x) J^\nu(0)|p\rangle \\
&= -\left(g^{\mu\nu} - \frac{q^\mu q^\nu}{q^2}\right)T_1 \\
&\quad + \left(p^\mu - q^\mu\frac{q\cdot p}{q^2}\right)\left(p^\nu - q^\nu\frac{q\cdot p}{q^2}\right)T_2 \,.
\end{aligned}
\tag{5.13}
$$

The first assumption which is required for the derivation of the sum rules is that the matrix element of the ST (if any) is a Lorentz scalar. It then follows that the seagull (if any) in the T* product is present only in the i,j components; see Exercise 2.7. It is to be recalled that this condition was met in our previous derivation.

The T_i satisfy fixed q^2 dispersion relations in ν. Such dispersion relations may be converted, by a change of variable, $\nu = -q^2/2\omega$, into dispersion relations in ω, for fixed q^2. Therefore T_i is taken to be a function of q^2 and ω. In order to derive the two sum rules it is necessary to assume that T_2 is unsubtracted.

$$
T_2(\omega, q^2) = \frac{2\omega^2}{\pi q^2}\int_0^1 d\omega'\,\frac{\tilde{F}_2(\omega', q^2)}{\omega'^2 - \omega^2} \,. \tag{5.14}
$$

To derive the ST sum rule, it is necessary to assume that the combination $T_L = T_1 + (\nu^2/q^2)T_2$ is unsubtracted.

$$
T_L(\omega, q^2) = \frac{\omega^2}{2\pi}\int_0^1 \frac{d\omega'}{\omega'^2}\,\frac{\tilde{F}_L(\omega', q^2)}{\omega'^2 - \omega^2} \,. \tag{5.15a}
$$

The Callan-Gross sum rule may be derived from a weaker hypothesis. It suffices to assume that T_1 requires at most one subtraction, as long as the subtraction does not grow with q^2.

$$T_1(\omega, q^2) = T(q^2) + \frac{1}{\pi}\int_0^1 d\omega' \, \frac{\omega' \widetilde{F}_1(\omega', q^2)}{\omega^2 - \omega'^2}$$

$$\lim_{q^2 \to \infty} T(q^2) < \infty \; . \tag{5.15b}$$

$T(q^2)$ is the value of T_1 at $\nu = 0$: $T(q^2) = T_1(\infty, q^2)$. Equation (5.15b) is certainly true if (5.14) and (5.15a) hold; it can be valid even when (5.15a) needs subtractions. The dispersive representations (5.14) and (5.15b) are consistent with conventional Regge lore which predicts that both $\widetilde{F}_2$ and $\omega \widetilde{F}_1$ possess a finite limit as $\omega \to 0$. Evidently (5.15a) is valid only if $\widetilde{F}_L$ vanishes sufficiently rapidly for small ω (large ν).

We now apply the BJL limit to selected components of $T^{*\mu\nu}$. First consider T^{*0i}. Since there is no seagull in this object, the T* product coincides with the T product. Therefore from (5.13) we have

$$\lim_{q_0 \to \infty} q_0 T^{0i}(q,p)$$

$$= -\int d^3x \, e^{-i\mathbf{q}\cdot\mathbf{x}} \langle p | [J^0(0,\mathbf{x}), J^i(0)] | p \rangle$$

$$= \lim_{q_0 \to \infty} \left[\frac{q_0^2 q^i}{q^2} T_1 + q_0 \left(p_0 - q_0 \frac{\nu}{q^2} \right) \left(p^i - q^i \frac{\nu}{q^2} \right) T_2 \right]$$

$$= q^i \lim_{q_0 \to \infty} T_1 + p^i \, \mathbf{p} \cdot \mathbf{q} \lim_{q_0 \to \infty} T_2 \; . \tag{5.16a}$$

Since the matrix element of the ST, by hypothesis, has no $\mathbf{p}$ dependence, $\lim_{q_0 \to \infty} T_2$ must be zero. The same conclusion can be drawn from (5.14). (In the BJL limit $\omega^2 \to \infty$, $\widetilde{F}_2(\omega', q^2) \to F_2(\omega')$.) Equation (5.16a) may be now rewritten as

$$- \int d^3x \, e^{-i\mathbf{q}\cdot\mathbf{x}} \langle p | [J^0(0,\mathbf{x}), J^i(0)] | p \rangle$$

$$= q^i \lim_{q_0 \to \infty} \left(T_1 + \frac{\nu^2}{q^2} T_2 \right) = q^i \lim_{q_0 \to \infty} T_L \; . \tag{5.16b}$$

The BJL limit of T_L is evaluated from the dispersion formula (5.15a), and the derivation of the ST sum rule is complete [6].

$$\lim_{q_0 \to \infty} T_L = -\frac{1}{2\pi}\int_0^1 \frac{d\omega'}{\omega'^2} F_L(\omega') . \tag{5.16c}$$

To derive the Callan-Gross sum rule, one begins with T^{*ij}. The first task is to extract the seagull and arrive at the T product so that the BJL theorem may be applied. It is sufficient for our purposes to examine T^{ij} in the rest frame of q [7]. Since the BJL limit of T_2 is zero, we have, from (5.13)

$$\lim_{q_0 \to \infty} T^{*ij}\Big|_{\mathbf{q}=0} = -g^{ij} \lim_{q_0 \to \infty} T_1\Big|_{\mathbf{q}=0} . \tag{5.17a}$$

Therefore the seagull, which is that portion of a T* product which does not vanish in the BJL limit, is

$$-g^{ij} \lim_{q_0 \to \infty} T_1\Big|_{\mathbf{q}=0} . \tag{5.17b}$$

According to the dispersive representation for T_1, the BJL limit of that object is $\lim\limits_{q_0 \to \infty} T(q^2)$. Defining an asymptotic expansion at large q^2 for the subtraction term

$$T(q^2) \approx a + b/q^2 , \tag{5.17c}$$

we learn from (5.17a) that the T product is

$$T^{ij}\Big|_{\mathbf{q}=0} = T^{*ij}\Big|_{\mathbf{q}=0} + g^{ij} a = -g^{ij} T_1\Big|_{\mathbf{q}=0} + p^i p^j T_2\Big|_{\mathbf{q}=0} + g^{ij} a . \tag{5.17d}$$

We now apply the BJL limit to $q_0^2 T^{ij}$.

$$\lim_{q_0 \to \infty} q_0^2 T^{ij}\Big|_{\mathbf{q}=0}$$

$$= -g^{ij} \lim_{q_0 \to \infty} q_0^2 (T_1 - a)\Big|_{\mathbf{q}=0} + p^i p^j \lim_{q_0 \to \infty} q_0^2 T_2\Big|_{\mathbf{q}=0}$$

$$= -i\int d^3x \langle p|[\dot{J}^i(0,\mathbf{x}), J^j(0)]|p\rangle . \tag{5.18a}$$

The BJL limits occurring in (5.18a) are evaluated from the dispersive representations of the respective function. We find

$$\lim_{q_0 \to \infty} q_0^2 T_2 \Big|_{\mathbf{q}=0} = -\frac{2}{\pi} \int_0^1 d\omega' F_2(\omega') ,$$

$$\lim_{q_0 \to \infty} q_0^2 (T_1 - a) \Big|_{\mathbf{q}=0} = b + \frac{4}{\pi} p_0^2 \int_0^1 d\omega' \omega' F_1(\omega') . \qquad (5.18b)$$

A division by p_0^2, which is then sent to ∞, completes the derivation of the Callan-Gross sum rule.

The present derivation is especially instructive, since it demonstrates explicitly that it is the BJL definition of the commutator that is relevant to a high energy sum rule. If there are subtractions beyond the ones assumed, the present method is incapable of yielding a sum rule; see Exercise 5.2.

5.4 Model Calculation

In a quark-vector gluon model, the above two sum rules, when coupled with canonical evaluations of the relevant commutators, predict that F_L vanishes. (Of course the ST cannot be evaluated canonically; however, as we shall discuss below, it is known that the non-canonical ST is a c number in this model, at least to low orders of perturbation theory.) As an explicit check of canonical reasoning, and of the sum rules, we evaluate F_L to lowest order perturbation theory — a non-vanishing result is found.

The model is described by the Lagrangian

$$\mathscr{L} = i\bar{\psi}\gamma_\mu \partial^\mu \psi - m\bar{\psi}\psi - \tfrac{1}{4}B^{\mu\nu}B_{\mu\nu} + \tfrac{1}{2}\mu^2 B^\mu B_\mu + g\bar{\psi}\gamma^\mu \psi B_\mu .$$
(5.19)

A vector meson B^μ of mass μ interacts with a fermion ψ of mass m through a Yukawa type coupling of strength g; $B^{\mu\nu} = \partial^\mu B^\nu - \partial^\nu B^\mu$. The current to which the meson couples is conserved, therefore the theory is renormalizable. All considerations of internal symmetry are being ignored. Since we are interested in high-energy limits, be they deep inelastic or BJL, it is possible to ignore the fermion mass. Therefore, to simplify all the calculations, we henceforth set m to zero.

To check the formal results, we need the cross section for the production of a meson by an off-mass-shell photon. The relevant one-boson amplitude, to lowest order perturbation theory, is given by the diagrammatic representation of Fig. 5-2. The wavy line is the photon with 4-momentum q. The solid line is the fermion, with incoming momentum p. The boson, represented by the dashed line, carries off 4-momentum r. By a standard and unambiguous application of the Feynman rule method for calculating amplitudes, the spin-averaged transverse and longitudinal cross section for the relevant process can be obtained. From these results, the $\widetilde{F}_i$ can be deduced. Passing to the deep inelastic limit, one finds

$$F_{\mathrm{L}}(\omega) \propto g^2\,\theta(1-\omega^2)\,\omega^2 \ . \tag{5.20}$$

Thus the conclusion of formal reasoning is not maintained: it is found that $F_{\mathrm{L}}(\omega)$ is non-vanishing. This is an unambiguous consequence of the dynamics. Note that neither sum rule diverges; $F_{\mathrm{L}}(\omega)$ is sufficiently regular at the origin.

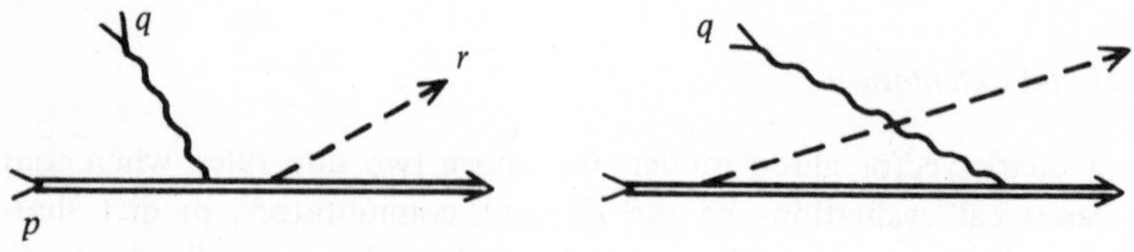

Fig. 5-2. Lowest order Feynman graphs for electroproduction in the model given by Eq. (5.19).

The other independent scaling function does not exist in this model. It is found that

$$\lim_{\substack{\nu\to\infty \\ \omega \text{ fixed}}} \widetilde{F}_2(\omega,\nu) \propto g^2 \log \nu/\mu^2 \ + \text{ finite terms} \ . \tag{5.21}$$

It might be objected that since scaling is not exhibited in this model, we have no business comparing sum rules which are derived with the help of the scaling hypothesis. This objection does not survive close scrutiny. To this order, one can unambiguously separate off the effects of the breakdown of the scaling hypothesis, and demonstrate that the failure of the formal reasoning is due to other causes. This is possible, since the longitudinal amplitude separately satisfies scaling, and no reference need be made to the transverse amplitude which violates scaling.

The failure of the ST sum rule is easy to understand. Examination of the explicit function $\widetilde{F}_L(\omega, q^2)$, as given by the perturbative calculation, indicates that the no subtraction hypothesis is wrong in the example. That is, even though the ST is indeed a c number, when calculated by the BJL limit (see below) the formula relating it to an integral over F_L was improperly derived [9, 10]. Therefore this failure is not due to anomalous commutators. Nevertheless, it is a particularly interesting example; there is no evidence in the deep inelastic limit of the need for a subtraction — the integral over F_L converges. We shall not discuss the ST sum rule any further since it is outside the scope of our main interest. The Callan-Gross sum rule is violated because the $[\dot{J}^i, J^j]$ ETC, when calculated by the BJL limit, as it must be in this application, does not have the form given by canonical reasoning (5.12). We now turn to a demonstration of this [11].

5.5 *Anomalous Commutators*

To compute the $[\dot{J}^i, J^j]$ ETC, to lowest order in the perturbation, we need to know the Compton amplitude $T^{*\mu\nu}$ to this order. The relevant unrenormalized graphs are those of Fig. 5-3. The physical amplitude is obtained by multiplying by Z^{-1}, the inverse of the wave-function renormalization constant. Although for the Callan-Gross application, only the forward Compton amplitude is relevant, we compute here the non-forward amplitude as well. Evidently we must also calculate the vertex function and the propagator to second order. In addition we shall need, in subsequent applications, the axial vertex function. The computation is performed by use of the Landau form for the boson propagator $D^{\mu\nu}(p) = -i(g^{\mu\nu} - p^\mu p^\nu/p^2)/(p^2 - \mu^2)$. We continue to keep the fermions massless. A convenient feature of this model is *finite* wave-function renormalization. Hence it is possible to speak of well defined unrenormalized fermion fields.

The following Green's functions are computed to lowest non-trivial order of perturbation theory.

Unrenormalized fermion propagator

$$G(p) = \int d^4x\, e^{ip\cdot x} \langle 0| T^* \psi(x) \bar{\psi}(0) |0\rangle \ . \tag{5.22}$$

Unrenormalized, improper vector vertex function

$$F^\mu(p,q) = \int d^4x\, d^4y\, e^{ip\cdot x}\, e^{-iq\cdot y} \langle 0| T^* \psi(x) \bar{\psi}(y) J^\mu(0) |0\rangle \ . \tag{5.23}$$

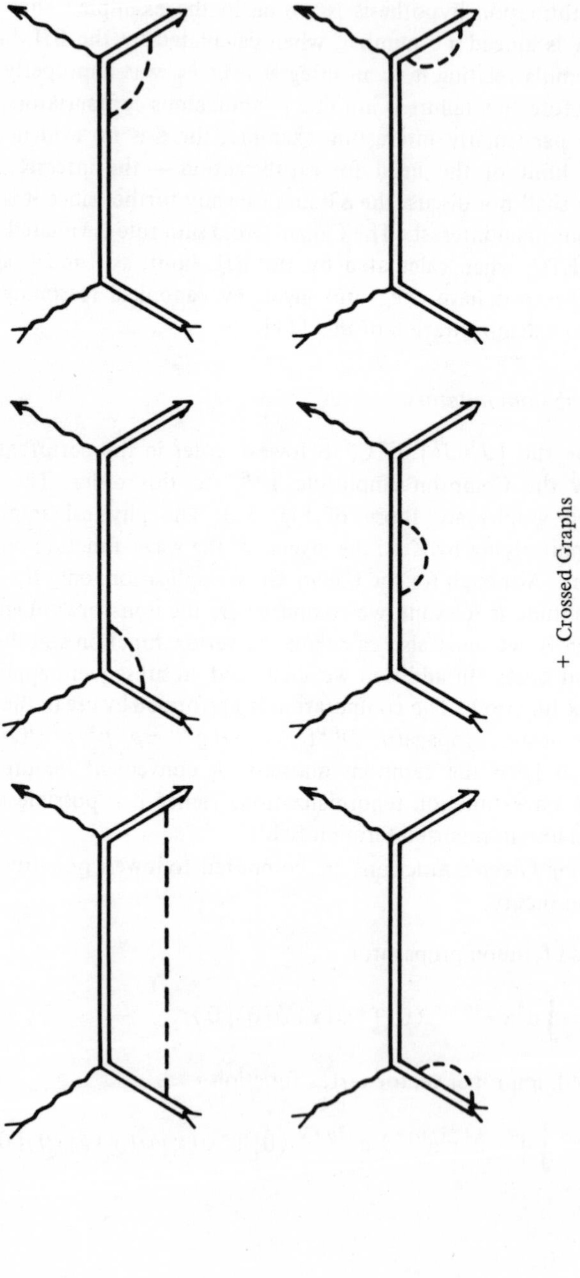

Fig. 5-3. Higher order corrections to the Compton amplitude.

Unrenormalized improper axial vector vertex function

$$F_5{}^{\mu}(p,q) = \int d^4x\, d^4y\, e^{ip\cdot x}\, e^{-iq\cdot y} \langle 0|T^* \psi(x)\, \bar\psi(y)\, J_5{}^{\mu}(0)|0\rangle\,. \tag{5.24}$$

Compton amplitude

$$T^{*\mu\nu}(q) = i\int d^4x\, e^{iq\cdot x} \langle p|T^* J^{\mu}(x)\, J^{\nu}(0)|p'\rangle\,. \tag{5.25}$$

Once these functions have been determined explicitly, several commutators can be computed by the BJL limit. They are not all relevant to the Callan-Gross sum rule, but they are sufficiently interesting to deserve mention. It turns out that none of the T* products possess seagulls; thus they are equivalent to T products, and the BJL theorem may be applied directly. The commutators which we determine are the following [12].

$$-i \lim_{p_0\to\infty} p_0 G(p) = \int d^3x\, e^{-i\mathbf{p}\cdot\mathbf{x}} \langle 0|[\psi(0,\mathbf{x}), \bar\psi(0)]_+|0\rangle\,. \tag{5.26}$$

$$-i \lim_{p_0\to\infty} p_0 F^{\mu}(p,q)\Big|_{q\text{ fixed}}$$

$$= \int d^3x\, d^4y\, e^{-i\mathbf{p}\cdot\mathbf{x}}\, e^{-iq\cdot y} \langle 0|T[\psi(0,\mathbf{x}), J^{\mu}(0)]\, \bar\psi(y)|0\rangle\,. \tag{5.27}$$

$$-i \lim_{q_0\to\infty} q_0\, T^{\mu\nu}(q) = i\int d^3x\, e^{-i\mathbf{q}\cdot\mathbf{x}} \langle p|[J^{\mu}(0,\mathbf{x}), J^{\nu}(0)]|p'\rangle\,. \tag{5.28}$$

$$- \lim_{q_0\to\infty} q_0^2\, T^{ij}(q)\Big|_{\substack{\mathbf{q}=0,\\ p=p'\\ \text{spin-averaged}}}$$

$$= i\int d^3x \langle p|[\dot J^{i}(0,\mathbf{x}), J^{j}(0)]|p\rangle\Big|_{\text{spin-averaged}} \tag{5.29}$$

The last commutator is the Callan-Gross commutator. No $1/q_0$ term is present in the forward amplitude, since the $[J^i, J^j]$ ETC between diagonal, spin-averaged states vanishes. The matrix elements of the axial current, calculated in (5.24), are required in an evaluation of the right-hand side of (5.28) for the ij component, which involves $\epsilon^{ijk} J_k{}^5$. When the limits are explicitly performed the following results are obtained [13].

$$[\psi(0,\mathbf{x}), \bar{\psi}(0)]_+ = \left(1 - \frac{3g^2}{32\pi^2}\right)\gamma^0 \delta(\mathbf{x}) , \tag{5.30}$$

$$[\psi(0,\mathbf{x}), J^0(0)] = \psi(0,\mathbf{x})\,\delta(\mathbf{x}) , \tag{5.31a}$$

$$[\psi(0,\mathbf{x}), J^i(0)] = \left(1 - \frac{g^2}{8\pi^2}\right)\gamma^0 \gamma^i \psi(0,\mathbf{x})\,\delta(\mathbf{x}) , \tag{5.31b}$$

$$\langle p|[J^0(0,\mathbf{x}), J^0(0)]|p'\rangle = 0 , \tag{5.32a}$$

$$\langle p|[J^0(0,\mathbf{x}), J^i(0)]|p'\rangle = 0 , \tag{5.32b}$$

$$\langle p|[J^i(0,\mathbf{x}), J^j(0)]|p'\rangle$$
$$= i\epsilon^{ijk}\left(1 - \frac{3g^2}{16\pi^2}\right)\langle p|J_5{}^k(0,\mathbf{x})|p'\rangle\,\delta(\mathbf{x}) , \tag{5.32c}$$

$$\int d^3x\, \langle p|[\dot{J}^i(0,\mathbf{x}), J^j(0)]|p\rangle \Big|_{\text{spin-averaged}}$$
$$= A(\delta^{ij}\mathbf{p}^2 - p^i p^j) + B\delta^{ij} + B'\mathbf{p}^2\delta^{ij} . \tag{5.33}$$

It is seen that most commutators are non-canonical; they have corrections which are interaction dependent. It is remarkable that all the non-canonical results are finite. Equation (5.30) indicates that the basic canonical (anti) commutator of the field variable with its conjugate momentum is not preserved. This is not particularly surprising; remember that the fields ψ are unrenormalized, and it is well known that, in general, the commutator of these fields ceases to exist due to infinite wave function renormalization Z.

It is only in our particular model, where Z is finite, that a well defined result is obtained, although it is noteworthy that the answer even when finite, is not the naive one. Equation (5.31b) shows that the field with the spatial components of the current density has a non-canonical commutator. Nevertheless the charge density with the field has the canonical commutator (5.31a). The $[J^0, J^0]$ and $[J^0, J^i]$ ETC are canonical. The latter result shows that here the ST has no fermion matrix element. This substantiates our previously made assertion that the ST term sum rule (5.8), if valid, predicts the vanishing of F_L in this model. Equation (5.32c) is the remarkable statement that the quark algebra of spatial components of currents is not maintained in perturbation theory. This has the consequence that certain radiative corrections to β decay, which were thought to be finite on the basis of the quark algebra, are, in fact, infinite [14]. Finally the term $B' \mathbf{p}^2 \delta^{ij}$ in (5.33) is non-canonical. The $\mathbf{p}^2$ factor in that term, allows it to contribute to the Callan-Gross sum rule. This spoils the canonical prediction that $F_L = 0$.

It is instructive to scrutinize the canonical derivation of the commutators given above, to pinpoint the precise step which is not validated by explicit calculation [15]. For definiteness we consider the $[\psi, J^i]$ ETC. One way to compute this commutator canonically is to exhibit the canonical expression for $J^i = \bar{\psi}\gamma^i\psi$ and then to use the canonical (anti) commutator relations for ψ with ψ and $\bar{\psi}$. Since we wish to minimize the use of the unreliable canonical formalism we prefer an alternate derivation which uses only reliable commutators, and which exposes the reason for the occurrence of an interaction dependent modification. For present purposes it is sufficient to consider only the term proportional to the δ function in the $[\psi, J^i]$ ETC. (Our above explicit calculation does not give evidence for the existence of gradient terms.) Thus we examine

$$\begin{aligned}
\int d^3y\, [\psi(t,\mathbf{x}), J^i(t,\mathbf{y})] &= \int d^3y\, y^i [\partial_j J^j(t,\mathbf{y}), \psi(t,\mathbf{x})] \\
&= \int d^3y\, y^i [\psi(t,\mathbf{x}), \dot{J}^0(t,\mathbf{y})] \\
&= \int d^3y\, y^i \frac{\partial}{\partial t}[\psi(t,\mathbf{x}), J^0(t,\mathbf{y})] - \int d^3y\, y^i [\dot{\psi}(t,\mathbf{x}), J^0(t,\mathbf{y})] \, .
\end{aligned} \tag{5.34a}$$

An integration by parts and the conservation of J^μ were used to pass from the first to the last expression in (5.34a). We now may use the canonical result for the $[J^0, \psi]$ ETC to evaluate the first of the two terms appearing in the last equality of (5.34a). Here the use of the canonical value is legitimate; explicit calculation did not cast any doubts on it. To determine the last occurring ETC in (5.34a), an expression for $\dot{\psi}$ is required. The equation of motion for ψ is by definition $i\gamma_\mu \partial^\mu \psi = m\psi + g\eta$; $\dot{\psi} = -\gamma^0\gamma^i\partial_i\psi - im\gamma^0\psi - ig\gamma^0\eta$. Here η is the source of the fermion field; in our model it is equal to $B_\mu\gamma^\mu\psi$, but we prefer to leave it unspecified for the moment. Equation (5.34a) is now seen to be equivalent to

$$\int d^3y\,[\psi(t,\mathbf{x}), J^i(t,\mathbf{y})]$$

$$= \int d^3y\,y^i\{\dot{\psi}(t,\mathbf{x})\delta(\mathbf{x}-\mathbf{y}) + \gamma^0\gamma^k\partial_k\psi(x)\delta(\mathbf{x}-\mathbf{y})$$

$$+ \gamma^0\gamma^k\psi(x)\partial_k\delta(\mathbf{x}-\mathbf{y}) + im\gamma^0\psi(x)\delta(\mathbf{x}-\mathbf{y})$$

$$+ ig\gamma^0[\eta(t,\mathbf{x}), J^0(t,\mathbf{y})]\} \quad . \tag{5.34b}$$

Performing the indicated time differentiation and using the equations of motion again, converts (5.34b) to

$$\int d^3y\,[\psi(t,\mathbf{x}), J^i(t,\mathbf{y})]$$

$$= \gamma^0\gamma^i\psi(x) + ig\gamma^0\int d^3y\,y^i\left\{[\eta(t,\mathbf{x}), J^0(t,\mathbf{y})] - \eta(x)\delta(\mathbf{x}-\mathbf{y})\right\} \quad . \tag{5.34c}$$

Thus it is seen that the canonical result is obtained only if the second term on the right-hand side of (5.34c) vanishes. This object *does* vanish if the canonical formula for $\eta(x)$ is taken: $\eta(x) = B_\mu(x)\gamma^\mu\psi(x)$; $[B_\mu(0,\mathbf{x})\gamma^\mu\psi(t,\mathbf{x}), J^0(t,\mathbf{y})] = B_\mu(x)\gamma^\mu\psi(x)\delta(\mathbf{x}-\mathbf{y}) = \eta(x)\delta(\mathbf{x}-\mathbf{y})$. However the expression $B_\mu(x)\psi(x)$ involves the product of two quantum mechanical operators at the same space-time point and is, in general, undefined. Evidently the "correct" form for $\eta(x)$ differs from the canonical one in such a way that the $[\eta, J^0]$ ETC is modified from its canonical form. It is easy to see that the following form for the $[\eta, J^0]$ ETC will reproduce the value given in (5.31b) for the $[\psi, J^i]$ ETC.

$$[\eta(t,\mathbf{x}), J^0(t,\mathbf{y})] = \eta(x)\delta(\mathbf{x}-\mathbf{y}) + \frac{i}{8\pi^2}g\gamma^i\psi(x)\partial_i\delta(\mathbf{x}-\mathbf{y}) \quad . \tag{5.35}$$

5.6 Discussion

We have demonstrated the occurrence of anomalous commutators in the quark model. The use of a vector gluon interaction is not important; similar results are obtained with scalar or pseudoscalar gluons. It is important to note that the local current algebra involving at least one time component has not been put into question. It therefore follows that vector Ward identities for Green's functions involving J^μ are not modified. This, of course, is a consequence of the fact that the symmetry implied by the conservation of J^μ can be maintained in perturbation theory. However the space-component algebra, as well as all higher commutators possess interaction dependent modifications. Such commutators are not used in derivations of Ward identities and low energy theorems. Rather they give rise, as we have seen, to high energy sum rules. It is these sum rules that are now found to be untrustworthy.

References

[1] For an excellent discussion of electroproduction kinematics see L. S. Brown, in *Lectures in Theoretical Physics* XII B, eds. K. Mahanthappa, and W. E. Brittin, Gordon and Breach, New York (1971).

[2] This hypothesis was enunciated by J. D. Bjorken, Phys. Rev. 179 (1969) 1547.

[3] This was derived by R. Jackiw, B. Van Royen and G. B. West, Phys. Rev. D2 (1970) 2473. See also L. S. Brown, Ref. 1 and J. M. Cornwall, D. Corrigan and R. E. Norton, Phys. Rev. Lett. 24 (1970) 1141.

[4] C. G. Callan, Jr. and D. J. Gross, Phys. Rev. Lett. 22 (1969) 156.

[5] The dispersive derivation for the ST sum rule is presented by R. Jackiw, R. Van Royen and G. B. West, Ref. 3. The Callan-Gross sum rule was derived in Ref. 4 by essentially the present method; the only difference being that the possibility of q-number ST was ignored.

[6] A technical comment: In order to avoid manipulations with the $\widetilde{F}_i$ in the unphysical, timelike domain of q^2, q_0 should be sent to $i\infty$ rather than to ∞. It can be shown that the BJL definition of the commutator holds in this limit as well.

[7] This does not take us out of the physical region of the $\widetilde{F}_i$, since q_0 can be imaginary; see Ref. 6.

[8] The treatment here follows R. Jackiw and G. Preparata, Phys. Rev. Lett. 22 (1969) 975, (E) 22 (1969) 1162 and Phys. Rev. 185 (1969) 1748. Identical results were found by S. L. Adler and Wu-Ki Tung, Phys. Rev. Lett. 22 (1969) 978 and Phys. Rev. D1 (1970) 2846.

[9] The "naive" derivation of the ST sum rule presumably fails due to improper interchange of limit with integral.

[10] J. M. Cornwall, D. Corrigan and R. E. Norton, Ref. 3, were able to derive a weaker ST sum rule, under hypotheses which are less stringent than the no

subtraction hypothesis employed here. However, in the case that $F_L(\omega)/\omega^2$ is regular at $\omega = 0$, their result is the same as the ST sum rule given in the text. Therefore the present model offers a counterexample to their argument as well. An analysis of this question has been given by A. Zee, Phys. Rev. D3, (1971) 2432.

[11] It is striking that the same model provides the same function, $F_L(\omega)$, as a counterexample to two common "regularity" assumptions: no subtraction in dispersion relations, validity of canonical reasoning. This suggests that there may be a relation between these two ideas. It would be most interesting to explore this relation.

[12] The presentation here is that of R. Jackiw and G. Preparata, Phys. Rev. 185 (1969) 1929.

[13] Equations (5.30), (5.31) and (5.33) were obtained by R. Jackiw and G. Preparata, Refs. 8 and 12. Simultaneously S. L. Adler and Wu-Ki Tung, Ref. 8, obtained (5.32) and (5.33). The current commutators (5.32) were previously studied in a different model by A. I. Vainshtein and B. L. Ioffe, Zh. Eksperim. i Toer. Fiz. Pisma v Redaktjiyu 6 (1967) 917 [English translation: Soviet Phys. JETP Lett. 6 (1967) 341].

[14] S. L. Adler and Wu-Ki Tung, Ref. 8.

[15] The development here is due to R. Jackiw and G. Preparata, Ref. 12.

6. Discussion of Anomalies in Current Algebra

6.1 *Miscellaneous Anomalies*

In addition to the results which were presented in the last two sections, various other commutators have been calculated in perturbation theory with the help of the BJL definition. We shall not discuss these calculations in detail, but merely quote three especially interesting conclusions.

(1) In the vector gluon model, which was introduced in the previous section, the connected matrix element of the ST between single boson states vanishes [1].

$$\langle B|[J^0(t,\mathbf{x}), J^i(t,\mathbf{y})]|B\rangle = 0 \ . \tag{6.1}$$

This, together with the previously quoted result that also the single fermion matrix element of the ST is zero, see (5.32b), strongly indicates that the ST is a c number. The relevant graph, whose high energy behavior yields (6.1), is the off-mass shell boson-boson scattering amplitude, Fig. 6-1. The result (6.1) is interesting because earlier calculations of this object, by methods other than the BJL definition, yielded a q-number ST, bilinear in the boson

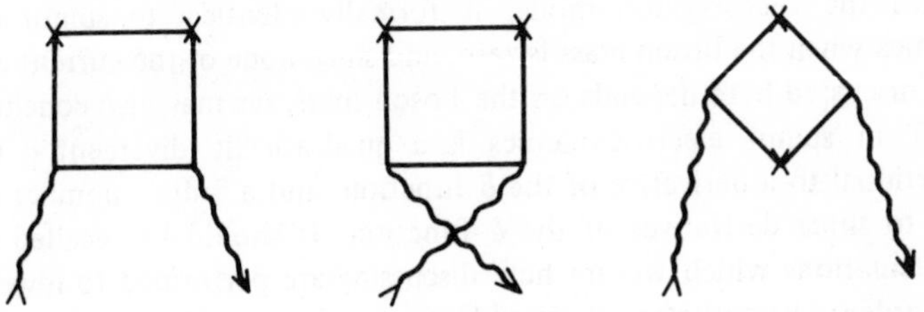

Fig. 6-1. Graphs from which the one-boson matrix element of the ST is determined.

field [2]. This shows, as was asserted previously, that alternate calculations of the ETC may give different results. Only the BJL definition has so far proven itself to be of physical interest.

(2) The vacuum expectation value of the ST in the same model has been shown to be a quadratically divergent term, proportional to the first derivative of the δ function. In addition, a finite term proportional to the triple derivative of the δ function has been uncovered [3].

$$[J^0(t,\mathbf{x}), J^i(t,\mathbf{y})] = iS_1 \partial^i \delta(\mathbf{x}-\mathbf{y}) + iS_3 \partial^i \nabla^2 \delta(\mathbf{x}-\mathbf{y}) \ . \tag{6.2}$$

S_1 is quadratically infinite; S_3 is well-defined. The relevant graph which yields (6.2) is the vacuum polarization, Fig. 6-2. The quadratic divergence of S_1 is correlated with the well known quadratic divergences of the relevant Feynman integral. The existence of the finite term involving S_3, is also a consequence of the presence of a dominant, quadratically divergent contribution. The triple derivative term violates both the operator analysis of the commutator, given in Section 2, (2.24), as well as the commonly accepted formula, derived in Exercise 3.2, for the vacuum expectation value of the ST. (That formula exhibits only one derivative of the δ function.) It is clear that the origin of the contradiction is the quadratic divergence of the $[J^0, J^i]$ ETC. The above two arguments proceed as if the commutator were well-defined. When it is divergent the arguments may be circumvented; see Exercise 6.1.

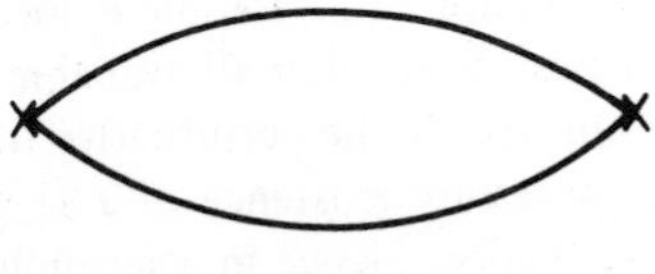

Fig. 6-2. Graph from which the vacuum expectation value of the ST is determined.

Since the vector-gluon model is formally identical to spinor electrodynamics when the boson mass is zero, and since none of the current commutators discussed here depends on the boson mass, we may also conclude that the ST in spinor electrodynamics is a quadratically divergent c number proportional to a derivative of the δ function, and a finite c number proportional to three derivatives of the δ function. It should be recalled that all the calculations which we are here discussing are performed to lowest non-trivial order in perturbation theory [4].

(3) The third discovery which we mention is the remarkable conclusion that the Jacobi identity for three spatial components of current densities is in general invalid in the quark model [5]. Explicit calculation shows that

$$\begin{aligned} &[J_a^{\mu}(t,\mathbf{x}),[J_b^{\nu}(t,\mathbf{y}),J_c^{\omega}(t,\mathbf{z})]] \\ &\quad\neq [[J_a^{\mu}(t,\mathbf{x}),J_b^{\nu}(t,\mathbf{y})],J_c^{\omega}(t,\mathbf{z})] \\ &\qquad + [J_b^{\nu}(t,\mathbf{y}),[J_a^{\mu}(t,\mathbf{x}),J_c^{\mu}(t,\mathbf{z})]] \\ &\qquad\qquad \mu,\nu,\omega = \text{spatial indices.} \end{aligned} \tag{6.3}$$

(A double commutator may also be defined by the BJL technique; see Exercise 6.2.) This anomaly, interesting in its own right, is relevant to the fact that it is possible to prove, with the help of the Jacobi identity, that the ST in the quark model is a q number [6]. Since we previously concluded that explicit calculation shows the ST to be a c number, we must now accept the violation of the Jacobi identity.

6.2 *Non-Perturbative Arguments for Anomalies*

Almost all the calculations which we have performed to substantiate the existence of commutator anomalies are based on perturbation theory. The important question arises whether these anomalies are peculiar to perturbation theory, or are they an essential feature of the complete theory? Although our knowledge of the structure of a complete local field theory is very limited, we may nevertheless assert that all available evidence points to the existence of anomalies even outside the perturbative framework.

In the first place, the necessary existence of a ST was established without reference to perturbation theory. Hence in a fermion theory, where a ST is not given by canonical reasoning, at least the ST anomaly must be present.

Furthermore the results quoted in subsection 6.1(3) above, indicate that this anomaly must be sufficiently singular so that either the ST is a q number, or the Jacobi identity is violated.

For other anomalies no one has constructed arguments which are as general as those for the ST. However it is clear that some of the anomalies must be present to *all* orders in perturbation theory. Consider for example the off mass shell pion decay amplitude $T(k^2)$. In renormalized perturbation theory this is a perfectly well defined and unambiguous quantity. Therefore if $T(0)$ does not vanish in lowest order, higher orders cannot cancel the first nonvanishing result, when the coupling constant is arbitrary. Even if the perturbative series does not converge, it is hard to see how the series can be interpreted to sum to zero for all values of the coupling. This argument, although valid to arbitrary order of perturbation theory, can be circumvented nevertheless, if it is asserted that there exists another solution to the theory which has no counterpart in perturbation theory. Such a point of view cannot be dismissed, however this speculation must be characterized as a dynamical assumption of the highest order. Therefore it remains true that the discovery of anomalies has shown current algebra to be dependent on unexpected dynamical assumptions.

For the anomalies associated with high-energy sum rules, like the Callan-Gross, lowest order calculations are less reliable indicators. For example, it is not impossible that $\widetilde{F}_L(\omega, \nu)$, which should be vanishing with $\nu \to \infty$ in the quark-gluon model, behaves for large ν as $g^2 \omega^2 \nu^{-ag^2}$, $a > 0$. In lowest order, this form is non-vanishing; however as $\nu \to \infty$, the complete function vanishes. However, higher order calculations of $\widetilde{F}_L(\omega, \nu)$ have been performed, and no evidence of such damping is found [7]. Of course since all orders have not been calculated, one is free to assume that this damping does occur, but again this must be characterized as a dynamical assumption.

Another class of evidence for anomalies derives from physically unrealistic, but solvable, models. It has been shown that the complete solutions of the Lee model [8] and of the Thirring model [9] possess commutator and divergence anomalies. It should be clear that the fundamental reason for the occurrence of anomalies in the renormalized theory is the fact that the underlying unrenormalized theory is divergent. No one knows with certainty whether these divergences are illusory, though all tentative evidence is that they are not. Indeed axiomatists who have examined and proven the existence of solutions to relativistic field theories, albeit in unrealistic two-dimensional space-time, have found that the kind of divergences which they rigorously show to be present, are accurately described by perturbation theory [10].

Some anomalies can perhaps disappear for a *particular* value of the coupling constant. It has been shown, for example, that in spinor electrodynamics, the canonical commutator of the unrenormalized electromagnetic potentials,

$$i[\dot{A}^i(t,\mathbf{x}) - \partial^i A^0(t,\mathbf{x}), A^j(t,\mathbf{y})] = -g^{ij}\,\delta(\mathbf{x}-\mathbf{y}) , \tag{6.4}$$

can be maintained in the complete theory if the unrenormalized charge satisfies an eigenvalue condition [11]. The condition is expressed by saying that a certain numerical function of the unrenormalized coupling constant must vanish; i.e., the coupling constant must be chosen to be a zero of this function. Unfortunately it is not known whether the relevant function possesses a zero — up to sixth order it does not. Moreover it is not likely that *all* anomalies would be removed even if the eigenvalue condition is met.

One might entertain the hope that the constraints of non-linear unitarity relations, which are not conveniently realized in perturbation theory, enforce sufficient damping on the high-energy behavior of various Green's functions, so that the various untoward results disappear in the complete solution. A preliminary investigation of this possibility has produced negative results: no mechanism for damping has been found [12].

The occurrence of anomalies in selected Ward identities may be understood in the following way. The "naive" Ward identities are a consequence of the relevant symmetries of the dynamics. In field theory, where the dynamics are described by a Lagrangian, these symmetries must be invariances of the Lagrangian. However, even when the Langrange function, expressed in terms of unrenormalized fields, possesses the appropriate invariance, the solution to the theory may not respect this symmetry. The reason is that the unrenormalized Lagrangian is undefined due to the infinities of the unrenormalized theory. A well defined theory emerges only after regularization. Thus the symmetry and the Ward identities are preserved only if the regulator method respects the invariance. Typically regulator procedures introduce large regulator masses into the problem. Therefore we should anticipate the possibility of the violation of symmetries which depend on the absence of mass terms, such as chirality, and scale invariance. (Scale invariance is discussed in the next section.)

6.3 Models without Anomalies

Can one somehow overcome these anomalies; that is, can a theory be set up where all results are naive? Only very limited investigations have been devoted to this question. The analysis, presented in the previous section in connection with the $[\psi, J^i]$ ETC, offers a hint how one may modify the theory so that this commutator retains its naive value. Evidently one must modify the dynamics so that the source of the fermion field, when commuted with J^0, no longer produces the gradient term as in (5.35), but coincides with the naive result.

Another example of such a modification can be given in reference to the failure of the Sutherland-Veltman theorem. By adding a direct interaction between the pion and the electromagnetic field of the form $\phi\, {}^*F^{\mu\nu}F_{\mu\nu}$, the non-zero value for the off mass shell decay amplitude for massless pions may be cancelled away, yielding the zero value as is required by the formal argument. Alternatively one may prescribe non-standard regulator techniques, which eliminate the non-zero coupling constant. All these modifications however probably lead to non-renormalizable perturbation series [13].

6.4 Discussion

It seems extremely probable, therefore, that if current algebra is derived from a local field theory, one cannot assume that canonical reasoning gives the correct results for current commutators and equations of motion. The anomalies are of two kinds: they may be sufficiently violent so that not only the canonical form of the commutators is destroyed, but also the Ward identities must be modified; other anomalies merely modify the commutators. The modified Ward identities lead to modified low energy theorems. It is likely that all instances of this have been discovered; they always involve the axial current and a catalog can be given [14]. The gentler anomalies which do not destroy naive Ward identities seem to occur in all commutators with the exception of the time-component algebra. Thus those results of current algebra which follow only from the $[J_a{}^0, J_b{}^0]$ ETC appear secure. The sum rules based on the space-component algebra, on the other hand, have been circumvented.

Ultimately the question whether our anomalies are present in nature, rather than in formal system, must be answered experimentally. The experimental predictions that have been obtained are rather limited. The Ward identity anomalies use PCAC and the possibility always remains that it is the latter idea, rather than the commutators that are at fault. Perhaps the experimental failure of the Sutherland-Veltman theorem is due to a rapid variator of $T(k^2)$ after all! [15]. Nevertheless some, model dependent, tests have been proposed, and it will be very interesting to see how they turn out; see Section 4.7.

The matrix elements of commutators can be measured experimentally, as the ST and Callan-Gross sum rules indicate. Unfortunately, with the exception of the Ward identity violations, the anomalies affect only those commutators which are already model dependent on the canonical level. Thus if F_L comes out to be non-vanishing experimentally, we still shall not know whether this is due to a non-canonical value for the $[\dot{J}^i, J^i]$ ETC, or whether the quark model, which canonically gives $F_L = 0$, is not the correct expression for the underlying dynamics.

The most positive consequence of the anomalies remains the fact that the Sutherland-Veltman theorem for π^0 decay can be circumvented without introducing mysteriously rapid variations. On the other hand, the anomalies do not seem relevant to other problematic predictions of current algebra and PCAC: $\eta \to 3\pi$ and $K_{\ell 3}$ decays [16].

Should it develop eventually that commutator anomalies must be rejected, we believe that this can be done only if current commutators are divorced from Lagrangian field theory. At the present time schemes are being developed to provide a non-Lagrangian basis for current algebra [17].

References

[1] D. G. Boulware and R. Jackiw, Phys. Rev. 186, (1969) 1442.

[2] R. A. Brandt, Phys. Rev. 166 (1968) 1795. Similarly Brandt calculates the anomalous commutators relevant to the Sutherland-Veltman theorem, see (4.36); and he obtains results different from those given by the BJL theorem; Phys. Rev. 180 (1969) 1490. Brandt's technique is that of split points. The current is defined from the fields by introducing a small separation ϵ: $J^\mu_\epsilon(x) = \bar{\psi}(x+\epsilon/2)\gamma^\mu \psi(x-\epsilon/2)$. The commutator is evaluated first for finite ϵ, and then ϵ is set to zero (ϵ is a 4 vector).

$$[J^\mu(t,\mathbf{x}), J^\nu(t,\mathbf{y})] = \lim_{\substack{\epsilon\to 0\\ \epsilon'\to 0}} [J^\mu_\epsilon(t,\mathbf{x}), J^\nu_{\epsilon'}(t,\mathbf{y})]$$

The reason for the discrepancy between this calculation and that of the BJL method has recently been explained by M. Chanowitz, Phys. Rev. D2 (1970) 3016. Chanowitz observes that, in the usual split-point calculations, the limit of the *time* components of ϵ and ϵ' is taken to zero first. It is then possible to use canonical commutators to evaluate the current commutators, where the currents are still defined with *spatially* separated points. Finally the spatial separation is set to zero. One the other hand, if the commutator is calculated first for unequal times, $\epsilon_0 = \epsilon'_0 \neq 0$; and then ϵ_0 and ϵ'_0 are sent to zero *after* the spatial separation has been eliminated, then the BJL result is obtained. Therefore, according to Chanowitz

$$[J^\mu(t,\mathbf{x}), J^\nu(t,\mathbf{y})]_{\mathrm{BJL}} = \lim_{\substack{\epsilon_0 \to 0 \\ \epsilon_0' \to 0}} \lim_{\substack{\epsilon \to 0 \\ \epsilon' \to 0}} [J^\mu_\epsilon(t,\mathbf{x}), J^\nu_{\epsilon'}(t,\mathbf{y})]$$

$$\neq \lim_{\substack{\epsilon \to 0 \\ \epsilon' \to 0}} \lim_{\substack{\epsilon_0 \to 0 \\ \epsilon_0' \to 0}} [J^\mu_\epsilon(t,\mathbf{x}), J^\nu_{\epsilon'}(t,\mathbf{y})] \, .$$

Note that our use of split-point techniques, (4.47), was for purposes of calculating the *divergence* of the axial current, and not commutators. Such applications have never been cast into doubt. The split-point technique has also been criticized by J. S. Bell and R. Jackiw, Nuovo Cimento 60 (1969) 47, and D. G. Boulware and R. Jackiw, Ref. 1.

[3] D. G. Boulware and R. Jackiw, Ref. 1; R. A. Brandt, Ref. 2, also obtained a third derivative term by use of the split-point method.

[4] The non-occurrence of q-number ST in spinor electrodynamics has also been noted by T. Nagylaki, Phys. Rev. 158 (1967) 1534 and D. G. Boulware and J. Herbert, Phys. Rev. D2 (1970) 1055.

[5] K. Johnson and F. E. Low, Progr. Theor. Phys. (Kyoto) Suppl. 37 – 38 (1966) 74. When all space-time indices select the time component: $\mu = \nu = \omega = 0$, no violation of the Jacobi identity has been found so far.

[6] F. Bucella, G. Veneziano, R. Gatto and S. Okubo, Phys. Rev. 149 (1966) 1268.

[7] R. Jackiw and G. Preparata, Phys. Rev. 185 (1969) 1748; S. L. Adler and Wu-Ki Tung, Phys. Rev. D1 (1970) 2846.

[8] J. S. Bell, Nuovo Cimento 47A (1967) 616.

[9] K. Wilson, Phys. Rev. D2 (1970) 1473; H. Georgi and J. Rawls, Phys. Rev. D3 (1971) 874; B. Schroer and J. Lowenstein, Phys. Rev. D3 (1971) 1981.

[10] For a summary, see, for example, K. Hepp in Proceedings of the 8th Symposium, N. Svartholm, ed. Interscience (John Wiley and Sons, New York, 1969).

[11] M. Gell-Mann and F. E. Low, Phys. Rev. 95 (1954) 1300; M. Baker and K. Johnson, Phys. Rev. 183 (1969) 1292.

[12] K. Bitar and N. Khuri, Phys. Rev. D3 (1971) 462.

[13] Nonstandard regulator techniques are discussed by J. S. Bell and R. Jackiw, Ref. 2. The observation that these methods may lead to infinities is due to S. L. Adler, Phys. Rev. 177 (1969) 2426. Non-minimal coupling is introduced by R. Jackiw and K. Johnson, Phys. Rev. 182 (1969) 1459.

[14] K. Wilson, Phys. Rev. 181 (1969) 1909; I. Gerstein and R. Jackiw, Phys. Rev. 181 (1969) 1955; W. Bardeen, Phys. Rev. 184 (1969) 1849; R. W. Brown, C.-C. Shih and B.-L. Young, Phys. Rev. 186 (1969) 1491; D. Amati, C. Bouchiat and J. L. Gervais, Nuovo Cimento 65A (1970) 55.

[15] R. A. Brandt and G. Preparata, Ann. Phys. (N. Y.) 61 (1970) 119 proposed a non-conventional interpretation of PCAC, which allows for such rapid variation.

[16] For both these processes, current algebra and PCAC predict a suppression of an appropriate quantity. Experimentally no such suppression occurs. For $\eta \to 3\pi$ see D. G. Sutherland, Nucl. Phys. B2 (1967) 433; for $K_{\ell 3}$ decays see C. G. Callan and S. B. Treiman, Phys. Rev. Lett. 16 (1966) 153. Recently K. Wilson has

provided a tentative explanation for the breakdown of PCAC in the η decay problem; see Phys. Rev. 179 (1969) 1499. R. A. Brandt and G. Preparata, Ref. 15, and Nuovo Cimento Lett. 4 (1970) 80 have proposed a non-conventional model of PCAC and SU(3) × SU(3) breaking, which escapes these problematical predictions. Their model has been criticized by M. Weinstein, Phys. Rev. D3 (1971) 481.

[17] K. Wilson, Ref. 16.

7. Approximate Scale Symmetry

7.1 Introduction

The experimental discovery of scaling in the electroproduction data has revived the long cherished notion, among symmetry-minded physicists, that at high energies masses can be ignored, and a new symmetry sets in: the symmetry of scale or dilatation invariance. Let us examine how the experimental observations may lead one to consider the topic of scale transformations.

As was stated in Section 5, the electroproduction measurements determine the commutator function.

$$C^{\mu\nu}(q,p) = \int \frac{d^4x}{(2\pi)^4} e^{iq\cdot x} \langle p|[J^\mu(x), J^\nu(0)]|p\rangle$$

$$= -\left(g^{\mu\nu} - \frac{q^\mu q^\nu}{q^2}\right)\widetilde{F}_1$$

$$+ \left(p^\mu - q^\mu \frac{p\cdot q}{q^2}\right)\left(p^\nu - q^\nu \frac{p\cdot q}{q^2}\right)\frac{\widetilde{F}_2}{p\cdot q} \; . \tag{7.1}$$

It can be verified that the $\widetilde{F}_i$ are dimensionless; see Exercise 7.1. (Recall that dimensions are measured in units of mass; $\hbar = c = 1$.) The $\widetilde{F}_i$ depend on the kinematical variables q^2 and ν. They also may be considered to be functions of whatever *dimensional* parameters determine the fundamental dynamics of the process. Examples of such additional parameters are masses of the "fundamental" particles that may be relevant. In addition it may be that the fundamental physical theory which governs natural phenomena is characterized by dimensional coupling constants. In that case these objects should also be included in the list of dimensional parameters on which the $\widetilde{F}_i$ can depend. However, the hypothesis which enables one to proceed from the experimental fact of scaling of the $\widetilde{F}_i$, to the theoretical considerations of scale invariance, is that there are *no* fundamental dimensional coupling constants. In that case the $\widetilde{F}_i$ depend only on q^2, ν and the masses; and since the functions are dimensionless, they must depend on these variables through dimensionless ratios: $\widetilde{F}_i = \widetilde{F}_i = (q^2/m^2, \nu/m^2, \ldots)$. Here m^2 is some mass in the problem, and the dots indicate possible other mass ratios. One says that m "sets the scale" for the problem.

When the kinematical variables q^2 and ν are large, it is not implausible to suppose that the mass dependence of $\widetilde{F}_i$ is unimportant. Hence the deep inelastic limit (q^2 and ν both large) is equivalent to the zero-mass limit. In that case $\widetilde{F}_i$ becomes a function only of the dimensionless ratio q^2/ν, and this is what is observed experimentally. When there are no dimensional parameters, other than the kinematical ones q^2 and ν, there is no scale against which q^2 and ν can be measured. Clearly a simultaneous rescaling of q and ν leaves q^2/ν unchanged. In position space this corresponds to rescaling the position variable x — in the absence of dimensional parameters such a rescaling leaves the theory invariant. Thus we are led to study the effects of the scale or dilatation transformation $x^\mu \to e^{-\rho} x^\mu$.

In the next subsection we shall derive the canonical theory of these transformations, as well as of the related conformal transformations. The currents that are conserved in the case of symmetry will be calculated. We shall then show how the canonical theory leads to Ward identities which may be used to derive theorems about the high energy behavior of Green's functions. Finally it will be demonstrated that in perturbation theory these results are false — there are anomalies which destroy them [1].

7.2 Canonical Theory of Scale and Conformal Transformations

A scale transformation of the coordinates x^μ takes x^μ into $x'^\mu = e^{-\rho} x^\mu$. This is to *coordinate* transformation, in the same class of transformations as a translation: $x^\mu \to x'^\mu = x^\mu + a^\mu$, or a Lorentz transformation $x^\mu \to x'^\mu = \Lambda^{\mu\nu} x_\nu$. It is also useful, for reasons which will presently be obvious, to consider conformal coordinate transformations defined by $x^\mu \to x'^\mu = (x^\mu - c^\mu x^2)(1 - 2cx + c^2 x^2)^{-1}$. Here ρ, a^μ, $\Lambda^{\mu\nu}$ and c^μ are parameters which specify the dilatation, translation, Lorentz transformation and conformal transformation, respectively. Infinitesimally these transformations are

translation $$\delta_T{}^\alpha x^\mu = g^{\alpha\mu} , \tag{7.2a}$$

Lorentz transformation $$\delta_L{}^{\alpha\beta} x^\mu = g^{\alpha\mu} x^\beta - g^{\beta\mu} x^\alpha , \tag{7.2b}$$

dilatation $\delta_D x^\mu = -x^\mu$, (7.2c)

conformal transformation $\delta_C{}^\alpha x^\mu = 2x^\alpha x^\mu - g^{\alpha\mu} x^2$. (7.2d)

In exhibiting the formulas (7.2) we have not included the infinitesimal parameters specifying the transformation; this practice will be followed throughout.

The set of 15 transformations given in (7.2) forms a 15 parameter Lie group, called the conformal group. This is a generalization of the 10 parameter Poincaré group, formed from the 10 transformations (7.2a) and (7.2b). By considering the combined action of various infinitesimal transformations in different orders, the Lie algebra of the group may be abstracted. Upon defining $P^\alpha, M^{\alpha\beta}, D$ and K^α to be respectively the generators of translations, Lorentz transformations, dilatations and conformal transformations, we find

$$i[P^\alpha, P^\beta] = 0 , \tag{7.3a}$$

$$i[M^{\alpha\beta}, P^\gamma] = g^{\alpha\gamma} P^\beta - g^{\beta\gamma} P^\alpha , \tag{7.3b}$$

$$i[M^{\alpha\beta}, M^{\mu\nu}] = g^{\alpha\mu} M^{\beta\nu} - g^{\beta\mu} M^{\alpha\nu} + g^{\alpha\nu} M^{\mu\beta} - g^{\beta\nu} M^{\mu\alpha} , \tag{7.3c}$$

$$i[D, P^\alpha] = P^\alpha , \tag{7.3d}$$

$$i[D, K^\alpha] = -K^\alpha , \tag{7.3e}$$

$$i[M^{\alpha\beta}, K^\gamma] = g^{\alpha\gamma} K^\beta - g^{\beta\gamma} K^\alpha , \tag{7.3f}$$

$$i[P^\alpha, K^\beta] = -2g^{\alpha\beta} D + 2M^{\alpha\beta} , \tag{7.3g}$$

$$i[D,D] = i[D, M^{\alpha\beta}] = i[K^\alpha, K^\beta] = 0 . \tag{7.3h}$$

The first three commutators define the Lie algebra of the Poincaré group. In offering (7.3d) to (7.3h) we make no claim that these commutators are realized in nature. They merely reflect the combination rules for the transformations. For if (7.3d) holds in nature, then it is also true that $e^{i\alpha D} P^2 e^{i\alpha D} = e^{2\alpha} P^2$. This implies that the mass spectrum is either continuous or all the masses vanish; neither eventuality is acceptable [3].

Before establishing the conditions under which a Lagrangian $\mathscr{L}$, which depends on the fields ϕ, is scale invariant or conformally invariant, we must decide how the fields ϕ transform under the dilatation and conformal transformations. For translations and Lorentz transformations the rules are the standard ones.

$$\delta_T{}^{\alpha}\phi(x) = i\,[P^{\alpha}, \phi(x)] = \partial^{\alpha}\phi(x)\,, \tag{7.4a}$$

$$\delta_L{}^{\alpha\beta}\phi(x) = i\,[M^{\alpha\beta}, \phi(x)] = (x^{\alpha}\partial^{\beta} - x^{\beta}\partial^{\alpha} + \Sigma^{\alpha\beta})\phi(x)\,. \tag{7.4b}$$

For the remaining operations, the following choice is consistent with (7.3), and we adopt it.

$$\delta_D\,\phi(x) = (d + x\cdot\partial)\phi(x)\,, \tag{7.4c}$$

$$\delta_C{}^{\alpha}\phi(x) = (2x^{\alpha}x^{\nu} - g^{\alpha\nu}x^2)\,\partial_{\nu}\phi(x) + 2x_{\nu}(g^{\nu\alpha}d - \Sigma^{\nu\alpha})\phi(x)\,. \tag{7.4d}$$

Here d is a constant; it is called the scale dimension of the field ϕ.

It is now easy, following the general discussion of Section 2, to establish the variation of $\mathscr{L}$.

$$\delta_D\mathscr{L} = \partial_{\nu}[x^{\nu}\mathscr{L}] - 4\mathscr{L} + \pi_{\mu}(d+1)\phi^{\mu} + \frac{\delta\mathscr{L}}{\delta\phi}d\phi\,, \tag{7.5a}$$

$$\delta_C{}^{\alpha}\mathscr{L} = \partial_{\nu}[(2x^{\alpha}x^{\nu} - g^{\alpha\nu}x^2)\mathscr{L}] + 2x^{\alpha}\left[\frac{\delta\mathscr{L}}{\delta\phi}d\phi + \pi_{\mu}(d+1)\phi^{\mu} - 4\mathscr{L}\right] + V^{\mu}\,. \tag{7.5b}$$

V^{μ} is called the field virial and is given by

$$V^{\mu} = \pi_{\alpha}[g^{\alpha\mu}d - \Sigma^{\alpha\mu}]\phi\,. \tag{7.6}$$

In deriving (7.5), Lorentz and translation invariance of $\mathscr{L}$ are used.

Examining (7.5a) we see that dilatation invariance requires that

$$-4\mathscr{L} + \pi_{\mu}(d+1)\phi^{\mu} + \frac{\delta\mathscr{L}}{\delta\phi}d\phi = 0\,. \tag{7.7}$$

If d is chosen to be $\frac{3}{2}$ for fermion fields, and 1 for boson fields, then the kinetic energy term of the Lagrangian satisfies (7.7). These values for d correspond to the natural dimensions of fields, in units of mass. It is easy to see that (7.7) requires that the scale dimension of $\mathscr{L}$ be 4; i.e., that there be no dimensional parameters in $\mathscr{L}$. Clearly mass terms violate (7.7). Examples of interactions that satisfy (7.7) are ϕ^4 or $\bar{\psi}\psi\phi$.

For conformal invariance, (7.5b) shows that *two* conditions must be met. Firstly, scale invariance must obtain; i.e., (7.7) must be true. This is already seen from (7.3g): if P^α, K^β and $M^{\alpha\beta}$ are symmetry generators, so also must be D. Secondly, the field virial must be a total divergence

$$V^\mu = \partial_\alpha \sigma^{\alpha\mu} \; . \tag{7.8}$$

It is remarkable that for all renormalizable theories (7.8) is true, even though scale invariance of course is broken. Equation (7.8) is also true in all theories involving spins $\leqslant 1$ without derivative coupling. This fact indicates that once a model for scale symmetry breaking has been adopted, one automatically is provided with a model for conformal symmetry breaking.

The currents associated with these transformations are now determined by the general techniques of Section 2. The *canonical* dilatation current $D_c^{\ \mu}$, and the *canonical* conformal current $K_c^{\ \alpha\mu}$ are

$$D_c^{\ \mu} = x_\alpha \theta_c^{\ \mu\alpha} + \pi^\mu d\phi \; , \tag{7.9a}$$

$$K_c^{\ \alpha\mu} = (2x^\alpha x_\nu - g_\nu^\alpha x^2)\theta_c^{\ \mu\nu} + 2x_\nu \pi^\mu (g^{\nu\alpha} d - \Sigma^{\nu\alpha})\phi - 2\sigma^{\alpha\mu} \; . \tag{7.9b}$$

Note that the currents are expressed in terms of the *canonical* energy momentum tensor $\theta_c^{\ \mu\alpha}$. We have argued in Section 2 that the Belinfante tensor $\theta_B^{\ \mu\alpha}$ has a greater significance than $\theta_c^{\ \mu\alpha}$, therefore (7.9) should be expressed in terms of it. Inserting the formula for $\theta_B^{\ \mu\alpha}$ in terms of $\theta_c^{\ \mu\alpha}$ and the appropriate superpotential, (2.17), one finds after some tedious steps

$$D_c^{\ \mu} = x_\alpha \theta_B^{\ \mu\alpha} + V^\mu + \partial_\beta X^{\beta\mu} \; , \tag{7.10a}$$

$$K_c^{\ \alpha\mu} = [2x^\alpha x_\nu - g_\nu^\alpha x^2]\,\theta_B^{\ \nu\mu} + 2x^\alpha \partial_\nu \sigma_+^{\ \mu\nu} - 2\sigma_+^{\ \alpha\mu} + \partial_\beta X^{\beta\mu\alpha} \; . \tag{7.10b}$$

Here $\sigma_+^{\ \mu\nu}$ is the symmetric part of $\sigma^{\mu\nu}$; $X^{\beta\mu}$ and $X^{\beta\mu\alpha}$ are objects which can be explicitly given — the only property that need concern us here is that both are anti-symmetric in β and μ. Hence they are superpotentials which may be dropped without loss of physical content. Thus we are left with the Belinfante dilatation and conformal current

$$D_B{}^{\mu} = x_{\alpha} \theta_B{}^{\mu\alpha} + V^{\mu} \ , \tag{7.11a}$$

$$K_B{}^{\alpha\mu} = [2x^{\alpha} x_{\nu} - g_{\nu}^{\alpha} x^2] \theta_B{}^{\nu\mu} + 2x^{\alpha} \partial_{\nu} \sigma_+{}^{\mu\nu} - 2\sigma_+{}^{\alpha\mu} \ . \tag{7.11b}$$

These currents may be further simplified by the following consideration. Instead of the Belinfante tensor, one may use, for discussions of translations and Lorentz transformations, the "new, improved energy-momentum tensor" $\theta^{\mu\alpha}$ [2]. This object is defined by adding to $\theta_B{}^{\mu\alpha}$ the following super-potential

$$\theta^{\mu\alpha} = \theta_B{}^{\mu\alpha} + \tfrac{1}{2} \partial_{\lambda} \partial_{\rho} X^{\lambda\rho\mu\nu} \ , \tag{7.12a}$$

$$X^{\lambda\rho\mu\nu} = g^{\lambda\rho} \sigma_+{}^{\mu\nu} - g^{\lambda\mu} \sigma_+{}^{\rho\nu} - g^{\lambda\nu} \sigma_+{}^{\mu\rho} + g^{\mu\nu} \sigma_+{}^{\lambda\rho}$$
$$- \tfrac{1}{3} g^{\lambda\rho} g^{\mu\nu} \sigma_+{}^{\alpha}{}_{\alpha} + \tfrac{1}{3} g^{\lambda\mu} g^{\rho\nu} \sigma_+{}^{\alpha}{}_{\alpha} \ . \tag{7.12b}$$

The additional superpoential does not destroy the conservation or symmetry properties of $\theta^{\mu\alpha}$, nor does it contribute to the translation or Lorentz generators. Hence $\theta^{\mu\alpha}$ may be used instead of $\theta_B{}^{\mu\alpha}$ as the Poincaré "current". For purposes of expressing the dilatation and conformal currents, $\theta^{\mu\alpha}$ is more convenient than $\theta_B{}^{\mu\alpha}$. In terms of $\theta^{\mu\alpha}$, the Belinfante dilatation and conformal currents are

$$D_B{}^{\mu} = x_{\alpha} \theta^{\mu\alpha} + \partial_{\beta} Y^{\beta\mu} \ , \tag{7.13a}$$

$$K_B{}^{\alpha\mu} = [2x^{\alpha} x_{\nu} - g_{\nu}^{\alpha} x^2] \theta^{\nu\mu} + \partial_{\beta} Y^{\beta\mu\alpha} \ . \tag{7.13b}$$

The total divergence terms in (7.13) are again superpotentials; they again may be dropped and we are left with the final expressions for the currents.

$$D^{\mu} = x_{\alpha} \theta^{\mu\alpha} \ , \tag{7.14a}$$

$$K^{\mu\alpha} = [2x^{\alpha} x_{\nu} - g_{\nu}^{\alpha} x^2] \theta^{\nu\mu} \ , \tag{7.14b}$$

In addition to providing a compact form for D^{μ} and $K^{\mu\alpha}$, the new improved tensor has another advantage. It can be shown that in renormalized perturbation theory, its matrix elements are less singular than those of $\theta_B{}^{\mu\nu}$ [2].

The divergence of these currents can be expressed in terms of the trace of $\theta^{\mu\nu}$.

$$\partial_\mu D^\mu = \theta^\mu_\mu \ , \tag{7.15a}$$

$$\partial_\mu K^{\mu\alpha} = 2x^\alpha \theta^\mu_\mu = 2x^\alpha \partial_\mu D^\mu \ . \tag{7.15b}$$

Thus we see that both scale and conformal invariance are broken by θ^μ_μ (in theories where $V^\mu = \partial_\alpha \sigma^{\alpha\mu}$). In the subsequent we shall always assume that θ^μ_μ is non-zero by virtue of the presence of mass terms in $\mathscr{L}$. Thus scale and conformal symmetries are broken, as they must be in order to avoid a physically absurd mass spectrum [3].

Explicit computation shows that V^μ and $\sigma^{\alpha\mu}$ are identically zero for spin $\frac{1}{2}$ and spin 1 fields. For spin zero fields, we have

$$V^\mu = \pi^\mu \phi = \phi^\mu \phi = \tfrac{1}{2} \partial^\mu \phi^2 \ ,$$

$$\sigma^{\alpha\mu} = \tfrac{1}{2} g^{\alpha\mu} \phi^2 \ . \tag{7.16}$$

An explicit formula for $\theta^{\mu\nu}$ may now be given.

$$\theta^{\mu\nu} = \theta_B^{\ \mu\nu} - \tfrac{1}{6} \sum_{\substack{\text{spin}\\ \text{zero}\\ \text{fields}}} (\partial^\mu \partial^\nu - g^{\mu\nu} \Box) \phi^2 \ . \tag{7.17}$$

(The unique role of spin zero has not as yet been understood.)

Although for purposes of transformation theory one may use *any* energy-momentum tensor: $\theta_c^{\ \mu\alpha}$, $\theta_B^{\ \mu\alpha}$ or $\theta^{\mu\alpha}$; our use of $\theta^{\mu\alpha}$, implies that we attach a unique physical significance to this object. As discussed in Section 2, this significance emerges when gravitational interactions are discussed. In the conventional gravity theory of Einstein general relativity, $\theta_B^{\ \mu\alpha}$ is the source of gravitons. If we wish to work with $\theta^{\mu\alpha}$, the theory must be changed so that $\theta^{\mu\alpha}$ is the source. This modification has been given. The modified theory is consistent with all the present tests of theory of relativity [4].

The charges $D(t)$ and $K^\alpha(t)$

$$D(t) = \int d^3x \, x_\mu \theta^{\mu 0}(x) \ , \tag{7.18a}$$

$$K^\alpha(t) = \int d^3x \, [2x^\alpha x_\nu - g^\alpha_\nu x^2] \, \theta^{\nu 0}(x) \ , \tag{7.18b}$$

are not Lorentz covariant in the absence of symmetry. Also they do not satisfy the algebra (7.3); see Exercise 7.2. Nevertheless it is true that $D(t)$ and $K^{\alpha}(t)$ effect the proper transformation on the fields even in the absence of symmetry; see Exercise 7.3.

$$i\,[D(t), \phi(x)] = \delta_{\mathrm{D}}\, \phi(x)\ , \tag{7.19a}$$

$$i\,[K^{\alpha}(t), \phi(x)] = \delta_{\mathrm{C}}{}^{\alpha} \phi(x)\ . \tag{7.19b}$$

7.3 Ward Identities and Trace Identities

We have assumed θ^{μ}_{μ} to be given in a Lagrangian model by mass terms, for example when fermions of mass m and bosons of mass μ are present, $\theta^{\mu}_{\mu} = m\bar{\psi}\psi + \mu^2\phi^2$. Evidently the divergence of the dilatation current is soft; the dimension of the operators occurring in θ^{μ}_{μ} is at most 3. At the present time there appear to be two interesting possibilities for applications of these ideas to physics. One may attempt to repeat the success of PCAC: dominate θ^{μ}_{μ} by scalar mesons and unravel the low energy dynamics of these mesons [5]. We shall not consider this point of view, but rather examine the second possible application: determination of high-energy behavior of Green's functions. For definiteness we study the two-point function, the renormalized propagator.

Extracting consequences of our hypotheses about scale and conformal symmetry breaking is most easily accomplished from Ward identities satisfied by matrix elements of D^{μ} and $K^{\mu\alpha}$. Since these currents are simply related to the energy momentum tensor, it is useful to consider Ward identities satisfied by matrix elements of $\theta^{\mu\nu}$. We now present a derivation of these.

In order to proceed we need to know the commutator of $\theta^{\mu\nu}$ with renormalized field ϕ of scale dimensionality d. Under very general hypothesis, one can show that

$$i\,[\theta^{00}(0,\mathbf{x}), \phi(0)] = \partial^0 \phi(0)\,\delta(\mathbf{x}) + \Sigma^{0i}\phi(0)\,\partial_i\,\delta(\mathbf{x})\ . \tag{7.20a}$$

$$i\,[\theta^{0i}(0,\mathbf{x}), \phi(0)] = \partial^i\phi(0)\,\delta(\mathbf{x}) - \frac{d}{3}\phi(0)\,\partial^i\,\delta(\mathbf{x})$$

$$+\frac{1}{2}\Sigma^{ij}\,\phi(0)\,\partial_j\,\delta(\mathbf{x})\ . \tag{7.20b}$$

These are the formal, canonical commutators. No statement is being made concerning their validity in perturbation theory. Since the commutator of

$\theta^{\mu\nu}$ with ϕ necessarily contains gradient terms, the T product of $\theta^{\mu\nu}$ with ϕ is not covariant. In order to arrive at the covariant T* product, a covariantizing seagull $\tau^{\mu\nu}$ must be added. Hence we are led to consider

$$F_{ij}{}^{\mu\nu}(p,q) = \int d^4x\, d^4y\, e^{iq\cdot x}\, e^{ip\cdot y} \langle 0|T^* \theta^{\mu\nu}(x)\,\phi_i(y)\,\phi_j(0)|0\rangle$$

$$= \int d^4x\, d^4y\, e^{iq\cdot x}\, e^{ip\cdot y} \langle 0|T^{\mu\nu}(x)\,\phi_i(y)\,\phi_j(0)|0\rangle$$

$$+ \tau_{ij}{}^{\mu\nu}(p,q)\ , \tag{7.21}$$

$$F_{ij}(p,q) = \int d^4x\, d^4y\, e^{iq\cdot x}\, e^{ip\cdot y} \langle 0|T\,\theta^{\mu}_{\mu}(x)\,\phi_i(y)\,\phi_j(0)|0\rangle\ . \tag{7.22}$$

In the above i, j label the fields: the labels may be space-time or internal indices. It is assumed that matrix elements of θ^{μ}_{μ} require no seagull. The covariantizing seagull $\tau_{ij}{}^{\mu\nu}$ may be explicitly constructed from the known commutators (7.20) by the method explained in Section 2. Once the seagull is determined, the Ward identity may be derived. We do not present the details here, but merely record the resulting Ward identity [6].

$$q_\mu F_{ij}{}^{\mu\nu}(p,q) = ip^\nu G(p) - i(p+q)^\nu G(p+q)$$

$$+ \frac{i}{2} q_\mu \Sigma_{ii'}{}^{\mu\nu}\, G_{i'j}(p+q) + \frac{i}{2} q_\mu \Sigma_{jj'}{}^{\mu\nu}\, G_{ij'}(p)\ . \tag{7.23}$$

Also a trace identity is obtained.

$$g_{\mu\nu} F_{ij}{}^{\mu\nu}(p,q) = F_{ij}(p,q) - id\,G_{ij}(p+q) - id\,G_{ij}(p)\ . \tag{7.24}$$

The terms in (7.24) additional to F_{ij} arise from the trace of $\tau_{ij}{}^{\mu\nu}$. In Eqs. (7.23) and (7.24) G_{ij} is the renormalized propagator.

$$G_{ij}(p) = \int d^4x\, e^{ip\cdot x} \langle 0|T\,\phi_i(x)\,\phi_j(0)|0\rangle\ . \tag{7.25}$$

The formulas (7.23) and (7.24) contain all the restrictions that the various space-time transformations (Lorentz, scale, and conformal) impose on the propagator. (Had we wished to study the n particle Green's function, we would consider the matrix element of $\theta^{\mu\nu}$ with n fields.) Once a model for scale symmetry breaking, e.g., mass terms is adopted, then one may deduce theorems about $G(p)$. We now show explicitly how these restrictions are contained in Eq. (7.23) and (7.24).

(A) Lorentz transformations. Differentiate (7.23) with respect to q_α and set q to zero. This gives

$$F_{ij}{}^{\alpha\nu}(p,0) = -ig^{\alpha\nu}\, G_{ij}(p) - ip^\nu \frac{\partial}{\partial p_\alpha} G_{ij}(p)$$

$$+ \frac{i}{2}\Sigma_{ii'}{}^{\alpha\nu}\, G_{i'j}(p) + \frac{i}{2}\Sigma_{ij'}{}^{\alpha\nu}\, G_{ij'}(p)\ . \tag{7.26}$$

Since $F_{ij}{}^{\alpha\nu}$ is symmetric in α and ν, we learn from (7.26) that

$$\left(p^\nu \frac{\partial}{\partial p_\alpha} - p^\alpha \frac{\partial}{\partial p_\nu}\right) G_{ij}(p) = \Sigma_{ii'}{}^{\alpha\nu}\, G_{i'j}(p) + \Sigma_{jj'}{}^{\alpha\nu}\, G_{ij'}(p)\ . \tag{7.27}$$

This is the trivial and well known constraint of Lorentz covariance.

(B) Scale transformations. Form the trace of (7.26). We have (suppressing indices)

$$g_{\mu\nu}\, F^{\mu\nu}(p,0) = -4i\, G(p) - ip^\alpha \frac{\partial}{\partial p^\alpha} G(p)\ . \tag{7.28a}$$

Combining (7.28a) with (7.24) at $q = 0$ leaves

$$F(p,0) = i(2d-4)\, G(p) - ip^\alpha \frac{\partial}{\partial p^\alpha} G(p)\ . \tag{7.28b}$$

This provides a constraint on $G(p)$, once $F(p,0)$ is known, i.e., once we have a model for scale-symmetry breaking.

(C) Conformal transformations. Differentiate (7.23) by

$$2\frac{\partial}{\partial q^\alpha}\frac{\partial}{\partial q^\nu} - g_{\alpha\nu}\frac{\partial}{\partial q^\beta}\frac{\partial}{\partial q_\beta}\ ,$$

and set $q = 0$. This gives

$$2 \frac{\partial}{\partial q_\alpha} g_{\mu\nu} F_{ij}{}^{\mu\nu}(p,q) \Big|_{q=0}$$

$$= -8i \frac{\partial}{\partial p_\alpha} G_{ij}(p) - 2i p_\beta \frac{\partial}{\partial p_\beta} \frac{\partial}{\partial p_\alpha} G_{ij}(p)$$

$$= i p^\alpha \frac{\partial}{\partial p^\beta} \frac{\partial}{\partial p_\beta} G_{ij}(p) + 2i \Sigma_{ii'}{}^{\alpha\beta} \frac{\partial}{\partial p^\beta} G_{i'j}(p) . \qquad (7.29a)$$

The left-hand side of (7.29a) may be evaluated from (7.24). After a rearrangement of terms, we are left with

$$2 \frac{\partial}{\partial q_\alpha} F_{ij}(p,q) \Big|_{q=0}$$

$$= i(2d-8) \frac{\partial}{\partial p_\alpha} G_{ij}(p) - 2i p_\beta \frac{\partial}{\partial p_\beta} \frac{\partial}{\partial p_\alpha} G_{ij}(p)$$

$$+ i p^\alpha \frac{\partial}{\partial p^\beta} \frac{\partial}{\partial p_\beta} G_{ij}(p) + 2i \Sigma_{ii'}{}^{\alpha\beta} \frac{\partial}{\partial p^\beta} G_{i'j}(p) . \qquad (7.29b)$$

This equation may be simplified by using (7.27) and (7.28b). First d is eliminated between (7.29b) and (7.28b). This gives

$$2 \frac{\partial}{\partial q_\alpha} F_{ij}(p,q) \Big|_{q=0} - \frac{\partial}{\partial p_\alpha} F_{ij}(p,0)$$

$$= i \frac{\partial}{\partial p^\beta} \left[p^\alpha \frac{\partial}{\partial p_\beta} G_{ij}(p) - p^\beta \frac{\partial}{\partial p_\alpha} G_{ij}(p) + 2 \Sigma_{ii'}{}^{\alpha\beta} G_{i'j}(p) \right] .$$

$$(7.29c)$$

Next we use (7.27)

$$2\frac{\partial}{\partial p_\alpha}F_{ij}(p,q)\bigg|_{q=0} - \frac{\partial}{\partial p_\alpha}F_{ij}(p,0)$$

$$= i\frac{\partial}{\partial p_\beta}\left[\Sigma_{ii'}{}^{\alpha\beta}\,G_{i'j}(p) - \Sigma_{jj'}{}^{\alpha\beta}\,G_{ij'}(p)\right] \; . \qquad (7.29d)$$

Equation (7.29d) determines the constraint on G which follows from a model for conformal-symmetry breaking.

7.4 *False Theorems*

The constraint equations (7.28b) and (7.29d) are, of course, without content as long as F, the symmetry breaking, is not specified. We now show that if a model for F is taken from the canonical result that θ^μ_μ is given only by mass terms; and furthermore, if the canonical value for d is accepted, theorems about G can be deduced, which however are contradicted by explicit calculation. We shall only discuss broken scale invariance; no further reference to conformal transformation will be made.

Consider for definiteness the propagator for a theory of spin zero fields with mass μ and a quartic self-interaction, $\lambda\phi^4$, which is scale invariant. The propagator may be written in the form

$$G(p) = \frac{i}{p^2}g(p^2/\mu^2) \; . \qquad (7.30)$$

We find from (7.28b) that g satisfies

$$\frac{p^2}{2}\,F(p,0) = \frac{p^2}{\mu^2}\,g'(p^2/\mu^2) + (1-d)\,g(p^2/\mu^2) \; . \qquad (7.31)$$

As $\mu^2 \to 0$, one might expect that the left-hand side above vanishes, since F is the matrix element of θ^μ_μ which formally is $\mu^2\phi^2$. On the right-hand side, this is equivalent to $p^2 \to \infty$. Hence we find

$$\lim_{p^2\to\infty} g(p^2/\mu^2) \propto (p^2/\mu^2)^{d-1} \; . \qquad (7.32)$$

Since $d = 1$ for boson fields, we further conclude that the boson propagator goes as $1/p^2$ for large p^2.

This result is manifestly false in perturbation theory where it is known that logarithmic terms are present in the asymptotic domain. Thus we must abandon the steps which lead from the true (by definition) Eq. (7.31) to the false result. Specifically we cannot conclude that $d = 1$ and that F vanishes with the mass.

7.5 *True Theorems*

Detailed calculation in perturbation theory *in lowest non trivial order of the interaction* yield the following conclusions. It remains possible to assert that F vanishes with the mass. However, d changes from its canonical value of 1. To exhibit the change in d, we consider the definition of that object.

$$i\,[D(0), \phi(0)] = i \int d^3x\, x_i\, [\theta^{0i}(0,\mathbf{x}), \phi(0)] = d\,\phi(0)\ . \tag{7.33}$$

The commutator is evaluated by the BJL prescription. Specifically an application of this technique to $F^{\mu\nu}(p,q)$ gives, by definition,

$$\lim_{q_0 \to \infty} \frac{\partial}{\partial q^i} q_0\, F^{0i}(p,q) \bigg|_{\mathbf{q}=0} = i d G(p)\ . \tag{7.34}$$

Hence the true value of d may be computed from the high-energy behavior of $F^{\mu\nu}$. Explicit calculation in lowest order gives

$$d = 1 + c\lambda^2\ , \tag{7.35}$$

where c is a well defined positive numerical constant. Substituting this value of d into Eq. (7.32) (which remains valid to lowest order, since F *does* vanish with the masses), we find

$$\lim_{p^2 \to \infty} g(p^2/\mu^2) \propto (p^2/\mu^2)^{c\lambda^2} \approx 1 + c\lambda^2 \log p^2/\mu^2\ . \tag{7.36}$$

Explicit calculation of the propagator to the same order, verifies (7.36) with precisely the same coefficient. Perturbative calcuations for several models have been performed, and the conclusion is always the same, *in lowest order*: although the scale breaking term vanishes with the masses, the dimension changes, and the resulting theorem about high-energy behavior is verified by comparison with explicit calculation of the propagator [7].

The situation to higher orders has been studied both by explicit calculations and by general analysis of the structure of Feynman graphs [8]. The conclusion is that d continues to migrate from the canonical value. However a new phenomenon sets in: F does not vanish with the masses; i.e., scale symmetry is broken by terms other than masses. This "anomalous", non-canonical scale breaking can be understood in the following way. In calculating matrix elements of $\theta^{\mu\nu}$ it is necessary that they be conserved. This is the requirement of Poincaré covariance. However, in specific calculations these matrix elements are *not* conserved, and conservation is achieved for example, by Pauli-Villars regularization. One defines $\theta_{\text{Reg}}{}^{\mu\nu} = \theta^{\mu\nu} - \theta_{\text{M}}{}^{\mu\nu}$, where $\theta_{\text{M}}{}^{\mu\nu}$ is formed from regulator fields $\bar{\phi}$ carrying mass M. Physical, conserved matrix elements are obtained by letting $M \to \infty$. Consider now the trace of $\theta_{\text{Reg}}{}^{\mu\nu}$, which breaks scale invariance. Evidently we have

$$g_{\mu\nu}\,\theta_{\text{Reg}}{}^{\mu\nu} = \mu^2\phi^2 - M^2\tilde{\phi}^2 \; . \tag{7.37}$$

Thus if matrix elements of $\tilde{\phi}^2$ behave as M^{-2} for large M, the regulator contribution to (7.37) survives, even in the physical limit $M \to \infty$. Specific calculation shows that $\tilde{\phi}^2$ does, indeed, behave in this fashion. Therefore even when μ^2 is zero, $g_{\mu\nu}\,\theta_{\text{Reg}}{}^{\mu\nu}$ does not vanish. (The analogy with the anomalous divergence of the axial current is clear.)

At the present time attention is being directed to the question of determining the precise form of the non-canonical scale-breaking terms. It is believed that they are absent if the coupling constant renormalization factor is finite. In perturbation theory this object is infinite; it is not yet known whether in the complete theory it can be finite. If it becomes established that local quantum field theory does not scale at high energy, one will have to turn to non-Lagrangian models to implement this physically attractive idea.

References

[1] The treatment here follows C. G. Callan, Jr., S. Coleman and R. Jackiw, Ann. Phys. (N. Y.) 59 (1970) 42 and S. Coleman and R. Jackiw, Ann. Phys. (N. Y.) 67 (1971) 552. For a less general discussion, and references to older literature, see G. Mack and A. Salam, Ann. Phys. (N. Y.) 53 (1969) 174.

[2] C. G. Callan, Jr., S. Coleman and R. Jackiw, Ref. 1.

[3] It is perhaps possible that these symmetries are exact on the Lagrangian level; yet a non-trivial mass spectrum is present because the vacuum is degenerate. This example of the Goldstone phenomenon has been studied by A. Salam and J. Strathdee, Phys. Rev. 184 (1969) 1750 and 1760. We shall not be considering this point of view.

[4] C. G. Callan, Jr., S. Coleman and R. Jackiw, Ref. 2; see also F. Gursey, Ann. Phys. (N. Y.) 24 (1963) 211 and R. Penrose, Proc. Roy. Soc. (London) 284A (1965) 159.

[5] This approach has been taken by M. Gell-Mann, Hawaii Summer School lectures (1969), and P. Carruthers, Phys. Rev. D2 (1970) 2265. The main obstacle to rapid advance in this direction seems to be the experimental fact that there are no low-lying scalar mesons.

[6] A detailed derivation of (7.23) and (7.24) is given in the first two papers of Ref. 1.

[7] These calculations are given by S. Coleman and R. Jackiw, Ref. 1. Similar calculations for the $\lambda\phi^4$ theory have been performed by K. Wilson, Phys. Rev. D2 (1970) 1478.

[8] S. Coleman and R. Jackiw, Ref. 1; K. Wilson, Ref. 7; C. G. Callan, Jr., Phys. Rev. D2 (1970) 1541; K. Symanzik, Commun. Math. Phys. 18 (1970) 227.

(A) EXERCISES

Exercise 2.1. By using only the canonical commutation relations (2.1), show that the space-time "charges", P^α, associated with the translation current $\theta_c^{\ \mu\alpha}$, $P^\alpha = \int d^3x\, \theta_c^{\ 0\alpha}(t, \mathbf{x})$, generate the correct transformation on the fields.

$$i\,[P^\alpha, \phi(x)] = \partial^\alpha \phi(x) = \delta_T^\alpha\, \phi(x) \ .$$

Exercise 2.2. Consider Lorentz transformations.

$$\delta_L^{\alpha\beta}\, \phi(x) = (x^\alpha \partial^\beta - x^\beta \partial^\alpha + \Sigma^{\alpha\beta})\, \phi(x) \ .$$

Here $\Sigma^{\alpha\beta}$ is the spin matrix appropriate to the field $\phi(x)$. It specifies the representation of Lorentz group under which $\phi(x)$ transforms. $\Sigma^{\alpha\beta} = 0$ for spin zero bosons; $\Sigma_{(ij)}^{\ \ \alpha\beta} = \frac{1}{2} i \sigma_{ij}^{\ \alpha\beta}$ for spin $\frac{1}{2}$ fermions; *ij* are the Dirac indices appropriate to a 4-component fermion field and $\sigma^{\alpha\beta} = [\gamma^\alpha, \gamma^\beta]/2i$; $\Sigma_{(\mu\nu)}^{\ \ \alpha\beta} = g_\mu^\alpha g_\nu^\beta - g_\nu^\alpha g_\mu^\beta$ for vector bosons ($\mu\nu$ are the space-time indices appropriate to a vector field). Under what conditions, on a translationally invariant $\mathscr{L}$, is the above a symmetry transformation of the theory? Show that the conserved, canonical space-time current appropriate to Lorentz transformation is

$$M_c^{\ \mu\alpha\beta}(x) = x^\alpha \theta_c^{\ \mu\beta}(x) - x^\beta \theta_c^{\ \mu\alpha}(x) + \pi^\mu(x)\Sigma^{\alpha\beta}\, \phi(x) \ .$$

By use of the canonical commutators (2.1), verify that the "charges"

$$M^{\alpha\beta} = \int d^3x\, M_c^{\ 0\alpha\beta}(t, \mathbf{x}) \ ,$$

generate the correct transformations on the fields.

$$i\,[M^{\alpha\beta}, \phi(x)] = \delta_L^{\alpha\beta}\, \phi(x) \ .$$

Exercise 2.3. Consider the Belinfante tensor $\theta_B^{\ \mu\alpha}$ defined in (2.17). Show that the Belinfante tensor is conserved when $\theta_c^{\ \mu\alpha}$ is. Next show that the charge P^α, defined in Exercise 2.1, are the same regardless whether they are constructed from $\theta_B^{\ \mu\alpha}$ or $\theta_c^{\ \mu\alpha}$. Therefore for purposes of describing translations, $\theta_B^{\ \mu\alpha}$ may be used instead of $\theta_c^{\ \mu\alpha}$. $\theta_B^{\ \mu\alpha}$ has additional advantages. With the help of the equations of motion, as well as the condition for Lorentz covariance, derived in Exercise 2.2, show that $\theta_B{}^{\mu\alpha}$ is symmetric in μ and α, while $\theta_c{}^{\mu\alpha}$ has this property only for spinless fields $\Sigma^{\alpha\beta} = 0$.

The Belinfante tensor takes on a significance over the canonical tensor in connection with the space-time current associated with Lorentz transformation. Consider

$$M_B{}^{\mu\alpha\beta}(x) = x^\alpha \theta_B{}^{\mu\beta}(x) - x^\beta \theta_B{}^{\mu\alpha}(x) .$$

Show that $M_B{}^{\mu\alpha\beta}(x)$ is conserved when $M_c{}^{\mu\alpha\beta}(x)$, defined in Exercise 2.2, possesses this property. Also show that $M_B{}^{\mu\alpha\beta}(x)$, leads to the same charges $M^{\alpha\beta}$ as $M_c{}^{\mu\alpha\beta}(x)$ does. Hence $M_B{}^{\mu\alpha\beta}$ can be used instead of $M_c{}^{\mu\alpha\beta}$ as the space-time current for Lorentz transformations; however $M_B{}^{\mu\alpha\beta}$ is much simpler than $M_c{}^{\mu\alpha\beta}$, since the former contains no explicit spin term $\pi^\mu \Sigma^{\alpha\beta}\phi$. The full significance of the Belinfante tensor emerges when one asserts that it is this energy-momentum tensor (rather than any other one) to which gravitons couple in the Einstein theory of gravity, general relativity.

Exercise 2.4. By use of canonical expressions for $\theta_c{}^{00}$ and $J_0{}^a$, as well as the canonical commutators, derive the equal-time commutator between these two operators, given in (2.18). Verify that the answer remains unchanged when $\theta_c{}^{00}$ is replaced by $\theta_B{}^{00}$.

Exercise 2.5. Consider the scalar field Lagrangian.

$$\mathscr{L}(x) = \tfrac{1}{2}\phi^\mu(x)\phi_\mu(x) + gx^2\phi(x) .$$

Construct the energy momentum tensor from the formula (2.11), and show that it is not conserved. Evaluate the commutator $[P^0(t), P^i(t)]$ and verify that it does not vanish. (When translation invariance holds this commutator vanishes.) Construct the Lorentz current for this model, and verify that it is conserved.

Exercise 2.6. Verify that (2.24) is the most general solution of (2.23b). Hint: Multiply (2.23b) by y_i and integrate over $\mathbf{y}$.

Exercise 2.7. Consider the ETC of electromagnetic currents which hold in scalar electrodynamics.

$$[J^0(0,\mathbf{x}), J^0(0)] = 0 ,$$

$$[J^0(0,\mathbf{x}), J^i(0)] = S(0)\,\partial^i\delta(\mathbf{x}) ,$$

$$[J^i(0,\mathbf{x}), J^i(0)] = 0 .$$

Here $S(y)$ is a scalar operator. Rewrite this commutator in a Lorentz covariant, but frame-dependent fashion by introducing the unit time-like vector n^μ. Consider also

$$\mathrm{T}J^\mu(x)J^\nu(0) \equiv \mathrm{T}^{\mu\nu}(x;n) .$$

Show that this is not covariant, and construct a covariantizing seagull $\tau^{\mu\nu}(x;n)$. Determine the seagull by requiring that $\mathrm{T}^{*\mu\nu}(x) = \mathrm{T}^{\mu\nu}(x;n) + \tau^{\mu\nu}(x;n)$ be conserved.

Exercise 3.1. Consider the free boson propagator $D(q) = i/(q^2 - m^2)$. By use of the BJL theorem verify the canonical commutation relations.

$$i[\phi(0,\mathbf{x}), \phi(0)] = i[\dot\phi(0,\mathbf{x}), \dot\phi(0)] = 0 ,$$

$$i[\dot\phi(0,\mathbf{x}), \phi(0)] = \delta(\mathbf{x}) .$$

Next consider the full propagator of renormalized fields. It may be written in the form

$$G(q) = i\int_0^\infty \mathrm{d}a^2 \, \frac{\rho(a^2)}{q^2 - a^2} .$$

The spectral function ρ can be shown to be non-negative. (A renormalized field $\tilde\phi$ is proportionally related to the unrenormalized field ϕ by $\phi = Z^{1/2}\tilde\phi$. In perturbation theory Z is cutoff dependent.) What can you deduce about the vacuum expectation of the canonical commutators in the complete theory?

Exercise 3.2. Consider the vacuum polarization tensor, which (formally) can be written as

$$\mathrm{T}^{*\mu\nu}(q) = \int \mathrm{d}^4x \, e^{iq\cdot x} \langle 0|\mathrm{T}^* J^\mu(x) J^\nu(0)|0\rangle$$

$$= (g^{\mu\nu} q^2 - q^\mu q^\nu)\int_0^\infty \mathrm{d}a^2 \, \frac{\sigma(a^2)}{q^2 - a^2} .$$

It can be shown that $\sigma(a^2)$ has a definite sign. Extract the T product from $\mathrm{T}^{*\mu\nu}(q)$ and calculate $\langle 0|[J^0(0,\mathbf{x}), J^i(0)]|0\rangle$ in terms of $\sigma(a^2)$. (Do not concern yourselves here with problems of convergence; these will be discussed in Exercise 6.1.)

Exercise 4.1. Show that

$$\Delta^{\mu}(a) = i\int \frac{d^4 r}{(2\pi)^4}\left[\frac{r^{\mu}+a^{\mu}}{([r+a]^2-m^2)^2} - \frac{r^{\mu}}{(r^2-m^2)^2}\right]$$

$$= \frac{-a^{\mu}}{32\pi^2} \; .$$

This verifies (4.19b).

Exericse 4.2. Show that

$$\Delta^{\alpha\mu\nu}(p,q\,|\,a)$$

$$= i\int \frac{d^4 r}{(2\pi)^4}\,\mathrm{Tr}\,\gamma^5\gamma^{\alpha}\Big\{[r_\beta\gamma^\beta + a_\beta\gamma^\beta + p_\beta\gamma^\beta - m]^{-1}$$

$$\times\; \gamma^{\mu}[r_\beta\gamma^\beta + a_\beta\gamma^\beta - m]^{-1}\gamma^{\nu}[r_\beta\gamma^\beta + a_\beta\gamma^\beta + q_\beta\gamma^\beta - m]^{-1}$$

$$- [r_\beta\gamma^\beta + p_\beta\gamma^\beta - m]^{-1}\gamma^{\mu}[r_\beta\gamma^\beta - m]^{-1}\gamma^{\nu}[r_\beta\gamma^\beta - q_\beta\gamma^\beta - m]^{-1}\Big\}$$

$$= \frac{-1}{8\pi^2}\,\epsilon^{\alpha\mu\nu\beta}\,a_\beta \; .$$

This verifies (4.21).

Exercise 4.3. Show that

$$i\int \frac{d^4 r}{(2\pi)^4}\,\mathrm{Tr}\,\gamma^5\gamma^{\alpha}\Big\{[r_\beta\gamma^\beta + q_\beta\gamma^\beta - m]^{-1}\gamma^{\nu}[r_\beta\gamma^\beta - p_\beta\gamma^\beta - m]^{-1}$$

$$- [r_\beta\gamma^\beta + p_\beta\gamma^\beta - m]^{-1}\gamma^{\nu}[r_\beta\gamma^\beta - q_\beta\gamma^\beta - m]^{-1}\Big\}$$

$$= \frac{1}{4\pi^2}\,\epsilon^{\alpha\mu\nu\beta}\,p_\mu q_\beta \; .$$

This verifies (4.25c).

Exercise 5.1. Compute canonically the symmetric part of

$$\int d^3x \, [\dot{J}^i(0, \mathbf{x}), J^j(0)] \ ,$$

in the quark-vector gluon model, where

$$i\gamma_\mu \partial^\mu \psi = m\psi - g\gamma^\mu \psi B_\mu \ .$$

Verify that the spin averaged, matrix element of that commutator, between diagonal proton states is of the form

$$A(\delta^{ij}\mathbf{p}^2 - p^i p^j) + B\delta^{ij} \ .$$

Exercise 5.2. Assume that the dispersion relation for T_L needs one subtraction, performed at $\nu = 0$, $(\omega = \infty)$. Show that the ST sum rule reduces in that instance to an uninformative relation between the subtraction term and the ST.

Exercise 6.1. When the spectral function $\sigma(a^2)$, relevant to the vacuum polarization tensor $T^{*\mu\nu}$, does not vanish as $a^2 \to \infty$, the dispersive representation for $T^{*\mu\nu}$ given in Exercise 3.2 will not converge. Assuming that

$$\lim_{a^2 \to \infty} \sigma(a^2) = A \ ,$$

$$\lim_{a^2 \to \infty} a^2 \Big(\sigma(a^2) - A \Big) = B \ ,$$

the following subtracted form for $T^{*\mu\nu}$ may be established

$$T^{*\mu\nu}(q) = \int d^4x \, e^{iq\cdot x} \langle 0 | T^* J^\mu(x) J^\nu(0) | 0 \rangle$$

$$= (g^{\mu\nu} q^2 - q^\mu q^\nu) \left[C + q^2 \int_{4m^2}^{\infty} da^2 \, \frac{\sigma(a^2)}{a^2(q^2 - a^2)} \right] \ .$$

Here C is a subtraction constant, and $4m^2$ is the appropriate threshold for the dispersive integral. Calculate the vacuum expectation value of the $[J^0, J^i]$ ETC, and show that a triple derivative of the δ function is present. Hint: Let $\tilde{\sigma}(a^2) = \sigma(a^2) - A - B/a^2$.

Exercise 6.2. Consider $T(p, q)$, defined by

$$T(p,q) = \int d^4x\, d^4y\, e^{ip\cdot x}\, e^{iq\cdot y} \langle \alpha | TA(x) B(y) C(0) | \beta \rangle ,$$

where α and β are arbitrary states. Show that

$$\lim_{q_0 \to \infty} \lim_{p_0 \to \infty} - p_0 q_0 T(p,q)$$

$$= \int d^3x\, d^3y\, e^{-i\mathbf{p}\cdot\mathbf{x}}\, e^{-i\mathbf{q}\cdot\mathbf{y}} \langle \alpha | [B(0,\mathbf{y}), [A(0,\mathbf{x}), C(0)]] | \beta \rangle .$$

What is the result if the limit is performed on opposite order,

$$\lim_{p_0 \to \infty} \lim_{q_0 \to \infty} - p_0 q_0 T(p,q) \; ?$$

Exercise 7.1. Show that the invariant functions $\widetilde{F}_i$, defined from

$$C^{\mu\nu}(q,p) = \int \frac{d^4x}{(2\pi)^4} e^{iq\cdot x} \langle p | [J^\mu(x), J^\nu(0)] | p \rangle$$

$$= -\left(g^{\mu\nu} - \frac{q^\mu q^\nu}{q^2} \right) \widetilde{F}_1$$

$$+ \left(p^\mu - q^\mu \frac{p\cdot q}{q^2} \right) \left(p^\nu - q^\nu \frac{p\cdot q}{q^2} \right) \frac{\widetilde{F}_2}{p\cdot q} ,$$

are dimensionless. The state $|p\rangle$ is normalized covariantly.

Exercise 7.2. Compute $i[D(t), P^\mu]$ and $i[D(t), M^{\mu\nu}]$, where $D(t)$ is the dilatation "charge".

Exercise 7.3. By use of canonical ETC, show that

$$i[D(t), \phi(x)] = \delta_D \phi(x) ,$$

$$i[K^\alpha(t), \phi(x)] = \delta_C{}^\alpha \phi(x) .$$

Here $D(t)$ and $K^\alpha(t)$ are dilatation and conformal "charges" respectively. Hint: Use the canonical formulas for the currents.

Exercise 7.4. Consider a scalar field Lagrangian $\mathscr{L}$ depending on ϕ and $\partial^\mu\phi$. What is the most general form for $\mathscr{L}$ which is scale invariant? What is the most general form that is conformly invariant? Derive $\theta_c{}^{\mu\nu}$ for this conformly invariant theory. Since this is a spinless theory, $\theta_c{}^{\mu\nu} = \theta_B{}^{\mu\nu}$. What is $\theta^{\mu\nu}$?

(B) SOLUTIONS

Exercise 2.1. It is clear that the argument of $\phi(x)$ is inessential. Hence we consider

$$i[P^\alpha, \phi(0)] = \int d^3y \, i[\theta_c{}^{0\alpha}(0,y), \phi(0)] \; .$$

(a) $\alpha = i$

$$\begin{aligned} i[P^i, \phi(0)] &= \int d^3y \, i[\pi^0(0,\mathbf{y})\,\phi^i(0,\mathbf{y}), \phi(0)] \\ &= \int d^3y \, i[\pi^0(0,\mathbf{y})\,\phi(0)]\;\phi^i(0,\mathbf{y}) \\ &= \int d^3y \, \delta(\mathbf{y})\,\phi^i(0,\mathbf{y}) \\ &= \phi^i(0) \; , \end{aligned}$$

(b) $\alpha = 0$

$$\begin{aligned} i[P^0, \phi(0)] &= \int d^3y \, i[\pi^0(0,\mathbf{y})\phi^0(0,\mathbf{y}) - \mathscr{L}(0,\mathbf{y}), \phi(0)] \\ &= \phi^0(0) + \int d^3y \left(\pi^0(0,\mathbf{y})\, i[\phi^0(0,\mathbf{y}), \phi(0)] \right. \\ &\qquad\qquad \left. - \, i[\mathscr{L}(0,\mathbf{y}), \phi(0)]\right) \; . \end{aligned}$$

We now show that the second term in the above is zero. To do this, we introduce the technique of functional differential with respect to $\phi(0)$ and $\pi^0(0)$. By definition,

$$\frac{\delta'\phi(0,\mathbf{x})}{\delta'\phi(0)} = \delta(\mathbf{x}) \; , \qquad \frac{\delta'\phi^i(0,\mathbf{x})}{\delta'\phi(0)} = \partial^i\,\delta(\mathbf{x}) \; ,$$

$$\frac{\delta'\pi^0(0,\mathbf{x})}{\delta'\phi(0)} = 0 \; , \qquad \frac{\delta'\phi(0,\mathbf{x})}{\delta'\pi^0(0)} = 0 \; ,$$

$$\frac{\delta'\partial^i\phi(0,\mathbf{x})}{\delta'\pi^0(0)} = 0 \; , \qquad \frac{\delta'\pi^0(0,\mathbf{x})}{\delta'\pi^0(0)} = \delta(\mathbf{x}) \; .$$

The quantity ϕ^0 is considered to be a functional of both π^0 and ϕ. For any functions F which depends on both ϕ and ϕ^μ, the chain rule holds

$$\frac{\delta' F(0,\mathbf{x})}{\delta'\phi(0)} = \frac{\delta F(0,\mathbf{x})}{\delta\phi}\,\delta(\mathbf{x}) + \frac{\delta F(0,\mathbf{x})}{\delta\phi^i}\,\partial^i\delta(\mathbf{x})$$

$$+ \frac{\delta F(0,\mathbf{x})}{\delta\phi^0}\,\frac{\delta'\phi^0(0,\mathbf{x})}{\delta'\phi(0)}$$

$$\frac{\delta' F(0,\mathbf{y})}{\delta'\pi^0(0)} = \frac{\delta F(0,\mathbf{x})}{\delta\phi^0}\,\frac{\delta'\phi^0(0,\mathbf{x})}{\delta'\pi^0(0)} \; .$$

Here δ indicates the ordinary variational derivative; δ' is the functional derivative. Applying this formalism, gives

$$\pi^0(0,\mathbf{y})\,i[\phi^0(0,\mathbf{y}),\phi(0)] - i[\mathscr{L}(0,\mathbf{y}),\phi(0)]$$

$$= \pi^0(0,\mathbf{y})\frac{\delta'\phi^0(0,\mathbf{y})}{\delta'\pi^0(0)} - \frac{\delta'\mathscr{L}(0,\mathbf{y})}{\delta'\pi^0(0)}$$

$$= \pi^0(0,\mathbf{y})\frac{\delta'\phi^0(0,\mathbf{y})}{\delta'\pi^0(0)} - \frac{\delta\mathscr{L}(0,\mathbf{y})}{\delta\phi^0}\,\frac{\delta'\phi^0(0,\mathbf{y})}{\delta'\pi^0(0)} = 0 \; .$$

The definition of π^0 was used: $\pi^0 = \delta\mathscr{L}/\delta\phi^0$.

(c) Note that we have established the stronger result

$$i[\theta_c^{0\alpha}(t,\mathbf{y}),\phi(t,\mathbf{x})] = \phi^\alpha(x)\,\delta(\mathbf{x}-\mathbf{y}) \; .$$

Exercise 2.2.

(a) $$\delta_L^{\alpha\beta}\mathcal{L} = \pi_\mu \delta_L^{\alpha\beta}\phi^\mu + \frac{\delta\mathcal{L}}{\delta\phi}\delta_L^{\alpha\beta}\phi$$

$$= \pi_\mu \partial^\mu \delta_L^{\alpha\beta}\phi + \frac{\delta\mathcal{L}}{\delta\phi}\delta_L^{\alpha\beta}\phi$$

$$= \pi_\mu \left[g^{\mu\alpha}\phi^\beta - g^{\mu\beta}\phi^\alpha + (x^\alpha\partial^\beta - x^\beta\partial^\alpha)\phi^\mu + \Sigma^{\alpha\beta}\phi^\mu\right]$$

$$+ \frac{\delta\mathcal{L}}{\delta\phi}(x^\alpha\partial^\beta\phi - x^\beta\partial^\alpha\phi) + \frac{\delta\mathcal{L}}{\delta\phi}\Sigma^{\alpha\beta}\phi \ .$$

By translation invariance, the above may be written as

$$\delta^{\alpha\beta}\mathcal{L} = x^\alpha\partial^\beta\mathcal{L} - x^\beta\partial^\alpha\mathcal{L} + \pi_\mu \Sigma^{\alpha\beta}\phi^\mu + \frac{\delta\mathcal{L}}{\delta\phi}\Sigma^{\alpha\beta}\phi + \pi^\alpha\phi^\beta$$

$$- \pi^\beta\phi^\alpha \ ,$$

$$= \partial_\mu [g^{\mu\beta}x^\alpha\mathcal{L} - g^{\mu\alpha}x^\beta\mathcal{L}] + \pi_\mu \Sigma^{\alpha\beta}\phi^\mu + \frac{\delta\mathcal{L}}{\delta\phi}\Sigma^{\alpha\beta}$$

$$+ \pi^\alpha\phi^\beta - \pi^\beta\phi^\alpha \ .$$

Hence the condition for Lorentz covariance of the theory is

$$\pi_\mu \Sigma^{\alpha\beta}\phi^\mu + \frac{\delta\mathcal{L}}{\delta\phi}\Sigma^{\alpha\beta}\phi = \pi^\beta\phi^\alpha - \pi^\alpha\phi^\beta \ .$$

(b) In the notation of Section 2,

$$\Lambda^{\mu\alpha\beta} = x^\alpha g^{\mu\beta}\mathcal{L} - x^\beta g^{\mu\alpha}\mathcal{L} \ ,$$

and the canonical conserved current is

$$M_c^{\mu\alpha\beta} = \pi^\mu \delta_L^{\alpha\beta}\phi - \Lambda^{\mu\alpha\beta}$$

$$= \pi^\mu (x^\alpha\phi^\beta - x^\beta\phi^\alpha + \Sigma^{\alpha\beta}\phi) - x^\alpha g^{\mu\beta}\mathcal{L} - x^\beta g^{\mu\alpha}\mathcal{L}$$

$$= x^\alpha\theta_c^{\ \mu\beta} - x^\beta\theta_c^{\ \mu\alpha} + \pi^\mu\Sigma^{\alpha\beta}\phi \ .$$

(c) Consider

$$i[M_c^{0\alpha\beta}(t,\mathbf{y}),\phi(t,\mathbf{x})]$$

$$= iy^\alpha[\theta_c^{\ 0\beta}(t,\mathbf{y}),\phi(t,\mathbf{x})] - iy^\beta[\theta_c^{\ 0\alpha}(t,\mathbf{y}),\phi(t,\mathbf{x})]$$

$$+ \ i[\pi^0(t,\mathbf{y})\Sigma^{\alpha\beta}\phi(t,\mathbf{y}),\phi(t,\mathbf{x})] \ .$$

The commutators with $\theta_c^{\ 0\alpha}$ are evaluated in Exercise 2.1. Hence

$$i[M_c^{0\alpha\beta}(t,\mathbf{y}),\phi(t,\mathbf{x})]$$

$$= (x^\alpha\partial^\beta\phi(x) - x^\beta\partial^\alpha\phi(x) + \Sigma^{\alpha\beta}\phi(x)\,\delta(\mathbf{x}-\mathbf{y}))$$

$$= \delta_L^{\alpha\beta}\phi(x)\,\delta(\mathbf{x}-\mathbf{y}) \ .$$

The desired result now follows.

Exercise 2.3.

(a) $$\theta_B^{\ \mu\alpha} = \theta_c^{\ \mu\alpha} + \tfrac{1}{2}\partial_\lambda X^{\lambda\mu\alpha} \ ,$$

$$X^{\lambda\mu\alpha} = -X^{\mu\lambda\alpha} = \pi^\lambda\Sigma^{\mu\alpha}\phi - \pi^\mu\Sigma^{\lambda\alpha}\phi - \pi^\alpha\Sigma^{\lambda\mu}\phi \ .$$

Since $X^{\lambda\mu\alpha}$ is explicitly anti-symmetric in $\lambda\mu$, $\partial_\lambda X^{\lambda\mu\alpha}$ does not contribute to the divergence of $\theta_B^{\ \mu\alpha}$ in μ, nor does it contribute to the charges $\int d^3x\,\theta_B^{\ 0\alpha}(x)$.

(b) $$\theta_B^{\ \mu\alpha} = \pi^\mu\phi^\alpha - g^{\mu\alpha}\mathcal{L}$$

$$+ \ \tfrac{1}{2}\partial_\lambda[\pi^\lambda\Sigma^{\mu\alpha}\phi - \pi^\mu\Sigma^{\lambda\alpha}\phi - \pi^\alpha\Sigma^{\lambda\mu}\phi] \ .$$

The only terms that are not explicitly symmetric in $\mu\alpha$ are

$$\pi^\mu\phi^\alpha + \frac{1}{2}\partial_\lambda[\pi^\lambda\Sigma^{\mu\alpha}\phi]$$

$$= \pi^\mu\phi^\alpha + \frac{1}{2}\partial_\lambda\pi^\lambda\,\Sigma^{\mu\alpha}\phi + \frac{1}{2}\pi^\lambda\,\Sigma^{\mu\alpha}\phi_\lambda \ ,$$

$$= \pi^\mu\phi^\alpha + \frac{1}{2}\frac{\delta\mathcal{L}}{\delta\phi}\,\Sigma^{\mu\alpha}\phi + \frac{1}{2}\pi^\lambda\,\Sigma^{\mu\alpha}\phi_\lambda \ .$$

The equations of motion have been used. Next use the constraint on $\mathscr{L}$ imposed by Lorentz covariance; see Exercise 2.2. We have

$$\pi^\mu \phi^\alpha + \tfrac{1}{2}\partial_\lambda[\pi^\lambda \Sigma^{\mu\alpha}\phi] = \pi^\mu\phi^\alpha - \tfrac{1}{2}\pi^\mu\phi^\alpha + \tfrac{1}{2}\pi^\alpha\phi^\mu$$
$$= \tfrac{1}{2}\pi^\mu\phi^\alpha + \tfrac{1}{2}\pi^\alpha\phi^\mu .$$

(c) $$M_{\mathrm{B}}{}^{\mu\alpha\beta} = x^\alpha \theta_{\mathrm{B}}{}^{\mu\beta} - x^\beta \theta_{\mathrm{B}}{}^{\mu\alpha}$$
$$= x^\alpha\theta_{\mathrm{c}}{}^{\mu\beta} - x^\beta\theta_{\mathrm{c}}{}^{\mu\alpha} + x^\alpha \tfrac{1}{2}\partial_\lambda X^{\lambda\mu\beta} - x^\beta\tfrac{1}{2}\partial_\lambda X^{\lambda\mu\alpha}$$
$$= M_{\mathrm{c}}{}^{\mu\alpha\beta} + \tfrac{1}{2}\partial_\lambda[X^{\lambda\mu\beta}x^\alpha - X^{\lambda\mu\alpha}x^\beta]$$
$$+ \tfrac{1}{2}X^{\beta\mu\alpha} - \tfrac{1}{2}X^{\alpha\mu\beta} - \pi^\mu\Sigma^{\alpha\beta}\phi .$$

From the explicit form for $X^{\lambda\mu\alpha}$, we see that the last three terms cancel. The total derivative term is the divergence in λ of a tensor which is anti-symmetric in $\mu\lambda$; hence that object does not contribute to the divergence in μ, nor to charges.

Exercise 2.4.

(a) $$i[\theta_{\mathrm{c}}{}^{00}(t,\mathbf{x}), J_0{}^a(t,\mathbf{y})] = i[\theta_{\mathrm{c}}{}^{00}(t,\mathbf{x}), \pi^0(t,\mathbf{y})T^a\phi(t,\mathbf{y})]$$
$$= \pi^0(t,\mathbf{y})T^a\, i[\theta_{\mathrm{c}}{}^{00}(t,\mathbf{x}), \phi(t,\mathbf{y})]$$
$$+ i[\theta_{\mathrm{c}}{}^{00}(t,\mathbf{x}), \pi^0(t,\mathbf{y})]\, T^a\phi(t,\mathbf{y}) .$$

The first commutator is evaluated as in Exercise 2.1. The second is expressed in terms of functional derivatives. The result is

$$i[\theta_{\mathrm{c}}{}^{00}(t,\mathbf{x}), J_0{}^a(t,\mathbf{y})] = \pi^0(t,\mathbf{y})T^a\phi^0(t,\mathbf{y})\,\delta(\mathbf{x}-\mathbf{y})$$
$$- \frac{\delta'\theta_{\mathrm{c}}{}^{00}(t,\mathbf{x})}{\delta'\phi(t,\mathbf{y})}\, T^a\phi(t,\mathbf{y}) .$$

The functional derivative is now calculated from the formula for $\theta_{\mathrm{c}}{}^{00}(t,\mathbf{x})$.

$$\frac{\delta'\theta_{\mathrm{c}}{}^{00}(t,\mathbf{x})}{\delta'\phi(t,\mathbf{y})} = \pi^0(t,\mathbf{x})\frac{\delta'\phi^0(t,\mathbf{x})}{\delta'\phi(t,\mathbf{y})} - \frac{\delta'\mathscr{L}(t,\mathbf{x})}{\delta'\phi(t,\mathbf{y})} ,$$

$$\frac{\delta'\mathscr{L}(t,\mathbf{x})}{\delta'\phi(t,\mathbf{y})} = \frac{\delta\mathscr{L}(t,\mathbf{x})}{\delta\phi}\delta(\mathbf{x}-\mathbf{y}) + \frac{\delta\mathscr{L}(t,\mathbf{x})}{\delta\phi^0}\frac{\delta'\phi^0(t,\mathbf{x})}{\delta'\phi(t,\mathbf{y})}$$

$$+ \frac{\delta\mathscr{L}(t,\mathbf{x})}{\delta\phi^i}\partial^i\delta(\mathbf{x}-\mathbf{y})$$

$$= \partial_\lambda\pi^\lambda(t,\mathbf{x})\delta(\mathbf{x}-\mathbf{y}) + \pi^0(t,\mathbf{x})\frac{\delta'\phi^0(t,\mathbf{x})}{\delta'\phi(t,\mathbf{y})}$$

$$+ \pi_i(t,\mathbf{x})\partial^i\delta(\mathbf{x}-\mathbf{y}) \ .$$

We have used the equations of motion and the definition of π^μ. It now follows that

$$\frac{\delta'\theta_c^{00}(t,\mathbf{x})}{\delta'\phi(t,\mathbf{y})} = -\partial_\lambda\pi^\lambda(t,\mathbf{x})\delta(\mathbf{x}-\mathbf{y}) - \pi_i(t,\mathbf{x})\partial^i\delta(\mathbf{x}-\mathbf{y}) \ ,$$

$$i[\theta_c^{00}(t,\mathbf{x}), J_0{}^a(t,\mathbf{y})]$$

$$= \pi^0(t,\mathbf{x})T^a\phi^0(t,\mathbf{x})\delta(\mathbf{x}-\mathbf{y}) + \partial_\lambda\pi^\lambda(t,\mathbf{x})T^a\phi(t,\mathbf{x})\delta(\mathbf{x}-\mathbf{y})$$

$$+ \pi^i(t,\mathbf{x})T^a\phi(t,\mathbf{y})\partial_i\delta(\mathbf{x}-\mathbf{y})$$

$$= \partial_0[\pi^0(t,\mathbf{x})T^a\phi(t,\mathbf{x})]\delta(\mathbf{x}-\mathbf{y})$$

$$+ \partial_i[\pi^i(t,\mathbf{x})T^a\phi(t,\mathbf{x})]\delta(\mathbf{x}-\mathbf{y})$$

$$+ \pi^i(t,\mathbf{x})T^a\phi(t,\mathbf{x})\partial_i\delta(\mathbf{x}-\mathbf{y})$$

$$= \partial^\mu J_\mu{}^a(x)\delta(\mathbf{x}-\mathbf{y}) + J_i^a(x)\partial^i\delta(\mathbf{x}-\mathbf{y}) \ .$$

In the last formula we have used the definition for the current.

$$J_\mu{}^a = \pi_\mu T^a\phi \ .$$

(b) The difference between θ_B^{00} and θ_c^{00} is

$$\tfrac{1}{2}\partial_\lambda X^{\lambda 00} = \tfrac{1}{2}\partial_i X^{i00} = \partial_i[\pi^0 \Sigma^{0i}\phi] \ .$$

Since Σ^{0i} and T^a commute, (they operate in different spaces) $J_0{}^a$ commutes with $\tfrac{1}{2}\partial_i X^{i00}$.

Exercise 2.5.

(a)
$$\begin{aligned}\theta_c^{\mu\nu} &= \pi^\mu \phi^\nu - g^{\mu\nu}\mathscr{L}\\ &= \phi^\mu \phi^\nu - g^{\mu\nu}\mathscr{L}\end{aligned}$$

$$\begin{aligned}\partial_\mu \theta_c^{\mu\nu} &= \partial_\mu \phi^\mu \phi^\nu + \phi^\mu \partial_\mu \phi^\nu - \partial^\nu \mathscr{L}\\ &= gx^2 \phi^\nu + \phi^\mu \partial_\mu \phi^\nu - \phi^\mu \partial^\nu \phi_\mu - 2gx^\nu \phi - gx^2 \phi^\nu\\ &= -\ 2gx^\nu \phi \ .\end{aligned}$$

We have used the equation of motion: $\partial_\mu \phi^\mu = gx^2$.

(b)
$$\begin{aligned}i[P^0(t), P^i(t)] &= \int d^3x\, d^3y\, i[\theta_c^{00}(t,\mathbf{x}), \theta_c^{0i}(t,\mathbf{y})]\\ &= \int d^3x\, d^3y\, i[\tfrac{1}{2}\phi^0(t,\mathbf{x})\phi^0(t,\mathbf{x}) - \tfrac{1}{2}\phi^j(t,\mathbf{x})\\ &\quad \times \phi_j(t,\mathbf{x}) - gx^2\phi(t,\mathbf{x}), \phi^0(t,\mathbf{y})\phi^i(t,\mathbf{y})]\\ &= \int d^3x\, d^3y \Big\{\phi^0(t,\mathbf{x})\phi^0(t,\mathbf{y})\frac{\partial}{\partial y_i}\delta(\mathbf{x}-\mathbf{y})\\ &\quad + \phi^j(t,\mathbf{x})\phi^i(t,\mathbf{y})\frac{\partial}{\partial x^j}\delta(\mathbf{x}-\mathbf{y})\\ &\quad + gx^2\phi^i(t,\mathbf{x})\delta(\mathbf{x}-\mathbf{y})\Big\}\\ &= \int d^3x \big\{-\phi^0(t,\mathbf{x})\partial^i\phi^0(t,\mathbf{x}) + \phi^j(t,\mathbf{x})\partial_j\phi^i(t,\mathbf{x})\\ &\quad + gx^2\phi^i(t,\mathbf{x})\big\}\end{aligned}$$

$$= -\tfrac{1}{2}\int d^3x\,\partial^i\left\{\phi^0(t,\mathbf{x})\phi^0(t,\mathbf{x}) - \phi^j(t,\mathbf{x})\phi_j(t,\mathbf{x})\right.$$

$$\left. - 2gx^2\,\phi(t,\mathbf{x})\right\} - g\int d^3x\,x^i\,\phi(t,\mathbf{x})\ .$$

The first term is a surface integral; it may be dropped. We are left with

$$i[P^0(t), P^i(t)] = -g\int d^3x\,x^i\,\phi(t,\mathbf{x})\ .$$

(c) Since $\Sigma^{\alpha\beta}$ is zero for spin zero fields,

$$M_c^{\ \mu\alpha\beta}(x) = x^\alpha\theta_c^{\ \mu\beta} - x^\beta\theta_c^{\ \mu\alpha}$$

$$\partial_\mu M_c^{\ \alpha\beta}(x) = \theta_c^{\ \alpha\beta} + x^\alpha\partial_\mu\theta_c^{\ \alpha\beta} - \theta_c^{\ \beta\alpha} - x^\beta\partial_\mu\theta_c^{\ \mu\alpha}$$

$$= x^\alpha\partial_\mu\theta_c^{\ \mu\beta} - x^\beta\partial_\mu\theta_c^{\ \mu\alpha}\ .$$

We have used the symmetry of $\theta_c^{\ \mu\nu}$. From (a) we have

$$\partial_\mu M_c^{\ \mu\alpha\beta}(x) = 2g(x^\beta x^\alpha - x^\alpha x^\beta) = 0\ .$$

Exercise 2.6.

$$[J_0^{\ b}(t,\mathbf{y}), J_k^a(t,\mathbf{z})]\,\partial^k\,\delta(\mathbf{x}-\mathbf{z}) + [J_0^{\ a}(t,\mathbf{x}), J_k^b(t,\mathbf{z})]\,\partial^k\,\delta(\mathbf{z}-\mathbf{y})$$

$$= -f_{abc}\,\delta(\mathbf{x}-\mathbf{y})\,J_k^c(t,\mathbf{z})\,\partial^k\,\delta(\mathbf{z}-\mathbf{x})\ .$$

Multiplying by y_i and integrating over $\mathbf{y}$ gives

$$[J_0^{\ a}(t,\mathbf{x}), J_i^b(t,\mathbf{z})]$$

$$= -f_{abc}\,x_i\,J_k^c(t,\mathbf{z})\,\partial^k\,\delta(\mathbf{z}-\mathbf{x})$$

$$- \int d^3y\,y_i\,[J_0^{\ b}(t,\mathbf{y}), J_k^a(t,\mathbf{z})]\,\partial^k\,\delta(\mathbf{x}-\mathbf{z})$$

$$= -f_{abc}\,J_i^c(t,\mathbf{z})\,\delta(\mathbf{z}-\mathbf{x}) - f_{abc}\,z_i\,J_k^c(t,\mathbf{z})\,\partial^k\,\delta(\mathbf{z}-\mathbf{x})$$

$$- \int d^3y\,y_i\,[J_0^{\ b}(t,\mathbf{y}), J_k^a(t,\mathbf{z})]\,\partial^k\,\delta(\mathbf{x}-\mathbf{z})\ .$$

Clearly this is of the form

$$[J_0{}^a(t,\mathbf{x}), J_i^b(t,\mathbf{z})]$$
$$= - f_{abc}\, J_i^c(t,\mathbf{z})\,\delta(\mathbf{z}-\mathbf{x}) + S_{ij}^{ab}(t,\mathbf{z})\,\partial^j\,\delta(\mathbf{y}-\mathbf{z}) \; .$$

Reinserting this solution into the starting equation, we have

$$f_{abc}\, J_k^c(t,\mathbf{z})\,\delta(\mathbf{y}-\mathbf{z})\,\partial^k\,\delta(\mathbf{x}-\mathbf{z}) + S_{kl}^{ba}(t,\mathbf{z})\,\partial^l\,\delta(\mathbf{y}-\mathbf{z})\,\partial^k\,\delta(\mathbf{x}-\mathbf{z})$$
$$- f_{abc}\, J_k^c(t,\mathbf{z})\,\delta(\mathbf{x}-\mathbf{z})\,\partial^k\,\delta(\mathbf{z}-\mathbf{y}) + S_{kl}^{ab}(t,\mathbf{z})\,\partial^l\,\delta(\mathbf{x}-\mathbf{z})\,\partial^k\,\delta(\mathbf{z}-\mathbf{y})$$
$$= - f_{abc}\, J_k^c(t,\mathbf{z})\,\delta(\mathbf{x}-\mathbf{y})\,\partial^k\,\delta(\mathbf{z}-\mathbf{x}) \; .$$

Since

$$-\delta(\mathbf{y}-\mathbf{z})\,\partial^k\,\delta(\mathbf{x}-\mathbf{z}) + \delta(\mathbf{x}-\mathbf{z})\,\partial^k\,\delta(\mathbf{z}-\mathbf{y}) = \delta(\mathbf{x}-\mathbf{y})\,\partial^k\,\delta(\mathbf{z}-\mathbf{x}) \, ,$$

we are left with

$$S_{kl}^{ba}(t,\mathbf{z})\,\partial^l\,\delta(\mathbf{y}-\mathbf{z})\,\partial^k\,\delta(\mathbf{x}-\mathbf{z}) + S_{lk}^{ab}(t,\mathbf{z})\,\partial^k\,\delta(\mathbf{x}-\mathbf{z})\,\partial^l\,\delta(\mathbf{z}-\mathbf{y})$$
$$= 0 \, ,$$

or

$$S_{kl}^{ba} = S_{lk}^{ab} \; .$$

Exercise 2.7. The covariant, but frame dependent formula, for the ETC is

$$[J^\mu(x), J^\nu(0)]\,\delta(x\cdot n)$$
$$= S(0)\,(n^\mu g^{\nu\alpha} + n^\nu g^{\mu\alpha})\,P_{\alpha\beta}\,\partial^\beta\,\delta^4(x) - \delta^4(x)\,n^\nu P^{\mu\alpha}\,\partial_\alpha S(0) \, .$$

Since there is a ST, $S^{\mu\nu\alpha}(n) = (n^\mu g^{\nu\alpha} + n^\nu g^{\mu\alpha})S(0)$, the T product is not covariant. Note that

$$S^{\mu\nu\alpha}(n) = S(0)\,\frac{\delta}{\delta n^\alpha}(n^\mu n^\nu) \; .$$

Hence a covariantizing seagull is $\tau^{\mu\nu}(x;n) = S(0) n^\mu n^\nu \delta^4(x)$, and

$$\mathrm{T}^{*\mu\nu}(x) = \mathrm{T}J^\mu(x)\,J^\nu(0) + n^\mu n^\nu S(0)\,\delta^4(x) + \tau^{\mu\nu}(x) \, ,$$

is covariant. The covariant seagull $\tau^{\mu\nu}(x)$ can be chosen to satisfy conservation of $\mathrm{T}^{*\mu\nu}(x)$. We have

$$\partial_\mu \mathrm{T}^{*\mu\nu}(x) = 0 = \delta(x\cdot n)n_\mu[J^\mu(x), J^\nu(0)] + n^\mu n^\nu S(0)\partial_\mu \delta^4(x)$$
$$+ \partial_\mu \tau^{\mu\nu}(x)$$

$$-\partial_\mu \tau^{\mu\nu}(x) = S(0)P^{\nu\beta}\partial_\beta \delta^4(x) + S(0)n^\nu n^\beta \partial_\beta \delta^4(x)$$
$$= S(0)\partial^\nu \delta^4(x) .$$

Therefore the choice $\tau^{\mu\nu}(x) = -g^{\mu\nu} S(0)\delta^4(x)$ assures conservation.

$$\mathrm{T}^{*\mu\nu}(x) = \mathrm{T}^{\mu\nu}(x) - P^{\mu\nu} S(0)\delta^4(x) .$$

Note that the seagull is present only in the ij component of $\mathrm{T}^{*\mu\nu}$.

Exercise 3.1.

(a) $$D(q) = \int d^4x\, e^{iq\cdot x} \langle 0|\mathrm{T}\phi(x)\phi(0)|0\rangle .$$

According to the BJL theorem and the canonical ETC, we have

$$\lim_{q_0\to\infty} q_0 D(q) = i\int d^3x\, e^{-i\mathbf{q}\cdot\mathbf{x}} \langle 0|[\phi(0,\mathbf{x}),\phi(0)]|0\rangle$$
$$= 0 ,$$

$$\lim_{q_0\to\infty} q_0^2 D(q) = -\int d^3x\, e^{-i\mathbf{q}\cdot\mathbf{x}} \langle 0|[\dot\phi(0,\mathbf{x}),\phi(0)]|0\rangle$$
$$= i ,$$

$$\lim_{q_0\to\infty} q_0^3\left[D(q) - \frac{i}{q_0^2}\right] = -i\int d^3x\, e^{-i\mathbf{q}\cdot\mathbf{x}} \langle 0|[\dot\phi(0,x),\phi(0)]|0\rangle$$

$$= -i\int d^3x\, e^{-i\mathbf{q}\cdot\mathbf{x}} \Big\{\partial_0 \langle 0|[\dot\phi(0,x),\phi(0)]|0\rangle$$
$$- \langle 0|[\dot\phi(0,x),\dot\phi(0)]|0\rangle\Big\} = 0 .$$

Since the explicit formula for $D(q)$ has the expansion $D(q) = (i/q_0^2) + O(1/q_0^4)$, all the BJL limits are satisfied.

(b) $G(q)$ formally has the expansion

$$G(q) = \frac{i}{q_0^2}\int_0^\infty da^2\, \rho(a^2) + 0\left(\frac{i}{q_0^4}\right).$$

Hence we deduce that

$$\langle 0 | [\tilde{\phi}(0, \mathbf{x}), \tilde{\phi}(0)] | 0 \rangle = 0 ,$$

$$\langle 0 | [\dot{\tilde{\phi}}(0, \mathbf{x}), \dot{\tilde{\phi}}(0)] | 0 \rangle = 0 ,$$

$$i \langle 0 | [\dot{\tilde{\phi}}(0, \mathbf{x}), \phi(0)] | 0 \rangle = \int_0^\infty da^2\, \rho(a^2)\, \delta(\mathbf{x}) .$$

If $\int_0^\infty da^2\, \rho(a^2)$ is divergent, as it is in perturbation theory, one cannot properly be speaking of the ETC for renormalized fields. On the other hand, if that quantity is finite, we find that

$$i \langle 0 | [\dot{\phi}(0, \mathbf{x}), \phi(0)] | 0 \rangle = \delta(\mathbf{x}) ,$$

$$\tilde{\phi} = Z^{-1/2} \phi ,$$

$$Z^{-1} = \int_0^\infty da^2\, \rho(a^2) .$$

$\rho(a^2)$ is non-negative. It has a contribution from the single particle state given by $\delta(a^2 - m^2)$. Hence $\rho(a^2) = \delta(a^2 - m^2) + \tilde{\rho}(a^2)$, $\tilde{\rho}(a^2) \geqslant 0$. Therefore

$$0 < Z^{-1} \leqslant 1 .$$

Exercise 3.2.

$$T^{*\mu\nu}(q) = (g^{\mu\nu} q^2 - q^\mu q^\nu) \int \frac{da^2\, \sigma(a^2)}{q^2 - a^2} ,$$

$$T^{*00}(q) = -\mathbf{q}^2 \int \frac{da^2\, \sigma(a^2)}{q^2 - a^2} \qquad T^{*00}(q) \xrightarrow[q_0 \to \infty]{} 0 ,$$

$$T^{*0i}(q) = -q^0 q^i \int \frac{da^2\, \sigma(a^2)}{q^2 - a^2} \qquad T^{*0i}(q) \xrightarrow[q_0 \to \infty]{} 0 ,$$

$$T^{*ij}(q) = (g^{ij} q^2 - q^i q^j) \int \frac{da^2\, \sigma(a^2)}{q^2 - a^2} \qquad T^{*ij}(q) \xrightarrow[q_0 \to \infty]{} g^{ij} \int da^2\, \sigma(a^2) .$$

Hence the seagull is only in the *ij* component. Therefore

$$T^{\mu\nu}(q) = T^{*\mu\nu}(q) - P^{\mu\nu} \int da^2\, \sigma(a^2) .$$

Finally

$$\lim_{q_0 \to \infty} q_0 T^{0i}(q) = i \int d^3x\, e^{-i\mathbf{q}\cdot\mathbf{x}} \langle 0 | [J^0(0,\mathbf{x}), J^i(0)] | 0 \rangle$$

$$= \lim_{q_0 \to \infty} -q_0^2 q^i \int \frac{da^2\, \sigma(a^2)}{q^2 - a^2} = -q^i \int da^2\, \sigma(a^2) .$$

This shows that

$$\langle 0 | [J^0(0,x), J^i(0)] | 0 \rangle = i \partial^i \delta(\mathbf{x}) \int da^2\, \sigma(a^2) .$$

Since $\sigma(a^2)$ has definite sign, the vacuum expectation value of the ST is non-vanishing. This provides an alternate argument for the necessary existence of ST. Note, however, that the derivation here is formal, in the sense that there is no guarantee that any of our steps are justified. Divergences in the integral over the spectral function will modify our result. In particular, there is no guarantee that only *one* derivative of the δ function is present; see Exercise 6.1. Since $\sigma(a^2)$ is proportional to the total lepton annihilation cross section, the present argument shows that the vacuum expectation value of the ST is a measureable quantity.

Exercise 4.1.

$$\Delta^\mu(a) = i \int \frac{d^4r}{(2\pi)^4} \left[\frac{r^\mu + a^\mu}{([r+a]^2 - m^2)^2} - \frac{r^\mu}{(r^2 - m^2)^2} \right]$$

$$= i \int \frac{d^4r}{(2\pi)^4} \left\{ \exp a^\nu \frac{\partial}{\partial r^\nu} - 1 \right\} \frac{r^\mu}{(r^2 - m^2)^2}$$

$$= i \int \frac{d^4 r}{(2\pi)^4} a^\nu \frac{\partial}{\partial r^\nu} \left\{ \left[1 + O(r^{-1}) \right] \frac{r^\mu}{(r^2 - m^2)^2} \right\} .$$

The integrand is a total derivative. If the integrand were sufficiently damped at large r, one would conclude by Gauss' theorem that the integral is zero. In the present instance, let us integrate over a surface $r = R$, by symmetric integration. Using the fact that

$$\int_{r=R} d^4 r\, \partial^\nu f(r) = i\, 2\pi^2 R^\nu R^2 f(R) ,$$

we have

$$\Delta^\mu(a) = -\frac{1}{8\pi^2} a^\nu \lim_{R\to\infty} R_\nu R^2 \left[1 + O(R^{-1}) \right] \frac{R^\mu}{(R^2 - m^2)^2} .$$

The limit is to be performed symmetrically.

$$\lim_{R\to\infty} \left\{ \text{odd number of factors of } R \right\} = 0 ,$$

$$\lim_{R\to\infty} \frac{R^\mu R^\nu}{R^2} = \tfrac{1}{4} g^{\mu\nu} .$$

Hence

$$\Delta^\mu(a) = -\frac{a^\mu}{32\pi^2} .$$

Exercise 4.2.

$$\Delta^{\alpha\mu\nu}(p, q\,|\,a)$$

$$= i \int \frac{d^4 r}{(2\pi)^4} \operatorname{Tr} \gamma^5 \gamma^\alpha \left\{ \exp a^\beta \frac{\partial}{\partial r^\beta} - 1 \right\} (\gamma_\delta r^\delta + \gamma_\delta p^\delta - m)^{-1}$$

$$\times\ \gamma^\mu (\gamma_\delta r^\delta - m)^{-1} \gamma^\nu (\gamma_\delta r^\delta + \gamma_\delta q^\delta - m)^{-1}$$

$$= i\int \frac{d^4 r}{(2\pi)^4}\, a^\beta \frac{\partial}{\partial r^\beta} \{[1 + O(r^{-1})] \operatorname{Tr} \gamma^5 \gamma^\alpha (\gamma_\delta r^\delta + \gamma_\delta p^\delta - m)^{-1}$$

$$\times\ \gamma^\mu (\gamma_\delta r^\delta - m)^{-1} \gamma^\nu (\gamma_\delta r^\delta + \gamma_\delta q^\delta - m)^{-1} \}$$

$$= -\ \frac{a^\beta}{8\pi^2} \lim_{R\to\infty}\ R_\beta R^2 [1 + O(R^{-1})]$$

$$\times\ [\operatorname{Tr} \gamma^5 \gamma^\alpha \gamma_\delta R^\delta \gamma_\epsilon R^\epsilon \gamma^\nu \gamma_\phi R^\phi R^{-6}\ +\ O(R^{-4})]\ .$$

The trace is simply $4\epsilon^{\alpha\omega\mu\nu} R_\omega / R^4$. Hence

$$\Delta^{\alpha\mu\nu}(p, q\,|\,a) = -\ \frac{a_\beta}{8\pi^2}\, \epsilon^{\alpha\beta\mu\nu}\ .$$

Exercise 4.3.

$$i\int \frac{d^4 r}{(2\pi)^4} \operatorname{Tr} \gamma^5 \gamma^\alpha \{[\gamma_\delta r^\delta + \gamma_\delta q^\delta - m]^{-1} \gamma^\nu [\gamma_\delta r^\delta - \gamma_\delta p^\delta - m]^{-1}$$

$$-\,[\gamma_\delta r^\delta + \gamma_\delta p^\delta - m]^{-1} \gamma^\nu [\gamma_\delta r^\delta - \gamma_\delta q^\delta - m]^{-1}\}$$

$$= i\int \frac{d^4 r}{(2\pi)^4} \operatorname{Tr} \gamma^5 \gamma^\alpha \left\{\exp (q-p)^\beta \frac{\partial}{\partial r^\beta}\ -\ 1\right\}$$

$$\times\ [\gamma_\delta r^\delta + \gamma_\delta p^\delta - m]^{-1} \gamma^\nu [\gamma_\delta r^\delta - \gamma_\delta q^\delta - m]^{-1}$$

$$= i\int \frac{d^4 r}{(2\pi)^4} (q-p)^\beta \frac{\partial}{\partial r^\beta} \{[1 + O(r^{-1})] \operatorname{Tr} \gamma^5 \gamma^\alpha [\gamma_\delta r^\delta + \gamma_\delta p^\delta - m]^{-1}$$

$$\times\ \gamma^\nu [\gamma_\delta r^\delta - \gamma_\delta q^\delta - m]^{-1}\}$$

$$= -\ \frac{(q-p)^\beta}{8\pi^2} \lim_{R\to\infty}\ R_\beta R^2 [1 + O(R^{-1})]$$

$$\times\ \frac{\operatorname{Tr} \gamma^5 \gamma^\alpha [\gamma_\delta R^\delta\ +\ \gamma_\delta p^\delta + m]\, \gamma^\nu [\gamma_\epsilon R^\epsilon - \gamma_\epsilon q^\epsilon - m]}{([R+p]^2 - m^2)([R-q]^2 - m^2)}\ .$$

The trace is $-4\epsilon^{\alpha\phi\nu\omega}R_\omega(q_\phi+p_\phi)+$ O(constant). Therefore the limit of the above is

$$\frac{(q-p)_\omega}{8\pi^2}\epsilon^{\alpha\phi\nu\omega}(q_\phi+p_\phi) = \frac{1}{4\pi^2}\epsilon^{\alpha\mu\nu\beta}\,p_\mu q_\beta \;.$$

Exercise 5.1.

$$J^i = \bar\psi\gamma^i\psi \;,$$

$$\dot J^i = \bar\psi\gamma^i\dot\psi + \text{h.c.} \;,$$

$$= -i\bar\psi\gamma^i\gamma^0 i\gamma^0\dot\psi + \text{h.c.} \;.$$

From equation of motion, $i\gamma_\mu\partial^\mu\psi = m\psi - gB_\mu\gamma^\mu\psi$, it follows that

$$i\gamma^0\dot\psi = -i\gamma^k\partial_k\psi + m\psi - gB_\mu\gamma^\mu\psi \;.$$

Hence

$$\dot J^i = -\bar\psi\gamma^i\gamma^0\gamma^k\partial_k\psi - im\bar\psi\gamma^i\gamma^0\psi + igB_\mu\bar\psi\gamma^i\gamma^0\gamma^\mu\psi + \text{h.c.} \;.$$

The commutator may now be evaluated with the help of the canonical commutators between the fileds. A long computation yields for the symmetric part of the commutator the following.

$$\int d^3x\,[\dot J^i(0,x), J^j(0)]$$
$$= \bar\psi(0)[\gamma^i\overleftrightarrow{\partial}^j + \gamma^j\overleftrightarrow{\partial}^i - 2g^{ij}\gamma^k\overleftrightarrow{\partial}_k - 2ig(\gamma^iB^j+\gamma^jB^i-2g^{ij}\gamma^kB_k)$$
$$-4img^{ij}]\,\psi(0) + \text{term anti-symmetric in } (i,j) \;.$$

To calculate the diagonal, spin-averaged proton matrix element of the above, we define

$$\langle p|\bar\psi\gamma^\mu\overleftrightarrow{\partial}^\nu\psi|p\rangle = g^{\mu\nu}iA_1 + p^\mu p^\nu iA_2 \;,$$

$$\langle p|\bar\psi\gamma^\mu B^\nu\psi|p\rangle = g^{\mu\nu}B_1 + p^\mu p^\nu B_2 \;,$$

$$\langle p|\bar\psi\psi|p\rangle = C \;.$$

It then follows that

$$\int d^3x \langle p|[\dot{J}^i(0,x), J^j(0)]|p\rangle = A(\delta^{ij}\mathbf{p}^2 - p^ip^j) + B\delta^{ij} ,$$

$$A = -2i(A_2 - 2gB_2) ,$$

$$B = 4i(A_1 - 2gB_1 + mC) .$$

Exercise 5.2. When the fixed q^2 dispersion relation in ν for T_L needs a subtraction, the dispersion relation in ω has the form

$$T_L(\omega, q^2) = T_L(q^2) + \frac{1}{2\pi}\int_0^1 d\omega' \frac{\tilde{F}_L(\omega', q^2)}{\omega'^2 - \omega^2} .$$

Here $T_L(q^2)$ is the subtraction, performed at $\nu = 0$ ($\omega = \infty$). This equation replaces (5.15a) in the text. It is evident that one may still establish that

$$-\int d^3x\, e^{-i\mathbf{q}\cdot\mathbf{x}} \langle p|[J^0(0,\mathbf{x}), J^i(0)]|p\rangle = \lim_{q_0\to\infty} q^i T_L .$$

However, from the dispersion relation, one now concludes only the uninformative relation

$$\lim_{q_0\to\infty} T_L(q^2, \omega) = T_L(\infty) .$$

Exercise 6.1.

$$T^{*\mu\nu}(q) = (g^{\mu\nu}q^2 - q^\mu q^\nu)T(q^2)$$

$$T(q^2) = C + q^2\int_{4m^2}^{\infty} da^2 \frac{\sigma(a^2)}{a^2(q^2 - a^2)} .$$

Define $\tilde{\sigma}(a^2) = \sigma(a^2) - A - B/a^2$, $\lim_{a^2\to\infty} a^2\tilde{\sigma}(a^2) = 0$. Therefore $T(q^2)$ can be represented by

$$T(q^2) = C + q^2\int_{4m^2}^{\infty} da^2 \frac{\tilde{\sigma}(a^2)}{a^2(q^2 - a^2)} + Aq^2\int_{4m^2}^{\infty} da^2 \frac{1}{a^2(q^2 - a^2)}$$

$$+ Bq^2\int_{4m^2}^{\infty} da^2 \frac{1}{a^4(q^2 - a^2)} ,$$

$$= \tilde{C} + \int_{4m^2}^{\infty} da^2 \frac{\tilde{\sigma}(a^2)}{q^2 - a^2} + A \log\left(1 - \frac{q^2}{4m^2}\right) + \frac{B}{q^2} \log\left(1 - \frac{q^2}{4m^2}\right),$$

$$\tilde{C} = C + \int_{4m^2}^{\infty} da^2 \frac{\tilde{\sigma}(a^2)}{a^2} + \frac{B}{4m^2} .$$

For large q_0

$$T(q^2) \xrightarrow[q_0 \to \infty]{} \tilde{C} + A \log\left(-\frac{q_0^2}{4m^2}\right) + \frac{B}{q_0^2} \log\left(-\frac{q_0^2}{4m^2}\right)$$

$$- \frac{A}{q_0^2}(4m^2 + \mathbf{q}^2) + \frac{1}{q_0^2} \int_{4m^2}^{\infty} da^2\, \tilde{\sigma}(a^2) .$$

Clearly $\tilde{C}$ is a covariant seagull; it may be dropped. Therefore it follows that the T product has the following form at large q_0.

$$q^0 \mathrm{T}^{0i}(q) = - q_0^2 q^i T(q^2) \xrightarrow[q_0^2 \to \infty]{} - q^i \left[A\, q_0^2 \log\left(-\frac{q_0^2}{4m^2}\right) \right.$$

$$\left. + B \log\left(-\frac{q_0^2}{4m^2}\right) - A(4m^2 + \mathbf{q}^2) + \int_{4m^2}^{\infty} da^2\, \tilde{\sigma}(a^2) \right].$$

It is thus seen that the terms proportional to one derivative of the δ function (q^i in momentum space) are the following: a quadratically divergent term, $A\, q_0^2 \log(-q_0^2/4m^2)$; a logarithmically divergent term, $B \log(-q_0^2/4m^2)$; and finite contributions, $-4m^2 A + \int_{4m^2}^{\infty} da^2\, \tilde{\sigma}(a^2)$. In addition to these, there is a *triple* derivative of a δ function ($q^i \mathbf{q}^2$) in momentum space) given by the well defined coefficient $-A$. Thus the existence of a quadratically divergent term in the $[J^0, J^i]$ ETC proportional to $\partial^i \delta(\mathbf{x})$, necessarily implies that an object proportional to $\partial^i \nabla^2 \delta(\mathbf{x})$ will be present with a finite coefficient.

Exercise 6.2.

$$\mathrm{T}(p, q) = \int d^4x\, d^4y\, e^{ip \cdot x} e^{iq \cdot y} \langle \alpha | \mathrm{T} A(x) B(y) C(0) | \beta \rangle ,$$

$$\lim_{p_0 \to \infty} p_0 \mathrm{T}(p, q) = i \int \mathrm{d}^3x \, \mathrm{d}^4y \, e^{-i\mathbf{p}\cdot\mathbf{x}} e^{iq\cdot y}$$

$$\times \langle \alpha | \mathrm{T}\, B(y)[A(0, \mathbf{x}), C(0)] | \beta \rangle ,$$

$$\lim_{q_0 \to \infty} q_0 \lim_{p_0 \to \infty} p_0 \mathrm{T}(p, q) = -\int \mathrm{d}^3x \, \mathrm{d}^3y \, e^{-i\mathbf{p}\cdot\mathbf{x}} e^{-i\mathbf{q}\cdot\mathbf{y}}$$

$$\times \langle \alpha | [B(0, \mathbf{y}), [A(0, \mathbf{x}), C(0)]] | \beta \rangle ,$$

$$\lim_{q_0 \to \infty} q_0 \mathrm{T}(p, q) = i \int \mathrm{d}^4x \, \mathrm{d}^3y \, e^{ip\cdot x} e^{-i\mathbf{q}\cdot\mathbf{y}}$$

$$\times \langle \alpha | \mathrm{T}\, A(x), [B(0, \mathbf{y}), C(0)] | \beta \rangle ,$$

$$\lim_{p_0 \to \infty} p_0 \lim_{q_0 \to \infty} q_0 \mathrm{T}(p, q) = -\int \mathrm{d}^3x \, \mathrm{d}^3y \, e^{-i\mathbf{p}\cdot\mathbf{x}} e^{-i\mathbf{q}\cdot\mathbf{y}}$$

$$\times \langle \alpha | [A(0, \mathbf{x}), [B(0, \mathbf{y}), C(0)]] | \beta \rangle .$$

Exercise 7.1.

$$C^{\mu\nu}(q, p) = \int \frac{\mathrm{d}^4x}{(2\pi)^4} e^{iq\cdot x} \langle p | [J^\mu(x), J^\nu(0)] | p \rangle .$$

The states are covariantly normalized.

$$\langle p | p' \rangle = 2p_0 \delta(\mathbf{p} - \mathbf{p}') .$$

Hence the dimension of the states is m^{-2}. The dimension of $[J^\mu(x), J^\nu(0)]$ is m^6. Finally d^4x is of dimension m^{-4}. Therefore $C^{\mu\nu}$ is dimensionless. Since the tensors which appear in the decomposition of $C^{\mu\nu}$ are also dimensionless, this property is shared by the $\widetilde{F}_i$.

Exercise 7.2.

(a) $$i[D(t), P^\mu] = i \int \mathrm{d}^3x \, x_\alpha [\theta^{\alpha 0}(x), P^\mu]$$

$$= -\int \mathrm{d}^3x \, x_\alpha \partial^\mu \theta^{\alpha 0}(x) .$$

Take $\mu = 0$.

$$i[D(t), P^0] = -\int d^3x\, x_\alpha \partial^0 \theta^{\alpha 0}(x) = \int d^3x\, x_\alpha \partial_i \theta^{\alpha i}(x)$$

$$= \int d^3x\, x_j \partial_i \theta^{ji}(x) = -\int d^3x\, \theta_i^i(x)$$

$$= \int d^3x\, \theta^{00}(x) - \int d^3x\, \theta_\mu^\mu(x)$$

$$= P^0 - \int d^3x\, \theta_\mu^\mu(x) .$$

We have used conservation of $\theta^{\mu\nu}$ and Gauss' theorem. Now take $\mu = i$.

$$i[D(t), P^i] = -\int d^3x\, x_\alpha \partial^i \theta^{\alpha 0}(x) = -\int d^3x\, x_j \partial^i \theta^{j0}(x)$$

$$= \int d^3x\, \theta^{i0}(x) = P^i .$$

It is seen that the algebra $i[D, P^\mu]$ is not satisfied for $\mu = 0$, when the dilatation current is not conserved, $\theta_\mu^\mu \neq 0$.

(b) $$i[D(t), M^{\mu\nu}] = i\int d^3x\, x_\alpha [\theta^{\alpha 0}(x), M^{\mu\nu}]$$

$$= -\int d^3x\, x_\alpha (x^\mu \partial^\nu \theta^{\alpha 0} + g^{\alpha\mu} \theta^{\nu 0} + g^{0\mu} \theta^{\alpha\nu})$$

$$- \{\mu \leftrightarrow \nu\} .$$

The commutator with $M^{\mu\nu}$ has been explicitly evaluated, since it is known that $\theta^{\alpha\beta}$ transforms as a second rank tensor under Lorentz transformations. Take first $\mu = 0$, $\nu = i$.

$$i[D(t), M^{0i}] = -\int d^3x \{x_\alpha x^0 \partial^i \theta^{\alpha 0}(x) + x^0 \theta^{i0}(x)$$

$$+ x_\alpha \theta^{\alpha i}(x) - x_\alpha x^i \partial_0 \theta^{\alpha 0}(x) - x^i \theta^{00}(x)\}$$

$$= -\int d^3x \{x_j x^0 \partial^i \theta^{j0}(x) + 2x^0 \theta^{i0}(x)$$

$$+ x_j \theta^{ji}(x) + x_\alpha x^i \partial_j \theta^{\alpha j}(x) - x^i \theta^{00}(x)\}$$

$$= - \int d^3x \left\{ x^0 \theta^{i0}(x) + x_j \theta^{ji}(x) - x^0 \theta^{0i}(x) - x_j \theta^{ij}(x) - x^i \theta^j_j(x) - x^i \theta^{00}(x) \right\}$$

$$= \int d^3x \, x^i \theta^\mu_\mu(x) .$$

We have used conservation of $\theta^{\mu\nu}$ and Gauss' theorem. Next we consider $\mu = i$, $\nu = j$.

$$i[D(t), M^{ij}] = - \int d^3x \, [x_\alpha x^i \partial^i \theta^{\alpha 0}(x) + x^i \theta^{j0}(x) - \{i \leftrightarrow j\}]$$

$$= - \int d^3x \, [-x^i \theta^{j0}(x) + x^i \theta^{j0}(x) - \{i \leftrightarrow j\}]$$

$$= 0 .$$

As expected from general principles, we find that $D(t)$ is not a Lorentz scalar when $\theta^\mu_\mu \neq 0$ scalar, though it is a rotational invariant.

Exercise 7.3.

(a) $$i[D(t), \phi(x)] = i \int d^3y \, y_\alpha [\theta_c^{\ 0\alpha}(t, \mathbf{y}), \phi(t, \mathbf{x})] + i \int d^3y \, [\pi^0(t, \mathbf{y}) d\phi(t, \mathbf{y}), \phi(t, \mathbf{x})] .$$

We have used the canonical formula for $D_c^{\ \mu}$; of course $D(t)$ is insensitive to which current is used. The $[\theta_c^{\ 0\alpha}, \phi]$ ETC has been evaluated in Exercise 2.1. We therefore have

$$i[D(t), \phi(x)] = \int d^3y \, y_\alpha \phi^\alpha(t, \mathbf{y}) \delta(\mathbf{x} - \mathbf{y}) + d\phi(t, \mathbf{x})$$

$$= x_\alpha \phi^\alpha(x) + d\phi(x) = \delta_D \phi(x) .$$

(b) $$i[K^\alpha(t), \phi(x)] = i \int d^3y \, [(2y^\alpha y_\nu - g^\alpha_\nu y^2) \theta_c^{\ 0\nu}(t, \mathbf{y}) + 2y_\nu \pi^0(t, \mathbf{y}) (g^{\nu\alpha} d - \Sigma^{\nu\alpha}) \phi(t, \mathbf{y}) - 2\sigma^{\alpha 0}(t, \mathbf{y}), \phi(t, \mathbf{x})] .$$

Again the canonical formula for the conformed current has been used. The commutators of ϕ with the first two terms in the expression for $K_c{}^{\alpha 0}$ are readily evaluated. To calculate the commutator of $\sigma^{\alpha 0}$ with ϕ, we observe that $\sigma^{\mu\nu}$ does not depend on derivatives of fields. The reason for this is that if $\sigma^{\mu\nu}$ were to depend on derivatives of fields then $\partial_\nu \sigma^{\mu\nu}$ would depend on second derivatives of the fields. This is impossible when $\mathscr{L}$ has no dependence on second derivatives, since

$$\partial_\nu \sigma^{\mu\nu} = V^\mu = \frac{\delta \mathscr{L}}{\delta \phi^\beta} [g^{\beta\alpha} d - \Sigma^{\beta\alpha}] \phi .$$

Therefore the commutator of ϕ with $\sigma^{\alpha 0}$ vanishes, and we are left with

$$\begin{aligned} i[K^\alpha(t), \phi(x)] &= \int d^3y [(2y^\alpha y_\nu - g_\nu^\alpha y^2)\phi^\nu(t, \mathbf{y}) + 2y_\nu(g^{\nu\alpha} d - \Sigma^{\nu\alpha})\phi(t, \mathbf{y})] \\ &= (2x^\alpha x_\nu - g_\nu^\alpha x^2)\phi^\nu(x) - 2x_\nu(g^{\nu\alpha} d - \Sigma^{\nu\alpha})\phi(x) \\ &= \delta_C{}^\alpha \phi(x) . \end{aligned}$$

Exercise 7.4.

(a) For the scale invariance we need $4\mathscr{L} = (d+1)\phi^\mu(\delta\mathscr{L}/\delta\phi^\mu) + d\phi(\delta\mathscr{L}/\delta\phi)$, $d = 1$. Since $\mathscr{L}$ is Poincaré invariant, it can depend on ϕ^μ only through the combination $\phi^\mu\phi_\mu$. Therefore we set, without loss of generality, $\mathscr{L}(\phi^\mu, \phi) = f(\phi^\mu\phi_\mu/\phi^4, \phi)\phi^4$. The above condition now gives a differential equation for f: $\partial f(x,y)/\partial y = 0$. Therefore the most general scale invariance Lagrangian is of the form $\mathscr{L}(\phi^\mu, \phi) = f(\phi^\mu\phi_\mu/\phi^4)\phi^4$, where f is an arbitrary function.

(b) For the Lagrangian to be conformally invariant, the field virial, $V^\mu = \pi^\mu\phi$, must be a total divergence. In the present case this requires

$$2\phi^\mu\phi f'\left(\frac{\phi^\alpha\phi_\alpha}{\phi^4}\right) = \partial_\alpha \sigma^{\alpha\mu} .$$

The quantity $\sigma^{\alpha\mu}$ is a function of ϕ and ϕ^μ. However, it cannot depend on ϕ^μ, for then the indicated differentiation on the right-hand side would produce second derivatives of ϕ, which are not present on the left-hand side. Thus

$$\sigma^{\alpha\mu} = g^{\alpha\mu}\,\sigma(\phi^2)\ ,$$

$$\partial_\alpha \phi^{\alpha\mu} = 2\sigma'(\phi^2)\phi^\mu\phi\ ,$$

and the condition on the field virial becomes

$$f'\left(\frac{\phi^\alpha\phi_\alpha}{\phi^4}\right) = \sigma'(\phi^2)\ .$$

It is now seen that f' must be constant, since $\sigma'(\phi^2)$ cannot depend on ϕ^α. We conclude therefore that the most general conformally invariant single boson Lagrangian is

$$\mathscr{L} = \frac{1}{2}\phi^\mu\phi_\mu - \frac{\lambda}{4!}\phi^4\ ,$$

$$\theta_c^{\ \mu\nu} = \phi^\mu\phi^\nu - g^{\mu\nu}\mathscr{L}\ , \qquad \sigma^{\mu\nu} = \frac{g^{\mu\nu}}{2}\phi^2\ ,$$

$$\theta^{\mu\nu} = \phi^\mu\phi^\nu - g^{\mu\nu}\mathscr{L} + \tfrac{1}{6}(g^{\mu\nu}\Box - \partial^\mu\partial^\nu)\phi^2\ .$$

TOPOLOGICAL INVESTIGATIONS OF QUANTIZED GAUGE THEORIES†

Roman Jackiw
Center of Theoretical Physics
Laboratory for Nuclear Science and Department of Physics
Massachusetts Institute of Technology
Cambridge, MA 02139, USA

†This is an updated version of the author's contribution in *Relativity, Groups and Topology II*, Les Houches 1983, eds., B. S. DeWitt and R. Stora (North-Holland, Amsterdam, 1984).

1. Introduction

A relativistic quantum field theory provides a dynamical framework for explaining a vast variety of physical phenomena: not only are all properties of bound and scattering states contained in the theory, but also processes in which particle number changes are described. The ultimate model that physicists are seeking is the unified theory, which must account for everything around us — a variety which is practically unlimited. It comes as no surprise therefore that realistic field theories are not exactly solvable, and we must resort to approximate but accurate methods of calculation — as already we do in simple classical mechanical or quantum mechanical problems involving two or more particles.

For a long time, the only available field theoretical approximation scheme was the Dyson-Feynman-Schwinger perturbation expansion, which is a generalization of the familiar quantum mechanical Born series. Since there is much in a quantum field theory which is not easily seen in such a series, we have looked for other approximations, and in the last decade the semi-classical method has been successfully developed and used to explore quantum field theory in novel ways.

The semi-classical approach begins by a study of the Heisenberg equations of motion for the quantum field operators, in the approximation that all quantum effects are ignored; i.e., one views the field equations as classical, nonlinear partial differential equations. Of course, such equations are still difficult to solve, and physicists have been engaged in much mathematical analysis — an activity which has put us in contact with mathematicians, who as it happens had come to similar equations for their own reasons which grew out of their own subject. This conjunction of interests between the two disciplines is gratifying because it has informed each field of the other's techniques, and in particular physicists have benefited from learning and using the mathematicians' topological analysis. With these topological ideas we have been able to go beyond the semi-classical approximation, and to prove exact quantum field theoretical results about various models of interest. Thus the development in quantum field theory that I am speaking about began with semi-classical approximations, passed through mathematical/topological analysis, and finally resulted in new insights into the structure of quantal systems.

In these lectures, I shall not discuss semi-classical approximations; this subject is close to ten years old and it is well described in the review literature [1] and in monographs [2]. Also, mathematical aspects of topology will not be treated in detail, beyond mentioning those facts that I shall need in my presentation. Again I refer you to other excellent sources [3].

Here I shall acquaint you with exact results about quantum field theory which have emerged from topological analysis. They are of three kinds. First, we have realized that there may exist parameters characterizing a quantum theory which are not apparent in the Lagrangian. Second, we have learned that some parameters of a Lagrangian (coupling constants, masses, etc.) cannot be arbitrary in a consistent quantum theory. The former results are the older; the latter are currently the subject of much active research. However, as we shall see, there is a relationship between the two, arising both from physical and from mathematical considerations. Finally, third, we have established that the interaction of gauge fields with fermions is strongly constrained. In particular, theories which appear gauge invariant and consistent in fact may lose gauge invariance when quantized, unless fermion content is carefully adjusted.

We shall need to develop considerable background before delving into the narrow topic of topological properties in quantum field theory. Thus, I shall begin my lectures by discussing gauge theories, their classical symmetries, the problem of quantization, and anomalies before coming to the main subject. I shall discuss Abelian and non-Abelian vector gauge theories, but not gravity theory, even though this School is primarily concerned with quantum gravity. Although my fundamental reason for ignoring gravitational models is that I am no expert about them — particle physics being my speciality — I can also assure you that everything that I shall say has an analogue in general relativity, which may also be viewed as a gauge theory. Thus you have the opportunity to learn the material in a simpler setting. Moreover, these days many of us are optimistic that a truly unified theory of natural phenomena is close at hand, so that particle-physics results should be of interest even to gravity students and researchers, and of course *vice versa*.

The focus on vector gauge theories reflects the circumstance that their topological properties are more intricate than those of other field theories (with the obvious exception of gravity) and that vector gauge theories are the crucial ingredient in all modern models for electromagnetic, weak and strong interactions.

The Exercises that are scattered throughout the text are intended to complement the lectures, and to introduce material beyond the scope of my presentation. I urge you to do them! Some solutions to the Exercises can be found in a separate section.

2. Description of Vector Gauge Theories

2.1 Action and Equations of Motion

We shall be dealing with field theories whose basic dynamical variables are vector fields $A^a_\mu(x)$, called potentials. The index μ refers to Minkowski space-time where the coordinate vector $x^\mu \equiv (t, \mathbf{r})$ is defined with metric $g_{\mu\nu} \equiv \mathrm{diag}(1, -1, -1, -1)$; we shall deal not only with the physical, four-dimensional world, but also with models in two and three dimensions — these are useful for physical and pedagogical reasons. Throughout, the velocity of light is set to unity; however $\hbar$ will be retained so that classical effects may be clearly separated from quantum ones. The index "a" on the potential labels internal degrees of freedom, and the theory is invariant against a (compact and in general non-Abelian) group of transformations operating on these degrees of freedom. The index a ranges over the dimension of the group: for SO(3) or SU(2) it goes from 1 to 3; for SU(3), it ranges to 8; while for the Abelian case a is single valued, and is suppressed. (For us, upper and lower internal symmetry indices are equivalent.)

From the potential (also called connection by the mathematically minded) the field strength $F^a_{\mu\nu}$ (curvature) is constructed by the formula

$$F^a_{\mu\nu} \equiv \partial_\mu A^a_\nu - \partial_\nu A^a_\mu + g f_{abc} A^b_\mu A^c_\nu \quad . \tag{2.1}$$

Here ∂_μ is the derivative with respect to x^μ, f_{abc} are completely anti-symmetric structure constants of the group, g is a coupling constant, and repeated indices are summed.

There exists a useful matrix notation for the gauge theory. Consider anti-hermitian representation matrices T^a for the Lie algebra of the invariance group,

$$[T^a, T^b] = f_{abc} T^c, \qquad (T^a)^\dagger = -T^a \quad , \tag{2.2}$$

normalized, for example, by

$$\mathrm{tr}\, T^a T^b = -\tfrac{1}{2}\delta_{ab} \quad , \tag{2.3}$$

and define a matrix-valued vector potential by

$$A_\mu \equiv gT^a A^a_\mu \quad , \tag{2.4a}$$

from which the components can be regained with the help of Eq. (2.3):

$$A^a_\mu = -\frac{2}{g} \operatorname{tr} T^a A_\mu \quad . \tag{2.4b}$$

[For SU(2), $T^a = \sigma^a/2i$, where the σ^a are Pauli matrices; for SU(3) the 3×3 Gell-Mann matrices $\lambda^a/2i$ are used.] The matrix-valued field strength

$$F_{\mu\nu} \equiv gT^a F^a_{\mu\nu} \quad , \tag{2.5a}$$

$$F^a_{\mu\nu} = -\frac{2}{g} \operatorname{tr} T^a F_{\mu\nu} \quad , \tag{2.5b}$$

is given by

$$F_{\mu\nu} = \partial_\mu A_\nu - \partial_\nu A_\mu + [A_\mu, A_\nu] \quad . \tag{2.6}$$

This compact notation encompasses the Abelian (Maxwell) theory (electrodynamics), where the matrices reduce to numbers, and the commutators vanish. When the matrix structure is present, the commutators are non-vanishing and the theory is called non-Abelian. [We emphasize that the commutators in Eq. (2.6) are considered only in the matrix space of the group generators; they are not quantum mechanical.]

Exercise 2.1. Convince yourself that any representation matrices, not only those of the defining representation, may be used for the matrix notation. Of course if their normalization differs from (2.3), the inversion formulas (2.4b) and (2.5b) are correspondingly modified.

The matrix and component notations will be used interchangeably; no confusion arises because the context leaves the convention unambiguous. For the Abelian theory, where the two collapse into one, I shall remain with the real "component" notation for potentials and fields.

We also introduce the gauge covariant derivative D_μ: operating on a matrix quantity $M = M^a T^a$ it gives

$$D_\mu M = \partial_\mu M + [A_\mu, M] \quad . \tag{2.7a}$$

In the gauge covariant derivative, the ordinary derivative is supplemented by a commutator, which however vanishes in the Abelian theory so that covariant and ordinary differentiation coincide. Equation (2.7a) may also be presented in component notation:

$$(D_\mu M)^a = \partial_\mu M^a + g f_{abc} A^b_\mu M^c \quad . \tag{2.7b}$$

The covariant derivative is distributive,

$$D_\mu (MM') = (D_\mu M)M' + M(D_\mu M') \quad , \tag{2.8a}$$

and when operating on functions rather than matrices it reduces to the ordinary derivative; for example

$$\partial_\mu (\mathrm{tr}\, MM') = \mathrm{tr}\,(D_\mu M)M' + \mathrm{tr}\, M(D_\mu M') \quad . \tag{2.8b}$$

Exercise 2.2. Show that

$$(D_\mu D_\nu - D_\nu D_\mu) M = [\, F_{\mu\nu}, M\,] . \tag{E2.1}$$

Exercise 2.3. Suppose A_μ is changed by an infinitesimal amount, $A_\mu \to A_\mu + \delta A_\mu$. Show that the first-order change in $F_{\mu\nu}$ is

$$\delta F_{\mu\nu} = D_\mu \delta A_\nu - D_\nu \delta A_\mu . \tag{E2.2}$$

Let us observe that from its definition (2.1) or (2.6) $F_{\mu\nu}$ satisfies an identity, called the Bianchi identity:

$$D_\alpha F_{\beta\gamma} + D_\beta F_{\gamma\alpha} + D_\gamma F_{\alpha\beta} = 0 \quad . \tag{2.9}$$

This is anti-symmetric in its three indices α, β, γ; hence in two dimensions it is vacuous. In higher dimensions, the formula may be presented more compactly by defining the dual field strength with the help of the totally anti-symmetric tensor appropriate to the dimensionality of space-time. In four dimensions, the dual is again a second-rank tensor,

$$*F^{\mu\nu} = \tfrac{1}{2}\, \epsilon^{\mu\nu\alpha\beta} F_{\alpha\beta} \quad , \tag{2.10a}$$

$$F_{\alpha\beta} = -\tfrac{1}{2}\, \epsilon_{\alpha\beta\mu\nu} *F^{\mu\nu} \quad , \qquad \epsilon^{0123} = 1 = -\epsilon_{0123} \quad , \tag{2.10b}$$

and the Bianchi identity (2.9) becomes

$$D_\mu {}^*F^{\mu\nu} = 0 \quad . \tag{2.11}$$

In three dimensions the dual is a vector,

$$ {}^*F^\mu = \tfrac{1}{2}\, \epsilon^{\mu\alpha\beta} F_{\alpha\beta} \quad , \tag{2.12a}$$

$$F_{\alpha\beta} = \epsilon_{\alpha\beta\mu} {}^*F^\mu \quad , \qquad \epsilon^{012} = 1 = \epsilon_{012} \quad , \tag{2.12b}$$

and the Bianchi identity (2.9) requires the dual field to be covariantly conserved,

$$D_\mu {}^*F^\mu = 0 \quad . \tag{2.13}$$

Finally in two dimensions, the dual is a scalar, which is unconstrained since (2.9) becomes vacuous.

$$ {}^*F = \tfrac{1}{2}\, \epsilon^{\mu\nu} F_{\mu\nu} \quad , \tag{2.14a}$$

$$F_{\mu\nu} = -\epsilon_{\mu\nu} {}^*F \quad , \qquad \epsilon^{01} = 1 = -\epsilon_{01} \quad . \tag{2.14b}$$

We have still to provide dynamical field equations which govern the space-time behavior of our variables. [The above Eqs. (2.9) — (2.13) are not dynamical; they are identities.] The equations of motion may be presented in terms of an action I, which is a functional of A_μ. The requirement that I be stationary against variations of A_μ gives Euler-Lagrange equations which are the field equations for A_μ.

The Yang-Mills theory results by taking the action to be the space-time integral of a Lagrangian (density) $\mathscr{L}_{YM}$, the latter being the simplest local invariant function constructed from $F_{\mu\nu}$ [4]:

$$I_{YM} = \int dx\, \mathscr{L}_{YM} \quad ,$$

$$\mathscr{L}_{YM} = \frac{1}{2g^2} \operatorname{tr} F^{\mu\nu} F_{\mu\nu} = -\frac{1}{4} F^{\mu\nu a} F^a_{\mu\nu} \quad . \tag{2.15}$$

The Yang-Mills field equations then read

$$\frac{\delta I_{\rm YM}}{\delta A^a_\mu} = 0 \Leftrightarrow D_\mu F^{\mu\nu} = 0 \quad , \tag{2.16a}$$

or in component from

$$\partial_\mu F^{\mu\nu a} + g f_{abc} A^b_\mu F^{\mu\nu c} = 0 \quad . \tag{2.16b}$$

We see that even in the absence of other matter couplings, the non-Abelian theory is non-trivial, nonlinear and interacting. (Further possible contributions to the gauge field action, even in the absence of matter fields, as well as interaction with matter fields, will be discussed later.)

It is clear that the non-Abelian Yang-Mills theory is a generalization, into non-commuting matrix-valued potentials and fields, of the Abelian Maxwell theory, which in Yang-Mills terminology we call an Abelian U(1) or SO(2) gauge theory. Notice however an important difference. The Maxwell theory's dynamics can be entirely formulated in terms of field strengths: the Maxwell equations are (the Abelian analogs of) (2.9) or (2.11) and (2.16), while the electromagnetic vector potentials are conventionally introduced to solve those Maxwell equations which coincide with the Bianchi identities, (2.9) or (2.11). Thus there is a historic prejudice that potentials are unphysical, secondary quantities and only field strengths are physically important. However, this attitude is unwarranted in view of modern developments. Already for electromagnetism, one knows that vector potentials are physically significant within quantum mechanics (Bohm-Aharanov effect), while the Yang-Mills theory cannot even be formulated without reference to the potentials, since the covariant derivatives, occurring in field equations, involved A_μ. Other similarities and differences between Maxwell and Yang-Mills theories will be mentioned later.

2.2 Symmetries

Next let us discuss invariance symmetries of the action (2.15); i.e. we consider transformations which take one solution of the field equations (2.16) into another.

The most important symmetry of the Yang-Mills gauge theory is its gauge invariance; indeed the particular form taken for the dynamics is chosen so that symmetry is respected. Let U be a position-dependent element of the group which transforms the internal symmetry degrees of freedom of the theory, and let U be given in the same representation as the matrices T^a. [For SU(2), U is a 2×2 unitary matrix with unit determinant; similarly for SU(3), where U is 3×3.] Then a gauge transformation of the vector potential matrix A_μ is defined as

$$A_\mu \to A_\mu^U \equiv U^{-1} A_\mu U + U^{-1} \partial_\mu U = A_\mu + U^{-1} D_\mu U \ , \quad U^\dagger = U^{-1} \ . \tag{2.17a}$$

This induces the following similarity transformation on the field strengths:

$$F_{\mu\nu} \to F_{\mu\nu}^U = U^{-1} F_{\mu\nu} U \ . \tag{2.17b}$$

One easily verifies that this is a symmetry transformation of the field Eq. (2.16), i.e., if A_μ, $F_{\mu\nu}$ solve them, so do $A_\mu^U, F_{\mu\nu}^U$ for arbitrary U. Correspondingly the action (2.15) is invariant, as also is the Lagrangian since $\mathscr{L}_{\text{YM}}$ involves a trace which is invariant against similarity transformations.

In the above, we have discussed finite gauge transformations. One may also consider infinitesimal ones. If U is written as

$$U = e^\Theta \ , \tag{2.18a}$$

and expanded in powers of Θ, which is anti-Hermitian,

$$U = I + \Theta + \ldots , \quad \Theta^\dagger = -\Theta \ , \tag{2.18b}$$

then the gauge transformation to first order in Θ reads

$$A_\mu \to A_\mu + \delta A_\mu \ , \qquad \delta A_\mu = D_\mu \Theta \ , \tag{2.19a}$$

$$F_{\mu\nu} \to F_{\mu\nu} + \delta F_{\mu\nu} \ , \qquad \delta F_{\mu\nu} = [F_{\mu\nu}, \Theta] \ , \tag{2.19b}$$

[Compare this with Eqs. (E2.1) and (E2.2).] Equation (2.19) becomes in component notation, $\Theta = \theta^a T^a$

$$\delta_\theta A_\mu^a = \frac{1}{g} \partial_\mu \theta^a + f_{abc} A_\mu^b \theta^c \ , \tag{2.20a}$$

$$\delta_\theta F_{\mu\nu}^a = f_{abc} F_{\mu\nu}^b \theta^c \ . \tag{2.20b}$$

These transformations are called "local" gauge transformations, because U and Θ are taken to be local matrix functions of the space-time coordinates.

When they are constant, the transformation reduces to a conventional internal symmetry transformation, also frequently described as a "global" gauge transformation. This is something of a misnomer; I prefer the name "rigid" gauge transformation.

One immediate and important consequence of local gauge invariance is the absence from the Lagrangian of a mass term for A_μ: $M^2 \operatorname{tr} A_\mu A^\mu$ is not locally gauge invariant.

The conserved Noether current j^μ for an invariant Lagrangian $\mathscr{L}$, which satisfies the invariance condition that must be established without use of equations of motion

$$\delta \mathscr{L} \equiv \frac{\partial \mathscr{L}}{\partial A^a_\nu}\, \delta A^a_\nu + \frac{\partial \mathscr{L}}{\partial \partial_\mu A^a_\nu}\, \partial_\mu \delta A^a_\nu = 0 \quad , \tag{2.21a}$$

is given by

$$j^\mu = \frac{\partial \mathscr{L}}{\partial \partial_\mu A^a_\nu}\, \delta A^a_\nu \quad . \tag{2.21b}$$

For local gauge transformations, (2.19) and (2.20), the conserved current is

$$j^\mu_\theta = \frac{2}{g^2} \operatorname{tr} F^{\mu\nu} D_\nu \Theta \quad , \tag{2.22}$$

while for rigid transformations, it reads

$$j^\mu_a = -f_{abc} F^{\mu\nu b} A^c_\nu \quad , \tag{2.23}$$

where I am using the component notation and have cancelled away the constant transformation parameter θ^a.

Exercise 2.4. With the help of the field equations (2.16) show that both j^μ_θ in (2.22) and j^μ_a in (2.23) are conserved. Hint: use (2.8b) and (E2.1). The conservation of these currents is of course expected, since the Lagrangian (2.15) is invariant against these transformations: $\delta_\theta \mathscr{L} = 0$.

Obviously non-Abelian gauge invariance is a generalization of familiar electromagnetic gauge invariance, where $U = \exp(i\theta/e)$ and both U and θ are functions, not matrices, thus reducing (2.17) and (2.19) to

$$A_\mu \to A_\mu + \frac{1}{e}\partial_\mu \theta \quad , \tag{2.24a}$$

$$F_{\mu\nu} \to F_{\mu\nu} \quad . \tag{2.24b}$$

(An electromagnetic coupling constant e is conventionally inserted, even though the free Maxwell theory is without interaction.)

However there are important differences between Abelian and non-Abelian gauge transformations: The electromagnetic gauge transformation (2.24) is the same both in its infinite and infinitesimal form, but for the Yang-Mills theory the two differ; compare (2.17) and (2.19). While iterating infinitesimal transformations produces finite transformations, not all finite non-Abelian transformations can be reached in this way. An equivalent statement is that there are finite gauge transformations U which cannot be continuously deformed to the identity I. This fact has far-reaching consequences for the topological structure of non-Abelian gauge theories, and I shall elaborate on this extensively in subsequent lectures.

Another important difference between the Abelian and non-Abelian theories is that the Yang-Mills field strength is not gauge invariant. As is seen from (2.17b), $F_{\mu\nu}$ transforms by a similarity transformation; or equivalently from (2.20b), $F^a_{\mu\nu}$ transforms according to the adjoint representation. We say that $F_{\mu\nu}$ is gauge covariant. This highlights once again the fact that field strengths are not fundamental to the theory. Indeed one cannot in general determine uniquely the potential, even up to gauge transformations, which gives rise to a specific field strength; i.e. gauge non-equivalent A_μ's can lead to the same $F_{\mu\nu}$ [5]. In spite of the more complicated way that gauge transformations operate in a non-Abelian gauge theory, we will insist that all physical quantities be gauge invariant, just as in the Abelian theory.

Exercise 2.5 Consider the charge constructed from the time components of the conserved current (2.23):

$$Q^a = \int \mathrm{d}r\, j^0_a(t, r) \quad . \tag{E2.3}$$

Show that Q^a is time-dependent, provided j_a falls off sufficiently rapidly at large r. Since j_a is not gauge invariant, the fall-off requirement restricts the large r behavior of gauge transformations. While the current j^μ_a has no simple gauge transformation properties, show that Q^a is gauge covariant against gauge transformations U which approach a definite angle-independent limit as r approaches infinity. Hint: use the time component of the Yang-Mills equation to express Q^a as an integral over a surface at spatial infinity.

Exercise 2.6. Show that no new charges, beyond Q^a, arise by integrating over space the time component of j^μ_θ, (2.22), provided Θ approaches an angle-independent limit at spatial infinity.

Gauge invariance of the theory may be used to set one space-time component of the vector potential to zero by appropriate choice of U and without loss of physical content. While this will be useful in our discussion of the quantized theory's Hamiltonian structure, it has a very dramatic effect on two-dimensional Yang-Mills theory: when one component of the vector potential is zero, the commutator $[A^\mu, A^\nu]$ vanishes and the theory becomes linear and trivial, just like two-dimensional, free electromagnetism.

Thus far I have discussed symmetries of Yang-Mills theory associated with transformations of internal degrees of freedom; of course there are also symmetries associated with transformations of space-time coordinates x^μ. The infinitesimal action of these transformations on coordinates

$$x^\mu \to x^\mu + \delta_f x^\mu \quad , \qquad \delta_f x^\mu = -f^\mu(x) \quad , \tag{2.25}$$

induces a transformation on potentials, which conventionally is given by a Lie derivative:

$$\delta_f A_\mu = f^\alpha \partial_\alpha A_\mu + (\partial_\mu f^\alpha) A_\alpha \equiv L_f A_\mu \quad . \tag{2.26}$$

A Lie derivative of a tensor $T^{\nu\ldots}_{\mu\ldots}$ with upper and lower indices is defined by

$$\begin{aligned} L_f T^{\nu\ldots}_{\mu\ldots} \equiv & f^\alpha \partial_\alpha T^{\nu\ldots}_{\mu\ldots} \\ & + (\partial_\mu f^\alpha) T^{\nu\ldots}_{\alpha\ldots} + \ldots \\ & - (\partial_\alpha f^\nu) T^{\alpha\ldots}_{\mu\ldots} - \ldots \quad , \end{aligned} \tag{2.27}$$

where the omitted terms are a repetition of the first, for each index. Note that in general $L_f A_\mu \neq g_{\mu\nu} L_f A^\nu$.

Exercise 2.7. Show that the transformation of the field strength which follows from that for the potential [cf. (E2.2)],

$$\delta_f F_{\mu\nu} = D_\mu \delta_f A_\nu - D_\nu \delta_f A_\mu \quad , \tag{E2.4}$$

is also given by the Lie derivative,

$$\delta_f F_{\mu\nu} = L_f F_{\mu\nu} \quad . \tag{E2.5}$$

For the Minkowski spaces on which our theories are defined, the coordinate transformations of space-time translation and Lorentz rotation are symmetry operations of the Yang-Mills theory. Together they form the Poincaré group, and their infinitesimal action is give by

$$\text{Translation } f^\mu = a^\mu \quad , \tag{2.28}$$

$$\text{Lorentz rotation } f^\mu = \omega^\mu{}_\nu x^\nu \quad , \qquad \omega^\mu{}_\nu = -\omega_\nu{}^\mu \, . \tag{2.29}$$

Here a^μ and $\omega^\mu{}_\nu$ are constant infinitesimal parameters of the transformation. Note that for these $L_f A_\mu = g_{\mu\nu} L_f A^\nu$ and the f^μ's are Killing vectors, i.e., they satisfy the Killing equation,

$$\partial_\mu f_\nu + \partial_\nu f_\mu = 0 \quad . \tag{2.30}$$

Exercise 2.8 The Lie bracket of two vector functions f^μ and g^μ is defined by

$$f^\alpha \partial_\alpha g^\mu - g^\alpha \partial_\alpha f^\mu \equiv h^\mu \quad . \tag{E2.6}$$

Show that the commutator of two coordinate transformations (2.25) satisfies

$$[\delta_f, \delta_g] = -\delta_h \quad , \tag{E2.7}$$

and the same is true for the transformations of the potentials (2.26). [N.B.: $\delta_f \partial_\alpha A_\mu \equiv \partial_\alpha \delta_f A_\mu \neq L_f(\partial_\alpha A_\mu)$.] Show that the Lie bracket of two Killing vectors is again a Killing vector. With (2.28) and (2.29) verify that Lie bracketing these transformations reproduces the commutators of the Poincaré group.

Under the Poincaré transformations (2.28) and (2.29), the Lagrangian changes by a total divergence,

$$\delta_f \mathscr{L}_{\text{YM}} = \partial_\mu \Omega_f^\mu \, , \tag{2.31a}$$

$$\Omega_f^\mu = f^\mu \mathscr{L}_{\text{YM}} \quad . \tag{2.31b}$$

The action changes only by surface terms so that the equations of motion are invariant; i.e., the Yang-Mills theory is Poincaré invariant.

The derivation of the conserved Poincaré currents is somewhat awkward. If one applies directly Noether's theorem to a Lagrangian $\mathscr{L}$ which is not invariant, but changes by a total derivative $\partial_\mu \Omega^\mu$,

$$\delta \mathscr{L} = \partial_\mu \Omega^\mu \quad , \tag{2.32a}$$

then the formula for the conserved current generalizes from (2.21) to [6]

$$j^\mu = \frac{\partial \mathscr{L}}{\partial \partial_\mu A^a_\nu} \delta A^a_\nu - \Omega^\mu \quad . \tag{2.32b}$$

[Just as in (2.21a), Eq. (2.32a) must be established without using equations of motion.] In the present case this produces a gauge non-invariant result, since δA^a_ν is not gauge covariant, and the Noether currents are not expressed in terms of the conserved, symmetric and gauge invariant energy-momentum tensor.

$$\begin{aligned} \theta^{\mu\nu} &= -F^{\mu\alpha a} F^{\nu a}_{\alpha} + \frac{1}{4} g^{\mu\nu} F^{\alpha\beta a} F^a_{\alpha\beta} \\ &= \frac{2}{g^2} \operatorname{tr}(F^{\mu\alpha} F^\nu_\alpha - \frac{1}{4} g^{\mu\nu} F^{\alpha\beta} F_{\alpha\beta}) \quad . \end{aligned} \tag{2.33}$$

Exercise 2.9. By using the equations of motion (2.16) prove that $\partial_\mu \theta^{\mu\nu} = 0$. Hint: Equation (2.8b) and the Bianchi identity (2.9) will simplify the calculation.

However, there exists a well known procedure for "improving" the Noether currents by adding to them superpotentials — divergences of anti-symmetric tensors $\partial_\nu X^{\mu\nu}$, $X^{\mu\nu} = -X^{\nu\mu}$ — which do not spoil current conservation, since $\partial_\mu \partial_\nu X^{\mu\nu} = 0$ [6]. (Ω^μ is not uniquely defined by the equation $\partial_\mu \Omega^\mu = \delta \mathscr{L}$!) With such improvement, the conserved Poincaré currents are

$$j^\mu_f = \theta^{\mu\nu} f_\nu \quad , \tag{2.34}$$

with f_ν given by (2.28) or (2.29). This is the so-called Bessel-Hagen form for the current [7].

Exercise 2.10. Find the superpotentials that must be added to the Noether current to arrive at (2.34). The possibility of finding these superpotentials is not accidental, but arises from the Lorentz invariance of the theory. For a general discussion see ref. [6].

By using some geometrical properties of gauge fields, one can arrive at (2.34) by an alternate, more elegant, method, which is particular to a gauge theory. Let us first observe that one can write $\delta_f A_\mu \equiv L_f A_\mu$ in terms of a gauge covariant quantity and an (infinitesimal) gauge transformation.

$$L_f A_\mu = f^\alpha \partial_\alpha A_\mu + (\partial_\mu f^\alpha) A_\alpha = f^\alpha F_{\alpha\mu} + D_\mu (f^\alpha A_\alpha) \quad . \tag{2.35}$$

Since the Yang-Mills theory is invariant against gauge transformations, an (infinitesimal) coordinate transformation supplemented by an (infinitesimal) gauge transformation, with gauge function $-f^\alpha A_\alpha$, is still a symmetry transformation. Therefore, we can define the action of coordinate transformations on gauge potentials by an alternate rule, which is gauge covariant [8].

$$\bar{\delta}_f A_\mu \equiv L_f A_\mu - D_\mu(f^\alpha A_\alpha) = f^\alpha F_{\alpha\mu} \ . \tag{2.36}$$

When Noether's theorem is used in conjunction with this gauge covariant transformation, the Bessel-Hagen current (2.34) is obtained immediately.

Exercise 2.11. Show that

$$\bar{\delta}_f F_{\mu\nu} \equiv D_\mu \bar{\delta}_f A_\nu - D_\nu \bar{\delta}_f A_\mu = f^\alpha D_\alpha F_{\mu\nu} + \partial_\mu f^\alpha F_{\alpha\nu} + \partial_\nu f^\alpha F_{\mu\alpha} \ . \tag{E2.8}$$

This may be called the gauge covariant Lie derivative. Hint: use the Bianchi identity.

Exercise 2.12. Show that the gauge covariant transformations on the potentials do not follow (E2.7), rather one has

$$[\bar{\delta}_f, \bar{\delta}_g]\, A_\mu = -\bar{\delta}_h A_\mu - D_\mu(f^\alpha g^\beta F_{\alpha\beta}) \ . \tag{E2.9}$$

For the finite version of these transformations see ref. [8].

In our discussion of symmetries no attention has been paid to the fact that one is dealing with quantum field operators, rather than classical c-number fields. There are two sources of differences between the two, which could invalidate the results here presented. First, there is the problem of operator ordering — in all our manipulations we ignored the quantum mechanical non-commutativity of the quantities with which we were working. Second, a difficulty specific to quantum field theories is the occurrence of infinities and hence the necessity of regularizing and renormalizing the theory. This in general produces further terms in the action, equations of motion etc. — the counter terms — which could spoil the symmetries that we have discussed. (When speaking of symmetries in the quantum field theory, we consider the parameters of the transformation — e.g. θ^a for gauge transformations, a^μ, ω^μ_ν for Poincaré transformations — to be c-numbers.)

An analysis of these questions requires a course in renormalization theory, well beyond the scope of my lectures. Let it suffice to state that the Yang-Mills model presented here can be quantized so that gauge invariance and Poincaré invariance are indeed preserved. However, as soon as additional couplings to other fields are included, it may not be possible to maintain the symmetries, and we shall discuss this below.

Whenever a classical theory possesses a symmetry, but there is no way of quantizing the theory to preserve that symmetry, we say that there are "anomalies" in the conservation equations for the symmetry currents. I am telling you, without proof, that for Yang-Mills theory there are no known Poincaré or gauge anomalies.

An example of anomalies is found in four-dimensional Yang-Mills theory, which on the classical level possesses a larger space-time symmetry than Poincaré invariance. Observe that the divergence of the Bessel-Hagen current (2.34) with arbitrary f^μ can be written as

$$\partial_\mu j_f^\mu = (\partial_\mu \theta^{\mu\nu}) f_\nu + \theta^{\mu\nu} \partial_\mu f_\nu = \tfrac{1}{2} \theta^{\mu\nu} (\partial_\mu f_\nu + \partial_\nu f_\mu) \quad . \tag{2.37}$$

In passing from the first equality to the second we have used the symmetry of $\theta^{\mu\nu}$ in (μ, ν) and the conservation of $\theta^{\mu\nu}$; see Exercise 2.9. For a Killing vector (2.30), in any number of dimensions, the term in parentheses vanishes; j_f^μ is conserved and this is the previously discussed Poincaré invariance. Maintaining this symmetry on the quantum level is equivalent to constructing a conserved, symmetric, renormalized energy-momentum tensor and this can indeed be done [9].

In four dimensions the formal (i.e. classical, not quantized and renormalized) energy-momentum tensor (2.33) is also trace-free in the (μ, ν) indices. Hence (2.37) may also be rewritten as

$$\partial_\mu j_f^\mu = \tfrac{1}{2} \theta^{\mu\nu} (\partial_\mu f_\nu + \partial_\nu f_\mu - \tfrac{1}{2} g_{\mu\nu} \partial_\alpha f^\alpha) \quad . \tag{2.38}$$

Consequently if the infinitesimal coordinate transformation f^μ satisfies the four-dimensional conformal Killing equation,

$$\partial_\mu f_\nu + \partial_\nu f_\mu - \tfrac{1}{2} g_{\mu\nu} \partial_\alpha f^\alpha = 0 \quad , \tag{2.39}$$

j_f^μ will be conserved, as a consequence of the theory's invariance against these further coordinate transformations. Vector functions solving (2.39) are called conformal Killing vectors; every Killing vector is a conformal Killing one, but one can find other conformal Killing vectors. The solutions of (2.39) beyond (2.28) and (2.29) correspond to dilatations and special conformal transformations:

$$\text{Dilatation } f^\mu = a x^\mu \quad , \tag{2.40}$$

$$\text{Special conformal transformation}\quad f^\mu = 2c\cdot x x^\mu - c^\mu x^2 \quad . \tag{2.41}$$

Here a and c^μ are infinitesimal parameters.

Exercise 2.13. Show that (2.28), (2.29), (2.40) and (2.41) are the only solutions to (2.39) in four dimensions. Is the same true in three dimensions, in two dimensions? The conformal Killing equation in d dimensions reads

$$\partial_\mu f_\nu + \partial_\nu f_\mu - \frac{2}{d} g_{\mu\nu}\partial_\alpha f^\alpha = 0 \quad . \tag{E2.10}$$

The finite version of these transformation is

$$\text{Dilatation}\quad x^\mu \to x'^\mu = e^{-a} x^\mu \quad . \tag{2.42}$$

$$\text{Special conformal transformation}\quad x^\mu \to x'^\mu = \frac{x^\mu + c^\mu x^2}{1 + 2c\cdot x + c^2 x^2} \tag{2.43}$$

One readily verifies that if the A_μ's are transformed by the rule (2.26) or (2.36), the Lagrangian changes as in (2.31) and Noether's theorem, combined with (2.36), again yields the Bessel-Hagen conserved current (2.34).

Exercise 2.14. Show that $\delta_f \mathscr{L}_{\text{YM}} = \partial_\mu \Omega^\mu_f$, $\Omega^\mu_f = f^\mu \mathscr{L}_{\text{YM}}$ when f^μ is a conformal Killing vector. Hint: use (E2.8) and (2.8b). From Noether's theorem derive the conserved current (2.34). Are the two- and three-dimensional Yang-Mills theories conformally invariant? If not, can one modify the transformation law for the vector potential, Eq. (2.26) or (2.36), to make the lower-dimensional theories conformally invariant? Examine separately the non-Abelian and Abelian models.

The transformations (2.40) and (2.41), or in finite form (2.42) and (2.43), together with the Poincaré transformations (2.28) and (2.29) form the fifteen-parameter four-dimensional conformal group, SO(4, 2), and the classical four-dimensional Yang-Mills theory is invariant against this large group of symmetry operations. A related fact is that the only parameter of the classical theory, the coupling constant g, is dimensionless, in units where $\hbar$ and the velocity of light are dimensionless.

Exercise 2.15. By expanding the finite special conformal transformation (2.43) for large c, show that up to terms of $O(c^{-2})$ it can be viewed as the following sequence of transformations: translation, improper Lorentz transformation, dilatation and coordinate inversion

$$x^\mu \to x'^\mu = x^\mu / x^2 \quad . \tag{E2.11}$$

This transformation cannot be constructed in infinitesimal form; but any conformally invariant theory is also inversion invariant. Show further that any finite special conformal transformation may be written as an inversion, followed by a translation, and then followed by another inversion.

Exercise 2.16. Show that the Lie bracket (E2.6) of two conformal Killing vectors is again a conformal Killing vector. Verify that Lie bracketing the various infinitesimal transformations of the conformal group reproduces the Lie algebra of the conformal groups; see ref [6] for the structure of this algebra.

The classical conformal symmetry cannot be maintained in the quantized version of the theory. Operationally this means that the renormalized energy-momentum tensor possesses a non-vanishing trace and there are anomalies in the dilatation and conformal currents: rather than the vanishing of (2.38), we get [10]

$$\partial_\mu j^\mu_f = \tfrac{1}{4}\,\partial_\alpha f^\alpha \theta^\mu_\mu \quad ; \tag{2.44}$$

since $\frac{1}{4}\,\partial_\alpha f^\alpha$ is nonzero for conformal Killing vectors, we see that the non-vanishing trace of $\theta^{\mu\nu}$ spoils the conformal symmetry.

The study of this anomalous breaking of conformal symmetry is of enormous practical significance in application to high-energy particle physics. If the symmetry were not broken, it is difficult to see how the theory could avoid being trivial, since it is unlikely that non-trivial scattering amplitudes can be constructed with only one dimensionless parameter. The anomalous response of a theory to conformal, more specifically scale, transformations, is the subject of the renormalization group [11], but neither this nor the trace anomalies involve any topological ideas, except when coupling to gravity is included. In flat space, the anomalous trace of the energy-momentum tensor is proportional to the Lagrangian, where the proportionality is established with the help of the renormalization group Gell-Mann-Low function [12]. Only in curved space do topologically interesting structures contribute to θ^μ_μ [13].

I shall, therefore, not elaborate on the subject of trace anomalies any further, beyond observing that this anomalous symmetry breaking is not unexpected in a quantum field theory. Conformal symmetry requires that there be no dimensional parameters. But a renormalization procedure necessarily introduces a dimensional quantity: the scale at which the theory is renormalized, and this breaks the scale symmetry and hence the conformal symmetry. More technically, field theoretical calculations must be regularized to avoid infinities and regularization is effected by introducing dimensional, conformal symmetry violating cutoff parameters, or by analytic continuation of the dimensionality of space-time, which also violates conformal symmetry, since the trace of $\theta_{\mu\nu}$ vanishes only in four dimensions.

In subsequent lectures, I shall discuss anomalies in great detail, in examples involving fermionic axial vector currents. These anomalies possess topologically non-trivial characteristics [14].

2.3 *Couplings to Matter Fields*

The pure Yang-Mills theory contains only vector gauge potentials. It is of interest to introduce couplings to other fields, the so-called matter fields. We consider multiplets of real scalar fields ϕ_n and spinor field ψ_n, which are taken to transform under a rigid (global) gauge transformation according to some definite representation of the Yang-Mills invariance group,

$$\delta\phi_n = -\tau^a_{nn'}\theta^a\phi_{n'} \qquad \delta\psi_n = -\tau^a_{nn'}\theta^a\psi_{n'} \qquad \theta^a \text{ constant} \quad . \tag{2.45}$$

Here the τ^a's are representation matrices for the Lie algebra; they need not coincide with the T^a representation matrices used earlier, nor need they be the same for the scalar and spinor fields, although they still satisfy the commutation relation (2.2).

$$[\tau^a, \tau^b] = f_{abc}\,\tau^c \quad . \tag{2.46}$$

For example, the scalar fields may transform according to the SU(N) adjoint representation,

$$\tau^a_{nn'} = f_{nan'} \quad , \tag{2.47a}$$

and the spinor fields according to the fundamental SU(N) representation,

$$\tau^a = \sigma^a/2\mathrm{i} \quad \text{for SU(2)} \ , \qquad \tau^a = \lambda^a/2\mathrm{i} \qquad \text{for SU(3)} \ , \ \text{ etc.} \tag{2.47b}$$

We then define a gauge covariant derivative of the matter fields by

$$(\mathscr{D}_\mu\phi)_n = \partial_\mu\phi_n + gA^a_\mu\tau^a_{nn'}\phi_{n'} \quad ,$$

$$(\mathscr{D}_\mu\psi)_n = \partial_\mu\psi_n + gA^a_\mu\tau^a_{nn'}\psi_{n'} \quad , \tag{2.48}$$

and one readily verifies that the covariant derivative responds covariantly to a local gauge transformation,

$$\delta(\mathscr{D}_\mu \phi)_n = -\tau^a_{nm}\theta^a(\mathscr{D}_\mu \phi)_m \quad ,$$

$$\delta(\mathscr{D}_\mu \psi)_n = -\tau^a_{nm}\theta^a(\mathscr{D}_\mu \psi)_m \quad . \tag{2.49}$$

Again a uniform matrix formalism may be used. In the notation that the matter multiplet (scalar or spinor) is a column vector in internal symmetry space, the covariant derivative is a matrix operator

$$\mathscr{D}_\mu = \partial_\mu + \mathscr{A}_\mu \quad , \tag{2.50a}$$

where the matrix potential $\mathscr{A}_\mu$ is now defined in the same representation as the matter field,

$$\mathscr{A}_\mu = g A^a_\mu \tau^a \quad . \tag{2.50b}$$

The matter fields transform under a finite gauge transformation $\mathscr{U}$, appropriate to their representation, according to the inverse rule,

$$\mathscr{U} = e^{\tau^a\theta^a} = I + \tau^a\theta^a + \ldots,$$

$$\phi \to \phi^{\mathscr{U}} = \mathscr{U}^{-1}\phi \quad , \quad \psi \to \psi^{\mathscr{U}} = \mathscr{U}^{-1}\psi \quad , \tag{2.51}$$

while the vector potential matrix in this representation, $\mathscr{A}_\mu$, transforms as in Eq. (2.17); see Exercise 2.1.

$$\mathscr{A}_\mu \to \mathscr{A}^{\mathscr{U}}_\mu = \mathscr{U}^{-1}\mathscr{A}_\mu\mathscr{U} + \mathscr{U}^{-1}\partial_\mu\mathscr{U} \quad . \tag{2.52}$$

Consequently the transformation law for the gauge covariant derivative is

$$\mathscr{D}_\mu\phi \equiv \partial_\mu\phi + \mathscr{A}_\mu\phi \to \partial_\mu\phi^{\mathscr{U}} + \mathscr{A}^{\mathscr{U}}_\mu\phi^{\mathscr{U}}$$

$$= \mathscr{U}^{-1}\mathscr{D}_\mu\psi \quad . \tag{2.53a}$$

and similarly for the Fermi fields.

$$\mathscr{D}_\mu\psi = \partial_\mu\psi + \mathscr{A}_\mu\psi \to \partial_\mu\psi^{\mathscr{U}} + \mathscr{A}^{\mathscr{U}}_\mu\psi^{\mu}$$

$$= \mathscr{U}^{-1}\mathscr{D}_\mu\psi \quad . \tag{2.53b}$$

Exercise 2.17. Prove that

$$[\mathscr{D}_\mu, \mathscr{D}_\nu] \ldots = \mathscr{F}_{\mu\nu} \ldots, \quad \mathscr{F}_{\mu\nu} = \partial_\mu \mathscr{A}_\nu - \partial_\nu \mathscr{A}_\mu + [\mathscr{A}_\mu, \mathscr{A}_\nu] \quad . \tag{E2.12}$$

The kinetic part of a gauge invariant matter Lagrangian for Dirac spinor and scalar fields is

$$\mathscr{L}_M = i\hbar\bar{\psi}\gamma^\mu \mathscr{D}_\mu \psi + \tfrac{1}{2}(\mathscr{D}^\mu \phi)(\mathscr{D}^\mu \phi) \quad , \tag{2.54}$$

$$I_M = \int dx \, \mathscr{L}_M \quad , \tag{2.54}$$

where I am using familiar Dirac theory quantities.

$$\bar{\psi} = \psi^\dagger \gamma^0 \quad , \quad \{\gamma^\mu, \gamma^\nu\} = 2g^{\mu\nu} \quad , \quad \bar{\gamma}^\mu \equiv \gamma^0 (\gamma^\mu)^\dagger \gamma^0 = \gamma^\mu \quad . \tag{2.55}$$

It is seen that local gauge invariance completely fixes the gauge potential — matter coupling. Of course there may also be further, pure matter, contributions to the Lagrangian, like a fermion mass term $m\bar{\psi}\psi$, boson mass term $\frac{1}{2}\mu^2\phi^2$, fermion-boson Yukawa couplings, and boson self-couplings governed by a potential $V(\phi)$. Provided these are invariant against rigid gauge transformations, local gauge invariance places no further constraints, and the parameters (masses, coupling constants) are arbitrary. But precisely because of their arbitrariness, they are unattractive to the theorist seeking a fundamental theory where everything is determined. Consequently, we shall not pay very much attention to these possible terms. (One can also construct locally gauge invariant matter-gauge field interaction with an arbitrary coupling constant, e.g. $\bar{\psi}\tau^\alpha \sigma^{\mu\nu}\psi F^a_{\mu\nu}$ with $\sigma^{\mu\nu} \equiv (1/2i)[\gamma^\mu, \gamma^\nu]$. This would not be renormalizable in four dimensions. Such interactions are called non-minimal and I shall not consider them.)

The field equation now becomes

$$(D_\mu F^{\mu\nu})^a = -\frac{\delta I_M}{\delta A^a_\nu} \equiv g J^\nu_a \quad , \tag{2.56}$$

where the matter current J^μ_a for the Lagrangian (2.54) is

$$J^\mu_a = -i\hbar\bar{\psi}\gamma^\mu \tau^a \psi - (\mathscr{D}^\mu \phi)\tau^a \phi \quad . \tag{2.57}$$

Note that with the definition (2.56) the current contains a factor of ħ, because the fermion Lagrangian does. Similarly fermion Noether symmetry currents have an ħ coefficient, since the canonical momentum is proportional to ħ. In addition to (2.56), there are also the matter field equations.

$$\frac{\delta I_M}{\delta\psi} = 0 \ , \qquad \frac{\delta I_M}{\delta\phi} = 0 \ . \tag{2.58}$$

In our example the fermion interactions are parity conserving and the current J^μ_a is a vector. One may also deal with parity non-conserving interactions involving γ_5 and axial vector currents. The γ_5 matrix is defined only in even dimensions, hence we consider two and four dimensions. (Three-dimensional fermions will be discussed in the last section.)

$$\text{two dimensions:} \quad \gamma_5 = \frac{1}{2i}\,\epsilon_{\alpha\beta}\gamma^\alpha\gamma^\beta = i\gamma^0\gamma^1 \ ,$$

$$\text{four dimensions:} \quad \gamma_5 = -\frac{1}{4!}\,\epsilon_{\alpha\beta\gamma\delta}\gamma^\alpha\gamma^\beta\gamma^\gamma\gamma^\delta = \gamma^0\gamma^1\gamma^2\gamma^3 \ ,$$

$$\bar\gamma_5 = \gamma_5 \ , \qquad (\gamma_5)^2 = -I \ . \tag{2.59}$$

In an even-dimensional non-Abelian theory a pure axial vector interaction with Dirac fermions would not be gauge invariant, but either of the two combinations $i\bar\psi\gamma^\mu\,\frac{1}{2}(1\pm i\gamma_5)\tau^a\psi A^a_\mu$ is an allowed interaction. The chiral combinations $\frac{1}{2}(1\pm i\gamma_5)$ are projection operators, so the above may also be written in terms of Weyl spinors.

$$\psi_\pm = \tfrac{1}{2}(1\pm i\gamma_5)\psi \ , \qquad i\gamma_5\,\psi_\pm = \pm\psi_\pm \ ; \tag{2.60a}$$

$$\bar\psi_\pm = \bar\psi\,\tfrac{1}{2}(1\mp i\gamma_5) \ , \qquad -\bar\psi_\pm\,i\gamma_5 = \pm\bar\psi_\pm \ ; \tag{2.60b}$$

$$\psi = \psi_+ + \psi_- \ . \tag{2.60c}$$

Since $\bar\psi_\pm\gamma^\mu\psi_\mp$ vanishes, we see that only one projection couples:

$$\bar\psi\gamma^\mu(\partial_\mu + \tfrac{1}{2}(1\pm i\gamma_5)\mathscr{A}_\mu)\psi = \bar\psi_\pm\gamma^\mu(\partial_\mu + \mathscr{A}_\mu)\psi_\pm + \bar\psi_\mp\gamma^\mu\partial_\mu\psi_\mp \ . \tag{2.61}$$

To preserve local gauge invariance, the right-handed spinor (+) must transform according to $\mathscr{U}^{-1}$ and the left-handed spinor (−) is a singlet, or *vice versa*. Also, there must be no couplings between right- and left-handed spinors; in particular there can be no fermion mass term: $\bar{\psi}_\pm \psi_\pm = 0$. In other words, the gauge group transformation (2.51) is defined separately for the right-handed Weyl spinors and for the left-handed Weyl spinors. Frequently one deletes spinors of one chirality entirely from the theory, and deals with massless Weyl spinors of definite chirality (handedness). Alternatively both may be present, transforming according to different representations of the gauge group. [If both transform by the same representation matrix, the vector theory (2.54) is regained.] Of course the field equations remain as in (2.56) and (2.58), with the current J^μ appropriately modified.

In an Abelian theory, a pure γ_5 vector coupling is allowed, but the fermions must be massless. This can be called axial spinor electrodynamics.

The matter current in a gauge theory satisfies an important constraint: it must be covariantly conserved. This follows when we differentiate covariantly the Yang-Mills equation (2.56). In matrix notation we find

$$D_\nu g J^\nu = D_\nu D_\mu F^{\mu\nu} = \tfrac{1}{2}[D_\nu D_\mu - D_\mu D_\nu]\, F^{\mu\nu} = \tfrac{1}{2}[F_{\mu\nu}, F^{\mu\nu}] = 0 \quad . \tag{2.62}$$

Indeed, in a gauge invariant theory the field equations (2.58) and the current definition (2.56) will imply covariant conservation of J^μ. However, once again I must remind you that all these manipulations are formal; they are valid on the classical level but must be re-examined in the quantum theory. As we shall see later, there are instances, involving chiral couplings, when the source current J^μ possesses an anomalous, non-vanishing divergence. In that case there are obstacles to constructing the quantum gauge theory, which we do not know how to overcome, even though classically there is no evidence for the problem. So the absence of anomalies in source currents is an important constraint on building quantum gauge field theory models.

Exercise 2.18. With the matter Lagrangian as in (2.54), show that the field equations (2.58) imply the covariant conservation of the current (2.57). Show also that $D_\mu J^\mu = 0$ is equivalent to the statement that the matter action is invariant against infinitesimal gauge transformations. Hint: recall the definition of J^μ in (2.56).

Exercise 2.19. The combined Yang-Mills matter Lagrangian is invariant under rigid gauge transformations. Using Noether's theorem, derive the conserved symmetry current j^μ_a. Show that in addition to its pure gauge field part (2.23) it now acquires a matter contribution given by J^μ_a. Show that the ordinary conservation of j^μ_a is equivalent to the covariant conservation of J^μ_a and the gauge field equation (2.56). Finally, show that the ordinary divergence of $F^{\mu\nu}_a$ is given by j^μ_a; hence the conservation of the latter is also required by the anti-symmetry of the field tensor.

Of course the interaction with matter preserves Poincaré invariance and this symmetry is maintained in the quantum theory for all cases of interest. (In two dimensions, with chiral couplings a conserved gauge invariant and local energy-momentum tensor cannot be constructed in the quantum theory; also the quantized model is beset by anomalies in the source current; more about this later.)

In four dimensions and if there are no dimensional parameters in the matter Lagrangian (mass terms, cubic scalar field self-couplings), conformal symmetry holds on the classical level, but just as for pure Yang-Mills, it is anomalously broken after quantization. The scale and special conformal transformation rules for four-dimensional scalar and spinor fields are given by

Dilatation transformation $f^\mu = ax^\mu$,

$$\delta_f \phi = f^\alpha \partial_\alpha \phi + a\phi \quad ,$$

$$\delta_f \psi = f^\alpha \partial_\alpha \psi + \tfrac{3}{2} a\psi \quad ; \tag{2.63}$$

Special conformal transformation $f^\mu = 2c\cdot x x^\mu - c^\mu x^2$,

$$\delta_f \phi = f^\alpha \partial_\alpha v + 2c\cdot x\phi \quad ,$$

$$\delta_f \psi = f^\alpha \partial_\alpha \psi + (3c\cdot x + ic_\alpha \sigma^{\alpha\beta} x_\beta)\psi \quad . \tag{2.64}$$

Note that these are not Lie derivatives; indeed that operator is not defined for spinor fields.

Exercise 2.20. Show that the gauge covariant coordinate transformations $\bar\delta_f\phi$ and $\bar\delta_f\psi$ are as above except that the ordinary derivatives are replaced by the appropriate covariant derivaties.

Exercise 2.21. Verify that the above field transformation rules satisfy the proper commutation relations (E2.7).

Exercise 2.22. For the above transformations, prove $\delta_f \mathscr{L}_{\mathrm{M}} = \partial_\mu \Omega^\mu_{f,\mathrm{M}}$. Note that when scalar fields are present $\Omega^\mu_{f,\mathrm{M}} \neq f^\mu \mathscr{L}_{\mathrm{M}}$. Hence, further improvement of the energy-momentum tensor is required to arrive at the current $\theta^{\mu\nu} f_\nu$; see ref. [6].

Finally let us comment on the role that all the fields which we have discussed play in physical theories. For strong interactions, one believes that Yang-Mills fields provide the "glue" that binds the quarks, described by Dirac spinor fields, in the hadrons. No observed particles are associated with elementary excitations of the gauge or Dirac fields [15]. These excitations are thought to be confined — a speculation which is well supported by various plausible arguments and approximate calculations, but no convincing proof has yet been given. For weak and electromagnetic interactions the vector potentials correspond to observed immediate vector bosons (W, Z) and photons that mediate these forces between quarks and leptons, which again are described by spinor fields coupling in chiral fashion to the gauge fields [16]. The W and Z are massive, yet local gauge invariance prohibits a conventional vector meson mass term in the Lagrangian — as mentioned earlier, $M^2 A_\mu^a A^{\mu a}$ is not gauge invariant. In order to circumvent this problem, model builders [17] have invoked the mechanism of spontaneous gauge symmetry breaking. One would like to see this breaking occur for dynamical reasons in a theory involving just gauge potentials and massless spinor fields, so that only one parameter — the gauge coupling constant — characterizes the theory. Unfortunately, no realistic and convincing model has been constructed in which this attractive speculation can be plausibly supported [18].

Failing at this, one arranges for the spontaneous breaking by introducing scalar fields whose gauge-invariant self-couplings are so chosen that the lower energy configuration is indeed non-symmetric; i.e., one uses the Goldstone-Higgs mechanism [17]. While there are many unattractive aspects to this procedure — the first being its *ad hoc* nature — it is phenomenologically successful [19], and this is where the subject stands today.

2.4 *Classical Gauge Fields*

While our aim is to discuss the quantized Yang-Mills theory, let us pause for a moment and examine the dynamical field equations in their classical setting. After all, the Maxwell theory, which is the antecedent and inspiration for Yang-Mills theory, was thoroughly investigated within classical physics, with results that are quite relevant physically even when quantum effects are ignored. Unfortunately, no such physical success can be claimed here, though much of mathematical interest has been achieved.

We consider first the sourceless equation in four dimensions,

$$D_\nu F^{\mu\nu} = 0 \quad . \tag{2.65}$$

In discussing solutions, it will be useful to characterize them by their energy, momentum and angular momentum. These expressions are of course determined by the energy-momentum tensor (2.33), and take familiar electromagnetic form in terms of the non-Abelian electric and magnetic fields.

$$E_a^i = F_{0i}^a ,$$

$$B_a^i = -\tfrac{1}{2}\epsilon^{ijk}F_{jk}^a \qquad \text{(three spatial dimensions)},$$

$$B_a = -\tfrac{1}{2}\epsilon^{ij}F_{ij}^a \qquad \text{(two spatial dimensions)}. \qquad (2.66)$$

(The discussion will be confined to the theory in three spatial dimensions.) The energy density is θ^{00},

$$\mathscr{E} = \tfrac{1}{2}(E_a^2 + B_a^2) \quad , \quad E = \int \mathrm{d}\boldsymbol{r}\,\mathscr{E} \quad , \qquad (2.67a)$$

the momentum density is the Poynting vector θ^{0i},

$$\mathscr{P} = \boldsymbol{E}_a \times \boldsymbol{B}_a \quad , \quad \boldsymbol{P} = \int \mathrm{d}\boldsymbol{r}\,\mathscr{P} \; , \qquad (2.67b)$$

and the angular momentum density $\epsilon^{ijk}x^j\theta^{0k}$ is given by

$$\mathscr{M} = \boldsymbol{r} \times (\boldsymbol{E}_a \times \boldsymbol{B}_a), \qquad \boldsymbol{M} = \int \mathrm{d}\boldsymbol{r}\,\mathscr{M} \; . \qquad (2.67c)$$

The first issue concerns the existence of regular solutions. If regular initial data is taken, will the solution evolve in a regular fashion, or will the nonlinearities produce singularities? This question has been answered: regular solutions to (2.65) do exist, and the same is true if one considers a larger system: scalar and spinor fields interacting with gauge fields [20]. However, physicists are not so interested in the general solution which depends on arbitrary initial data, but rather in specific solutions which reflect some physically interesting situation. For example, in the Maxwell theory we are interested in plane wave solutions.

Let us note that any Maxwell solution is a solution of the Yang-Mills equation, when one makes the *Ansatz* that the space and internal symmetry degrees of freedom decouple. If one forms $A_\mu^a(x) = \eta^a A_\mu(x)$ with η^a constant and $A_\mu(x)$ satisfying the Maxwell equation, then $A_\mu^a(x)$ is a solution to the Yang-Mills equation, which we shall call "Abelian".

It is interesting to see whether there are plane wave solutions in the non-Abelian theory, which are not Abelian. By "plane wave", we shall mean a configuration of finite energy density ($0 < \mathscr{E} < \infty$), of constant direction for the Poynting vector [$\mathscr{P}(x) = \hat{\mathscr{P}} \, |\mathscr{P}(x)|$ with $\hat{\mathscr{P}}$ constant], and with magnitude of the Poynting vector equal to the energy density ($|\mathscr{P}| = \mathscr{E}$). Such solutions have been constructed [21], but unlike their Maxwell analogs, they do not seem to have any physical significance. Certainly, if gauge quanta are confined, one cannot make a coherent superposition of them to construct an observable plane wave. Alternatively, one may view the Maxwell waves as quantum mechanical wave functions for the photon. However, the non-Abelian plane waves solve a nonlinear equation; they cannot be superposed to form other solutions, and it is hard to see how they can be used as wave functions.

Another class of solutions, more appropriate to nonlinear field theories, are the celebrated solitons, which *do* have a quantum meaning — they are the starting point of a semi-classical description of coherently bound quantum states [22]. A soliton should be a static solutions, have finite energy and be stable in the sense that small perturbations do not grow exponentially in time. However, one proves with virial theorems that no such solution exists in the pure Yang-Mills theory in four, three or two dimensions [23].

Another tack that one can take is that of symmetry. Recall that the classical Yang-Mills theory in four dimensions possesses conformal SO(4, 2) symmetry. One may seek solutions invariant under the maximal compact subgroup, i.e. SO(4) × SO(2). This solution has been constructed [24]; it is called a "meron". But again no physical significance has been attached to it, or to its generalization which possesses the smaller compact invariance symmetry group, SO(4) [25].

There are many other solutions to (2.65) that have been found [26], and while their discoverers invariably highlight some unique characteristic, no physical application has been given thus far — although doubtlessly they are mathematically interesting.

In the above discussion, I have touched on a subject which is worth elaborating upon. I spoke of the SO(4) × SO(2) invariant meron solution. But what exactly does one mean by "invariance" in a gauge theory? Let us approach this question first by considering a scalar field $\phi(x)$. We say that a given functional form for ϕ is invariant against translations in the a^μ direction if there is no dependence on $x \cdot a$, or equivalently if $a^\mu \partial_\mu \phi = 0$. Similarly, ϕ is rotationally invariant if it depends only on $|r|$ and not on angles, i.e., if $\epsilon^{ijk} r^i \partial_k \phi = 0$. In both cases we see that the derivative is a Lie derivative (2.27) with respect to an infinitesimal coordinate transformation against which the ϕ field configuration is invariant. Consequently we define an arbitrary tensor field $T^{\nu\cdots}_{\mu\cdots}$ to be invariant against an arbitrary coordinate transformation, infinitesimally given by f^μ, when

$$L_f T^{\nu\cdots}_{\mu\cdots} = 0 \Leftrightarrow \text{invariant tensor field.} \tag{2.68}$$

However, for a gauge potential A_μ, this definition is too restrictive since we are interested in coordinate invariance of gauge invariant quantities, and not necessarily of a gauge variant object like the potential. Therefore we extend (2.68) by saying that a given gauge potential A_μ is invariant against a coordinate transformation not only if the Lie derivative annihilates it, but more generally if the Lie derivative acting on A_μ produces an infinitesimal gauge transformation.

$$L_f A_\mu = D_\mu W_f \Leftrightarrow \text{invariant gauge potential.} \tag{2.69a}$$

(This condition refers to arbitrary gauge potential configurations, not necessarily solutions to Yang-Mills equations.) One may transform (2.69a) by recalling the earlier result that a Lie derivative of A_μ supplemented by a gauge transformation with gauge function $-f^\alpha A_\alpha$ equals $f^\alpha F_{\alpha\mu}$; see Eq. (2.35). Hence an equivalent test for coordinate invariance is

$$f^\alpha F_{\alpha\mu} = D_\mu \Phi_f \Leftrightarrow \text{invariant gauge field.} \tag{2.69b}$$

This is a gauge covariant test, because it is applied to the gauge covariant field strength and not to the gauge-variant potential.

Aside from giving a convenient criterion for deciding whether some gauge field configuration is invariant under coordinate transformations, the formalism allows for the construction of the most general invariant gauge fields: given f^μ, one solves (2.69a) for A_μ with W_f arbitrary [27].

Also, these ideas have been used in the following "physical" ways. Observe that Φ_f in (2.69b) is a gauge covariant Lorentz scalar field. Hence we see that an invariant gauge field always has some components that involve scalar fields. If one considers a pure gauge theory in $4 + n$ dimensions, and dimensionally reduces to four dimensions, by asserting that the potentials are independent of the n additional dimensions, i.e., that they are invariant against coordinate transformations in those n directions, then some components of the $(4 + n)$-dimensional $F_{\mu\nu}$ survive in four dimensions as scalar fields. It has been suggested that the scalar fields necessary for spontaneous symmetry breaking in the Weinberg-Salam model [17] might arise in this fashion from dimensional reduction of a higher-dimensional pure Yang-Mills model [28]. However, as with all attempts to replace the *ad hoc* symmetry-breaking procedures based on scalar fields with something more natural (is dimensional reduction "natural"?), this idea, though promising at first, has not reproduced all the phenomenologically necessary details.

Another observation is interesting: the quantity Φ_f, defined in (2.69b), has physical significance. Consider a particle moving freely in space — there will be constants of motion C_f associated with that motion as a consequence of the (free) dynamics being invariant against coordinate transformations f^μ, e.g. energy and momentum as a consequence of time and space translation invariance, angular momentum as a consequence of rotational invariance, etc. Now consider the same particle moving in some background gauge field. In general, an arbitrary background field will break the invariance and the constants of motion will disappear. However if the background is itself invariant, the constants of motion will remain, but their form will be modified by a contribution from the field: a term proportional to Φ_f must be added to C_f. This explains the frequently noted fact that particles moving in prescribed gauge fields have unexpected contributions to their constants of motion [29].

Exercise 2.23. The above remarks apply to an Abelian theory as well. Consider a charged particle moving non-relativistically in a magnetic monopole field $\boldsymbol{B} = g\hat{\boldsymbol{r}}/r^2$ according to the Lorentz force law,

$$m\ddot{\boldsymbol{r}} = e\dot{\boldsymbol{r}} \times \boldsymbol{B} \quad . \tag{E2.13}$$

Show that the conserved angular momentum $\boldsymbol{J}$ has an unexpected contribution beyond $\boldsymbol{r} \times m\dot{\boldsymbol{r}}$,

$$\boldsymbol{J} = \boldsymbol{r} \times m\dot{\boldsymbol{r}} - eg\hat{\boldsymbol{r}} \quad . \tag{E2.14}$$

Next consider an infinitesimal rotation $f^i = \epsilon^{ijk} r^j \omega^k$ ($\boldsymbol{\omega}$ is the rotation parameter). Compute $f^i F_{ij}$ with the above magnetic field and show that it is a gradient of a scalar Φ. Compare with (E2.14). (Note: here g is the monopole strength, not the gauge theory coupling constant.)

Returning now to classical solutions, let us take note of another category that has been studied: Yang-Mills fields with prescribed external delta-function sources, i.e., solutions to (2.65) that are singular, with the singularity giving rise to a delta function which is interpreted as a localized source. Again all sorts of interesting phenomena have been found, but physical relevance is obscure, presumably because it is not sensible to approximate quark sources (which should be confined) by a classical localized source [30]. I mention briefly some salient results:

(1) The Coulomb potential, being a Maxwell solution, also solves the Yang-Mills equation with a delta-function point source. However, when

the source strength exceeds a critical magnitude, the solution becomes unstable [31].

(2) The same source can produce more than one, gauge non-equivalent, solution as the source strength is increased beyond a critical value, i.e. there is a bifurcation phenomenon [32, 33].

(3) Some static solutions are stable, without minimizing the energy. They are stabilized by gyroscopic forces analogous to those operating in a spinning top [33, 34].

(4) The Dirac magnetic point monopole, which solves the Abelian theory, but with a singular vector potential containing "strings", may be represented in a non-Abelian theory by a solution which is regular (aside from the singularity at the location of the monopole). For example in SU(2), the following describes a point monopole at the origin [35]:

$$A_a^0 = 0, \qquad A_a^i = \frac{1}{g}\,\epsilon^{aij}\,\frac{\hat{r}^j}{r} \quad . \tag{2.70a}$$

The above solves (2.65) and is SU(2) gauge equivalent to

$$A_a^0 = 0\ , \quad A_{1,2} = 0\ , \quad A_3 = A_{\text{Dirac}}\ , \quad \nabla \times A_{\text{Dirac}} = \frac{1}{g}\,\frac{\hat{r}}{r^2}\ . \tag{2.70b}$$

Note that the potential in (2.70a) is not manifestly rotationally symmetric since it mixes spatial degrees of freedom with internal degrees of freedom — but it is rotationally symmetric according to the criterion (2.69), as it should be since a point magnetic monopole is a rotationally symmetric object.

Exercise 2.24. Find an explicit formula for A_{Dirac}; i.e., solve the equation

$$\nabla \times A_{\text{Dirac}} = g\,\frac{\hat{r}}{r^2} \quad . \tag{E2.15}$$

Hint: Use radial coordinates. Find a gauge transformation that transform (2.70a) to (2.70b). Verify that A_{Dirac} as well as the configuration in (2.70a) are rotationally invariant in the sense (2.69a). (Note: here g is the monopole strength, not the gauge theory coupling constant.)

Example (4) leads to the final and most important category of solutions I shall mention. Now dynamical sources J^μ are included and the source current is given by scalar fields as in (2.57), while the scalar fields satisfy their own dynamical equation as in (2.58). It has been understood that whenever the gauge group is simple and the dynamics of the scalar potential is such that the symmetry group is spontaneously broken to one with a U(1) factor, a *smooth* monopole solution can exist, i.e., the solution (2.70a) acquires a form factor, which vanishes at the origin and approaches unity at large distances. Thus the configuration is regular at the origin, but far away from the origin it still appears as a monopole. Also one finds a smooth, self-consistent solution for the scalar field. Everything is static; the classical energy is finite; there is good reason to believe that the solution is stable — one is speaking of a soliton in three spatial dimensions and consequently of a true quantum state which is being described semi-classically [36].

Exercise 2.25. Call $a(r)$ the form factor occurring in the monopole soliton; i.e., consider the configuration (2.70a) with a further factor $a(r)$. Perform the gauge transformation determined in Exercise 2.24 and establish how (2.70b) is changed.

After the initial discovery of this smooth monopole solution in a SU(2) gauge theory with scalar field in the adjoint representation breaking the symmetry to U(1) — the original 't Hooft-Polyakov monopole [37] — there has been much study of this curious structure [38]. Multi-monopole generalizations and monopoles with electric charge — dyons — have been found. In addition to the obvious mathematical fascination, there is physical interest as well because current speculative "Grand Unified Theories" (which one hopes will unify the already unified weak and electromagnetic interactions with the strong interaction) are precisely of the type that support monopoles. They are based on a simple group (that is why they are unified); the symmetry must break spontaneously (Nature doesn't exhibit this unity); a U(1) factor must survive (electromagnetism exists). Interest was further spurred by the reported observation of a monopole [39].

Nevertheless no definite physical role has yet been found for monopoles and dyons in our present understanding of Nature. Conclusive experimental evidence for their existence is lacking, and their theoretical implications are problematical: cosmological models cannot easily accomodate them [40]; it appears that monopoles cause proton decay [41], yet the proton appears to be stable. Thus we do not know at the present time whether Grand Unified Theory is wrong, or whether its consequences are being improperly applied, or whether new experiments will bring everything into line. Suffice it to say, that monopoles remain important, if only in a negative way, by providing important constraints on model building.

Although particle theory has not yet absorbed the monopole soliton into a consistent phenomenology, solitons in other branches of physics have led to important new insights, especially in lower-dimensional systems that are realized in condensed matter, for example vortices in superconductors and soliton-induced charge fractionization in one-dimensional polymers, like polyacetylene [42].

While analysis of the monopole and other (lower dimensional) soliton solutions uses topological methods, this subject concerns classical field theory, and I refer you to the literature for further discussion [38].

There is one more class of solutions, which I shall describe later. These do not solve the Yang-Mills equations (2.65) in Minkowski space, but rather in Euclidean space, and are called instantons (pseudoparticles). In fact instantons solve the Euclidean self-duality equation

$$ {}^*F^{\mu\nu} = \pm F^{\mu\nu} \quad , \tag{2.71} $$

and then (the Euclidean-space analog of) (2.65) follows by the Bianchi identity. [In Minkowski space, the proportionality constant in (2.71) must be $\pm i$.] Of all the solutions, the instantons have interested mathematicians most; for physicists they give a semi-classical understanding of some of the topological effects that are present in Yang-Mills theory. This will be explained in a subsequent lecture.

3. Quantization

I now come to the problem of quantizing our Yang-Mills theory. It is important to develop this subject carefully, because as we shall see, the topological subtleties of theory can be uncovered in the quantization process. Since Yang-Mills theory is gauge invariant, we expect that there will be complications with a straightforward approach to the canonical formalism, as there already are in Maxwell theory. It turns out that because of local gauge invariance, we are dealing with a constrained canonical system, and therefore I shall first exemplify and solve a constrained quantum mechanical problem, which we can all understand easily.

3.1 Constrained Quantum Mechanical Example

Consider a two-body mechanical system, governed by a Hamiltonian for one-dimensional motion.

$$ H = \frac{p_1^2}{2m_1} + \frac{p_2^2}{2m_2} + V(q_1 - q_2) \quad . \tag{3.1} $$

Obviously, the total momentum is a constant of motion, as a consequence of the translational invariance of the interaction.

$$P = p_1 + p_2 \quad , \quad \frac{i}{\hbar}\,[H, P] = 0 \quad . \tag{3.2}$$

It is useful therefore to pass to center-of-mass coordinates, which are also canonical.

$$Q = \frac{m_1 q_1 + m_2 q_2}{m_1 + m_2} \quad , \quad q = q_1 - q_2 \quad , \quad p = \frac{m_2 p_1 - m_1 p_2}{m_1 + m_2} \tag{3.3}$$

and the Hamiltonian in terms of the new canonical variables reads

$$H = \frac{P^2}{2M} + \frac{p^2}{2\mu} + V(q) \equiv \frac{P^2}{2M} + H_{\mathrm{CM}}(p, q) \quad ,$$

$$M = m_1 + m_2 \quad , \quad \frac{1}{\mu} = \frac{1}{m_1} + \frac{1}{m_2} \quad . \tag{3.4}$$

It is now clear that H does not depend on Q — that is why P commutes with it — so H and P can be simultaneously diagonalized. The states are of the form

$$\Psi(Q, q) = \frac{1}{\sqrt{2\pi}}\, e^{iKQ}\, \psi(q) \quad , \tag{3.5}$$

where $\hbar K$ is an eigenvalue of P and $\psi(q)$ is an eigenfunction of H_{CM}.

Suppose now that for some reason one is instructed to append to our theory the requirement that "physical" states have zero total momentum. One cannot satisfy this requirement by setting the operator $P = p_1 + p_2$ to zero; we cannot have $p_1 = -p_2$ since this would violate the commutation relations satisfied by p_i and q_i. However, we can enforce the requirement by demanding that P acting on physical states is zero.

$$P\Psi_{\mathrm{phys}} = 0 \quad . \tag{3.6}$$

This means that Ψ_{phys} does not depend on the variable conjugate to P — the same variable which is absent from H — and the general solution to (3.6) is an arbitrary function depending only on q, $\Psi_{\mathrm{phys}} = \psi(q)$. This wavefunction is governed by the Hamiltonian H_{CM}, and we can say that the constraint has been solved.

The one disadvantage of the procedure is that physical states are no longer normalizable on the full Hilbert space; clearly the integral $\int dQ dq (\Psi^* \Psi)_{phys}$ will be infinite since Ψ_{phys} does not depend on Q. This reflects the physical fact that our constraint insures that the probability of the total momentum vanishing is 1, and correspondingly the probability of finding nonzero total momentum vanishes. The solution to this difficulty is trivial: do not normalize with respect to Q. Yet one must guard against a formal contradiction: consider the expectation value of the canonical P, Q commutator between physical states,

$$\langle \text{physical} | [Q, P] | \text{physical} \rangle = i\hbar \langle \text{physical} | \text{physical} \rangle \quad .$$

The left-hand side vanishes since P annihilates physical states, while the right-hand side does not. But the contradiction disappears when one remembers that it is illegitimate to take expectation values between non-normalizable states.

It may however not be clear how to solve the constraints, and one may wish to use a functional integral involving all the variables. If there were no constraints, the integral would take the form

$$\langle x_i'; t' | x_i; t \rangle = \int \mathscr{D}p(\tau) \mathscr{D}q_i(\tau) \exp \frac{i}{\hbar} \int_t^{t'} d\tau \left[p_i \dot{q}_i - H \right] \, , \tag{3.7}$$

where the functional integration is over all $p_i(\tau)$ and all $q_i(\tau)$ such that $q_i(t) = x_i$ and $q_i(t') = x_i'$. To enforce the constraint, one must insert a functional delta function of $p_1(\tau) + p_2(\tau) \equiv P(\tau)$. But recall that the functional integral (without constraints) is derived by breaking up the time interval $t' - t$ into small steps of size Δt, and also representing the propagation kernel $\langle x_i'; t' | x_i; t \rangle$ by multiple ordinary integrals,

$$\langle x_i'; t' | x_i; t \rangle =$$

$$\int d\bar{p}_1 \, d\bar{x}_1 \ldots \langle x_i'; t' | \bar{p}_1; t' - \Delta t \rangle \langle \bar{p}_1; t' - \Delta t | \bar{x}_1; t' - 2\Delta t \rangle$$
$$\times \langle \bar{x}_1; t' - 2\Delta t | \cdots | x_i; t \rangle \quad . \tag{3.8}$$

To be sure, inserting momentum delta functions will enforce the constraint, but the $\bar{x}$ integrals will diverge, since the amplitudes do not depend on the center-of-mass coordinate. This can be remedied by inserting another functional delta function in the variable conjugate to the constraint, setting it to any arbitrary value Q_0. This is legitimate, since physical, zero-momentum states do not depend on Q. So the functional integral for the constrained problem reads

$$\langle x_i';t'|x_i;t\rangle = \int \mathscr{D}p_i(\tau)\,\mathscr{D}q_i(\tau)\delta(P)\delta(Q-Q_0)\exp\frac{i}{\hbar}\int_t^{t'} d\tau[p_i\dot{q}_i - H] \quad . \tag{3.9}$$

Finally, if it is not clear how to identify the canonical variable Q, we may use a delta function of an arbitrary function of Q and q, provided we recall the formula

$$\delta(Q-Q_0) = \delta(f(Q,q))\frac{\partial}{\partial Q}f(Q,q) \quad , \tag{3.10a}$$

which functionally is promoted to

$$\delta(Q(\tau)-Q_0) = \delta(f(Q(\tau),q(\tau)))\det\frac{\delta f(Q(\tau),q(\tau))}{\delta Q(\tau')} \quad . \tag{3.10b}$$

[We assume that f vanishes only at Q_0, and $(\partial/\partial Q)f$ is non-vanishing and positive there.] When it is recognized that $\delta f(Q(\tau), q(\tau))/\delta Q(\tau')$ is also the Poisson bracket between the constraint and f, we arrive, heuristically, at Faddeev's formula for the functional integral appropriate to a constrained quantum system [43]:

$$\begin{aligned}\langle x_i';t'|x_i;t\rangle &= \int \mathscr{D}p_i(\tau)\,\mathscr{D}q_i(\tau)\delta(p_1+p_2)\delta(f) \\ &\times \det\{p_1+p_2, f\}\exp\frac{i}{\hbar}\int_t^{t'} d\tau[p_i\dot{q}_i - H] \quad .\end{aligned} \tag{3.11}$$

The first delta function enforces the constraint; the second involves an arbitrary function, hence it is called "a choice of gauge". The exponent is recognize to be the classical action, in terms of canonical variables.

Exercise 3.1. In the above quantum mechanical example the constraint that the total moment vanish is imposed "by hand"; it did not arise from a gauge principle. One may also construct a gauge invariant quantum mechanical example. Consider the Lagrangian

$$L = \frac{1}{2} m_1 (\dot{q}_1 + eA)^2 + \frac{1}{2} m_2 (\dot{q}_2 + eA)^2 - V(q_1 - q_2) \quad . \tag{E3.1}$$

Show that L is invariant under the time-dependent translation $\delta q_i(t) = a(t)$, provided the "gauge potential" $A(t)$ is also transformed. Verify that "Gauss' law" (the equation obtained by varying A) enforces the constraint, and that in the "Weyl gauge" ($A = 0$) the dynamical equations reduce to those of the above example. Show that the Lagrangian (E3.1) may be derived from (3.11), when the constraint delta function is represented by an exponential integral over A.

Exercise 3.2. Show that the Lagrangian describing motion of a point particle in a plane,

$$L = \frac{1}{2} m (\dot{r}^i + eA \epsilon^{ij} r^j)^2 - V(r) \quad , \qquad i = 1, 2 \quad , \tag{E3.2}$$

is gauge invariant against time-dependent rotations, $\delta r^i = \epsilon^{ij} r^j \omega(t)$, provided the "gauge potential" A is gauge transformed. Verify that "Gauss' law" enforces the vanishing of the angular momentum. What are the dynamical equations in the "Weyl gauge"? Solve the constraint and derive the unconstrained Hamiltonian.

3.2 *Quantizing a Yang-Mills Theory*

Let us now turn to the Yang-Mills theory, which is governed by the Lagrangian $\mathscr{L}_{\text{YM}}$. (Since a gauge theory is trivial in two dimensions, we take the dimensionality to be three or greater. We shall use a notation appropriate to the four-dimensional theory, but an identical development can be given in any number of dimensions, with the modification that the magnetic field is not a space vector, but an anti-symmetric tensor, except in two spatial dimensions, where it is a scalar.)

$$\mathscr{L}_{\text{YM}} = \frac{1}{2}(E_a^2 - B_a^2) \quad ,$$

$$E_a = -\dot{A}_a - \nabla A_a^0 - g f_{abc} A_b^0 A_c \quad ,$$

$$B_a = \nabla_a \times A_a - \frac{1}{2} g f_{abc} A_b \times A_c \quad . \tag{3.12}$$

Our task is to build a Hamiltonian scheme, which will give rise to the Yang-Mills equations. These I record one again in non-covariant form. The time component of (2.16a) — the non-Abelian Gauss' law — reads

$$(\boldsymbol{D}\cdot\boldsymbol{E})_a = 0 \quad , \tag{3.13}$$

and the space component — the non-Abelian Ampère law — is

$$(D^0\boldsymbol{E})_a = (\boldsymbol{D}\times\boldsymbol{B})_a \quad . \tag{3.14}$$

(In the space vector notation, the covariant derivative $\boldsymbol{D}$ has components D_i.)

The first problem that is encountered in passing to a Hamiltonian description arises from the fact that $\mathscr{L}_{\mathrm{YM}}$ does not depend on $\dot{A}_a^0$; thus there is no momentum conjugate to A_a^0. To remedy this, we use our gauge freedom to set A_a^0 to zero. This choice of gauge is called the "Weyl gauge" because Weyl publicized its use it in electrodynamics [44]. Then (3.12)-(3.14) reduce to

$$\mathscr{L}_{\mathrm{YM}} = \tfrac{1}{2}(\dot{\boldsymbol{A}}_a^2 - \boldsymbol{B}_a^2) \quad , \qquad \boldsymbol{E}_a = -\dot{\boldsymbol{A}}_a \quad ,$$

$$\boldsymbol{B}_a = \nabla\times\boldsymbol{A}_a - \tfrac{1}{2}g f_{abc}\boldsymbol{A}_b\times\boldsymbol{A}_c \quad , \tag{3.15}$$

and

$$(\boldsymbol{D}\cdot\boldsymbol{E})_a = 0 \quad , \tag{3.16}$$

$$\dot{\boldsymbol{E}}_a = (\boldsymbol{D}\times\boldsymbol{B})_a \quad . \tag{3.17}$$

Now the Lagrangian lends itself to a canonical transcription into a Hamiltonian. The dynamical variable is $\boldsymbol{A}_a$, its canonical momentum is $\dot{\boldsymbol{A}}_a = -\boldsymbol{E}_a$, and the Hamiltonian becomes

$$H = \frac{1}{2}\int \mathrm{d}\boldsymbol{r}(\boldsymbol{E}_a^2 + \boldsymbol{B}_a^2) \quad , \tag{3.18}$$

which is also the total energy; see (2.67a). The non-vanishing canonical equal time commutator

$$[E_a^i(\boldsymbol{r}), A_b^j(\boldsymbol{r}')] = i\hbar\delta_{ab}\delta^{ij}\delta(\boldsymbol{r}-\boldsymbol{r}') \quad , \tag{3.19}$$

implies

$$\dot{\boldsymbol{A}}_a = \frac{i}{\hbar}[H, \boldsymbol{A}_a] = -\boldsymbol{E}_a \quad , \tag{3.20a}$$

$$\dot{\boldsymbol{E}}_a = \frac{i}{\hbar}[H, \boldsymbol{E}_a] = (\boldsymbol{D} \times \boldsymbol{B})_a \quad , \tag{3.20b}$$

and we see that the Hamiltonian equations reproduce Ampère's law (3.17) and the definition $\boldsymbol{E}_a$ in terms of $\dot{\boldsymbol{A}}_a$. However, Gauss' law (3.16) has as yet not emerged, because it is a fixed-time constraint between canonical variables. (Since we are developing a fixed-time Schrödinger picture for the quantum field theory, the time argument of the operators is suppressed.)

Let us for the moment ignore the absence of Gauss' law, and observe that we have arrived at a completely consistent quantum field which, however, does not yet coincide with the Lorentz-invariant Yang-Mills theory, since we do not have Gauss' law. Certainly we cannot simply set $(\boldsymbol{D} \cdot \boldsymbol{E})_a$ to zero; this operator does not commute with the canonical variables.

We observe that the Lagrangian (3.15) possesses a Noether symmetry in which $\boldsymbol{A}_a$ changes infinitesimally according to

$$\delta \boldsymbol{A}_a = -\frac{1}{g}(\boldsymbol{D}\theta)_a \quad , \tag{3.21}$$

where θ^a is a c-number, space dependent but time independent, function. Of course this is recognized as the residual local gauge invariance in the Weyl gauge: $A_a^0 = 0$ is preserved by time-independent gauge transformations. But now we view it as an ordinary continuous symmetry, and Noether's theorem gives the conserved charge:

$$Q_\theta = \int \mathrm{d}\boldsymbol{r}\, \boldsymbol{\Pi}_a \cdot \delta \boldsymbol{A}_a = \frac{1}{g}\int \mathrm{d}\boldsymbol{r} \boldsymbol{E}_a \cdot (\boldsymbol{D}\theta)_a \quad . \tag{3.22}$$

Since θ^a is arbitrary, we also know that $(1/g)\int \mathrm{d}\boldsymbol{r}(\boldsymbol{D}\cdot\boldsymbol{E})_a \theta^a$ is conserved, and so also is $(\boldsymbol{D}\cdot\boldsymbol{E})_a$; a fact which may be verified by an explicit commutation with H. Note that $(\boldsymbol{D}\cdot\boldsymbol{E})_a$ is not zero, since Gauss' law is not one of our operator equations. Thus we recognize that $(\boldsymbol{D}\cdot\boldsymbol{E})_a$ is the time-dependent generator of infinitesimal time-independent gauge transformations, with the space-dependent c-number parameter θ^a stripped away; we call it G_a.

$$G_a = -\frac{1}{g}\,(\boldsymbol{D}\cdot\boldsymbol{E})_a \quad . \tag{3.23}$$

It satisfies the following commutation relations:

$$\frac{i}{\hbar}[G_a(\boldsymbol{r}), A_b(\boldsymbol{r}')] = \frac{1}{g}\,\delta_{ab}\nabla\delta(\boldsymbol{r}-\boldsymbol{r}') + f_{abc}A_c(\boldsymbol{r})\delta(\boldsymbol{r}-\boldsymbol{r}') \ ,$$

$$\frac{i}{\hbar}[G_a(\boldsymbol{r}), E_b(\boldsymbol{r}')] = f_{abc}E_c(\boldsymbol{r})\delta(\boldsymbol{r}-\boldsymbol{r}') \quad . \tag{3.24}$$

These show that G_a does indeed generate infinitesimal gauge transformations. The Hamiltonian is gauge invariant and G_a commutes with it, so G_a is time independent,

$$\dot{G}_a = \frac{i}{\hbar}[H, G_a] = 0 \quad . \tag{3.25}$$

The commutators of different G_a follow the Lie algebra of the gauge group,

$$\frac{i}{\hbar}[G_a(\boldsymbol{r}), G_b(\boldsymbol{r}')] = f_{abc}G_c(\boldsymbol{r})\delta(\boldsymbol{r}-\boldsymbol{r}') \quad . \tag{3.26}$$

We now see how to impose Gauss' law: the operator G_a is *not* set to zero; rather one demands that physical states be annihilated by it.

$$G_a(\boldsymbol{r})|\text{physical}\rangle = 0 \quad . \tag{3.27}$$

One may consider the states to be realized in a Schrödinger representation as functionals of A_a. Thus (3.27) becomes a functional differential equation satisfied by physical state functional $\Psi_{\text{phys}}(A)$,

$$\left(\boldsymbol{D}\cdot\frac{\delta}{\delta \boldsymbol{A}}\right)_a \Psi_{\text{phys}}(A)$$

$$= (\boldsymbol{\nabla}\delta_{ab} - g f_{acb}A_c)\cdot\frac{\delta}{\delta \boldsymbol{A}_b}\,\Psi_{\text{phys}}(A) = 0 \ . \tag{3.28}$$

while the Hamiltonian eigenvalue equation reads

$$\int \mathrm{d}r \left\{ -\frac{\hbar^2}{2} \frac{\delta^2}{\delta A_a^2} + \frac{1}{2} B_a^2 \right\} \Psi(A) = E\Psi(A) \quad . \tag{3.29}$$

Eqs. (3.28) and (3.29) correspond to the Yang-Mills quantum field theory, where only those solutions of the latter which also satisfy the former are physical. Equation (3.26) shows that the constraints close on commutation; hence (3.28) is integrable, at least locally.

We shall still need to examine the gauge transformation properties of our theory more closely, but let us postpone this and first discuss the constraint (3.28). The conservation of G_a, i.e., the fact that it commutes with H, means that H does not depend on certain combinations of the dynamical variables A_a. Correspondingly the constraint $G_a \Psi_{\text{phys}}(A) = 0$ forces the state functional to be independent of these quantities. As a consequence, the physical states are not normalizable, and one should guard against contradictory statements that would arise if expectation values of the commutators in (3.24) are taken between physical states.

One may proceed by solving the constraint (3.28), i.e., by finding the most general functional satisfying Gauss' law, and then deriving the effective Schrödinger equation for the unconstrained functional. For the Abelian theory this is trivial to do, since one can immediately identify the variable conjugate to $\nabla \cdot \boldsymbol{E}$; it is essentially the longitudinal component of $\boldsymbol{A}$. Hence taking the wave functional to depend only on the transverse, but not on the longitudinal, components of $\boldsymbol{A}$ solves the constraint. This is equivalent to setting $\nabla \cdot \boldsymbol{A}$ to zero, and the conventional electrodynamic Coulomb gauge emerges naturally with our approach to the Maxwell theory. Details are in the Exercises.

Exercise 3.3. Show that the most general solution to the Abelian version (3.28) in three spatial dimensions is a functional that depends on the transverse components of A, but not on the longitudinal ones. Show that a solution to the Abelian version of (3.29) is the functional

$$\Psi_0(A) = \exp\left[-\frac{1}{2\hbar} \int \mathrm{d}r \mathrm{d}r' \, A^i(r) \, G^{ij}(r, r') \, A^j(r') \right] \quad . \tag{E3.3}$$

with

$$G^{ij}(r, r') = (-\nabla^2 \delta^{ij} + \partial_i \partial_j) \int \frac{\mathrm{d}k}{(2\pi)^3} \, \mathrm{e}^{ik \cdot (r - r')} \, \frac{1}{k}$$

$$= -\frac{2}{\pi^2 R^4} (\delta^{ij} - 2\hat{R}^i \hat{R}^j) \quad , \qquad R = r - r' \quad . \tag{E3.4}$$

The singularity at $R = 0$ is treated with the principal-value prescription. The energy eigenvalue

$$E = \hbar \int \mathrm{d}r \int \frac{\mathrm{d}k}{(2\pi)^3}\, k \;, \tag{E3.5}$$

is the conventional infinite vacuum energy, hence (E3.3) is the ground state wave functional. Note that the constraint is automatically satisfied, and that (E3.3) may also be written as a functional of gauge invariant quantities,

$$\Psi_0(A) = \exp\left[-\frac{1}{4\pi^2\hbar}\int \mathrm{d}r\mathrm{d}r'\, \frac{B(r)\cdot B(r')}{|r-r'|^2}\right] . \tag{E3.6}$$

Observe that $\Psi_0(A)$ does not depend on the longitudinal component of A, hence it cannot be functionally integrated over all components of A. What is the wave functional for the one-photon state with momentum p, and what is its energy eigenvalue? What is the two-photon wave functional?

Exercise 3.4. In the Maxwell theory with an external, static, c-number charge density $\rho(r)$, Gauss' law reads

$$\nabla\cdot E = \rho \;. \tag{E3.7}$$

Solve the constraint and show that physical states involve an arbitrary functional of the transverse part of A, times a phase factor depending on ρ and on the longitudinal part of A. Show that in the effective Schrödinger equation for the transverse functional, the energy eigenvalue includes the Coulomb energy.

For the non-Abelian theory the task is more difficult; in particular $\nabla\cdot A_a$ is *not* conjugate to $\nabla\cdot E_a$ and the Coulomb gauge does not arise naturally. In fact the constraint has been solved [45], but the effective Hamiltonian is complicated and does not lend itself to a power series expansion in g, since $1/g$ terms are present. One may understand these inverse powers of the coupling constant by recognizing that the solution of the constraint treats the non-Abelian gauge group exactly, while in the limit $g = 0$ the Yang-Mills theory is no longer invariant under a non-Abelian group of transformations. Hence, removing the non-Abelian gauge degrees of freedom exactly, which is what solving the constraint equation amounts to, prevents one from taking the limit $g = 0$.

While it would be useful to understand the dynamics of the unconstrained Hamiltonian, we have not yet succeeded in doing so. Consequently one remains with the constrained formalism and uses, for example, a path integral formulation.

Before deriving the functional integral representation of the Yang-Mills quantum theory, we re-examine the gauge transformations of the theory.

3.3 θ Angle

In addition to the infinitesimal gauge transformations, one may also perform finite, time-independent gauge transformations, U. These leave the Lagrangian (3.15) and the Hamiltonian (3.18) invariant, and are implemented by the unitary operator $\mathscr{G}_U$.

$$
\begin{aligned}
&\boldsymbol{A} \rightarrow \mathscr{G}_U \boldsymbol{A} \mathscr{G}_U^{\dagger} = U^{-1} \boldsymbol{A} U - U^{-1} \nabla U \quad , \\
&\boldsymbol{E} \rightarrow \mathscr{G}_U \boldsymbol{E} \mathscr{G}_U^{\dagger} = U^{-1} \boldsymbol{E} U \quad , \\
&\boldsymbol{B} \rightarrow \mathscr{G}_U \boldsymbol{B} \mathscr{G}_U^{\dagger} = U^{-1} \boldsymbol{B} U \quad , \\
&[H, \mathscr{G}_U] = 0 \quad .
\end{aligned}
\tag{3.30}
$$

Clearly the effect of $\mathscr{G}_U$ on all states (physical and unphysical) is to gauge transform the argument,

$$
\mathscr{G}_U \Psi(\boldsymbol{A}) = \Psi(\boldsymbol{A}^U) \equiv \Psi(U^{-1} \boldsymbol{A} U - U^{-1} \nabla U) \quad . \tag{3.31}
$$

Since $\mathscr{G}_U$ and H commute, we may choose $\Psi(\boldsymbol{A})$ to be an eigenstate of $\mathscr{G}_U$, for a given U, with an eigenvalue which is a phase, since $\mathscr{G}_U$ is unitary:

$$
\mathscr{G}_U \Psi(\boldsymbol{A}) = \mathrm{e}^{-\mathrm{i}\theta_U} \Psi(\boldsymbol{A}) \quad . \tag{3.32}
$$

The question now is whether physical states, i.e., those annihilated by G_a, are truly invariant against finite gauge transformations or only phase invariant. Let us observe that no physical principle will be violated if θ_U is nonzero, since the probability density $\Psi^*(\boldsymbol{A})\Psi(\boldsymbol{A})$ is gauge invariant.

One might suppose that $\mathscr{G}_U$ could be represented by exponentiating the infinitesimal generator (3.22),

$$
\mathscr{G}_U = \exp\left(\frac{\mathrm{i}}{\hbar} Q_\theta\right)
$$

with $U = \mathrm{e}^{\Theta}$ and $\Theta = \theta^a T^a$, where the exponential operator is defined by its power series. If this were the case, then $\mathscr{G}_U$ would leave physical states invariant, because G_a and hence Q_θ annihilates them. (An integration by parts is needed for G_a to act on the state; we assume no surface terms arise, for otherwise certainly $\mathscr{G}_U$ will not leave the state invariant. A precise statement about large-distance behavior of gauge functions will be given presently.) However, it is easy to see that in three-dimensional space there are finite gauge transformations that are not obtained by operating with the exponential operator.

Consider the following quantity, defined in three-space:

$$W(A) = \frac{-1}{4\pi^2}\int d\boldsymbol{r}\,\epsilon^{ijk}\,\mathrm{tr}\left(\frac{1}{2}A^i\partial_j A^k - \frac{1}{3}A^iA^jA^k\right)$$

$$= -\frac{1}{16\pi^2}\int d\boldsymbol{r}\,\epsilon^{ijk}\,\mathrm{tr}\left(F_{ij}A_k - \frac{2}{3}A_iA_jA_k\right) . \qquad (3.33)$$

I shall show that $W(A)$ is not gauge invariant, yet is is invariant against gauge transformations implemented by $\exp(iQ_\theta/\hbar)$. Observe that $W(A)$ has the property that

$$\frac{\delta W(A)}{\delta A_a^i} = \frac{g^2}{8\pi^2}B_a^i \quad . \qquad (3.34)$$

Consequently

$$e^{iQ_\theta/\hbar}W(A)e^{-iQ_\theta/\hbar}$$

$$= W(A) + \frac{i}{\hbar}[Q_\theta, W(A)] + \ldots$$

$$= W(A) + \int d\boldsymbol{r}\,\frac{i}{\hbar}[Q_\theta, \boldsymbol{A}_a(\boldsymbol{r})]\cdot\frac{g^2}{8\pi^2}\boldsymbol{B}_a(\boldsymbol{r}) + \ldots$$

$$= W(A) - \frac{g}{8\pi^2}\int d\boldsymbol{r}(\boldsymbol{D}\theta)_a\cdot\boldsymbol{B}_a + \ldots$$

$$= W(A) + \frac{g}{8\pi^2}\int d\boldsymbol{r}\,\theta_a(\boldsymbol{D}\cdot\boldsymbol{B})_a + \ldots$$

$$= W(A) \quad , \qquad (3.35)$$

since the Bianchi identity states that the covariant divergence of $\boldsymbol{B}$ is zero. (In order to drop surface terms, we have assumed that θ^a goes to a constant at spatial infinity and the $\boldsymbol{B}$ falls faster than $1/r^2$, i.e., there are no magnetic monopoles; see also below.) Thus $W(A)$ is gauge invariant against gauge transformations implemented by the exponential operator $\exp(\mathrm{i}Q_\theta/\hbar)$.

The gauge transformation of $W(A)$ may also be computed directly:

$$\begin{aligned} \mathscr{G}_U W(A)\mathscr{G}_U^{-1} &= W(A^U) \\ &= W(A) - \frac{1}{8\pi^2}\int \mathrm{d}r\epsilon^{ijk}\,\partial_i\, tr(\partial_j U U^{-1} A^k) \\ &+ \frac{1}{24\pi^2}\int \mathrm{d}r\epsilon^{ijk}\,\mathrm{tr}(U^{-1}\partial_i U U^{-1}\partial_j U U^{-1}\partial_k U) \quad . \end{aligned} \tag{3.36}$$

To evaluate the gauge change in $W(A)$, we must impose a boundary condition on U so that integrals in (3.36) not diverge. We shall assume that U tends to a constant matrix at spatial infinity, and without loss of generality this may be taken to be $\pm I$. Also $\boldsymbol{A}$ is assumed to fall faster than $1/r$. Note that both these statements are consistent with the earlier requirement that θ^a approach a constant and $\boldsymbol{B}$ decrease faster than $1/r^2$. With these, the middle integral in (3.36) vanishes, but the last remains. It occurs only for a non-Abelian theory and makes no reference to the potentials since it depends on only the non-Abelian gauge transformations,

$$\begin{aligned} \omega(U) &\equiv \frac{1}{24\pi^2}\int \mathrm{d}r\epsilon^{ijk}\,\mathrm{tr}(U^{-1}\partial_i U U^{-1}\partial_j U U^{-1}\partial_k U) \\ &= W(-U^{-1}\nabla U) \quad . \end{aligned} \tag{3.37}$$

This integration may be performed for a definite U, and one finds a non-vanishing results; see Exercise 3.5.

However, it is not necessary to perform the explicit evaluation. We recognize that the gauge functions U, with large-distance asymptotes $\pm I$, provide a mapping of the three-sphere S_3 (which is equivalent to our three-space once the points at infinity have been identified) into the gauge group. Such mappings fall into disjoint homotopy classes, labeled by the integers, and gauge functions belonging to different classes cannot be deformed continuously into each other. In particular, only those in the zero class are deformable to the identity. This fact is expressed by the mathematical statement that

$$\Pi_3\,(\text{non-Abelian compact gauge group}) \\ = (\text{group of all integers under addition}) \equiv \mathscr{Z}\,. \tag{3.38}$$

Furthermore, $\omega(U)$ is an analytic expression for the integer which labels U's homotopy class. It is called the "winding number" of the gauge transformation.

Thus we see that $W(A)$ is not gauge invariant against homotopically non-trivial gauge transformations; rather it changes by U's winding number.

$$W(A) \to W(A^U) = W(A) + n_U \\ n_U \equiv \omega(U)\;. \tag{3.39}$$

We now recognize that $\mathscr{G}_U$ may be represented by the exponential only when U belongs to the trivial homotopy class and is deformable to the identity. (For these θ^a vanishes at $r = \infty$.) Correspondingly physical states are gauge invariant against these gauge transformations. But homotopically non-trivial gauge transformations that are not deformable to the identity are not implemented by the exponential operator, and physical states are only phase invariant. If $\mathscr{G}_n$ is the unitary operator that implements a representative gauge transformation U_n belonging to the nth homotopy class, then we have

$$\mathscr{G}_n|\text{physical}\rangle = e^{-in\theta}\,|\text{physical}\rangle\;, \\ \left.\begin{aligned} \mathscr{G}_0 &= e^{iQ_\theta/\hbar} = I \\ \mathscr{G}_n\,\mathscr{G}_m &= \mathscr{G}_{n+m} \end{aligned}\right\}\ \text{on physical states.} \tag{3.40}$$

[It should be clear that θ in (3.40) is distinct from the gauge parameter θ^a.] The topologically non-trivial gauge transformations are called "large", while the trivial ones are called "small."

This is the origin of the famous vacuum angle in gauge theories [46, 47]; we see that its presence is established without any approximation but rather by carefully following the response of a non-Abelian gauge theory (in four-dimensional space-time) to the large gauge transformations which are topologically richer than anything in electrodynamics [46]. We emphasize that in two spatial dimensions, all gauge transformations are small.

Exercise 3.5. Consider the SU(2) gauge group, and evaluate (3.37) with

$$U = e^{\theta^a T^a} \quad , \qquad T^a = \sigma^a/2i \quad . \tag{E3.8}$$

Show that the integrand in (3.37) may be written as a total divergence, so that the volume integral for $\omega(U)$ may be recast as an integral over the surface at infinity, provided θ^a is regular in the interior.

$$\omega(U) = \frac{1}{16\pi^2}\int dS^i \epsilon^{ijk} \epsilon_{abc}\, \hat{\theta}^a \partial_j \hat{\theta}^b \partial_k \hat{\theta}^c (\sin|\theta| - |\theta|) \quad ,$$

$$|\theta| = \sqrt{\theta^a\theta^a} \quad , \qquad \hat{\theta}^a = \theta^a/|\theta| \quad . \tag{E3.9}$$

Exercise 3.6. For an arbitrary SU(2) gauge transformation parametrized as in (E3.8), the condition $U \xrightarrow[r\to\infty]{} \pm I$ sets the requirement that $|\theta| \xrightarrow[r\to\infty]{} 2\pi n$. Thus the general formula (E3.9) reduces to

$$\omega(U) = -\frac{n}{8\pi}\int dS^i \epsilon^{ijk} \epsilon_{abc}\, \hat{\theta}^a \partial_j \hat{\theta}^b \partial_k \hat{\theta}^c \quad . \tag{E3.10}$$

By parametrizing the unit three-vector $\hat{\theta}^a$ as

$$\hat{\theta}^1 = \sin\psi \cos\phi \quad , \qquad \hat{\theta}^2 = \sin\psi \sin\phi \quad , \qquad \hat{\theta}^3 = \cos\psi \quad , \tag{E3.11}$$

and the surface of the two-sphere at $r = \infty$ by angles α and β on which ψ and ϕ depend, show that

$$\omega(U) = -\frac{n}{4\pi}\int_0^{2\pi} d\beta \int_0^{\pi} d\alpha \sin\psi \left(\frac{\partial\psi}{\partial\alpha}\frac{\partial\phi}{\partial\beta} - \frac{\partial\psi}{\partial\beta}\frac{\partial\phi}{\partial\alpha} \right) \quad . \tag{E3.12}$$

The quantity in parentheses is the Jacobian, apart from sign, of the transformation from (α, β) to (ψ, ϕ). Hence the above integral is the integer that counts the (signed) number of times (ψ, ϕ) range over their two-sphere as (α, β) range over theirs. This shows quite generally, that $\omega(U)$ is an integer.

Verify the above analysis by considering the gauge function

$$\theta^a = \hat{r}^a f(r), \qquad f(0) = 0 \quad , \qquad f(\infty) = 2\pi n \quad . \tag{E3.13}$$

Evaluate (3.37), (E3.9) and (E3.12). (It is important to appreciate that here one is not compactifying R_3 to S_3. Even though the points at infinity can be identified as far as the gauge function U is concerned — by hypothesis U tends to a uniform angle-independent limit — the Lie-algebra valued quantity Θ does not possess a uniform limit; $\Theta \xrightarrow[r\to\infty]{} -i\boldsymbol{\sigma}\cdot\hat{\mathbf{r}}\pi n$. Thus with the asymptote (E3.13), the Lie-algebra cannot be defined on a compact manifold, and this is why a total derivative expression for the integrand $\omega(U)$ can be given. On a compact manifold no such formula is available. The generalization of (E3.9) to an arbitrary group on R_3 is given in ref. [48]).

Exercise 3.7. Prove that $\exp[\pm(8\pi^2/\hbar g^2)\, W(A)]$ solves the non-Abelian functional Schrödinger equation (3.29) with zero eigenvalue. Regrettably this remarkable solution diverges for large A and does not seem to have physical meaning [49].

3.4 *Functional Integral Formulation*

The Yang-Mills quantum theory can be transcribed into a functional integral formulation, which has the advantage of exhibiting the vacuum angle in an unambiguously gauge- and Lorentz-invariant fashion. In order to derive the functional integral, we prefer to work with states which are invariant even against large gauge transformations. Thus we inquire whether we can modify our physical wave functionals so that they are gauge invariant. This is easy to do. Recall from Eq. (3.39) that $W(A)$ shifts by n under a gauge transformation in homotopy class n. Hence

$$\Phi \equiv e^{iW(A)\theta}\,\Psi_{\rm phys} \,, \tag{3.41}$$

continues to be annihilated by G_a, and is also gauge invariant against all gauge transformations, large and small. However, Φ satisfies a Schrödinger equation more complicated than (3.29), which follows from (3.34) and (3.41).

$$\int dr \left[\frac{1}{2}\left(\frac{\hbar}{i}\frac{\delta}{\delta A_a} - \frac{\hbar\theta g^2}{8\pi^2} B_a\right)^2 + \frac{1}{2} B_a^2\right] \Phi = E\Phi \,. \tag{3.42}$$

The path integral for this Hamiltonian is given by analogy with the constrained quantum mechanical system; see (3.11).

$$Z = \int \mathscr{D}E_a \mathscr{D}A_a\, \delta(G_a)\,\delta(\chi_b) \det\{G_a, \chi_b\}$$
$$\times \exp \frac{-i}{\hbar} \int dx \left\{ E_a \cdot \dot{A}_a + \frac{1}{2}\left(E_a + \frac{\hbar\theta g^2}{8\pi^2} B_a\right)^2 + \frac{1}{2} B_a^2 \right\}. \tag{3.43a}$$

Here χ_b is an arbitrary gauge choice, taken to depend on $\boldsymbol{A}_a$. Since G_a generates gauge transformations, the Poisson bracket is just an infinitesimal gauge transformation of χ_b, with "parameter" θ^a stripped off; we represent it by $\delta_a \chi_b$. The constraint delta function may be written as a (functional) phase integral, with an integration variable which we call A_a^0. Thus (3.43a) becomes

$$Z = \int \mathscr{D}A^a_\mu \, \mathscr{D}\boldsymbol{E}_a \, \delta(\chi_b) \det(\delta_a \chi_b) \exp \frac{-i}{\hbar} \times \int dx \left\{ \boldsymbol{E}_a \cdot \dot{\boldsymbol{A}}_a - A_a^0 (\boldsymbol{D} \cdot \boldsymbol{E})_a + \frac{1}{2}\left(\boldsymbol{E}_a + \frac{\hbar \theta g^2}{8\pi^2} \boldsymbol{B}_a\right)^2 + \frac{1}{2} \boldsymbol{B}_a^2 \right\} . \tag{3.43b}$$

Next the Gaussian $\boldsymbol{E}_a$ integral is performed, leaving an expression that may be written in invariant notation.

$$Z = \int \mathscr{D}A^a_\mu \, \delta(\chi_b) \det(\delta_a \chi_b) \exp \frac{i}{\hbar} \int dx \mathscr{L},$$

$$\mathscr{L} \equiv \frac{1}{2g^2} \operatorname{tr} F^{\mu\nu} F_{\mu\nu} - \frac{\hbar\theta}{16\pi^2} \operatorname{tr} {}^*F^{\mu\nu} F_{\mu\nu} \quad . \tag{3.43c}$$

Note that even though in our derivation the gauge condition χ depends only on $\boldsymbol{A}$, the result holds for arbitrary χ; hence we may allow χ to depend also on A^0 provided the determinant is correspondingly adjusted [50].

Aside from the familiar gauge fixing delta function and gauge compensating determinant, we have arrived at the functional integral formulated in terms of the gauge- and Lorentz-invariant Yang-Mills action, but with an additional term contributing to the Lagrangian as a consequence of the angle θ. This gauge invariant term does not contribute to the equations of motion because it is a total divergence (of a gauge variant quantity).

$$\mathscr{P} \equiv - \frac{1}{16\pi^2} \operatorname{tr} {}^*F^{\mu\nu} F_{\mu\nu} = \partial_\mu \mathscr{C}^\mu \quad , \tag{3.44a}$$

$$\mathscr{C}^{\mu} = -\frac{1}{16\pi^2}\,\epsilon^{\mu\alpha\beta\gamma}\,\mathrm{tr}(F_{\alpha\beta}A_{\gamma} - \tfrac{2}{3}A_{\alpha}A_{\beta}A_{\gamma}) \tag{3.44b}$$

Also it does not contribute to the energy-momentum tensor; $\theta^{\mu\nu}$ in (2.33) remains conserved. However, tr $^*F^{\mu\nu}F_{\mu\nu}$ does affect the canonical formalism because it depends on time derivatives of the canonical variables. Observe that $W(A)$ is given by the spatial integral of $\mathscr{C}^0$, with the time dependence suppressed,

$$W(A) = \int \mathrm{d}r\,\mathscr{C}^0(A) \quad . \tag{3.45}$$

Exercise 3.8. Derive the energy-momentum tensor (2.33) for the Lagrangian in (3.43c) by using Noether's theorem and an appropriate improvement.

The form of the θ term in the Lagrangian shows that θ is Lorentz and gauge invariant. However, since tr $^*F^{\mu\nu}F_{\mu\nu}$ is odd under P and T reflection symmetries, the θ term is P and T [or CP] violating. Moreover, we see now that an independent argument for the existence of a θ-parameter in four-dimensional Yang-Mills theory is the fact that tr $^*F^{\mu\nu}F_{\mu\nu}$ could have been added to the original Lagrangian without affecting classical dynamics, which is entirely determined by equations of motion. However, such a term modifies quantum dynamics which depends on the Lagrangian and on the action, as is seen, for example, from the functional integral formulation.

Exercise 3.9. Starting with the Lagrangian of (3.43c) in the Weyl gauge ($A^0 = 0$) show that the Hamiltonian is conventional,

$$H = \frac{1}{2}\int \mathrm{d}r\,(E_a^2 + B_a^2) \quad . \tag{E3.14}$$

Derive the canonical form for the Hamiltonian in terms of canonical variables and compare with (3.42).

The conclusion therefore is that a four-dimensional Yang-Mills quantum theory is characterized not only by its coupling constant g, but also by a hidden parameter θ, which enters on the quantum level and involves tr$^*F^{\mu\nu}F_{\mu\nu}$. (This effect does not occur in three space-time dimensions, but we shall see later that there too an unexpected parameter characterizes the Yang-Mills theory.)

The novel addition to the action has a well-established place in mathematics: $\mathscr{P} \equiv -(1/16\pi^2)$ tr $^*F^{\mu\nu}F_{\mu\nu}$ is called the "Pontryagin density" and $\mathscr{C}^{\mu}$ — the vector whose divergence equals the Pontryagin density — is the "Chern-Simons secondary characteristic class".

For well-defined classical potentials, the Pontryagin index

$$q \equiv \int \mathrm{d}x\, \mathscr{P} \tag{3.46}$$

is a topological invariant; it does not change under local variations of the potentials.

$$\delta q = -\frac{1}{8\pi^2}\int \mathrm{d}x\, \mathrm{tr}\, {}^*F^{\mu\nu}\,\delta F_{\mu\nu}$$

$$= -\frac{1}{4\pi^2}\int \mathrm{d}x\, \mathrm{tr}\, {}^*F^{\mu\nu} D_\mu\, \delta A_\nu = 0 \quad . \tag{3.47}$$

No surface terms arise in the integration by parts, which together with the Bianchi identity is needed to pass from the second to the third equalities, since δA_ν is arbitrary, and therefore may be taken to be localized. According to (3.44) q is given by the large-distance properties of the gauge potential — another hallmark of a topological quantity.

$$q = \int \mathrm{d}S_\mu\, \mathscr{C}^\mu \quad . \tag{3.48}$$

If we consider gauge potentials that tend to a pure gauge at large distances in all four directions,

$$A_\mu \xrightarrow[x\to\infty]{} U^{-1}\partial_\mu U \quad , \tag{3.49a}$$

then q may be represented in terms of U, by substituting into the expression (3.44b) for $\mathscr{C}^\mu$ the asymptotic form of A_μ, (3.49a).

$$q = \frac{1}{24\pi^2}\int \mathrm{d}S_\mu\, \epsilon^{\mu\alpha\beta\gamma}\, \mathrm{tr}\,(U^{-1}\partial_\alpha U U^{-1}\partial_\beta U U^{-1}\partial_\gamma U) \quad . \tag{3.49b}$$

Finally, note also that q is a geometrical invariant; even in curved space-time no factors of the metric are needed to make (3.46) a world scalar [51].

These topological remarks apply to well-behaved *classical* potentials which are all that mathematicians are concerned with. They are not directly relevant to quantum operator fields, nor to functional integrals, where the integration ranges over irregular field configurations. Most frequently the

the Pontryagin index is used for classical potentials defined on Euclidean four-space, and we shall meet it again when we discuss the semi-classical picture for the θ-angle. However, even on spaces with Minkowski signature q plays a role; for non-Abelian gauge fields that describe a 't Hooft-Polyakov monopole, q coincides with the monopole strength [52].

In Euclidean space q is an integer for regular potentials. This is seen by compactifying R_4 to S_4, and recognizing that the limit (3.49a) defines a gauge function U on S_3, the boundary of S_4. Again because Π_3 (gauge group) $= \mathscr{Z}$, the U's fall into integer-labeled homotopy classes and (3.49b) is an analytic expression for the winding number. Alternatively, one may work on R_4 (or even in Minkowski space) and present the limit (3.49a) in the following way.

Assume that in three directions, A_μ goes rapidly to zero, faster than $1/r$, where r is the modulus of the three-vector. In the fourth direction (time or Euclidean time) at negative infinity take A_μ to vanish, but at positive infinity to tend to a pure gauge. From (3.44) and (3.46) we have

$$q = \int_{-\infty}^{\infty} \mathrm{d}x^0 \int \mathrm{d}\mathbf{r}\, \partial_0 \mathscr{C}^0 + \int_{-\infty}^{\infty} \mathrm{d}x^0 \int \mathrm{d}\mathbf{r}\, \nabla \cdot \mathscr{C} \quad . \tag{3.50a}$$

The last term is converted to a three-surface integral and may be dropped; the x^0 integration in the remaining integral is trivial to perform, with the contribution from negative infinity vanishing. Thus

$$q = \int \mathrm{d}\mathbf{r}\, \mathscr{C}^0 \Big|_{x^0 = \infty} \quad . \tag{3.50b}$$

However, as already noted in (3.45), the integral (3.50b) coincides with $W(A)$, hence at $x^0 = \infty$ it is the winding number of the gauge transformation to which A tends.

The above discussion of the Pontryagin density is restricted to four-dimensional space-time, which is where the θ-angle of Yang-Mills theory appears. However, as a topological/mathematical object it can be defined in any even-dimensional space, and always it is a divergence of a vectorial Chern-Simons secondary characteristic. For example, in two dimensions, for an Abelian gauge theory,

$$\mathscr{P}_2 = \frac{-\mathrm{i}}{2\pi} {}^*F = \frac{-\mathrm{i}}{4\pi} \epsilon^{\mu\nu} F_{\mu\nu} \ , \qquad \mathscr{C}_2^\mu = \frac{-\mathrm{i}}{2\pi} \epsilon^{\mu\nu} A_\nu \ , \tag{3.51}$$

while in six dimensions

$$\mathscr{P}_6 = \frac{i}{384\pi^3} \epsilon^{\alpha\beta\gamma\delta\epsilon\phi} \operatorname{tr} F_{\alpha\beta} F_{\gamma\delta} F_{\epsilon\phi} \quad ,$$

$$\mathscr{C}_6^{\mu} = \frac{i}{192\pi^3} \epsilon^{\mu\alpha\beta\gamma\delta\epsilon} \operatorname{tr}(F_{\alpha\beta} F_{\gamma\delta} A_\epsilon - F_{\alpha\beta} A_\gamma A_\delta A_\epsilon + \tfrac{2}{5} A_\alpha A_\beta A_\gamma A_\delta A_\epsilon) \quad . \tag{3.52}$$

We shall discuss some physical consequences of the two-dimensional Pontryain density below, while the six-dimensional quantities will appear in our discussion of chiral anomalies.

3.5 Semi-Classical Picture for the Vacuum Angle

The emergence of a phase in the response of a physical state to a finite symmetry transformation is reminiscent of the quantum mechanical Bloch momentum associated with the wave function of a particle in a periodic potential $V(q)$, of the type pictured in Fig. 1. Even though the Hamiltonian $H = p^2/2m + V(q)$ is invariant under the shift $q \rightarrow q + a$, the wave function acquires a phase: $\psi(q + a) = e^{i\theta} \psi(q)$, where θ is proportional to the Bloch momentum. Moreover, even though the classical zero-energy configuration is infinitely degenerate ($q^{(n)} = na$; $n = 0, \pm 1, \pm 2, \ldots$), quantum mechanical tunnelling between classical minima lifts the degeneracy and produces a band spectrum $E(\theta)$. This is the physical situation in a crystal.

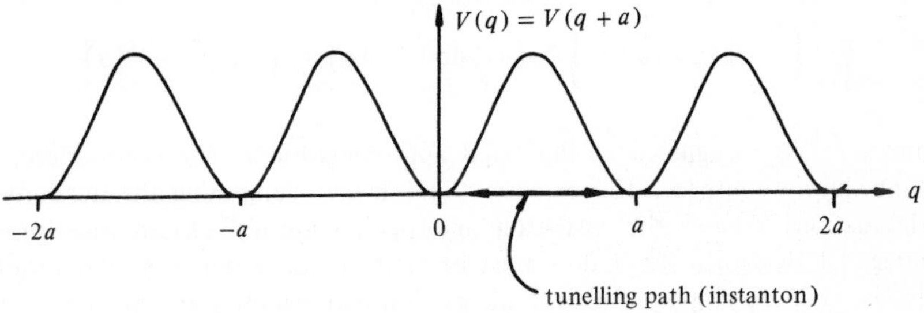

Fig. 1. Periodic potential whose quantum mechanics is analogous to the quantized Yang-Mills theory, with its gauge freedom with respect to infinitesimal gauge transformations removed. Zero-energy configurations $q^{(n)} = an$, $n = 0, \pm 1, \pm 2, \ldots$, are analogous to $A^{(n)} = -U_n^{-1} \nabla U_n$ and shifting by a is analogous to gauge transforming by U_1. Tunneling is discovered by finding an imaginary-time classical solution (instanton) that follows the tuneling path.

For our gauge field theory, we recognize that gauge transforming by a gauge function U_1, belonging to the first homotopy class, is analogous to shifting by a, and the analog of the infinite number of classical zero-energy configurations are the pure gauge potentials,

$$A^{(n)} = -U_n^{-1} \nabla U_n \quad , \tag{3.53}$$

for which $\boldsymbol{E}$ and $\boldsymbol{B}$ are zero, and hence the energy vanishes. (It is understood that the gauge freedom associated with homotopically trivial gauge transformations is completely fixed by the imposition of Gauss' law.)

How can we recognize if there is tunnelling in the gauge theory, which would close the analogy with the crystal and produce a band spectrum? It is well known to chemists and condensed-matter physicists that semi-classical evidence for tunnelling is obtained by solving the classical equations of motion, not in real time, but in imaginary time $\tau = it$, and by finding a solution which interpolates as τ passes from $-\infty$ to $+\infty$ between adjacent classical minima. For the ground state, these should have vanishing (imaginary time) energy, $-\frac{1}{2}m\dot{q}^2 + V(q) = 0$. Moreover, a semi-classical formula for the tunnelling probability amplitude Γ is gotten by dominating the functional integral, continued to imaginary time, with the (imaginary time) solution, which these days is called an "instanton"; i.e., $\Gamma \propto e^{-I/\hbar}$ where I is the classical (imaginary time) action $I = \int_{-\infty}^{\infty} d\tau [\frac{1}{2}m\dot{q}^2 + V(q)]$ evaluated for the instanton. The zero-energy instanton satisfies $\dot{q} = \pm(2V(q)/m)^{1/2}$, hence

$$I = \int_{-\infty}^{\infty} d\tau 2V(q) = \int dq (d\tau/dq) 2V(q) = \int dq \sqrt{2mV(q)} \quad ,$$

and $e^{-I/\hbar}$ is recognized as the WKB approximation to the (zero-energy) tunnelling amplitude. (Higher corrections involve computing the quadratic fluctuations around the instanton and performing a Gaussian functional integral.) Evidently the action must be finite to get a non-vanishing result [53]. [Let me emphasize that in the exact functional integral, one integrates over configurations, not merely solutions, and that finite-action solutions and configurations give a vanishingly small contribution. The infinite-action configurations are so much more numerous, that even though each one gives zero for $e^{-I/\hbar}$, their large number (entropy) ensures a finite answer. It is only in the semi-classical approximation that finite action solutions are relevant.]

These ideas carry over to the Yang-Mills theory [46]. The imaginary time theory becomes defined in Euclidean space; the imaginary time energy is $\frac{1}{2}\int d\boldsymbol{r}(-\boldsymbol{E}_a^2 + \boldsymbol{B}_a^2)$; and the Euclidean action becomes

$$I = \int dx \left(-\frac{1}{2g^2} \operatorname{tr} F^{\mu\nu} F_{\mu\nu} + \frac{i\hbar\theta}{16\pi^2} \operatorname{tr} {}^*F^{\mu\nu} F_{\mu\nu} \right) . \tag{3.54}$$

(The topological term retains its factor of i in the continuation to Euclidean space, since it is a world scalar.) Zero energy is assured if $\boldsymbol{E} = \pm\boldsymbol{B}$ or in covariant notation

$${}^*F^{\mu\nu} = \pm F^{\mu\nu} . \tag{3.55}$$

As mentioned earlier, solutions to (3.55) automatically satisfy the Euclidean-space Yang-Mills equation, by virtue of the Bianchi identity.

Since the action must be finite, we may expect the potentials to be sufficiently regular so that Pontryagin index is an integer, $q = N \neq 0$, and the solution is called $|N|$ instanton solution. The action then becomes

$$I_N = \frac{8\pi^2 |N|}{g^2} - iN\hbar\theta . \tag{3.56}$$

For the physical application we need the smallest action, hence $|N| = 1$ and the two cases $N = \pm 1$ add coherently. Thus the semi-classical tunnelling amplitude is

$$\Gamma \propto \exp(-8\pi^2/\hbar g^2) \cos\theta . \tag{3.57}$$

(Instantons with $|N| \geqslant 2$ give exponentially subdominant contributions.) Also the quadratic fluctuations have been evaluated [54].

Finally, one may represent schematically the ground state wave functional by analogy with the tight-binding approximation of crystal physics. The quantum state should be a superposition of wave functionals, $\Psi_n(\boldsymbol{A})$, each peaked in $\boldsymbol{A}$-function space near the classical zero-energy configuration $-U_n^{-1} \nabla U_n$. They must be gauge invariant against homotopically trivial gauge transformations, but the non-trivial ones shift n.

$$\mathcal{G}_n \Psi_{n'} = \Psi_{n'+n'} . \tag{3.58}$$

The ground state band functional then is

$$\Psi(A) = \sum_n e^{in\theta}\, \psi_n(A) \quad , \tag{3.59}$$

and the band energy (density) has the form $\alpha + \beta \cos\theta$ [46, 47].

Let me emphasize that the existence of the vacuum angle does not rely on instantons; it is an exact statement. Instantons provide an approximate method for calculating the consequences of the angle. This is just as with the periodic potential in quantum mechanics: Bloch-Floquet theory makes exact statements, and the tight-binding approximation provides approximate analysis.

I shall describe the physical importance of these results when I consider a more realistic model: Yang-Mills theory coupled to fermions.

Explicit Euclidean-space instantons have been found. For the SU(2) theory the self-dual one-instanton potential is [55]

$$A^\mu = \frac{-2i}{x^2 + \lambda^2}\, \alpha^{\mu\nu} x_\nu \quad , \tag{3.60}$$

where the self-dual 2×2 matrices $\alpha^{\mu\nu}$ are defined by

$$\alpha^{\mu\nu} = {}^*\alpha^{\mu\nu} = \frac{1}{4i}(\bar{\alpha}^\mu \alpha^\nu - \bar{\alpha}^\nu \alpha^\mu) \quad , \qquad \alpha^\mu = (-i\boldsymbol{\sigma}, I) \quad ,$$

$$\bar{\alpha}^\mu = (\alpha^\mu)^\dagger = (i\boldsymbol{\sigma}, I) \quad , \tag{3.61}$$

and the ensuing field strength is

$$F^{\mu\nu} = \frac{4i\lambda^2 \alpha^{\mu\nu}}{(\lambda^2 + x^2)^2} \quad . \tag{3.62}$$

The configurations depend on five parameters: the instanton "size" λ, and four parameters specifying the location, here set at zero. The solution is invariant against SO(5) rotations [in the sense (2.69)] which form the maximal compact subgroup of the Euclidean space SO(5, 1) conformal group — the symmetry group for classical Euclidean-space Yang-Mills theory [56]. The anti-self-dual solution can be gotten from the above by replacing $\alpha^{\mu\nu}$ by the anti-self-dual matrices $\bar{\alpha}^{\mu\nu}$.

$$\bar{\alpha}^{\mu\nu} = -*\bar{\alpha}^{\mu\nu} = \frac{1}{4i}(\alpha^{\mu}\bar{\alpha}^{\nu} - \alpha^{\nu}\bar{\alpha}^{\mu}) \quad . \tag{3.63}$$

Although only the $|N| = 1$ solutions are physically important, both physicists and mathematicians have been fascinated by the multi-instanton configurations. It has been shown that for SU(2) the $|N|$-instanton solution depends on $8|N| - 3$ parameters, which can be interpreted as $5|N|$ size and position parameters, and $3|N| - 3$ parameters specifying relative orientations in the 3-parameter SU(2) group space [57]. The most general instanton solution which can be simply written has been found by physicists. The self-dual solution is [58]

$$A^{\mu} = i\bar{\alpha}^{\mu\nu}\partial_{\nu}\ln\rho \quad , \qquad \rho = \sum_{i=1}^{N+1} \frac{\lambda_i^2}{(x - y_i)^2} \quad . \tag{3.64}$$

It is regular — the singularities are gauge artifacts — closed under the conformal group, and depends on $5N + 4$ gauge invariant parameters for $N \geqslant 3$, on thirteen parameters for $N = 2$, and on five for $N = 1$. [Some of the parameters in (3.64) are gauge artifacts.] While the field strength is complicated, the Pontryagin density is given by the elegant formula

$$-\frac{1}{16\pi^2}\,\mathrm{tr}\,{*F^{\mu\nu}}F_{\mu\nu} = \Box\Box\ln\rho \quad , \tag{3.65}$$

and the Pontryagin index is N [59]. Evidently, for $N \geqslant 3$ (3.64) is not the most general solution. Mathematicians have given a constructive procedure, which involves (many!) finite steps, for constructing any of the most general $8|N| - 3$ parameter solutions [60]; however, no closed expression analogous to (3.64) is available.

Exercise 3.10. Show that the (Euclidean space version of) of the energy-momentum tensor (2.33) vanishes on self-dual and anti-self-dual field strengths.

4. Fermion Interactions and Anomalies

4.1 Quantized Fermions

To be realistic, a Yang-Mills theory must be supplemented by other fields, coupling to it in a gauge invariant manner. While scalar fields are used to ensure spontaneous symmetry breaking, I shall here discuss only the interaction with fermion fields, which represent quarks and leptons in realistic models. I have already indicated how these interactions should be constructed to preserve gauge invariance; but the presentation was classical. It is a fascinating and important fact that in general, quantum mechanics puts further constraints on these couplings — what appears gauge invariant and consistent on the classical level, often fails to be so on the quantum level. Thus we come to the subject of anomalies, but before exploring this, let me briefly recall the structure of a gauge invariant quantum field theory with fermions.

In the presence of fermion sources, the Gauss' law constraint is modified to

$$(D \cdot E)_a = \rho_a \quad . \tag{4.1}$$

Here ρ_a is given by the time component of the fermion source current J^μ_a,

$$J^\mu_a = -\frac{\delta I_{\mathrm{M}}}{g \delta A^a_\mu} = -\hbar \mathrm{i}\, \bar{\psi} \gamma^\mu \tau^a \psi \quad , \qquad \rho_a \equiv g J^0_a \quad . \tag{4.2}$$

In the Weyl gauge, the generator of infinitesimal, space-dependent gauge transformations (with its infinitesimal parameter θ^a "stripped off") is now

$$G_a = -\frac{1}{g}(D \cdot E)_a + \frac{1}{g}\rho_a \quad , \tag{4.3a}$$

while the full generator is

$$Q_\theta = \frac{1}{g}\int \mathrm{d}r \,\{ E_a \cdot (D\theta)_a + \rho_a \theta^a \} \quad . \tag{4.3b}$$

Note that the gauge field part and fermion part commute and may be considered independent operators. Gauss' law is ensured by demanding that G_a annihilates physical states,

$$G_a \,|\text{physical}\rangle = 0 \quad , \tag{4.4a}$$

or that finite homotopically trivial gauge transformations, generated by the exponential operator, leave the state invariant,

$$\mathrm{e}^{\mathrm{i}Q_\theta/\hbar} |\text{physical}\rangle = |\text{physical}\rangle \quad . \tag{4.4b}$$

Let me be more specific about the nature of the quantum states in this gauge boson-fermion theory. In the same Schrödinger representation used earlier, states are functionals of the vector potential, and also they are "infinite-component" column vectors in the fermionic Hilbert space. The static gauge transformations operate on the dynamical (gauge-potential) variables, changing A into $A^U \equiv U^{-1} A U - U^{-1} \nabla U$ as before, and also the components in the fermion Hilbert space are mixed by a gauge transformation. As a consequence, states are no longer invariant (or phase invariant) against gauge transforming only A; one must also transform the fermionic gauge degrees of freedom. Moreover, the finite version of Gauss' law in the presence of fermions may be presented as

$$\left\{\exp \frac{\mathrm{i}}{\hbar g} \int \mathrm{d}\boldsymbol{r} \boldsymbol{E}_a \cdot (\boldsymbol{D}\theta)_a\right\} \Psi_{\text{phys}}(A)$$

$$= \Psi_{\text{phys}}(A^U) \underset{\text{Gauss' law}}{=} \left\{\exp\left[-\frac{\mathrm{i}}{\hbar g} \int \mathrm{d}\boldsymbol{r} \rho_a \theta^a\right]\right\} \Psi_{\text{phys}}(A)$$

$$= {}^{\mathrm{F}}\mathscr{G}_U^{-1}\, \Psi_{\text{phys}}(A) \quad , \tag{4.5}$$

where ${}^{\mathrm{F}}\mathscr{G}_U$ is the fermionic operator which implements the (topologically trivial) gauge transformation U. Of course, after a topologically non-trivial gauge transformation (as in three spatial dimensions) the states acquire a phase. (Similar remarks apply when scalar fields provide dynamical sources, but we shall not discuss this here.)

In a functional integral description of fermionic interactions, the generating functional (3.43) becomes

$$Z = \int \mathscr{D}\psi\, \mathscr{D}\bar{\psi} \mathscr{D} A_\mu^a$$

$$\exp \frac{\mathrm{i}}{\hbar} \int \mathrm{d}x \left\{ \frac{1}{2g^2} \operatorname{tr} F^{\mu\nu} F_{\mu\nu} - \frac{\hbar\theta}{16\pi^2} \operatorname{tr} {}^*F^{\mu\nu} F_{\mu\nu}\right.$$

$$\left. + \mathrm{i}\hbar \bar{\psi} \gamma^\mu (\partial_\mu + \mathscr{A}_\mu) \psi \right\} . \tag{4.6a}$$

(The topological θ term is present only in four dimensions.) Gauge fixing and compensating factors are included in the measure and the fermionic integration may be performed to give

$$Z = \int \mathscr{D}A_\mu^a \det(\partial\!\!\!/ + A\!\!\!/) \exp \frac{i}{\hbar} \int dx \left\{ \frac{1}{2g^2} \operatorname{tr} F^{\mu\nu} F_{\mu\nu} - \frac{\hbar\theta}{16\pi^2} \operatorname{tr} {}^*F^{\mu\nu} F_{\mu\nu} \right\} . \tag{4.6b}$$

Here we have assumed that there is just one species of massless Dirac fermions; if there are N, the determinant is raised to the Nth power. A possible mass term would change the argument of the determinant to $\partial\!\!\!/ + A\!\!\!/ + im$. Furthermore if the fermions are chiral, then there are half as many independent components in ψ and each species of fermions contributes a square root of the determinant to (4.6b).

4.2 *Gauge Invariance and Fermion Couplings*

The issue of gauge invariance is therefore the issue whether the (regulated) det $(\partial\!\!\!/ + A\!\!\!/)$ is gauge invariant or not. [For chiral fermions one is concerned with $\det^{1/2}(\partial\!\!\!/ + A\!\!\!/)$.] If the relevant homotopy classes of the gauge transformation are trivial, it suffices to check infinitesimal gauge invariance; if not then finite transformations must be examined as well. Note that the functional determinant is a space-time quantity, so in four space-time dimensions, for example, one is concerned with Π_4, in contrast to the Hamiltonian formulation which is at fixed time and involves homotopy groups of one dimension less — Π_3 for three spatial dimensions.

We first consider infinitesimal gauge transformations; that is, we inquire what is the value of $(D_\mu \delta/\delta A_\mu)_a \det(\partial\!\!\!/ + A\!\!\!/)$, which is also recognized as the covariant divergence of the source current operator, since

$$\frac{\delta}{\delta A_\mu^a} \det(\partial\!\!\!/ + A\!\!\!/) = \frac{-ig}{\hbar} \int \mathscr{D}\psi \mathscr{D}\bar{\psi} J_a^\mu \exp \frac{i}{\hbar} \int dx i\hbar \bar{\psi} (\partial\!\!\!/ + A\!\!\!/) \psi .$$

Thus we regain the statement that for gauge invariance of a quantum gauge theory, the source current must be covariantly conserved as an operator, not only as a classical field, and this brings us to the subject of chiral anomalies, which describe the situation that in a quantum field theory classical current conservation equations are frequently modified.

4.3 Chiral (or Axial Vector) Anomalies

The fact that classical symmetries need not survive quantization is now a well-established, but still poorly understood fact. The whole subject of symmetry plays a much more vigorous role in a quantum theory than it ever did in classical dynamics. For aesthetic and practical reasons, we prefer theories with a high degree of symmetry, but because Nature is asymmetric, we must also account for this symmetry breaking. The oldest and most primitive idea for symmetry breaking is that of "approximate" symmetries. One supposes that there are terms in the Lagrangian that violate the symmetry, but they are "small". More refined is the concept of spontaneous symmetry breaking, introduced by W. Heisenberg in condensed-matter physics, and extended by him as well as by J. Goldstone and Y. Nambu to the particle-physics domain. Here the dynamical equations are completely symmetric, but energetic considerations of stability indicate that the ground state is asymmetric. As is well known, these ideas are realized in the modern theory of low-energy processes involving pseudo-scalar mesons, principally the pion and in the unified models for weak, electromagnetic as well as (speculatively) strong interactions. Anomalous breaking of symmetries — the third, most subtle mechanism — arises from quantum mechanical effects, in a way whose fundamental origin remains obscure. Certainly there are no energetic or stability considerations as in spontaneous breaking. Our only clue comes from perturbation theory: there does not exist a regularization procedure which respects the anomalously broken symmetries. In addition to the scale and conformal symmetries, on whose anomalous breaking I have already commented, it is the chiral fermion symmetries that are anomalously broken and that possess a rich topological structure, even in flat space. Both symmetries are dimension-specific hence dimensional regularization breaks them. Both symmetries rely on zero-mass fields, hence Pauli-Villars regularization breaks them as well.

Nevertheless, there is good reason to believe that anomalies are not obscure consequences of problems with perturbation theory, but reflect a deep fact about Nature, which when understood will surely illuminate a whole complex of related ideas: chirality, spontaneous mass generation, spontaneous symmetry breaking and the reasons for parity violation. Moreover, as I shall show in a two-dimensional example, the occurrence of anomalies can be established from general principles, with no recourse to perturbation theory. In that example, the anomaly will also be responsible for spontaneous mass generation. In higher dimensions, we do not have such explicit construction of the anomalies, but we shall present non-perturbative/topological arguments for the existence of some of them.

Thus we expect that anomalies are a true aspect of quantum mechanical Nature, and their prevalence in all branches of physics gives support for this: consequences of scale and chiral anomalies are widely proliferated in particle physics [61]; scale (= trace) anomalies are widely studied in gravity theory [13], even though chiral anomalies have not yet had an impact there [62]; finally in condensed-matter physics, scale anomalies lead to the understanding of critical phenomena [63] and chiral anomalies are just beginning to enter the field [64].

Although the subject is contemporary [65], the idea that symmetries may be broken by quantum effects possesses a prehistory. Before the neutrino hypothesis, some speculated that the β-decay spectrum indicates a quantum mechanical violation of energy conservation. Also before gauge invariant Pauli-Villars regularization was developed for quantum electrodynamics, there was some question whether electromagnetic gauge invariance could be maintained in a quantum field theory. Both puzzles were ultimately resolved, and symmetries were maintained, but the idea that quantum effects can eliminate a classical conservation law has survived and is realized in the anomaly phenomenon.

Since anomalies arise from the unavoidable infinities of relativistic and local field theory, specifically when fermions are involved (Dirac's negative energy sea is one example) the resulting formulas for the anomalies reflect the ambiguities which arise when infinities are regulated. Consequently, there is a certain amount of arbitrariness in the expressions. In perturbation theory, the source of ambiguity comes from the fact that the renormalization rules for perturbation theory allow adjusting the value of any graph by arbitrary local functions of the coordinates (polynomials in the momenta). These local terms can also modify current divergences.

The arbitrariness is somewhat limited when it is realized that three different types of currents are under discussion:

(1) The most important current in a gauge theory is the source current J^μ_a to which gauge fields couple. If it is possible, the regularization and renormalization scheme must be chosen in such a way that this current be covariantly conserved. Moreover, all other physical quantities must be gauge invariant, and the regularization procedure must define them in a gauge-invariant fashion. If it is impossible to maintain source current conservation, the theory loses gauge invariance, and it is therefore rejected [66].

(2) There may also be in the model Noether symmetry currents j^μ which are classically conserved, but are not coupled to gauge fields. Their regularized version should be defined consistently with the gauge principle, but if they

fail to be conserved, the theory need not be rejected, although its symmetry will be reduced.

(3) Finally there may be "partially" conserved currents, whose formal divergences are "small", reflecting an "approximate" symmetry. The anomalies associated with these currents are the most arbitrary, since it may not be possible to separate unambiguously a quantum addition to a nonzero divergence which is already present classically. Nevertheless the regularization should be gauge invariant.

Here I shall discuss mainly chiral anomalies of the first two categories. Before examining them in realistic four-dimensional models, let us look at some two-dimensional Abelian models where much the same phenomena can be seen in a simpler setting where results can be established without perturbation theory. (There are no anomalous divergences of chiral currents in three- or any other odd-dimensional space-time; however, there are other chiral anomalies in odd-dimensional field theories; see section 5.2.)

In two dimensions, Dirac fields are two-component objects (aside from any further degrees of freedom associated with internal symmetry) and Dirac matrices may be chosen to be Pauli matrices:

$$\gamma^0 = \sigma^1 \quad , \quad \gamma^1 = i\sigma^2 \quad , \quad \gamma_5 = -i\sigma^3 \quad . \tag{4.7}$$

A peculiar property of these two-dimensional matrices, which leads to all our results, is that the axial vector is dual to the vector.

$$\epsilon^{\mu\nu}\gamma_\nu = i\gamma^\mu\gamma_5 \quad , \quad \epsilon^{01} = 1 = -\epsilon_{01} \quad . \tag{4.8}$$

Hence the axial vector current $j_5^\mu \equiv \hbar\bar{\psi} i\gamma^\mu\gamma_5\,\psi$ is dual to the vector current $j^\mu \equiv \hbar\bar{\psi}\gamma^\mu\psi$,

$$j_5^\mu = \epsilon^{\mu\nu} j_\nu \quad , \tag{4.9}$$

But now it follows that in a dynamically non-trivial theory, both currents cannot be conserved. To see this, consider the vacuum correlation function of two vector currents, whose most general, Poincaré-invariant, form is

$$\langle j^\mu(x) j^\nu(y)\rangle = g^{\mu\nu}\Pi_1(x-y) - \frac{\partial^\mu\partial^\nu}{\Box}\Pi_2(x-y)$$

$$+\left(\frac{\partial^\mu\epsilon^{\nu\alpha}\partial_\alpha}{\Box} + \frac{\partial^\nu\epsilon^{\mu\alpha}\partial_\alpha}{\Box}\right)\Pi_3(x-y) \quad . \tag{4.10a}$$

The axial vector-vector correlation function is determined by the above

$$\langle j_5^\mu(x) j^\nu(y)\rangle = \epsilon^{\mu\alpha}\langle j_\alpha(x) j^\nu(y)\rangle$$
$$= \epsilon^{\mu\nu}\,\Pi_1(x-y) - \epsilon^{\mu\alpha}\,\frac{\partial_\alpha\partial^\nu}{\Box}\,\Pi_2(x-y)$$
$$-\left(g^{\mu\nu} - 2\,\frac{\partial^\mu\partial^\nu}{\Box}\right)\Pi_3(x-y) \quad . \tag{4.10b}$$

Vector current conservation requires $\Pi_1 - \Pi_2 = 0$, and $\Pi_3 = 0$, but axial vector conservation would be obtained only if $\Pi_2 = 0$ and $\Pi_3 = 0$. The two are incompatible for a non-trivial theory.

Detailed calculation reveals that the one-loop graph contributing to the two-current correlation function cannot have its logarithmic divergence regulated so that both vector and axial vector vertices are conserved [67]. Moreover, when these currents arise in an Abelian gauge theory with massless fermions — if the coupling is vector-like one is speaking of the well-known Schwinger model of two-dimensional massless electrodynamics [68] — the functional determinant may be explicitly computed. Three couplings to Dirac fermions may be considered: vector, pseudovector, and chiral, giving rise to the following determinants:

$$\Delta_1(A) = \det(\partial\!\!\!/ + ie A\!\!\!/) \quad ,$$
$$\Delta_2(A) = \det(\partial\!\!\!/ - e\gamma_5 A\!\!\!/) \quad ,$$
$$\Delta_3(A) = \det(\partial\!\!\!/ + ie\,\tfrac{1}{2} A\!\!\!/(1 \pm i\gamma_5)) \quad . \tag{4.11}$$

By virtue of (4.8), the last two determinants may be rewritten in the same form as the first, since $i\gamma^\mu\gamma_5 A_\mu = \gamma^\mu\{-\epsilon_{\mu\nu}A^\nu\}$ and $\frac{1}{2}\gamma^\mu(1 \pm i\gamma_5)A_\mu = \gamma^\mu\{\frac{1}{2}(g_{\mu\nu} \mp \epsilon_{\mu\nu})A^\nu\}$. Note that $\frac{1}{2}(g^{\mu\nu} \mp \epsilon^{\mu\nu})$ is a projection operator; only one component of A^μ couples. Since the first determinant is known [68], closed forms can be given for all three:

$$-i\hbar\ln\Delta_1(A) = \frac{\hbar e^2}{2\pi}\int dx\,dy$$
$$\times\ A^\mu(x)\left[g_{\mu\nu}a_1 - \frac{\partial_\mu\partial_\nu}{\Box}\right]\delta(x-y)A^\nu(y) \quad , \tag{4.12}$$

$$-i\hbar \ln \Delta_2(A) = \frac{\hbar e^2}{2\pi} \int dx\,dy$$

$$\times \; A^\mu(x) \left[g_{\mu\nu} a_2 - \frac{\partial_\mu \partial_\nu}{\Box} \right] \delta(x-y) A^\nu(y) \; , \qquad (4.13)$$

$$-i\hbar \ln \Delta_3(A) = \frac{\hbar e^2}{8\pi} \int dx\,dy A^\mu(x) \left[g_{\mu\nu} - 2\, \frac{\partial_\mu \partial_\nu}{\Box} \right.$$

$$\left. \mp \frac{\partial_\mu \epsilon_{\nu\alpha} \partial^\alpha}{\Box} \mp \frac{\partial_\nu \epsilon_{\mu\alpha} \partial^\alpha}{\Box} \right] \delta(x-y) A^\nu(y) \; . \qquad (4.14)$$

Here a_1 and a_2 are coefficients of local terms which are undetermined by the one-loop graph — they may be fixed at will. For the vector case, gauge invariance dictates that a_1 be set to unity. The determinant is then gauge invariant; the vector current is the source current and it is conserved. But the axial vector current — a Noether current for the axial vector symmetry of massless fermions and obtained by operating with $\epsilon^{\mu\nu} \delta / \delta A_\nu$ on the determinant — possesses an anomaly [67].

$$-i\hbar \ln \Delta_1(A) = -\frac{\hbar e^2}{4\pi} \int dx F^{\mu\nu} \frac{1}{\Box} F_{\mu\nu} = \frac{\hbar e^2}{2\pi} \int dx \, {}^*F \frac{1}{\Box} {}^*F \; ,$$

$$\partial_\mu J^\mu = 0, \qquad \partial_\mu j_5^\mu = -\frac{\hbar e}{2\pi} \epsilon^{\mu\nu} F_{\mu\nu} = -\frac{\hbar e}{\pi} {}^*F \; . \qquad (4.15)$$

(Here and below, the factor $\hbar$ in the anomalous divergence is present because the current is defined with that factor as well. Also $\hbar$ reflects the "one-loop" nature of the anomaly.) A similar gauge-invariant expression for the determinant emerges in the second case with a_2 set to unity. A conserved axial vector current exists, but now the anomaly is in the vector current — the Noether fermion number current:

$$-i\hbar \ln \Delta_2(A) = -\frac{\hbar e^2}{4\pi} \int dx F^{\mu\nu} \frac{1}{\Box} F_{\mu\nu} = \frac{\hbar e^2}{2\pi} \int dx \, {}^*F \frac{1}{\Box} {}^*F \; ,$$

$$\partial_\mu J_5^\mu = 0 \; , \qquad \partial_\mu j^\mu = \frac{\hbar e}{2\pi} \epsilon^{\mu\nu} F_{\mu\nu} = \frac{\hbar e}{\pi} {}^*F \; . \qquad (4.16)$$

In both cases, the anomalously non-conserved current may be defined gauge invariantly, and the anomaly is gauge invariant, given by twice the Pontryagin density. Finally, in the third case, the local term cannot be adjusted at will because the kernel must have the projection properties of $(g_{\mu\nu} \mp \epsilon_{\mu\nu})$ so that only one component of A_μ be present in (4.14). This is assured by the expression exhibited, which however is not gauge invariant. Rather one finds

$$-\mathrm{i}\hbar \ln \Delta_3(A) = \frac{\hbar e^2}{4\pi} \int \mathrm{d}^2 x \left({}^*F \frac{1}{\Box} {}^*F \mp {}^*F \frac{1}{\Box} \partial_\mu A^\mu - \frac{1}{2} A^2 \right),$$

$$\partial_\mu J^\mu = \mp \frac{\hbar e}{4\pi} ({}^*F \mp \partial_\mu A^\mu) = \frac{\hbar e}{4\pi} (g^{\mu\nu} \mp \epsilon^{\mu\nu}) \partial_\mu A_\nu \quad . \tag{4.17}$$

The anomaly is not even gauge invariant, though it can be written as a total divergence. Since here the source current is not conserved, the quantized gauge theory has lost gauge invariance [66, 69].

Exercise 4.1. Verify from the explicit formulas for the appropriate determinants that the anomalous divergence equations are given by (4.15), (4.16) and (4.17).

Let us observe that in the first two (consistent) examples the anomaly is expressed by the two-dimensional Pontryagin density (3.51). Moreover, one may also show that the massless vector meson spontaneously acquires a mass without the intervention of scalar symmetry breaking fields. A straightforward argument, applied to the vector Schwinger model (first example above), makes use of the topological quantity that emerges in connection with the axial vector anomaly. Consider the gauge field equation

$$\partial_\mu F^{\mu\nu} = e J^\nu \quad . \tag{4.18a}$$

By contracting with $\epsilon_{\nu\alpha}$ this becomes

$$\partial^\mu {}^*F = e j^\mu_5 \quad . \tag{4.18b}$$

Taking a second divergence and using the anomalous conservation equation for j^μ_5 yields a free equation for $^*F$ which shows explicitly that the gauge field is massive [70]. (Recall that the two-dimensional coupling constant has the dimension of mass, in units where $\hbar$ and the velocity of light are dimensionless.)

$$\Box {}^*F = e\partial_\mu j_5^\mu = -\frac{\hbar e^2}{\pi} {}^*F \quad . \tag{4.18c}$$

While we shall see another topological mechanism for vector meson mass generation in three space-time dimensions, no similarly elegant result has yet been established in four dimensions.

If in the models with chiral couplings we drop the components of the Dirac field that do not participate in the interaction, we can write the respective fermionic Lagrangians $\mathcal{L}_+ (\mathcal{L}_-)$ solely in terms of right (left) spinors coupled to $\frac{1}{2}(g_{\mu\nu} - \epsilon_{\mu\nu})A^\nu$ whose only non-vanishing component is $A^+ = (A^0 + A^1)/\sqrt{2}$ [respectively, $\frac{1}{2}(g_{\mu\nu} + \epsilon_{\mu\nu})A^\nu$ with non-vanishing component $A^- = (A^0 - A^1)/\sqrt{2}$].

$$\mathcal{L}_\pm = \hbar\bar{\psi}_\pm (i\partial\!\!\!/ - eA\!\!\!/)\psi_\pm \quad . \tag{4.19}$$

The fermion determinants, obtained from (4.14), are respectively

$$-i\hbar \ln \Delta_3^+(A) = -\frac{\hbar e^2}{4\pi}\int A^+ \frac{\partial_+}{\partial_-} A^+ \quad , \tag{4.20a}$$

$$-i\hbar \ln \Delta_3^-(A) = -\frac{\hbar e^2}{4\pi}\int A^- \frac{\partial_-}{\partial_+} A^- \quad . \tag{4.20b}$$

On the other hand, the vector theory (Schwinger model) is described by a fermionic Lagrangian which is the sum of left and right Lagrangians.

$$\mathcal{L}_V = \mathcal{L}_+ + \mathcal{L}_- \quad . \tag{4.21}$$

In this model, the conservation of the vector current, as dictated by gauge invariance, and the anomalous non-conservation of the axial vector current, see (4.15), indicate that separately the right and left currents are not conserved, but their sum is. Why this should be the case, in spite of the fact that the formal Lagrangian in (4.21) exhibits no interaction between left and right spinors, is the surprise of the chiral anomaly. The puzzle is resolved when it is appreciated that the gauge invariant effective Lagrangian given by the fermion determinant (4.12) with $a_1 = 1$, is not merely the sum of left and right terms, neither of which is separately gauge invariant nor is their sum. Rather, to insure gauge invariance a contact term which couples left to right must be added.

$$-i\hbar \ln \Delta_1(A) = -i\hbar \ln \Delta_3^+(A)\Delta_3^-(A) + \frac{\hbar e^2}{2\pi^2}\int A^+ A^- \quad . \qquad (4.22)$$

This shows how the anomaly in the axial Noether current is forced by demanding vector gauge invariance. Also, we see that, contrary to assertions in the literature [62], the determinant for Dirac fermions is not merely the product of determinants for Weyl fermions of each chirality.

Finally, we remark that very similar results hold in two-dimensional non-Abelian fermionic models, since also for these the fermionic determinant may be evaluated, not explicitly in terms of the vector potentials as in (4.12) and (4.14), but rather in terms of non-local matrix functionals of the vector potentials defined by $A_\pm = g_\pm^{-1}\,\partial_\pm\, g_\pm$, where $g_\pm$ are group elements [71].

Turning now to anomalies in four dimensions, the following summarizes the results of the last decade's research [72]. While it is impossible to evaluate the functional determinant exactly, the anomalous graphs have been identified — they are the reflection non-symmetric triangle graphs, involving one or three axial vertices [73]. No other graphs introduce new structures, except that one may need to adjust them so that the anomalous divergence possesses a preferred form [74]. Consequently purely vectorial gauge theories, like quantum chromodynamics are gauge invariant. However, when the fermions are massless, the gauge invariant, group singlet axial vector Noether current j_5^μ is not conserved. Rather it satisfies

$$\begin{aligned}\partial_\mu j_5^\mu &= \frac{\hbar}{8\pi^2}\,\mathrm{tr}\,{}^*\mathscr{F}^{\mu\nu}\mathscr{F}_{\mu\nu}\\ &= \frac{\hbar}{8\pi^2}\,\partial_\mu\,\epsilon^{\mu\alpha\beta\gamma}\,\mathrm{tr}\,(\mathscr{F}_{\alpha\beta}\mathscr{A}_\gamma - \tfrac{2}{3}\mathscr{A}_\alpha\mathscr{A}_\beta\mathscr{A}_\gamma)\\ &= \frac{\hbar}{4\pi^2}\,\partial_\mu\,\epsilon^{\mu\alpha\beta\gamma}\,\mathrm{tr}\,(\mathscr{A}_\alpha\partial_\beta\mathscr{A}_\gamma + \tfrac{2}{3}\mathscr{A}_\alpha\mathscr{A}_\beta\mathscr{A}_\gamma) \quad . \qquad (4.23)\end{aligned}$$

For one fermion in the fundamental representation, this is twice the Pontryagin density apart from $-\hbar$. Since the anomaly is a total divergence, [see (3.44)] a conserved axial vector current can be defined, but it is not gauge invariant:

$$\tilde{j}_5^\mu = j_5^\mu + 2\hbar\mathscr{C}^\mu \Leftrightarrow \partial_\mu \tilde{j}_5^\mu = 0 \quad . \qquad (4.24a)$$

The charge constructed from the conserved current is time independent and invariant against small gauge transformations. Under a large gauge transformation, it changes by ħ times twice the winding number of the transformation, since the anomalous addition to the charge is twice $\hbar W(A)$, see (3.45), whose gauge transformation properties are established in (3.39),

$$Q_5 = \int \mathrm{d}r\, j_5^0 \ , \qquad \tilde{Q}_5 = \int \mathrm{d}r (j_5^0 + 2\hbar \mathscr{C}^0) = Q_5 + 2\hbar W(A) \ ,$$

$$\mathscr{G}_U \tilde{Q}_5 \mathscr{G}_U^\dagger = \tilde{Q}_5 + 2\hbar n_U \ . \tag{4.24b}$$

With vector gauge couplings and massless fermions there are also non-singlet axial vector currents j_{5a}^μ which classically are covariantly conserved. (They are not Noether currents since they are not "ordinarily" conserved.) However, the covariant conservation law acquires an anomaly upon quantization,

$$\partial_\mu j_{5a}^\mu + g f_{abc} A_\mu^b j_{5c}^\mu = \frac{\mathrm{i}\hbar}{8\pi^2} \operatorname{tr} \tau^a {}^*\mathscr{F}^{\mu\nu} \mathscr{F}_{\mu\nu} \ . \tag{4.25}$$

This is the obvious generalization of (4.23); but the anomaly is not a total divergence.

There are also anomalies in source currents for gauge fields when axial vector couplings are present. While there are no purely axial ($\mathrm{i}\gamma^\mu \gamma_5$) non-Abelian theories, chiral couplings ($\frac{1}{2}(1 \pm \mathrm{i}\gamma_5)\gamma^\mu$) can occur if the fermions are massless. For these the functional determinant responds to an infinitesimal gauge transformation in a non-trivial way. We give here the preferred result for a simple group [66, 74]; for direct product groups there is some ambiguity in handling the individual factors; see Exercise 4.2.

$$(D_\mu J_\pm^\mu)_a = \pm \frac{\mathrm{i}\hbar}{24\pi^2} \partial_\mu \epsilon^{\mu\alpha\beta\gamma} \operatorname{tr} \tau^a (\mathscr{A}_\alpha \partial_\beta \mathscr{A}_\gamma + \tfrac{1}{2} \mathscr{A}_\alpha \mathscr{A}_\beta \mathscr{A}_\gamma)$$

$$= \pm \frac{\mathrm{i}\hbar g^2}{24\pi^2} D_{abc} \partial_\mu \epsilon^{\mu\alpha\beta\gamma} (A_\alpha^b \partial_\beta A_\gamma^c + \tfrac{1}{4} g f_{cde} A_\alpha^b A_\beta^d A_\gamma^e) ,$$

$$D_{abc} \equiv \operatorname{tr} \tau^a \, \tfrac{1}{2} \{\tau^b, \tau^c\} \ . \tag{4.26}$$

If (4.26) is non-vanishing, the fermionic determinant is not gauge invariant and the gauge theory is rejected. Note that the anomalous divergence is not gauge covariant, which highlights once again that gauge invariance has been lost. In particular (4.26) does not have the same structure as (4.23) with an additional τ^a matrix inside the trace, nor is it of the form (4.25).

One may understand the total derivative structure of the anomaly (4.26) in the following way. Recall from Exercise 2.19 that the conservation of the Noether current j^μ arising from the rigid gauge invariance of the Lagrangian is equivalent to the Yang-Mills field equation and the covariant conservation of the source current J^μ. If the latter is not covariantly conserved, j^μ is not conserved. However, if $D_\mu J^\mu$ can be expressed as a total divergence, then j^μ may be modified and a conserved current may be defined. This indicates that the anomaly (4.26) leads to a breakdown of the local gauge symmetry, but not of the rigid gauge symmetry.

Exercise 4.2. Consider the SU(2) × U(1) group, with a singlet vector potential a_μ, as well as a triplet of SU(2) gauge potentials A^a_μ, coupling chirally to a massless fermion doublet. The determinant is

$$\Delta(A) = \det{}^{1/2}\left(\partial\!\!\!/ + g\,\frac{\sigma^a}{2i}\,A\!\!\!/_a + \frac{g'}{2i}\,a\!\!\!/\right) \quad . \tag{E4.1}$$

Compute the divergence of the singlet current using (4.26), and observe that it is not SU(2) gauge invariant. Is it U(1) gauge invariant? Compute also the divergence of the triplet current and show that it is neither SU(2) nor U(1) gauge covariant.

Show that one may modify the (unspecified) definition of $-i\hbar \ln \Delta(A)$ which gives (4.26) by adding a local term, proportional to

$$\int dx\, \epsilon^{\mu\alpha\beta\gamma}\,\epsilon_{abc}\, a_\mu A^a_\alpha A^b_\beta A^c_\gamma \quad ,$$

so that the singlet current anomalous divergence is now given by an SU(2) × U(1) gauge invariant expression. What is the (modified) divergence of the SU(2) current? Alternatively, can the triplet current divergence be SU(2) × U(1), SU(2) or U(1) gauge covariant? If so, what is the modified divergence of the singlet current?

Exercise 4.3. Derive the commutator algebra of the operators $(D_\mu\, \partial/\delta A_\mu)_a$ and compare with (3.26). By applying this commutator to $\det(\partial\!\!\!/ + A\!\!\!/)$, find an integrability condition that the anomalous divergence of J^μ_a must satisfy [75]. Verify that (4.26) satisfies this condition, but that (4.25) does not. This condition is called the Wess-Zumino consistency conditon.

Let us explain in detail the reason why anomalies (4.25) and (4.26) differ in form. In the latter $J^\mu_\pm$ is given by the gauge variation of a functional — the determinant for fermions of positive or negative chirality — and as a consequence its divergence satisfies the Wess-Zumino integrability condition, see Exercise 4.3. In the former j^μ_5 is not the variation of anything; the Noether current j^μ_5 does not result from a variation, since nothing couples to it. Thus, its anomalous divergence need not satisfy the Wess-Zumino condition, and one checks that (4.25) does not, see Exercise 4.3. (Statements in the literature that an anomaly "must" satisfy the Wess-Zumino condition [61, 75, 76] are inaccurate; recently, this confusion has been extensively elucidated [77].) On the other hand, since (4.25) arises in the consistent, gauge invariant theory of QCD, it should be gauge invariant, as indeed it is, while the gauge non-invariant formula (4.26) arises in a theory which has lost gauge invariance.

In spite of the quite different physical settings for the two anomalies, (4.25) and (4.26), there is a mathematical relationship between them. Observe the identity

$$\partial_\mu \epsilon^{\mu\alpha\beta\gamma}(\mathscr{A}_\alpha \partial_\beta \mathscr{A}_\gamma + \tfrac{1}{2}\mathscr{A}_\alpha \mathscr{A}_\beta \mathscr{A}_\gamma)$$

$$+ D_\mu \epsilon^{\mu\alpha\beta\gamma}(\mathscr{A}_\alpha \partial_\beta \mathscr{A}_\gamma + \partial_\beta \mathscr{A}_\gamma \mathscr{A}_\alpha + \tfrac{3}{2}\mathscr{A}_\alpha \mathscr{A}_\beta \mathscr{A}_\gamma)$$

$$= \frac{3}{2} {}^*\mathscr{F}^{\mu\nu} \mathscr{F}_{\mu\nu} \tag{4.27}$$

This means that if we add to $J^\mu_\pm$

$$\Delta J^\mu_\pm \equiv \pm \frac{i\hbar}{24\pi^2} \epsilon^{\mu\alpha\beta\gamma} \operatorname{tr}(\mathscr{A}_\alpha \partial_\beta \mathscr{A}_\gamma + \partial_\beta \mathscr{A}_\gamma \mathscr{A}_\alpha + \tfrac{3}{2}\mathscr{A}_\alpha \mathscr{A}_\beta \mathscr{A}_\gamma) \tag{4.28}$$

we shall obtain currents whose covariant divergence is $(\pm i\hbar/16\pi^2)^*\mathscr{F}^{\mu\nu}\mathscr{F}_{\mu\nu}$ [78]. Since j^μ_5 is the difference of the right chiral current and the left current we recognize in this way the formula (4.25).

The meaning of this manipulation is the following. Regardless whether a current is gauge source current or a Noether current, the same Feynman diagrams describe its matrix elements. However, different local terms, which are not determined by the diagrams, are appropriate in the two cases, and the above addition $\Delta J^\mu_\pm$ reflects the difference between the local terms contributing to $J^\mu_\pm$ and those in j^μ_5. Moreover, $\Delta J^\mu_\pm$ will not in general be integrable with respect to $\mathscr{A}^\mu$, and that is why j^μ_5 will not be the variation of anything, nor will its divergence satisfy the Wess-Zumino condition.

The quantity D_{abc} in (4.26) must vanish, if chiral couplings are to be gauge invariant [66]. Two cases may be distinguished. It may be that for all representations of the group, $D_{abc} = 0$; these are called "safe" groups and they include SU(2) but no other special unitary groups, all orthogonal groups except SO(6) ≈ SU(4), and all symplectic groups. On the other hand even if the group is not safe, like SU(N), with $N > 2$, it may still be that for some particular representations D_{abc} vanishes [79]. This then gives a useful limitation on the allowed fermion representations.

For an Abelian gauge theory, an axial coupling is classically allowed if the fermions are massless — we are speaking of axial electrodynamics. But again, the axial vector current is not conserved, owing to the three axial current triangle graph [73]; so axial quantum electrodynamics is not gauge invariant.

One believes that the coefficients of the anomalies are not modified by radiative corrections; a theorem proven for the vector Abelian theory — electrodynamics [81]. Although the proof is technical, depending on details of the renormalization procedure, the idea is simple. If one regulates the photons, all divergences of the theory are removed, save those associated with graphs containing fermion loops without photon insertions. But regulating photons does not interfere with chiral symmetry, so chiral anomalies should arise only from lowest-order fermion loops, with no further corrections. (Correspondingly, scaling anomalies associated with the trace of the energy-momentum tensor do possess corrections [10, 12] because any regulator violates scale invariance.) Presumably the result holds in the non-Abelian theory as well, but it may very well be regularization dependent. For example in a supersymmetric theory, one should treat bosons and fermions on equal footing, so regulating only boson lines would be inappropriate [82].

I shall postpone discussing the behaviour of the functional fermion determinant under finite gauge transformations. Suffice it to say here that physically interesting theories, like quantum chromodynamics with vector couplings and unified models with chiral couplings, are invariant against finite gauge transformations as soon as they are infinitesimally invariant.

4.4 Some Physical Consequences of Anomalies

We have seen that in two dimensions the axial anomaly is responsible for generating a vector meson mass. Moreover, in physical four-dimensional theories, the chiral anomaly has many applications [61]. We discuss here the most important ones.

The Glashow-Weinberg-Salam unified theory of electro-weak interactions utilizes chiral SU(2) × U(1) couplings. Since that is not a safe group, one must insure the fermions lie in safe representations. This can be done, provided quarks and leptons balance in number. Thus the requirement that the standard model be anomaly-free leads to the prediction that for every observed lepton there should exist a quark [66]. The prediction has thus far been verified; most recently the discovery of the τ lepton was soon followed by evidence for the "bottom" quark. Now we are anxiously awaiting word about the τ neutrino and the "top" quark.

Once the chiral source currents for the electro-weak gauge fields are conserved, the baryon number current acquires an anomaly [83] [just as in the two-dimensional example the fermion number current is anomalous with gauge-invariant axial-vector interactions; see (4.16)].

$$\partial_\mu j^\mu = \frac{\hbar}{8\pi^2} \operatorname{tr} {}^*\mathcal{F}^{\mu\nu}\mathcal{F}_{\mu\nu} , \qquad (4.29)$$

Since the divergence of the baryon number current is proportional to $\operatorname{tr} {}^*\mathcal{F}^{\mu\nu}\mathcal{F}_{\mu\nu}$, whenever that quantity is sizeable in a quantum process one may expect "topological" baryon decay (which should not be confused with "ordinary" baryon decay in Grand Unified Theories). Two mechanisms for topological baryon decay are known. The first involves tunnelling, and in a semi-classical description instantons are the dominant field configurations. For these, $\operatorname{tr} {}^*\mathcal{F}^{\mu\nu}\mathcal{F}_{\mu\nu}$ clearly is sizeable, but the tunnelling rate, being exponentially small, is negligible [84]. However, as mentioned earlier, the Pontryagin index of an 't Hooft-Polyakov monopole is also nonzero [52]; therefore one expects baryon decay in the presence of a monopole. While the magnitude of this process is still controversial, there are arguments that it is large [41]. Nevertheless, practical significance is obscured by the absence of any experimental evidence for monopoles, other than just one reported sighting [39].

Although I have not discussed anomalies in "partially conserved" currents, one physical effect should be mentioned, since historically it opened the subject [65]. The hypothesis of partial conservation of flavor SU(2) axial vector currents (PCAC) implies, in the absence of anomalies, that a massless neutral pion cannot decay into two photons [85]. But the physical pion does decay, with a width of about 7.9 eV. This cannot be accounted for by the finite mass of the physical pion. However, taking into account the anomaly in the axial vector current of the type (4.21), which arises from electromagnetic couplings, one obtains a non-vanishing result [86], that depends on the number of quark colors. Excellent agreement (about 10% too small) with the experimental number is gotten for three colors. (The remaining discrepancy is attributed to finite mass effects of the pion.) Therefore the anomaly provides an experimental determination of the number of colors.

For the final application, let us return to the four-dimensional vectorial Yang-Mills theory, and inquire how the addition of fermions affects the vacuum angle and the semi-classical picture of tunnelling. We discuss first massless Dirac fermions, whose axial vector Noether current is anomalously non-conserved, according to (4.23). For definiteness we consider a SU(2) gauge theory with one fermion doublet, and we return to the Hamiltonian formalism.

From (4.24) one sees that a conserved current does exist, and the time-independent charge $\tilde{Q}_5$ is composed of two pieces; a gauge invariant contribution coming from the fermions, and an anomalous term constructed from gauge potentials, which we recognize to be twice $\hbar W(A)$, defined in (3.33). As mentioned earlier, $\tilde{Q}_5$ is not invariant against homotopically non-trivial, static gauge transformations; rather $\tilde{Q}_5$ shifts by twice the winding number of the gauge function. The commutator algebra of the three operators H, $\tilde{Q}_5$ and $\mathscr{G}_n$ is

$$[H, \tilde{Q}_5] = 0 \quad , \qquad [H, \mathscr{G}_n] = 0, \qquad [\mathscr{G}_n, \tilde{Q}_5] = 2\hbar n\, \mathscr{G}_n \ . \tag{4.30}$$

The three cannot be simultaneously diagonalized, and gauge invariance requires that physical states be eigenstates of $\mathscr{G}_n$. But this means that $\tilde{Q}_5$ acts as a lowering operator for θ.

$$\exp\left(\frac{i}{\hbar}\,\theta'\tilde{Q}_5\right)\Psi_\theta = \Psi_{\theta-2\theta'} \quad . \tag{4.31}$$

Energy eigenvalues of H, which commutes with $\tilde{Q}_5$, can no longer depend on θ; tunnelling is suppressed, and the entire energy band collapses to one level. Physical, gauge and chiral invariant quantities cease to depend on θ. Moreover, chiral symmetry is spontaneously broken because states are not chirally invariant. However, this spontaneous breaking does not derive from energetic stability reasons as in the Goldstone-Nambu mechanism, but rather it occurs because of the axial vector anomaly [46].

The same results may be seen in a functional integral formulation [47]. The generalization of (3.43) to include fermions is (gauge fixing terms suppressed)

$$Z_\theta = \int \mathscr{D}\psi\,\mathscr{D}\bar{\psi}\,\mathscr{D}A_\mu^a \exp\left(\frac{i}{\hbar}\int dx\,\mathscr{L}\right) \ , \tag{4.32}$$

where

$$\mathcal{L} = \frac{1}{2g^2} \operatorname{tr} F^{\mu\nu} F_{\mu\nu} - \frac{\hbar\theta}{16\pi^2} \operatorname{tr} {}^*F^{\mu\nu} F_{\mu\nu} + i\hbar \bar{\psi} \gamma^\mu (\partial_\mu + \mathcal{A}_\mu) \psi \quad .$$

When the fermionic integration variable is redefined by a chiral transformation,

$$\psi \to e^{-\theta' \gamma_5} \psi \quad , \quad \bar{\psi} \to \bar{\psi} e^{-\theta' \gamma_5} \quad , \tag{4.33}$$

it would appear that the functional integral is left invariant. However, the axial anomaly indicates that this is not so; rather Z_θ changes according to

$$Z_\theta \to Z_{\theta - 2\theta'} \quad . \tag{4.34}$$

The detailed reason for non-invariance of the superficially invariant integral in (4.32) has been traced to the singular nature of the fermion measure [87]. Physical quantities cannot be affected by changing integration variables but such changes modify θ, so we must conclude that in the presence of massless fermions the angle is not a physical parameter and can be set to zero.

In a semi-classical treatment, the suppression of tunnelling is recognized after the fermion integration is performed. Then (4.32) leaves

$$Z_\theta = \int \mathcal{D} A_\mu^a \det(\partial\!\!\!/ + \mathcal{A}\!\!\!/)$$
$$\times \exp \frac{i}{\hbar} \int dx \left\{ \frac{1}{2g^2} \operatorname{tr} F^{\mu\nu} F_{\mu\nu} - \frac{\hbar\theta}{16\pi^2} {}^*F^{\mu\nu} F_{\mu\nu} \right\} \quad . \tag{4.35}$$

When this integral is continued to Euclidean space, and dominated by a tunnelling instanton configuration, it vanishes because the Euclidean Dirac operator $(\partial\!\!\!/ + \mathcal{A}\!\!\!/)$ possesses a zero eigenvalue [88], thus forcing the determinant to vanish. Indeed, the number of zero modes of the Dirac operator is counted by the Pontryagin index; a fact which relates the axial vector anomaly to the topological Atiyah-Singer index theorem [89].

Exercise 4.4. Solve the four-dimensional Euclidean Dirac equation in the instanton field (3.60) and find the expression for the zero mode eigenfunction.

But physical fermions are not massless, and the θ angle in the physical quantum chromodynamical model remains observable. This now presents a problem. Recall that θ is a *CP* violating parameter, but the stringent experimental limits on the neutron's electric dipole moment require that it be practically zero. (More accurately, the limit is $\theta \lesssim 10^{-9}$ [90].) Yet there is no known, physically acceptable principle which insures the vanishing of θ. In particular, setting it to zero *ab initio* would not help for the following reason. Although the origin of fermion masses remains obscure, we expect that spontaneous symmetry breaking is responsible. In that context, the fermion mass matrix would arise in the quantum chromodynamical Lagrangian pointing in an arbitrary *CP* direction, i.e., its form would be $\bar{\psi} M_1 \psi + \bar{\psi} \gamma_5 M_2 \psi$. In order to isolate *CP* violating effects, one needs to remove the term involving M_2. This may be achieved by a chiral redefinition of the Fermi fields, which formally leaves the rest of the Lagrangian invariant, but actually — because of the axial vector anomaly — induces a tr $^*\mathscr{F}^{\mu\nu} \mathscr{F}_{\mu\nu}$ term, giving rise to a vacuum angle. Thus what is needed is a principle that would insure that the "initial" value of θ be precisely cancelled by this chiral redefinition — but such a principle is missing.

We are facing a problem not unlike that of the cosmological constant in gravity theory. The same general principles of invariance and renormalizability which select the kinetic part of the Lagrangian, allow the additional constant. Experimental observation, however, requires it to be zero. But setting the constant to zero initially does not help because spontaneous symmetry breaking gives rise to it anew. Of course one difference is that the cosmological constant modifies the classical theory, while the vacuum angle is a quantum effect.

Fortunately there is also good news for quantum chromodynamics from the θ angle and the associated phenomena. For a long time it appeared that the theory possesses too much symmetry to be phenomenologically acceptable, since it was not realized that $\widetilde{Q}_5$ is gauge variant. This symmetry predicts that there would be a particle degenerate with the pion, and no such particle exists [91]. Now we recognize this to be a false prediction. The so-called U(1) problem has dissolved [92]!

5. Quantization Constraints on Physical Parameters

We have seen that quantum mechanics and gauge invariance constrain the structure of a consistent quantum field theory: the theory must be anomaly free, and if there is a possibility of anomalies, fermions must transform according to an anomaly-free representation of the gauge group. In this section, I shall show that in special circumstances the constraints are much more exacting: they require that some parameters of the theory be quantized.

Quantization of physical quantities goes back to the foundations of modern physics, and we can now identify three different reasons for quantization. The first and oldest is the usual quantum mechanical one: dynamical quantities like energy and angular momentum are eigenvalues of Hermitian operators, whose spectrum in general is not continuous. In the last decade a second framework for quantization was developled — the quantization of soliton and instanton number. Unlike the first example, which is quantum mechanical and invokes $\hbar$, this is classical and arises for topological reasons from the requirement of finite energy or action [1, 2]. The third quantization, which is the subject of the remainder of my lectures, requires parameters describing dynamics (masses, coupling constants) to be quantized for both quantum mechanical and topological reasons arising from gauge invariance.

Although in field theory this has been discovered only recently [93, 94], the first example is in fact more than fifty years old, and concerns the Dirac monopole. I shall begin by reviewing that classic result, with special emphasis on questions of gauge invariance, so that we may first understand the mechanism in a relatively simple setting.

5.1 Dirac Point Monopole

Consider a particle of charge e and mass m moving (non-relativistically) in the field of a point magnetic monopole of strength g. The classical dynamical equations are given by the Lorentz law:

$$m\boldsymbol{r} = e\dot{\boldsymbol{r}} \times \boldsymbol{B} \quad , \qquad \boldsymbol{B} = g\hat{\boldsymbol{r}}/r^2 \quad , \qquad \nabla \cdot \boldsymbol{B} = 4\pi g \delta(\boldsymbol{r}) \quad . \tag{5.1}$$

They are seen to be entirely gauge invariant, well-defined and can be solved for any value of e and g. However, as Dirac showed, quantum mechanics is consistent only when $eg = \hbar n/2$, n any integer. To understand this, let us examine the Lagrangian which governs this motion.

$$L = \tfrac{1}{2} m\dot{\boldsymbol{r}}^2 + e\dot{\boldsymbol{r}} \cdot \boldsymbol{A} \quad , \qquad \nabla \times \boldsymbol{A} = \boldsymbol{B} \quad . \tag{5.2}$$

Observe that the Lagrangian is not gauge invariant; rather, under a static gauge transformation $\boldsymbol{A} \to \boldsymbol{A} - e^{-1}\nabla\theta$ it changes by a total time derivative,

$$L \to L - \dot{\boldsymbol{r}} \cdot \nabla\theta = L - \frac{\mathrm{d}}{\mathrm{d}t}\theta \quad , \tag{5.3a}$$

and as a consequence the action $I \equiv \int_{t_2}^{t_1} \mathrm{d}tL$ changes by endpoint contribution to θ,

$$I \to I - \theta \Big|_{\text{endpoints}} \quad . \tag{5.3b}$$

The fact that the Lagrangian is not gauge-invariant and that the action changes by the endpoint contributions is irrelevant for classical physics, which is entirely determined by the dynamical equations of motion. But quantum mechanics does care about the Lagrangian and the action, as is seen for example in a functional integral formulation which involves $\exp(\mathrm{i}I/\hbar)$. Thus we need to examine carefully the invariance properties of the action; if it changes by ΔI under a gauge transformation, the requirement of gauge invariance in the quantum theory may still be met, provided the change is an integral multiple of $2\pi\hbar$.

For the monopole problem, one cannot restrict gauge transformations to vanish at the endpoints, because the vector potentials that describe a monopole are not rotationally covariant. Since the physical results must be rotationally symmetric — the monopole has no preferred direction — the asymmetric response of the gauge potential to a rotation must be compensated by a gauge transformation, in the sense of Eq. (2.69) and Exercise 2.23, and these gauge transformations do not vanish at the endpoints of the motion; see Exercise 5.1.

To evaluate ΔI explicitly, consider for example, the action integrated over a closed path P for which $\boldsymbol{r}(t_1) = \boldsymbol{r}(t_2) \equiv \boldsymbol{R}$. This contributes to the quantum mechanical propagation amplitude $\langle \boldsymbol{R}; t_2 | \boldsymbol{R}; t_1 \rangle$. Only the interaction part of the Lagrangian need concern us, and its action may be written as $e\oint_P \mathrm{d}\boldsymbol{r} \cdot \boldsymbol{A}$. Now imagine that this path lies on a closed surface, as in Fig. 2. By Stokes' theorem the line integral over P may be converted to an integral over the surface enclosed by P, provided the integrand is non-singular. But a vector potential $\boldsymbol{A}$ which gives rise to a magnetic monopole field is necessarily singular owing to the presence of a Dirac string, whose location is gauge dependent, and can be confined to one hemisphere. Therefore if the path integral is cast on the surface S_1, the vector potential must be presented in a gauge $\boldsymbol{A}_1$, with the string rendered harmless by being relegated to other surface,

$$e\oint_P \mathrm{d}\boldsymbol{r} \cdot \boldsymbol{A} = e\int_{S_1} \mathrm{d}\boldsymbol{S} \cdot (\nabla \times \boldsymbol{A}_1) = e\int_{S_1} \mathrm{d}\boldsymbol{S} \cdot \boldsymbol{B} \quad . \tag{5.4a}$$

Alternatively, the integral may cast over the surface S_2 provided we are

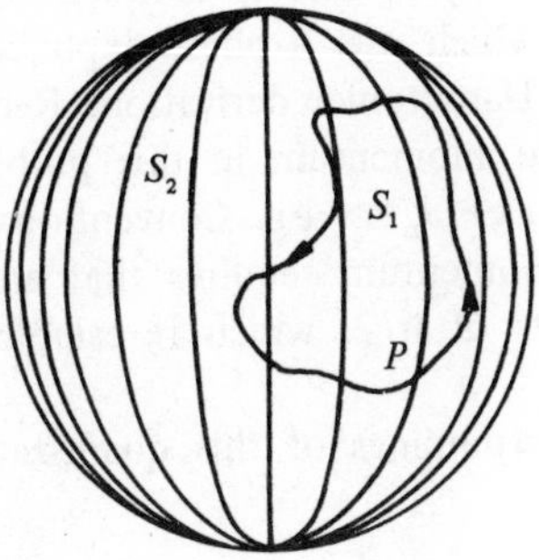

Fig. 2. Integration of the point monopole action. The line integral of A along P may be convered to a surface integral over S_1 or S_2, provided the Dirac gauge potential is non-singular on the surface in question. The difference between the two surface integrals equals the total magnetic flux through the entire surface $S_1 \cup S_2$.

in a gauge A_2, with the string passing through S_1,

$$e\oint_P \mathrm{d}\boldsymbol{r} \cdot \boldsymbol{A} = -e\int_{S_2} \mathrm{d}\boldsymbol{S} \cdot (\nabla \times \boldsymbol{A}_2) = e\int_{S_2} \mathrm{d}\boldsymbol{S} \cdot \boldsymbol{B} \quad . \tag{5.4b}$$

(The minus sign arises from the change of orientation.) The difference between the two expressions is the change of the action, and by Gauss' theorem may be written as a volume integral of the magnetic field's divergence,

$$\Delta I = e\int_{S_1 \cup S_2} \mathrm{d}\boldsymbol{S} \cdot \boldsymbol{B} = e\int_V \mathrm{d}\boldsymbol{r}\, \nabla \cdot \boldsymbol{B} = 4\pi e g \quad . \tag{5.5}$$

The requirement that ΔI be an integral of a multiplet of $2\pi\hbar$ gives the Dirac quantization condition.

Exercise 5.1. Exhibit a magnetic monopole vector potential $\boldsymbol{A}_1$ with its Dirac string along the positive z-axis, and $\boldsymbol{A}_2$ with the Dirac string along the negative z-axis; see Exercise 2.24. Show that $\boldsymbol{A}_1$ and $\boldsymbol{A}_2$ are related by a gauge transformation.

Evaluate $e\int_{t_1}^{t_2} \mathrm{d}t\, \dot{\boldsymbol{r}} \cdot \boldsymbol{A}$ for both cases along the path

$$\boldsymbol{r}(t) = R\,(\cos 2\pi(t_2 - t)/(t_2 - t_1),\, \sin 2\pi(t_2 - t)(t_2 - t_1),\, 0) \quad , \tag{E5.1}$$

and compare the two answers.

The above discussion is appropriate to a functional integral formulation of the quantized theory, which uses a phase exponential of the action. It is also instructive to give a Hamiltonian derivation. Recall from Exercise 2.23 that the conserved angular momentum in this problem possesses a radial component: $\boldsymbol{J} = \boldsymbol{r} \times m\dot{\boldsymbol{r}} - eg\hat{\boldsymbol{r}}, J_r = -eg$. Conventional quantum mechanical quantization of angular momentum requires that any one component be quantized in integral units of $\hbar/2$, which re-establishes the quantization condition.

The mathematical underpinnings of this quantization make use of the fact that

$$\Pi_1(U(1)) = \mathscr{Z} \quad . \tag{5.6}$$

One is dealing with Π_1, because the gauge functions depend on one variable — time — and provide a mapping into the gauge group, here U(1).

5.2 *Three-Dimensional Gauge Theories*

Consider a gauge theory in three-dimensional space-time (or in Euclidean three-space — below I shall explain the physical significance of the latter). It is a curious fact of three-dimensional kinematics that an unconventional term may be added to the Yang-Mills Lagrangian, which gives a mass to the vector fields, yet the equations of motion are gauge covariant [95],

$$I = \int \mathrm{d}x \mathscr{L} \quad ,$$

$$\mathscr{L} = \frac{1}{2g^2} \operatorname{tr} F^{\mu\nu} F_{\mu\nu} - \frac{\mu}{2g^2} \epsilon^{\mu\nu\alpha} \operatorname{tr}(F_{\mu\nu} A_\alpha - \tfrac{2}{3} A_\mu A_\nu A_\alpha) \ . \tag{5.7}$$

The field equation involves the dual field.

$$\frac{\delta I}{\delta A^a_\mu} = 0 \Leftrightarrow D_\mu F^{\mu\nu} + \frac{1}{2} \mu \epsilon^{\nu\alpha\beta} F_{\alpha\beta} = D_\mu F^{\mu\nu} + \mu {}^*F^\nu = 0 \ . \tag{5.8}$$

Evidently the parameter μ has the dimension of mass. The Abelian "Maxwell" theory, or the linear non-interacting approximation to the non-Abelian model, may of course be solved exactly, and one finds that the field excitations are indeed massive, with mass μ and spin, which in two spatial dimensions is a scalar, equal to $\mu/|\mu| = \pm 1$ [93].

Exercise 5.2. By operating on (5.8) with $\epsilon_{\alpha\beta\nu} D^\beta$ derive the second-order equation,

$$(D^\alpha D_\alpha + \mu^2)^* F_\mu = \epsilon_{\mu\alpha\beta} [{}^*F^\alpha, {}^*F^\beta] \quad , \tag{E5.2}$$

which shows the excitations to be massive.

The fact that excitations possess only one spin value, + 1 or − 1 depending on the sign of μ, indicates that reflection symmetries — parity and time reversal — are violated, as is already seen from the occurrence of the anti-symmetric tensor in the Lagrangian. In two-dimensional space, a parity transformation changes the sign of one coordinate; changing both would be equivalent to a rotation.

Parity:

$$A^0(t, r) \to A^0(t, r'), \qquad A^1(t, r) \to - A^1(t, r'), \qquad A^2(t, r) \to A^2(t, r') ,$$
$$r \equiv (x, y), \qquad r' \equiv (-x, y) \quad . \tag{5.9a}$$

Time reversal:

$$A^0(t, r) \to A^0(-t, r), \qquad \mathbf{A}(t, r) \to - \mathbf{A}(-t, r) \quad . \tag{5.9b}$$

Taking a doublet of models, one with mass μ and the other with $-\mu$, and defining the reflection transformations to include a field exchange, produces a reflection invariant system.

We recognize of course that the mass term in the action is a topological quantity; it coincides with $8\pi^2\mu/g^2$ times the Chern-Simons expression $W(A)$ introduced in (3.33). So we know that the mass action

$$I_{CS} \equiv - \frac{\mu}{2g^2} \int dx\, \epsilon^{\mu\nu\alpha} \operatorname{tr}(F_{\mu\nu} A_\alpha - \tfrac{2}{3} A_\mu A_\nu A_\alpha)$$
$$= 8\pi^2 \frac{\mu}{g^2} W(A) \quad , \tag{5.10}$$

is not gauge invariant against homotopically non-trivial, large three-dimensional gauge transformations of a non-Abelian gauge group whose Π_3 is $\mathscr{Z}$. Rather

I_{CS} changes by $8\pi^2\mu/g^2$ times the integral winding number of the gauge transformation. Consequently the requirement that the phase exponential of the action be gauge invariant enforces a quantization condition on the parameters of the theory [93],

$$4\pi\mu/g^2 = \hbar n \quad . \tag{5.11}$$

(A three-dimensional gauge coupling constant has dimension [mass]$^{1/2}$, in units where $\hbar$ and the velocity of light are dimensionless.)

Although the above argument is given for the Minkowski-space theory, the same result emerges in a Euclidean formulation. In Euclidean space, the Chern-Simons term acquires a factor of i, so that $e^{-I/\hbar}$ still changes by a phase after a gauge transformation. This is a consequence of the fact that $W(A)$ is a world scalar on arbitrary manifolds, without additional metric tensor factors, which also implies that the energy-momentum tensor for the theory (5.7) is given by the conventional formula (2.33).

Exercise 5.3. Show that (2.33) remains conserved with the field Eq. (5.8). Derive (2.33) by applying Noether's theorem, properly improved, to $\mathcal{L}$ in (5.7). Show that the Hamiltonian remains conventional,

$$H = \frac{1}{2}\int \mathrm{d}r(E_a^2 + B_a^2) \quad , \quad B_a = -\frac{1}{2}\,\epsilon^{ij} F_{ij}^a \quad , \tag{E5.3}$$

but the topological mass term reappears when H is expressed in canonical variables.

Of course an Abelian theory possesses a trivial Π_3, and does not require quantized parameters; indeed, in the absence of matter couplings it is free and the massive nature of the excitation — the "photon" — may be explicitly established. We shall not dwell on this model beyond noting the following fact, which puts into evidence topologically non-trivial effects here as well. Consider the Abelian field equations in the presence of a static, external charge density ρ, and total charge $Q = \int \mathrm{d}r\rho$. The time component of the field equation — Gauss' law for this theory — reads

$$\nabla \cdot E - \mu B = \rho \ . \tag{5.12a}$$

(Recall that the two-dimensional magnetic field is a scalar.) When this equation is integrated over all space, the E field is integrated over the circle at infinity C_∞. However, since fields are massive, they drop off exponentially at large distance, so one is left with

$$-\frac{Q}{\mu} = \int \mathrm{d}rB = \int \mathrm{d}r\, \nabla \times A = \oint_{C_\infty} \mathrm{d}r \cdot A \quad . \tag{5.12b}$$

Thus we see that the total charge measures the total flux through our two-dimensional space, and the magnetic potential is long-range

$$\boldsymbol{A} \xrightarrow[r \to \infty]{} -\frac{Q}{2\pi\mu} \nabla \tan^{-1}(y/x) \quad . \tag{5.13}$$

This brings out an analogy with the three-dimensional Higgs model: both describe massive gauge fields, the present theory by a topological mechanism, the Higgs model by spontaneous symmetry breaking; both give rise to topological vortex excitations [96].

The argument for quantizing parameters in the non-Abelian theory was presented in terms of the space-time action, appropriate to a functional integral formulation of the quantum theory. But just as in the analysis of the four-dimensional vacuum angle and of the quantum mechanical Dirac monopole, one may also give a Hamiltonian derivation [97], based on the response of our system to spatial, two-dimensional gauge transformations, which we restrict to approach $\pm I$ at spatial infinity. Since Π_2 of the groups with which we are dealing is trivial, there is only one homotopy class for the two-dimensional gauge functions U — they are all small — and they may be implemented by exponentiating the infinitesimal generator, which for our theory is no longer $\int d\boldsymbol{r}\, \boldsymbol{\Pi}_a \cdot \delta \boldsymbol{A}_a$, because the Lagrangian is not gauge invariant, rather it changes by a total derivative under an infinitesimal gauge transformation; see (2.32). We find

$$Q_\theta = -\frac{1}{g}\int d\boldsymbol{r}\,[\boldsymbol{\Pi}_a \cdot (\boldsymbol{D}\theta)_a - \tfrac{1}{2}\mu(\nabla\times \boldsymbol{A}_a)\theta^a] \quad . \tag{5.14}$$

$\boldsymbol{\Pi}_a$ is not $-\boldsymbol{E}_a$, owing to the derivative terms present in the mass term:

$$\Pi^i_a = -E^i_a + \tfrac{1}{2}\mu\epsilon^{ij}A^j_a \quad . \tag{5.15}$$

Nevertheless, the integrand of (5.14), when integrated by parts, coincides with the time component of the field equation,

$$G_a \equiv \frac{1}{g}(\boldsymbol{D}\cdot\boldsymbol{\Pi})_a + \frac{\mu}{2g}\nabla\times \boldsymbol{A}_a = -\frac{1}{g}\left\{(\boldsymbol{D}\cdot\boldsymbol{E})_a - \mu B_a\right\} \quad , \tag{5.16}$$

and this must vanish, when operating on physical states, in order to satisfy Gauss' law. (Since all gauge transformations are trivial, we may take θ^a to vanish at large distance, and no surface terms interfere in an integration by parts.)

$$G_a \, |\text{physical}\rangle = 0 \quad . \tag{5.17}$$

Correspondingly, the finite transformation must leave the state invariant.

$$e^{iQ_\theta/\hbar} \, |\text{physical}\rangle = |\text{physical}\rangle \quad . \tag{5.18}$$

(There is no θ angle in two spatial dimensions!)

The quantization condition arises because Eqs. (5.17) and (5.18) cannot be satisfied without it. This is seen as follows. We realize (5.17) in a Schrödinger picture, where (5.17) becomes a functional differential equation,

$$G_a \Psi_{\text{phys}}(A) = \left\{ \frac{\hbar}{gi} \left(D \cdot \frac{\delta}{\delta A} \right)_a + \frac{\mu}{2g} \nabla \times A_a \right\} \Psi_{\text{phys}}(A) = 0 \; . \tag{5.19}$$

While the algebra of the generators G_a still closes as in (3.26), this merely insures that Eq. (5.19) may be consistently integrated locally. But whether a global solution exists, which according to (5.18) must satisfy

$$\exp \int d\boldsymbol{r} \theta^a \left\{ \frac{1}{g} \left(D \cdot \frac{\delta}{\delta A} \right)_a + \frac{i\mu}{2\hbar g} \nabla \times A_a \right\} \Psi_{\text{phys}}(A)$$
$$= \Psi_{\text{phys}}(A) \quad , \tag{5.20}$$

remains an open question.

Exercise 5.4. Verify that the commutator algebra of G_a satisfies (3.26).

That there may be a problem with satisfying (5.20) arises from the following topological consideration. The statement that Π_3 is non-trivial also implies that the space of gauge functions defined on two-space is not simply connected. For consider any U defined on three-space which is not deformable to the identity. We now view U as providing a family of matrices depending on the spatial two-vector $\boldsymbol{r}$ and on a parameter τ. As $\tau \to \pm\infty$, $U \to I$; as a function of τ, our family describes a loop which starts and ends at the identity. Yet this loop cannot be shrunk to a point, since by hypothesis U is not deformable to the identity!

In order to make the obstruction to (5.20) explicit, we first evaluate the effect of the exponential operator on an arbitrary function of A. For definiteness, we work with the SU(2) group, and find

$$\left(\exp \frac{i}{\hbar}\int dr\theta^a G_a\right) \Psi(A) = \left(\exp \frac{i}{\hbar}\ \frac{8\pi^2\mu}{g^2}\ \Omega(\theta)\right) \dot{\Psi}(A^U) \ , \tag{5.21}$$

where

$$U = \exp \frac{\sigma^a\theta^a}{2i} \ ,$$

$$\Omega(\theta) \equiv \frac{1}{8\pi^2}\int dr\epsilon^{ij}\ \mathrm{tr}\,(\partial_1 U U^{-1} A^j)$$

$$+ \frac{1}{16\pi^2}\int dr\epsilon^{ij}\epsilon^{abc}\hat{\theta}^a\partial_i\hat{\theta}^b\partial_j\hat{\theta}^c(|\theta| - \sin|\theta|) \ ,$$

$$\theta = \sqrt{\theta^a\theta^a} \ , \qquad \hat{\theta}^a = \theta^a/|\theta| \ . \tag{5.22}$$

Exercise 5.5. Derive (5.21). Procedure: replace θ^a by $t\theta^a$; call

$$\Psi_t \equiv \left(\exp \frac{i}{\hbar}\int drt\theta^a G_a\right) \Psi(A) \ , \tag{E5.4}$$

differentiate with respect to t and obtain a differential equation for Ψ_t. Show that the right-hande side of (5.21) with Ω given in (5.22) solves the equation.

The exponentiated Gauss' law (5.18) or (5.20) demands that in our theory physical wave functionals not be gauge invariant; rather there is also a phase change:

$$\Psi_{\mathrm{phys}}(A^U) = \left(\exp \frac{-i}{\hbar}\ \frac{8\pi^2\mu}{g^2}\Omega(\theta)\right)\Psi_{\mathrm{phys}}(A) \ . \tag{5.23}$$

This complicated transformation law reflects a general quantum mechanical phenomenon. Whenever a symmetry transformation on canonical variables changes the Lagrangian by a total time derivative of a function, that function appears as a phase in the transformation law for the quantum mechanical state.

In our case the transformation law (3.36) for the Lagrange density (5.7) implies that under a spatial gauge transformation the Lagrangian changes according to

$$L = \int \mathrm{d}r \mathscr{L} \to L + \frac{\mu}{g^2} \int \mathrm{d}r \epsilon^{\mu\alpha\beta} \partial_\mu \operatorname{tr}(\partial_\alpha U U^{-1} A_\beta)$$

$$+ \frac{\mu}{3g^2} \int \mathrm{d}r \epsilon^{\alpha\beta\gamma} \operatorname{tr}(U^{-1} \partial_\alpha U U^{-1} \partial_\beta U U^{-1} \partial_\gamma U) . \quad (5.24\text{a})$$

The integrand in the last integral may be written as a total divergence; see Exercise 3.5. Dropping the two-space surface terms in (5.24a) leaves

$$L \to L - \frac{\mathrm{d}}{\mathrm{d}t} \left[\frac{8\pi^2 \mu}{g^2} \Omega(\theta) \right] . \quad (5.24\text{b})$$

This provides an alternative derivation of (5.23).

Consider next a "loop" of two-dimensional gauge functions parametrized by a homotopy parameter τ, such that $U_{\tau = \pm\infty} = I$, but the loop cannot be shrunk to a point. We have in mind a large gauge function defined in three-space, where the third coordinate is identified with τ, as explained above. We follow (5.23) as τ varies from $-\infty$ to $+\infty$. At the end points, U is the identity, so for consistency the phase factor must be unity, or

$$\frac{8\pi^2 \mu}{g^2} \Omega(\theta) \Big|_{\tau = -\infty}^{\tau = +\infty} = \frac{8\pi^2 \mu}{g^2} \int_{-\infty}^{\infty} \mathrm{d}\tau \frac{\partial}{\partial \tau} \Omega(\theta) = 2\pi\hbar n . \quad (5.25)$$

But comparing the integral in (5.25) to (5.24b) and reasoning as in the derivation of that equation, we recognize the integral in (5.25) to be just $\omega(U)$. Since $\omega(U)$ is an integer, (5.25) gives the quantization condition; if it is not met Gauss' law cannot be satisfied.

With an eye towards further application, let me summarise the key reasons for quantizing the mass term. First, we have the non-trivial homotopic structure of three-dimensional gauge transformations: $\Pi_3 = \mathscr{Z}$. Second, there is a term in the action which is invariant against small gauge transformations but not against large ones. Consequently, unless a quantization condition is enforced, the functional integral is not gauge invariant. Alternatively, in a Hamiltonian formulation, the non-trivial structure of Π_3 implies that the space of two-dimensional gauge functions is not simply connected. Unless the mass term is quantized, the differential equation which implements the modified Gauss' law (the modification arises from the mass term) cannot be globally integrated.

We have thus learned that a gauge theory in three dimensions is not specified just by a coupling constant; also there is an integer which gives the magnitude of a possible topological mass,

$$I = \frac{1}{2g^2}\int dx \, \mathrm{tr} F^{\mu\nu} F_{\mu\nu} + 2\pi\hbar n W(A) \quad . \tag{5.26}$$

Finally let me remark that the non-Abelian gauge theory without a mass term is infrared divergent in perturbation theory. Since one is expanding in a coupling constant g with the dimension of $[\text{mass}]^{1/2}$, momentum-space graphs in a high perturbative order involve high powers of g divided by a high power of the momentum. When these are inserted in a still higher order graph, an infrared divergence will ensue, even for off-mass-shell Green's functions. A re-summation procedure which heals these infrared divergences has been devised; one finds non-analyticity in the coupling constant [98]. Also the massless model has been analyzed non-perturbatively [99].

Let us now couple fermions to our massive gauge theory. In three-dimensional space-time the Dirac algebra is realized with 2×2 Pauli matrices, as in two dimensions — see (4.7) — with also $\gamma^2 = i\sigma^3$. There is no γ_5 matrix, since no matrix anti-commutes with all three Pauli matrices. Reflection transformations operate as follows.

$$\text{Parity:} \qquad \psi(t, r) \to \sigma^2 \psi(t, r') \quad ,$$

$$r \equiv (x, y) \quad , \qquad r' \equiv (-x, y) \quad . \tag{5.27a}$$

$$\text{Time reversal:} \qquad \psi(t, r) \to \sigma^3 \psi(-t, r) \quad . \tag{5.27b}$$

One sees that a mass term $m\bar{\psi}\psi$ is reflection non-invariant.

Coupling massless fermions to the Abelian theory without a topological mass produces perturbative infrared divergences, for the same reasons as explained above for the massless Yang-Mills model, and again a re-summation procedure, healing these divergences at the expense of coupling constant non-analyticity, has been discussed [98]. Massive fermions coupled to topologically massive Abelian fields give rise to a super-renormalizable theory which is entirely free of perturbative divergences: the masses provide an infrared cut-off; Lorentz and gauge invariance suffice to remove the ultraviolet infinities. In this respect the model is very interesting, but since it seems devoid of any topological interest, I shall not discuss it further [100].

An interesting topological effect arises for massless fermions coupling to the non-Abelian theory [101]. Again for definiteness we take the SU(2) group, and one doublet of fermions in the fundamental representation. (For the moment we are not concerned with the gauge field action, and are not committed to the presence or absence of the Chern-Simons mass term.) The fermion determinant is

$$\Delta(A) = \det(\not\partial + \not A) \quad , \tag{5.28}$$

and we inquire whether this is gauge invariant. There are no three-dimensional anomalies in current divergences, hence $\Delta(A)$ is certainly invariant against infinitesimal transformations. However, since Π_3 (SU(2)) is $\mathscr{Z}$, we still must check the homotopically non-trivial ones. The surprising result is that $\Delta(A)$ is not invariant against large gauge transformations U, but changes sign when U belongs to an odd homotopy group.

One may establish the non-invariance of $\Delta(A)$ without an explicit calculation of the determinant. We consider a family of vector potentials $A_\mu(x, \tau)$, labelled by a homotopy parameter such that A_μ vanishes at $\tau = -\infty$ and is a pure gauge $U^{-1}\partial_\mu U$ at $\tau = +\infty$, with U belonging to the first homotopy class. (U has unit winding number.) Of course at intermediate values of τ, A_μ is not a pure gauge, since the gauge function U cannot be deformed to the identity. It will be shown that the determinant at $\tau = +\infty$ has opposite sign from that at $\tau = -\infty$; hence it is not gauge invariant [102].

To effect the calculation, it is convenient to pass to Euclidean space and to double the size of the Dirac matrices to 4×4.

$$\Delta(A) = \Delta^{1/2}_{(4)}(A) = \det{}^{1/2}(\not\partial + \not A)_{(4)} \quad . \tag{5.29}$$

The subscript (4) reminds us that we are dealing with the usual 4×4 γ-matrices, but still only with three of them since we remain in three-dimensional space. A determinant of an operator is given by the product of the operator's eigenvalues. The spectrum of $(\not\partial + \not A)_{(4)}$ is symmetric about zero, because there exists a 4×4 matrix that anti-commutes with the three 4×4 Dirac matrices occurring in that operator. (We assume that at $\tau = \pm\infty$ there is no zero mode, so the determinant is non-vanishing.) By definition we take $\Delta^{1/2}_{(4)}(A)$ to be the product of positive eigenvalues at $\tau = -\infty$, and we follow these eigenvalues at τ passes $+\infty$. If an odd number crosses the axis, the determinant changes sign.

To see that with the chosen potential, $A_\mu(x, \tau)$, precisely one eigenvalue crosses, we imbed the operator $(\partial + A)_{(4)}$ in one higher dimension — the fourth dimension is τ, but A_4 is still zero — $\gamma^4 \partial_\tau + (\partial + A)_{(4)}$. It is now recognized that the vector potential we are using is exactly the four-dimensional instanton (3.60), in the gauge $A_4 = 0$.

Exercise 5.6. Perform a gauge transformation which takes the instanton (3.60) into the gauge $A_4 = 0$. Verify that the remaining three components vanish for $\tau = -\infty$, and become a pure, homotopically non-trivial, gauge at $\tau = +\infty$.

The four-dimensional Dirac equation in the instanton field has one normalizable zero mode [88]; see Exercise 4.4.

$$[\gamma^4 \partial_\tau + (\not\partial + \not A)_{(4)}]\Psi_0 = 0 \quad . \tag{5.30}$$

This means that one eigenvalue $\lambda(\tau)$ of $(\not\partial + \not A)_{(4)}$ must have the kink shape, changing sign as τ passes from $-\infty$ to $+\infty$, so that $[\gamma^4 \partial_\tau + \lambda(\tau)] f(\tau)$ possesses a τ-normalizable solution, and this completes the argument.

In order to insure a gauge-invariant theory with fermions, we must proceed in one of two ways. We may double the number of fermion species, so that the determinant gets squared and no longer changes sign under a large gauge transformation. Alternatively, if we wish to remain with one fermion, there must be a parity-violating Chern-Simons term in the gauge field action, with half-integer coefficient, so that the complete boson plus fermion action is gauge invariant.

In a sense the required Chern-Simons term can be found in $\Delta(A)$ when that determinant is regulated in a parity non-invariant fashion by introducing massive, parity-violating Pauli-Villars regulator fermions. [In the above proof of gauge non-invariance, the implicit regularization does not use regulator fields. Rather one has in mind for example zeta-function regularization of $\Delta^{1/2}_{(4)}(A)$, which does not introduce parity violation.] In fact one may then compute $\Delta(A)$, but of course not for arbitrary A_μ, but for those that correspond to constant, non-vanishing Abelian $F_{\mu\nu}$ — this is the three-dimensional analog of the four-dimensional Euler-Heisenberg calculation of the fermionic determinant for constant electromagnetic fields. [$\Delta(A)$ may also be evaluated for constant, non-Abelian $F_{\mu\nu}$.] The result is [103]

$$-i\hbar \ln \Delta_{PV}(A)\Big|_{F_{\mu\nu}\ \text{constant} \neq 0} = \hbar\pi W(A) \quad . \tag{5.31}$$

The parity-violating Chern-Simons term is induced by the parity-violating regularizing procedure. Thus the Pauli-Villars regulated determinant $\Delta_{PV}(A)$ is gauge invariant, but loses coordinate reflection symmetry.

This is analogous to the chiral anomaly story in even dimensions: while there are good *a priori* reasons for their existence, an explicit calculation encounters them only as a consequence of the regularization procedure. It is especially noteworthy that in the present instance an internal symmetry anomaly is correlated with the (spontaneous?) breakdown of the space-time reflection symmetry. This hints at a deep connection between discrete space-time transformations and gauge theory anomalies, which gives further hope that *P* and *CP* violation will be understood in this context. Note also that the Chern-Simons term is induced by fermions already in the Abelian three-dimensional theory. This again is as in even dimensions: Abelian models are topologically trivial, yet they support structures which in the non-Abelian setting become topologically non-trivial. Anomalies go beyond topology; they serve to expose topological intricacies, but are present even in topologically trivial circumstances!

Let me assess the physical import of three-dimensional gauge theories. Certainly they provide fascinating examples of the interrelation between quantum mechanics, gauge invariance, topology and discrete space-time symmetries. However, they may also be relevant to actual physical phenomena, since a field theory in three dimensions gives a phenomenological description of four-dimensional physics at infinite temperature. I shall not dwell here on details of this dimensional reduction, but I shall describe briefly some aspects of the temperature formalism [104].

Finite-temperature formulations of field theory proceed in the well known Martin-Schwinger manner. Amplitudes of interest are computed by the same "momentum-space" rules as at zero temperature, except that the energy variable is discrete, rather than continuous [105]. Specifically, a boson propagator $D(p)$ becomes

$$D(p) = \mathrm{i}/(p_0^2 - \boldsymbol{p}^2 - m^2) \quad , \qquad p_0 = \mathrm{i}\pi(2n)T \quad , \tag{5.32a}$$

while a fermion propagator $S(p)$ is

$$S(p) = \mathrm{i}/(\gamma^0 p^0 - \boldsymbol{\gamma}\cdot\boldsymbol{p} - m) \quad , \qquad p_0 = \mathrm{i}\pi(2n+1)T \quad , \tag{5.32b}$$

where n is any integer and T is (proportional to) the temperature. Furthermore an integral over virtual four-"momenta" becomes a sum over the integers n, and an integration over spatial three-momenta:

$$\int \frac{\mathrm{d}p}{(2\pi)^4} \to \mathrm{i}T \sum_{n=-\infty}^{\infty} \int \frac{\mathrm{d}\boldsymbol{p}}{(2\pi)^3} \quad . \tag{5.33}$$

Consider now the contribution of a propagator to some perturbation theoretic diagram. With bosons, we have

$$\mathrm{i}T \sum_{n=-\infty}^{\infty} \int \frac{\mathrm{d}\boldsymbol{p}}{(2\pi)^3} \; \lambda \; \frac{\mathrm{i}}{-4n^2\pi^2T^2 - \boldsymbol{p}^2 - m^2} \; \lambda \ldots \quad .$$

Here λ is the coupling constant governing the vertices that are connected by the propagator; the summation over the discrete "energy" is weighted by $\mathrm{i}T$; the dots signify further terms. In the high-T limit, all modes with $n \neq 0$ decouple, since our expression vanishes, owing to the temperature dependence of the denominator — the modes behave like very heavy particles. Only the $n = 0$ mode escapes this fate, and we conclude that, as $T \to \infty$, the above approaches

$$\int \frac{\mathrm{d}\boldsymbol{p}}{(2\pi)^3} \; \lambda\sqrt{T} \; \frac{1}{\boldsymbol{p}^2 + m^2} \; \lambda\sqrt{T} \; \ldots \quad .$$

This is exactly what one would find in a field theory on a Euclidean space of one dimension less, with an effective coupling constant which is $\sqrt{T}$ times the physical coupling. Moreover, it is further seen that the fermion contribution is subdominant, since the energy modes in the fermion propagator never vanish.

A position-space version of the above argument relies on the fact that finite-temperature field theory is formulated over a finite temporal interval which is purely imaginary and extends from 0 to $1/\mathrm{i}T$. Fields are defined over this interval to be periodic for bosons and anti-periodic for fermions. As T becomes large, the interval shrinks to zero, and the temporal dimension disappears from the problem, as do the anti-periodic Fermi fields.

From this we conclude that the graphical perturbation series for a simple theory — e.g., involving spin-$\frac{1}{2}$ fields and self-interacting spinless fields — coincides at sufficiently high temperature with the graphical series of a theory possessing the same structure of bosonic interactions, with all fermions deleted, and everything evaluated in a Euclidean space diminished by one dimension. Since four-dimensional renormalizable theories become super-renormalizable in three dimensions, our effective field theory is less divergent than the full, finite-temperature one. This reflects the fact that zero-temperature renormalization is sufficient to remove ultra-violet infinities from a finite-temperature theory [105]. Also three-dimensional coupling constants have different dimensions than the corresponding four-dimensional ones. This comes about, as we have seen, from the circumstance that the effective couplings acquire powers of T in the high-temperature domain.

The above is the most superficial inference that one can draw from perturbation theory. However, as we move away from the simplest models, new questions arise. Upon considering a four-dimensional theory with a four-component gauge field, we realize that a three-dimensional vector theory involves a three-component field and we wonder how to take the temporal field component of the four-dimensional theory into account in a high-temperature reduction. An answer requires detailed calculation. What is found in quantum electrodynamics and quantum chromodynamics is that the propagator of the temporal (electric) component acquires in lowest-order perturbation theory, a contribution which acts as a mass term that increases linearly with temperature. This is just the Debye screening length arising from the charged particle plasma which is present at finite temperature. The effect is that the temporal component of the gauge potential decouples, and one is left in three dimensions with a truly three-dimensional gauge theory [104]. For QED, this is a non-interacting theory without dynamical interest (the fermions decouple!); for QCD (or pure Yang-Mills theory) we are still facing a three-dimensional non-Abelian gauge theory, with highly non-trivial dynamics.

However, perturbative analysis beyond lowest order encounters new difficulties. The problem is that we are dealing with massless gauge fields, whose propagator diverges at zero momentum. Thus in four-dimensional, finite-temperature perturbation theory we encounter integrals of the form $\int d^3 p / \boldsymbol{p}^2$ arising from the zero mode in the energy summation. Alternatively and equivalently we meet the same integral in the perturbative analysis of a three-dimensional gauge theory at zero temperature. While in lowest order the $p = 0$ singularity is integrable, it becomes enhanced in higher orders, integrals are infrared divergent, and a direct perturbative approach is vitiated.

This is the central problem with perturbation theory for massless theories at finite temperature in four (or fewer) dimensions and at zero temperature in three (or fewer) dimensions. While a complete understanding of these models is still lacking, it is generally agreed that the infrared divergences arise only in perturbation theory, but the complete theory is free from inconsistencies.

In particular we cannot use finite-temperature perturbation theory for the four-dimensional non-Abelian gauge theory to derive the effective three-dimensional Lagrangian which should govern the infinite-temperature tail. Moreover, the four-dimensional theory has a rich topological structure — the θ angle coming from the tr $^*F^{\mu\nu}F_{\mu\nu}$ term in the Lagrangian (3.43c) — which is not seen in perturbative theory.

It is in this context that a physical role for the topological mass term presents itself: We conjecture that the effective three-dimensional Lagrangian, which governs the infinite-temperature tail of quantum chromodynamics, is not merely of the conventional massless Yang-Mills form, but also there is a quantized topological mass term.

There is no derivation of this fact; but neither can it be falsified, since naive perturbation theory does not exist and we do not have sufficient control over the formalism to extract non-perturbative behavior. Confronting such a hiatus, we invoke the principle of "naturalness" to aid in constructing the effective Lagrangian. The three-dimensional effective Lagrangian should possess all terms with quantum numbers of the four-dimensional theory, whose high-temperature asymptote is being described. According to this criterion, the gauge-invariant mass should be present, since its reflection non-invariance mirrors the reflection non-invariance of the four-dimensional θ vacua. Indeed, we that there is an intimate mathematical relation between the two, through the Chern-Simons secondary characteristic class; see (3.44), (3.45) and (5.26). But it is not known at present whether this mathematical relationship can be the basis for a physical derivation.

If we accept the gauge-invariant mass as a proper term in the effective Lagrangian which summarizes high-temperature behavior of physical non-Abelian gauge theories, we get another bonus, beyond infrared regularity. Owing to the quantization condition (5.11), the mass becomes evaluated in terms of the coupling constant and if we recall the connection between the three- and four-dimensional coupling constants, we find

$$\mu = n\hbar g^2/4\pi = n\hbar\lambda^2 T/4\pi \quad . \tag{5.34}$$

The integer structure of μ is most fascinating. A non-vanishing mass presumably arises from a non-vanishing θ, and discontinuities in the former for the three-dimensional model are suggestive of different phases in the latter for the four-dimensional theory. That different values of θ correspond to different phases has been occasionally suggested. Clearly it would be most satisfying if more understanding of these speculative ideas could be obtained [106].

Perturbation theory for gauge invariant massive Yang-Mills theory has been performed [93]. Though much more complicated than in the Abelian case, the one-loop results have been evaluated. It appears that infrared divergences are absent as in the Abelian theory; ultraviolet finiteness should also be guaranteed by the super-renormalizable interaction. So probably this too is a finite theory.

Another, much more vague, speculation makes reference to mass generation for gauge fields. A mechanism for this must be found if gauge fields are to correspond to the electro-weak intermediate bosons. The present approach, in terms of scalar fields and the Higgs mechanism, is frequently viewed as preliminary and unsatisfactory; the alternative of replacing Higgs fields by fermion pair bound states [107] has not succeeded phenomenologically [18]. Thus one may wonder if topological mechanisms, which are known to produce gauge-field masses in two and three dimensions, have an analogue in four space-time dimensions, thus providing a topological solution to this outstanding problem.

Finally we mention that fermion quantum electrodynamics in three-dimensional space-time is relevant when describing planar systems; e.g. charged particles moving in external crossed magnetic and electric fields. This is the physical situation for the Hall effect, and there are attempts to use the fermion-induced current, $j^\mu \propto (\delta / \delta A_\mu) W(A) \propto \epsilon^{\mu\alpha\beta} F_{\alpha\beta}$, for an analysis of this effect [108].

Chern-Simons terms may be added to the action of any gauge theory in an odd number of dimensions $2d - 1$. The procedure is to construct the Pontryagin class $\mathscr{P}_{2d}$ in one higher even dimension $2d$, write it as a total derivative, $\mathscr{P}_{2d} = \partial_\mu \mathscr{C}^\mu_{2d}$, take the $2d$ component of $\mathscr{C}^\mu_{2d}$ and integrate it over the $2d - 1$ space of interest, having suppressed the $2d$ coordinate. However, only in three dimensions does the Chern-Simons expression involve field bilinears leading to a mass.

Since higher-dimensional, particularly seven- and eleven-dimensional field theories are of interest in supergravity, and also in Kaluza-Klein theory of dimensional reduction, Chern-Simons terms may be relevant here [109].

One can also insert a topological mass term in three-dimensional gravity, since that model may be viewed as a vector gauge theory based on the Lorentz SO(2, 1) group [93]. Ordinarily three-dimensional gravity is without propagating excitations, because Einstein's field equations imply that the space is flat [110]. However, with a toplogical mass term, non-trivial dynamics take place, and there is one massive excitation with spin + 2 or −2, depending on the sign of the mass. But there is a difference between the space-time theory and its Euclidean space formulation. In the former the "gauge group" is the non-compact SO(2, 1), whose Π_3 is trivial. Hence mass is not quantized. For the latter, the group becomes the compact SO(3) with a non-vanishing Π_3, so a quantized mass is required. The resolution of this puzzle has not been given. Perhaps is it not true that gravity theories can be continued to imaginary time, i.e., to Euclidean space.

Exercise 5.7. Consider the quantum mechanical gauge theory of two-dimensional rotations in Exercise 3.2. Add a Chern-Simons term with an arbitrary coefficient to that Lagrangian. Since the gauge group is SO(2) = U(1), and the t space is one-dimensional, Π_1 (U(1)) = $\mathscr{Z}$ is relevant. Show from an action argument that this implies a quantization of the coefficient. Construct the Hamiltonian theory and re-derive the quantization condition from angular-momentum quantization.

Exercise 5.8. Consider the fermion quantum mechanical Lagrangian given by $L = i\hbar a^\dagger (d/dt + ieA)a$, where the a and the $a^\dagger$ satisfy the usual fermionic anti-commutation relations. Show that the theory is formally gauge invariant. Integrate out the fermions and evaluate the resulting determinant. Show that it is not gauge invariant against homotopically non-trivial U(1) gauge transformations, which occur since Π_1 (U(1)) = $\mathscr{Z}$.

5.3 Four-Dimensional Gauge Theories

An example of parameter quantization in four-dimensional theories is also known [111]. Consider an SU(2) gauge theory, with N massless Weyl fermion doublets, each in the fundamental (doublet) representation. Because

$$\Pi_4\,(\mathrm{SU}(2)) = (\text{group of } \{0, 1\} \text{ under modulo 2 addition}) \equiv \mathscr{Z}_2 \ , \tag{5.35}$$

there is one homotopically non-trivial class of four-dimensional SU(2) gauge functions, which cannot be deformed to the identity. Therefore, the functional fermion determinant,

$$\Delta^N(A) \equiv \det{}^{N/2}(\partial\!\!\!/ + A\!\!\!/) \ , \tag{5.36}$$

which is invariant against infinitesimal gauge transformations since SU(2) is anomaly-free, must still be checked for invariance against finite, large gauge transformations belonging to the non-trivial homotopy class. (The square root occurs because one is dealing with Weyl fermions, which have half as many components as Dirac fermions.) The check has been performed, and one finds that the action is not gauge invariant, but changes according to

$$-\mathrm{i}\hbar \ln \Delta^N(A) \to -\mathrm{i}\hbar \ln \Delta^N(A) + \pi N \hbar \quad . \tag{5.37}$$

Hence gauge invariance requires that N, the number of fermion species, be even.

The analysis of the fermion determinant $\Delta(A)$ proceeds in Euclidean space in the same way as three-dimensional example. One considers a potential $A_\mu(x, \tau)$ which interpolates between zero and a pure non-trivial gauge as τ ranges from $-\infty$ to $+\infty$. To determine any change in the determinant, the flow of eigenvalues $\lambda(\tau)$ of the four-dimensional Dirac operator $(\not\partial + \not A)$ is followed. If any cross zero, the five-dimensional Dirac operator $\gamma_5 \partial_\tau + (\not\partial + \not A)$ will possess a normalizable zero mode. That such a mode does indeed exist can be established, which then shows that $\Delta(A)$ changes sign.

The details of this argument are not explicit, but make use of advanced mathematical techniques (mod 2 index theory) and will not be presented here [111]. Rather a Hamiltonian derivation of the result that N must be even will be given. The argument is especially interesting not only for its elegant simplicity, but also because it relates the restriction on N to the chiral anomaly, which plays no role in the Lagrangian discussion [112].

Chiral SU(2) currents $J^\mu_a = -\hbar \bar\psi \gamma^\mu \frac{1}{2} \sigma^\alpha \psi$, constructed from Weyl fermions, are covariantly conserved, as they must be since SU(2) is anomaly-free and $\Delta(A)$ is invariant against infinitesimal gauge transformations. However, the singlet chiral number current, $j^\mu = \hbar \bar\psi \gamma^\mu \psi$, which is not a gauge current but a Noether symmetry current, here carries the anomaly, with half the coefficient in Eq. (4.23), because the Weyl fields have only two independent components:

$$\partial_\mu j^\mu = \frac{\hbar}{16\pi^2} \operatorname{tr} {}^*F^{\mu\nu} F_{\mu\nu} \quad . \tag{5.38}$$

[Compare with the two-dimensional example in Eq. (4.16).] As a consequence, the conserved chiral fermion number operator

$$N_\mathrm{f} = \int \mathrm{d}r (j^0 + \hbar N \mathscr{C}^0) \quad , \tag{5.39a}$$

is not gauge invariant, but changes under a three-dimensional gauge transformation belonging to the nth homotopy class according to

$$N_\mathrm{f} \to N_\mathrm{f} + \hbar n N \quad . \tag{5.39b}$$

As with our other Hamiltonian arguments, the relevant question concerns whether Gauss' law can be satisfied.

$$G_a \,|\text{physical}\rangle = 0 \quad . \tag{5.40}$$

Here G_a includes a contribution from the fermion charge density — compare with (4.3). Because the three-dimensional gauge function space is not simply connected, owing to Π_4 (SU(2)) being non-trivial — compare with the discussion in Section 5.2 — it is not certain that (5.40) can be globally satisfied. Thus we need to examine the condition that states must satisfy when (5.40) is exponentiated. For finite three-dimensional gauge transformations U in the trivial Π_3 class, (5.40) implies

$$\left\{\exp \frac{-\mathrm{i}}{\hbar g} \int \theta^a (\boldsymbol{D}\cdot\boldsymbol{E})_a\right\} \Psi_{\text{phys}}(A) \underset{\text{Gauss' law}}{=} {}^F\mathscr{G}_U^{-1}\, \Psi_{\text{phys}}(A) \quad , \tag{5.41}$$

where ${}^F\mathscr{G}_U$ is the operator which implements the gauge transformation on the fermion degrees of freedom, compare (4.5).

Obviously if (5.41) is met, ${}^F\mathscr{G}_U$ must be uniquely defined operator, taking the same form for equal U's. Consider now the following gauge function $U(r;\tau)$, labelled by the homotopy parameter τ, which varies from 0 to π.

$$U(r;\tau) = \mathrm{e}^{\mathrm{i}\sigma_3\tau}\, \mathrm{e}^{\mathrm{i}\boldsymbol{\sigma}\cdot\hat{\boldsymbol{r}} f(r)}\, \mathrm{e}^{-\mathrm{i}\boldsymbol{\sigma}\cdot\hat{\boldsymbol{r}} f(r)} \quad ,$$

$$f(0) = 0 \quad , \qquad f(\infty) = \pi \quad . \tag{5.42}$$

Although U has the factor $U_1 \equiv \mathrm{e}^{\mathrm{i}\boldsymbol{\sigma}\cdot\hat{\boldsymbol{r}} f(r)}$, which is a gauge transformation in the first homotopy class — see Exercise 3.5 — it also contains the inverse transformation, hence U belongs to the trivial class. Moreover, U contains factors $\mathrm{e}^{\pm\mathrm{i}\sigma_3\tau}$, which are rigid (global) isospin gauge transformations, and are realized on the fermion degrees of freedom by $\exp(\pm\mathrm{i}\tau/\hbar)2T_3$, where T_3 is the third component of the fermion isospin operator, with eigenvalues $\pm\hbar/2$. Since the eigenvalues of $2T_3$ measure the total of "up" fermions minus the total of "down" fermions, $2T_3$ differs from N_{f} by $\hbar$ times an even integer.

Let us now inquire how the fermion operator ${}^F\mathscr{G}_U$ varies as τ ranges from 0 to π. At $\tau = 0$, $U = I$ and we set ${}^F\mathscr{G}_U$ to the identity operator.

$${}^F\mathscr{G}_I = I \quad . \tag{5.43a}$$

To follow the evolution of ${}^F\mathscr{G}_U$ as τ approaches π, where U is again I, we represent ${}^F\mathscr{G}_U$ as a product of factors corresponding to the factors in U,

$$ {}^F\mathscr{G}_U = \mathrm{e}^{-2\mathrm{i}\tau T_3/\hbar}\, \mathscr{G}_{U_1}\, \mathrm{e}^{2\mathrm{i}\tau T_3/\hbar}\, \mathscr{G}_{U_1}^{-1} \quad . \tag{5.43b} $$

Here $\mathscr{G}_{U_1}$ implements the homotopically non-trivial gauge transformation. At $\tau = \pi$,

$$ \exp \pm \mathrm{i}\pi 2T_3/\hbar = (-I)^{2T_3/\hbar} = (-I)^{N_\mathrm{f}/\hbar} \quad . $$

But $\mathscr{G}_{U_1} N_\mathrm{f}\, \mathscr{G}_{U_1}^{-1} = N_\mathrm{f} + \hbar N$, hence

$$ {}^F\mathscr{G}_U \Big|_{\tau=\pi} = {}^F\mathscr{G}_I = (-I)^{N_\mathrm{f}/\hbar}\,(-I)^{N + N_\mathrm{f}/\hbar} = (-I)^N \quad , \tag{5.43c} $$

which contradicts (5.43a), unless N is even.

We see that the U(1) anomaly interferes with the SU(2), anomaly-free, gauge theory. This hints at a mathematical connection between the two, which has not yet been developed.

A similar conclusion applies to all SP(N) gauge theories [SP(1) = SU(2)], since Π_4 (SP(N)) = $\mathscr{Z}_2$. Unfortunately, physically interesting models do not use these groups, but only for these is Π_4 non-trivial. Therefore, no phenomenological application of this mathematically interesting result is available. Let us recall that this gauge variance of the determinant has a counterpart in three dimensions. But there, the problem could be cured by adding a Chern-Simons gauge field mass term by hand to the Yang-Mills Lagrangian or by finding it in the fermionic determinant through parity violating regularization procedures. No such option is available here, since we know no local four-dimensional structure constructed from gauge fields, which could compensate for the gauge non-invariance of $\Delta^N(A)$, when N is odd [113].

5.4 *Discussion*

There are additional examples of parameter quantization in field theory. In nonlinear σ-models coupled to gravity, one finds Newton's contant taking on quantized values [114]. Also nonlinear σ-models in four dimensions, which summarize SU(3) × SU(3) current algebra, together with its anomalies in the partially conserved currents, contain a term related to the five-dimensional Chern-Simons quantity (3.52). This is called the Wess-Zumino contribution to the action [75], and is further described in Section 6. Owing to a topological argument, the coefficient of the Wess-Zumino term is quantized since Π_5 (SU(3)) = $\mathscr{Z}$, thus fixing the coefficient of the chiral anomaly at an arbitrary integer [115]. Of course this is not a new result, since the magnitude of the anomaly was known previously from quark-model calculations. Nevertheless, it is remarkable that general arguments can give this information, without perturbative calculation.

In two dimensions the fermion determinant $\det(\partial\!\!\!/ + A\!\!\!/)$ may be evaluated even in non-Abelian theories [116]. With conserved vector gauge currents, there are anomalies in the axial vector Noether currents, compare with (4.15). The non-local formula for the two-dimensional $\det(\partial\!\!\!/ + A\!\!\!/)$ involves a Chern-Simons-type term in three dimensions, responsible for the anomalies. This is the Wess-Zumino action appropriate to the model in question, and its explicit coefficient is consistent with Π_3 being non-trivial.

6. Conclusions

While the discussion of topological effects was carried out separately for different dimensions, there exist unifying ideas and a unified formalism which are relevant to all dimensions, as is already seen in the relationship between the even-dimensional Chern-Pontryagin density and the odd-dimensional Chern-Simons density. In this concluding section, we give some of these unified formulations which, although producing no new physics, make use of elegant mathematics that may yet lead to physical discoveries.

6.1 Abelian Functional Gauge Structures in Non-Abelian Theories

Let us recall that a (non-relativistic) particle, of mass m and charge e, moving in an external magnetic field $\boldsymbol{B}$ is governed by a Hamiltonian which does not see the magnetic field when it is written in terms of velocities, $\mathbf{v}$.

$$H = \tfrac{1}{2} m \mathrm{v}^2 \quad . \tag{6.1}$$

The way in which H differs from the free-particle Hamiltonian is that the canonical momentum $\boldsymbol{p}$ is not merely $m\mathbf{v}$, rather one has

$$\boldsymbol{p} = m\mathbf{v} + e\boldsymbol{A}(r) \quad ,$$

$$\nabla \times \boldsymbol{A} = \boldsymbol{B} \quad . \tag{6.2}$$

Alternatively, one need not introduce the gauge variant potential A; a gauge invariant statement is that the commutator (Poisson bracket in the classical, canonical formulation) of two velocity components is non-vanishing, defining the field strength.

$$[v^i, v^j] = i \frac{e\hbar}{m^2} \epsilon^{ijk} B^k(r) \quad . \tag{6.3}$$

Commutators involving the coordinate r remain conventional. In this way, the Lorentz equations of motion (5.1) are regained as gauge invariant operator equations.

$$\dot{r} = \frac{i}{\hbar}[H, r] = v \quad , \qquad \dot{v} = \frac{i}{\hbar}[H, v] = \frac{e}{m} v \times B \quad . \tag{6.4}$$

But now we recognize that the three-dimensional Hamiltonian (describing a gauge theory in four-dimensional space-time) corresponding to the Yang-Mills Lagrangian with a θ term (3.43c), as well as the two-dimensional Hamiltonian (describing a gauge theory in three-dimensional space time) corresponding to the topologically massive Lagrangian (5.7) behave in an anologous fashion. When written in terms of A and E [the latter is like a (negative) field velocity since in the Weyl gauge it equals $-\dot{A}$] the Hamiltonian does not see the θ-term, nor the topological mass term; see Eqs. (E3.14) and (5.4). These reappear when E is expressed in terms of the canonical momentum $\Pi = \partial\mathcal{L}/\partial\dot{A}$. In the two cases, we have respectively

$$-E^i_a = \Pi^i_a - \frac{\theta}{8\pi^2} B^c_a \quad , \quad \text{(three-space)} \quad , \tag{6.5}$$

and

$$-E^i_a = \Pi^i_a - \frac{\mu}{2} \epsilon^{ij} A^j_a \quad . \quad \text{(two-space)} \quad . \tag{6.6}$$

One may therefore think of this circumstance as describing a non-Abelian gauge field with coordinates A and velocities $-E$, moving in an *external* U(1) *functional* gauge potential (connection) $\mathfrak{A}$ for which a formula in terms of the coordinates (A) is in the two respective cases

$$\mathfrak{A} = \frac{\theta}{8\pi^2} B^i_a = \frac{\theta}{8\pi^2} \epsilon^{ijk} \left(\partial_j A^k_a - \frac{g}{2} f_{abc} A^j_b A^k_c \right) \quad , \tag{6.7}$$

$$\mathfrak{A} = \frac{\mu}{2} \epsilon^{ij} A_a^j \quad . \tag{6.8}$$

The functional U(1) field strength (curvature) corresponding to these connections is deduced from the commutator of two velocities, compare (6.3). One finds, respectively

$$[E_a^i(\boldsymbol{x}), E_b^j(\boldsymbol{y})] = 0 \quad , \quad \text{(three space)} \quad , \tag{6.9}$$

$$[E_a^i(\boldsymbol{x}), E_b^j(\boldsymbol{y})] = -\mathrm{i}\hbar\mu\delta_{ab}\epsilon^{ij}\delta(\boldsymbol{x}-\boldsymbol{y}) \qquad \text{(two-space)} \ . \tag{6.10}$$

In the first case, the curvature vanishes, the connection $(\theta/8\pi^2)B_a^i$ is a pure gauge — a result consistent with (3.34), which also identifies the functional "gauge-function(al)" as the integrated Chern-Simons term. In the second case, the connection $(\mu/2)\epsilon^{ij}A_a^j$ corresponds to a "constant" curvature. The fact that the strength of this curvature is quantized, brings out once again the analogy between the Dirac monopole and the topological mass term [117].

One wonders whether there are other forms of non-Abelian dynamics that appear as field motion in an external U(1) functional connection [118].

6.2 *Cochains, Cocycles and the Coboundary Operation*

The behavior of gauge transformations in an anomalous gauge theory, as well as in a consistent gauge theory with a Chern-Simons term, can be given a unified description in terms of cocyles [119]. In order to present this, we begin by summarizing the role of cocyles in representation theory.

Consider a transformation g which belongs to a group of transformations. Suppose g acts on variables q according to a definite rule,

$$q \xrightarrow[g]{} q^g \quad , \tag{6.11}$$

and the group composition law is

$$g_1 g_2 = g_{12} \quad . \tag{6.12}$$

Next, consider functions $\Psi(q)$ defined on q, and represent the group action on these functions by an operator $U(g)$. In the simplest case, we have

$$U(g)\Psi(q) = \Psi(q^g) \quad , \tag{6.13}$$

and the composition law for the operators follows that of the group.

$$U(g_1)U(g_2) = U(g_{12}) \quad . \tag{6.14}$$

However, various generalizations are possible.

The first generalization consists of allowing a phase in (6.13).

$$U(g)\Psi(q) = e^{i2\pi\omega_1(q;g)}\,\Psi(q^g) \quad . \tag{6.15}$$

Imposing (6.14) shows that ω_1 must satisfy

$$\Delta\omega_1 \equiv \omega_1(q^{g_1};g_2) - \omega_1(q;g_{12}) + \omega_1(q;g_1) = 0 \ \text{(mod integer)} \ . \tag{6.16}$$

A quantity depending on the variable q and one group element is called a 1-cochain. When also it satisfies (6.16) it is called 1-cocycle. It may be that ω_1 can be written as

$$\omega_1(q;g) = \alpha_0(q^g) - \alpha_0(q) \equiv \Delta\alpha_0 \quad , \tag{6.17}$$

for some quantity of α_0. It is easy to show that then (6.16) is indeed satisfied. When (6.17) holds, the 1-cocycle is called trivial, and may be removed by rewriting (6.15) as

$$e^{i2\pi\alpha_0(q)}\,U(g)e^{-i2\pi\alpha_0(q)}\,e^{i2\pi\alpha_0(q)}\,\Psi(q) = e^{i2\pi\alpha_0(q^g)}\,\Psi(q^g) \quad , \tag{6.18}$$

i.e., by defining new functions

$$e^{i2\pi\alpha_0(q)}\,\Psi(q) \equiv \Psi'(q) \quad , \tag{6.19a}$$

and new, conjugated operators

$$e^{i2\pi\alpha_0(q)}\,U(g)\,e^{-i2\pi\alpha_0(q)} \equiv U'(q) \quad , \tag{6.19b}$$

the primed quantities satisfy the simple rules (6.13) and (6.14).

Another generalization occurs when the composition law (6.14) is relaxed and a phase is permitted to occur, so that one is dealing with a projective representation.

$$U(g_1)U(g_2) = e^{i2\pi\omega_2(q;g_1,g_2)}U(g_{12}) \quad . \tag{6.20}$$

Associativity of the triple product

$$\big(U(g_1)U(g_2)\big)\,U(g_3) = U(g_1)\big(U(g_2)U(g_3)\big) \quad , \tag{6.21}$$

imposes the condition

$$\begin{aligned}\Delta\omega_2 \equiv\ & \omega_2(q^{g_1};g_2,g_3) - \omega_2(q;g_{12},g_3) \\ & + \omega_2(q;g_1,g_{23}) - \omega_2(g;g_1,g_2) = 0 \ (\text{mod integer}) \quad , \end{aligned} \tag{6.22}$$

and the 2-cochain $\omega_2(q;g_1,g_2)$ is called a 2-cocyle since it satisfies (6.23). (It depends on two group elements.) Once again, if ω_2 can be written as Δ of some 1-cochain α_1,

$$\omega_2(q;g_1,g_2) = \alpha_1(q^{g_1},g_2) - \alpha_1(q;g_{12}) + \alpha_1(q;g_1) = \Delta\alpha_1 \quad , \tag{6.23}$$

ω_2 is trivial, then (6.22) is satisfied and (6.20) may be presented as

$$\begin{aligned} e^{-i2\pi\alpha_1(q;g_{12})}\,U(g_{12}) &= e^{-i2\pi\alpha_1(q;g_1)}\,e^{-i2\pi\alpha_1(q^{g_1};g_2)}\,U(g_1)U(g_2) \\ &= e^{-i2\pi\alpha_1(q;g_1)}\,U(g_1)e^{-i2\pi\alpha(q;g_2)}\,U(g_2) \quad , \end{aligned} \tag{6.24}$$

i.e., the modified operators

$$e^{-i2\pi\alpha_1(q;g)}\,U(g) \equiv U'(g) \quad , \tag{6.25}$$

satisfy the naive composition law (6.14).

The next generalization, involving 3-cocycles, arises when the representation behaves truly anomalously in that the operators implementing the transformation do not associate [120]. This means that they cannot be well-defined linear operators on a vector or Hilbert space, because such operators associate, by definition. If in (6.21) a phase is allowed,

$$\big(U(g_1)U(g_2)\big)\,U(g_3) = e^{i2\pi\omega_3(q;g_1,g_2,g_3)}\,U(g_1)\,\big(U(g_2)U(g_3)\big)\,, \tag{6.26}$$

consistency of four-fold products requires that the 3-cochain $\omega_3(q;g_1,g_2,g_3)$ satisfy the 3-cocyle condition.

$$\begin{aligned}\Delta\omega_3 \equiv\ & \omega_3(q^{g_1};g_2,g_3,g_4) - \omega_3(q;g_{12},g_3,g_4)\\ & + \omega_3(q;g_1,g_{23},g_4) - \omega_3(q;g_1,g_2,g_{34})\\ & + \omega_3(q;g_1,g_2,g_3) = 0 \ \text{(mod integer)}\ .\end{aligned} \tag{6.27}$$

A 3-cocyle is trivial if it can be written as Δ of a 2-cochain α_2.

$$\begin{aligned}\omega_3(q;g_1,g_2,g_3) =\ & \alpha_2(q^{g_1};g_2,g_3) - \alpha_2(q;g_{12},g_3)\\ & + \alpha_2(q;g_1,g_{23}) - \alpha_2(q;g_1,g_2) = \Delta\alpha_2\ .\end{aligned} \tag{6.28}$$

Next, we examine the implication of all this for the infinitesimal, algebraic relations when the transformation group is a continuous Lie group, and the finite transformation is expressed in terms of infinitesimal generators. The group element is represented by $g = e^{\theta^a T^a}$, where θ^a are the infinitesimal parameters. The occurrence of a 2-cocyle, as in (6.20), manifests itself in the infinitesimal formulation by the fact that the Lie algebra of the generators does not follow the Lie algebra of the group; rather, there is an extension [119]. Moreover, with a 3-cocyle, the Jacobi identify fails [120].

Let me now set down some mathematical terminology. Quantities that depend on n group elements are called n-cochains. The Δ operation, which has been presented for n = 0, 1, 2, 3, is called the coboundary operation and can be given a general definition.

$$\begin{aligned}\Delta\omega_n \equiv\ & \omega_n(q^{g_1};g_2,\ldots,g_{n+1}) - \omega_n(q;g_{12},g_3,\ldots,g_{n+1}) + \ldots\\ & + (-1)^m\,\omega_n(q;g_1,\ldots,g_{m\,m+1},\ldots,g_{n+1}) + \ldots\\ & + (-1)^{n+1}\,\omega_n(q;g_1,\ldots,g_n)\ .\end{aligned} \tag{6.29}$$

Evidently, operating on an n-cochain, Δ creates an $(n+1)$-cochain, and one sees that $\Delta^2 = 0$. An n-cochain which can be written as Δ of an $(n-1)$-cochain is an n-coboundary, while a cochain ω_n satisfying $\Delta\omega_n = 0$ (mod integer) is n-cocycle, which is trivial if it is a coboundary i.e., if $\omega_n = \Delta\omega_{n-1}$.

6.3 *Cocycles in Quantum Mechanics*

While the structures introduced above have a general mathematical setting, they possess particular significance in quantum mechanics which naturally concerns itself with unitary operators $U(g)$ that implement transformations g of dynamical variables g on the wave functions $\Psi(q)$. Our principal interest here is quantum gauge field theory, where q corresponds to the spatial components of the vector potential $\boldsymbol{A}$ — the dynamical variable in a canonical/Hamiltonian description — while the states are (in a Schrödinger representation) wave functionals of $\boldsymbol{A}$, $\Psi(\boldsymbol{A})$, and the group elements g are local gauge functions depending on the space-time variables. However, before delving into the quantum field theoretic application of these ideas, it is useful to exemplify them in the much simpler context of quantum mechanics of point particles. Indeed, quantum gauge field theory behaves exactly in the same way, except that one is dealing with gauge transformations, whose effect ultimately must be unobservable, while in the quantum mechanical examples discussed below the transformations describe actual changes in the physical system.

6.3.1 *1-Cocycle*

A 1-cocycle occurs in quantum mechanics whenever one is dealing with a transformation which is a symmetry operation of the action, but not of the Lagrangian. Specifically, if we consider a transformation defined by

$$q \to F(q) \quad , \tag{6.30a}$$

or in infinitesimal form

$$\delta q = f(q) \quad , \tag{6.30b}$$

which does not leave the Lagrangian L invariant, but changes L by a total derivative,

$$L \to L - \frac{d}{dt}\chi \quad , \tag{6.31a}$$

$$\delta L = -\frac{d}{dt}\chi \quad , \tag{6.31b}$$

then Noether's theorem gives the infinitesimal generator as

$$C = \frac{\partial L}{\partial \dot{q}}\,\delta q + \chi = p\delta q + \chi \quad , \tag{6.32}$$

and the unitary operator effecting the finite transformation

$$U = e^{iC/\hbar} \quad , \tag{6.33}$$

acts on wave functions with a 1-cocycle, which is just the quantity that appears in the finite transformation of the Lagrangian (apart from a factor of $\hbar^{-1}$).

$$U\Psi(q) = e^{i2\pi\omega}\,\Psi(F(q)) \quad , \quad 2\pi\omega \equiv \chi/\hbar \quad . \tag{6.34}$$

Moreover, if the cocyle is trivial, $\omega = \Delta\alpha$, then it can be removed by adjusting the phase of the wave function, as in (6.19). A phase change in a wave function corresponds to a canonical transformation, which in general changes the Lagrangian by a total time-derivative. In the present case, the new Lagrangian will read

$$L' = L + \frac{d}{dt}\,2\pi\hbar\alpha \quad , \tag{6.35}$$

and L' will have the property that it is invariant under the transformation.

These remarks are well-illustrated by the example of a Galilean transformation, $q \to q - \mathrm{v}t$, in a free theory governed by the Lagrangian $L = (m/2)\dot{q}^2$. We have $L \to L - m(d/dt)(q\mathrm{v} - \frac{1}{2}\mathrm{v}^2 t)$ and $\delta q = -\mathrm{v}t$, $\delta L = -m(d/dt)q\mathrm{v}$. The constant motion $C = -m\dot{q}\,\mathrm{v}t + mq\mathrm{v}$ gives rise to the unitary operator $U(\mathrm{v}) = e^{iC/\hbar} = e^{-i\mathrm{v}(pt - mq)/\hbar}$ whose effect on wave functions $\psi(q)$ can be easily evaluated with the Baker-Hausdorff lemma: $U(\mathrm{v})\psi(q) = e^{i2\pi\omega_1(q;\,\mathrm{v})}\,\psi(q - \mathrm{v}t)$. One recognizes the 1-cocycle $2\pi\omega_1(q;\,\mathrm{v}) = m/\hbar(q\mathrm{v} - \frac{1}{2}\mathrm{v}^2 t)$, which is also seen in the finite change of L and satisfies the 1-cocycle condition,

$\Delta\omega_1 = \omega_1(q - v_1 t; v_2) - \omega_1(q; v_1 + v_2) + \omega_1(q; v_1) = 0$. Moreover, the 1-cocycle is trivial since it can be written as $\omega_1(q; v) = \alpha_0(q - vt) - \alpha_0(q) = \Delta\alpha_0$, $2\pi\alpha_0 = -mq^2/2\hbar t$. To remove the 1-cocycle, we redefine the wave functions as in (6.19a), $e^{-imq^2/2\hbar t}\psi(q) = \psi'(q)$. To see how the Lagrangian changes, we first compute the Hamiltonian relevant to ψ' from the Schrödinger equation for ψ.

$$i\hbar \frac{\partial}{\partial t}(e^{imq^2/\hbar t}\psi') = \frac{p^2}{2m}(e^{imq^2/2\hbar t}\psi')$$

$$\Rightarrow i\hbar \frac{\partial}{\partial t}\psi' = \left(\frac{1}{2m}\left(p + \frac{mq}{t}\right)^2 - \frac{mq^2}{2t^2}\right)\psi' .$$

Hence, the new Hamiltonian $H' = (p + mq/t)^2/2m - mq^2/2t^2$ leads to the new Lagrangian

$$L' = m\left(\frac{1}{2}\dot{q}^2 - \frac{\dot{q}q}{t} + \frac{q^2}{2t^2}\right) = \frac{m}{2}\dot{q}^2 - \frac{d}{dt}\frac{mq^2}{2t}$$

$$= L + \frac{d}{dt}2\pi\hbar\alpha_0 \quad ,$$

and one verifies that L' is indeed invariant against Galilean transformations.

6.3.2 2-Cocycle

A quantum mechanical 2-cocycle arises in the representation of translations on phase space [coordinates and momenta]. The translation generators are $\boldsymbol{r}$ and $\boldsymbol{p}$, whose Lie algebra is non-commutative, possessing a central extension — the Heisenberg commutator $i[p^i, q^j] = \hbar\delta^{ij}$. The finite translations are implemented by $U(\boldsymbol{a}, \boldsymbol{b}) = e^{i(\boldsymbol{a}\cdot\boldsymbol{p} + \boldsymbol{b}\cdot\boldsymbol{q})/\hbar}$ which compose as follows.

$$U(\boldsymbol{a}_1, \boldsymbol{b}_1)U(\boldsymbol{a}_2, \boldsymbol{b}_2) = e^{i(\boldsymbol{a}_1\cdot\boldsymbol{b}_2 - \boldsymbol{a}_2\cdot\boldsymbol{b}_1)/\hbar}\,U(\boldsymbol{a}_1 + \boldsymbol{a}_2, \boldsymbol{b}_1 + \boldsymbol{b}_2)$$

These operators also serve as coherent state creation operators [121].

6.3.3 3-Cocycle

A 3-cocycle thus far has not been seen in quantum field theory, but quantum mechanics makes use of it when translations are represented on configuration space q in the presence of magnetic sources, specifically a magnetic monopole. Conventionally, the finite translation operator is taken to be $e^{i\boldsymbol{a}\cdot\boldsymbol{p}/\hbar}$, and no cocycles occur in this representation. However, in the presence of an external magnetic field $\boldsymbol{B}$, a gauge invariant canonical momentum does not exist, rather we use the gauge invariant velocity operator $\mathbf{v}$. Since $m\mathbf{v}$ satisfies the same commutation relation with $\boldsymbol{r}$ and $\boldsymbol{p}$, we may define

$$U(a) = e^{i a\cdot m\mathbf{v}/\hbar} \quad , \tag{6.36}$$

as the translation operator. However, it does not represent the Abelian translation group trivially, since the components of $\mathbf{v}$ do not commute; see (6.3).

$$[\mathrm{v}^i, \mathrm{v}^j] = i\,\frac{e\hbar}{m^2}\,\epsilon^{ijk} B^k \quad . \tag{6.37}$$

Moreover, the triple commutator

$$[\mathrm{v}^1, [\mathrm{v}^2, \mathrm{v}^3]] + [\mathrm{v}^2 [\mathrm{v}^3, \mathrm{v}^1]] + [\mathrm{v}^3, [\mathrm{v}^1, \mathrm{v}^2]] = \frac{e\hbar^2}{m^3}\nabla\cdot \boldsymbol{B} \ , \tag{6.38}$$

is non-vanishing in the presence of a magnetic sources, i.e., when $\nabla\cdot\boldsymbol{B}\neq 0$. In particular, for a point monopole strength g, located at $\boldsymbol{r}_0$ the divergence of $\boldsymbol{B}$ is

$$e\,\nabla\cdot\boldsymbol{B} = 4\pi g e\,\delta(\boldsymbol{r}-\boldsymbol{r}_0) \quad . \tag{6.39}$$

When the Jacobi identity fails, we anticipate the occurrence of a 3-cocycle, and it remains to understand why the finite quantities (6.36) do not associate in the presence of magnetic sources [120].

Before proceeding, let me discuss the numerial coefficient in (6.39). According to Dirac and as explained in Section 5.1, a consistent quantum dynamics for the monopole requires that eg be quantized in half integer units of $\hbar$. Hence, the coefficient in (6.39) is in fact $2\pi n\hbar$. For the moment, let us ignore this, and remain with an arbitrary value for eg.

To recognize the non-associativity, we form $U(\boldsymbol{a}_1)U(\boldsymbol{a}_2)$ and find from (6.36) and (6.37)

$$U(\boldsymbol{a}_1)U(\boldsymbol{a}_2) = e^{ie\Phi/\hbar}\,U(\boldsymbol{a}_1+\boldsymbol{a}_2) \quad , \tag{6.40}$$

where Φ is the outward [direction $\boldsymbol{a}_1 \times \boldsymbol{a}_2$] flux through the triangle with vertices $(\boldsymbol{r}, \boldsymbol{r}+\boldsymbol{a}_1, \boldsymbol{r}+\boldsymbol{a}_2)$. We set $e\Phi/\hbar$ to $-2\pi\alpha_2(\boldsymbol{r};\boldsymbol{a}_1 \cdot \boldsymbol{a}_2)$, but note that α_2 is not a 2-cocycle, as long as $\nabla \cdot \boldsymbol{B} \neq 0$, for then a globally defined vector potential cannot be defined. (If a well-defined vector potential exists, α_2 can be written as a coboundary of a 1-cochain constructed from the integrals of $\boldsymbol{A}$ along the edges of the triangle. Of course, the existence of $\boldsymbol{A}$ is tantamount to a divergence-free $\boldsymbol{B}$, so that the Jacobi identity is not violated.) Consider now three translations in non-coplanar directions $\boldsymbol{a}_1$, $\boldsymbol{a}_2$, $\boldsymbol{a}_3$; see Fig. 3. Forming the triple product as in (6.26), we find for the left-hand side

$$\begin{aligned}\big(U(\boldsymbol{a}_1)U(\boldsymbol{a}_2)\big)\,U(\boldsymbol{a}_3) &= \mathrm{e}^{-\mathrm{i}2\pi\alpha_2(\boldsymbol{r};\boldsymbol{a}_1,\boldsymbol{a}_2)}\,U(\boldsymbol{a}_1+\boldsymbol{a}_2)U(\boldsymbol{a}_3)\\ &= \mathrm{e}^{-\mathrm{i}2\pi\left(\alpha_2(\boldsymbol{r};\boldsymbol{a}_1,\boldsymbol{a}_2)+\alpha_2(\boldsymbol{r};\boldsymbol{a}_1+\boldsymbol{a}_2,\boldsymbol{a}_3)\right)}\,U(\boldsymbol{a}_1+\boldsymbol{a}_2+\boldsymbol{a}_3)\ .\end{aligned} \tag{6.41a}$$

For the right-hand side we have

$$\begin{aligned}&\mathrm{e}^{\mathrm{i}2\pi\omega_3(\boldsymbol{r};\boldsymbol{a}_1,\boldsymbol{a}_2,\boldsymbol{a}_3)}\,U(\boldsymbol{a}_1)\big(U(\boldsymbol{a}_2)U(\boldsymbol{a}_3)\big)\\ &= \mathrm{e}^{\mathrm{i}2\pi\omega_3(\boldsymbol{r};\boldsymbol{a}_1,\boldsymbol{a}_2,\boldsymbol{a}_3)}\,U(\boldsymbol{a}_1)\mathrm{e}^{-\mathrm{i}2\pi\alpha_2(\boldsymbol{r};\boldsymbol{a}_2,\boldsymbol{a}_3)}\,U(\boldsymbol{a}_2+\boldsymbol{a}_3)\\ &= \mathrm{e}^{\mathrm{i}2\pi\omega_3(\boldsymbol{r};\boldsymbol{a}_1,\boldsymbol{a}_2,\boldsymbol{a}_3)}\,\mathrm{e}^{-\mathrm{i}2\pi\alpha_2(\boldsymbol{r}+\boldsymbol{a}_1;\boldsymbol{a}_2,\boldsymbol{a}_3)}\,U(\boldsymbol{a}_1)U(\boldsymbol{a}_2+\boldsymbol{a}_3)\\ &= \mathrm{e}^{\mathrm{i}2\pi\left(\omega_3(\boldsymbol{r};\boldsymbol{a}_1,\boldsymbol{a}_2,\boldsymbol{a}_3)-\alpha_2(\boldsymbol{r}+\boldsymbol{a}_1;\boldsymbol{a}_2,\boldsymbol{a}_3)-\alpha_2(\boldsymbol{r};\boldsymbol{a}_1,\boldsymbol{a}_2+\boldsymbol{a}_3)\right)}\,U(\boldsymbol{a}_1+\boldsymbol{a}_2+\boldsymbol{a}_3)\end{aligned} \tag{6.41b}$$

Hence, we see that

$$\begin{aligned}\omega_3(\boldsymbol{r};\boldsymbol{a}_1,\boldsymbol{a}_2,\boldsymbol{a}_3) &= \alpha_2(\boldsymbol{r}+\boldsymbol{a}_1;\boldsymbol{a}_2,\boldsymbol{a}_3)-\alpha_2(\boldsymbol{r};\boldsymbol{a}_1+\boldsymbol{a}_2,\boldsymbol{a}_3)\\ &\quad+\alpha_2(\boldsymbol{r};\boldsymbol{a}_1,\boldsymbol{a}_2+\boldsymbol{a}_3)-\alpha_2(\boldsymbol{r};\boldsymbol{a}_1,\boldsymbol{a}_2)\ .\end{aligned} \tag{6.42}$$

The sum in (6.42) is $-(e/2\pi\hbar)$ times the total flux emerging from the tetrahedron formed from three vectors $\boldsymbol{a}_i$ with one vertex at $\boldsymbol{r}$. For a monopole it is $-(e/2\pi\hbar)\times 4\pi g = -(2eg/\hbar)$, while more generally

$$\omega_3 = -\frac{e}{2\pi\hbar}\int \mathrm{d}\boldsymbol{S}\cdot\boldsymbol{B} = -\frac{e}{2\pi\hbar}\int \mathrm{d}\boldsymbol{r}\,\nabla\cdot\boldsymbol{B}\ . \tag{6.43}$$

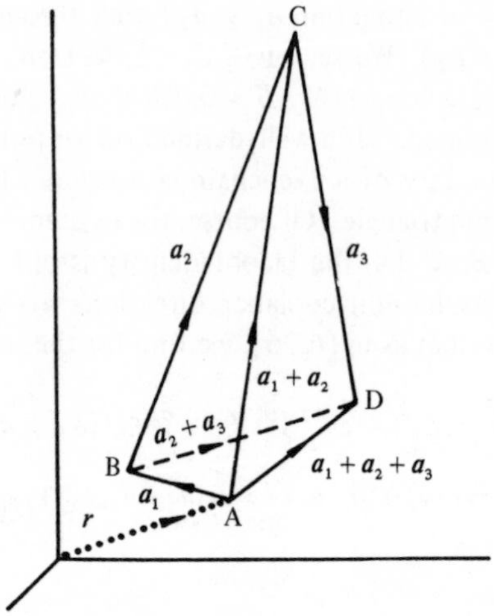

Fig. 3 Tetrahedron at point $\boldsymbol{r}$ defined by three translations $\boldsymbol{a}_i$. The 3-cocycle is proportional to the flux out of the tetrahedron.

In conclusion, we see that whenever $\nabla \cdot \boldsymbol{B} \neq 0$, there occurs a 3-cocycle in the representation of translations. According to (6.42), it is trivial, in the sense of (6.28); nevertheless, it is present. Moreover, for infinitesimally small $\boldsymbol{a}_i$, the Jacobi identity fails precisely by an amount $\nabla \cdot \boldsymbol{B}$.

We may now also understand the Dirac quantization condition. In order to have a proper quantum mechanical formalism, translations must commute. This may be achieved provided ω_3 is an integer n so that the 3-cocycle is invisible in the exponent, $e^{i2\pi n} = 1$. This then requires that (i) $\nabla \cdot \boldsymbol{B}$ be localized as in (6.39), so that its volume integral retain integrality for arbitrary $\boldsymbol{a}_i$, and (ii) eg be quantized in integer units of $\hbar/2$, so that ω_3 is an integer. Thus, we see that conventional quantum mechanics can accommodate magnetic sources only if they are concentrated at a point and if their strength is quantized. Notwithstanding associativity of finite translations, the infinitesimal generators do not associate at one point, the location of the monopole. However this does not interfere with a realization on a Hilbert space, because an isolated point — the location of the monopole — may be always removed.

6.4 Cocycles in Quantum Field Theory

We now turn to the gauge-field theoretic application of these ideas. Let us observe that in this context the transformation group becomes the local gauge group with position-dependent gauge functions U; the variables on which the group acts are the vector potentials A; and the function Ψ are wave functionals in the Schrödinger representation. Moreover, we show that Chern-Simons terms, their gauge transforms and other related quantities give rise to cochains, cocycles and the coboundary operation in a natural way [119].

Let us recall Eq. (3.44), which we record here once again, renaming the Chern-Simons four-vector $\mathscr{C}^\mu$ as $\Omega_0^\mu(A)$,

$$\mathscr{P} \equiv -\frac{1}{32\pi^2} \operatorname{tr} \epsilon^{\mu\nu\alpha\beta} F_{\mu\nu} F_{\alpha\beta} = \partial_\mu \Omega_0^\mu$$

$$\Omega_0^\mu(A) = -\frac{1}{8\pi^2} \operatorname{tr} \epsilon^{\mu\alpha\beta\gamma} (A_\alpha \partial_\beta A_\gamma + \tfrac{2}{3} A_\alpha A_\beta A_\gamma) \quad . \tag{6.44}$$

The following facts should be noted.

(1) The Chern-Simons Ω_0^μ density naturally lives in one dimension lower than the Chern-Pontryagin $\mathscr{P}$, in the following sense. Owing to the ϵ symbol, picking one component of Ω_0^μ, say $\mu = \mu_0$, forces the remaining components of the derivatives and gauge potentials occurring in (6.44) to be other than μ_0 — i.e., they belong to the three-dimensional subspace of the four-dimensional space that is complementary to μ_0. As a consequence, without loss of information, we may suppress one dimension and write the Chern-Simons density as a 3-dimensional object.

$$\Omega_0(A) = -\frac{1}{8\pi^2} \operatorname{tr} \epsilon^{\alpha\beta\gamma} (A_\alpha \partial_\beta A_\gamma + \tfrac{2}{3} A_\alpha A_\beta A_\gamma) \quad . \tag{6.45}$$

(2) The passage from $\mathscr{P}$ to Ω_0 is ambiguous. More specifically, Ω_0 is gauge dependent, while $\mathscr{P}$ is gauge invariant. When $\int d^4x\, \mathscr{P}$ is non-zero, Ω_0 will possess gauge dependent singularities, and/or slow long-range fall-off.

(3) $\Omega_0(A)$ may be called a 0-cochain — it depends on A, but on no U's and its change under a gauge transformation involves the coboundary operation.

$$\Omega_0(A^U) - \Omega_0(A) \equiv \Delta\Omega_0 \quad . \tag{6.46}$$

(4) The integral of Ω_0 over the appropriate space – here three-space —

is a 0-cocycle, since

$$\omega_0(A) \equiv \int d^3x\,\Omega_0(A) \quad , \tag{6.47a}$$

$$\Delta\omega_0 = \int d^3x(\Omega_0(A^U) - \Omega_0(A)) = 0 \text{ (mod integer)} \quad . \tag{6.47b}$$

(5) The explicit form for $\Delta\Omega_0$ may be read off (3.36).

$$\Delta\Omega_0 = \frac{1}{8\pi^2}\,\mathrm{tr}\,\epsilon^{\alpha\beta\gamma}\partial_\alpha(a_\beta A_\gamma) + \frac{1}{24\pi^2}\,\mathrm{tr}\,\epsilon^{\alpha\beta\gamma}a_\alpha a_\beta a_\gamma$$

$$a_\alpha \equiv \partial_\alpha U U^{-1} \quad . \tag{6.48}$$

$\Delta\Omega_0$ is also a total divergence of a quantity which is ambiguous and/or singular. That the first term on the right-hand side in (6.48) is a total derivative is manifest; that the same is true of the second term is demonstrated for SU(2) in Exercise 3.5 and 3.6, and for other groups in Ref. [48]. Thus, we write

$$\Delta\Omega_0 = \partial_\mu \Omega_1^\mu(A; U) \quad , \tag{6.49}$$

and recognize that owing to the ϵ symbol, $\Omega_1^\mu(A; U)$ again lives in one lower dimension, i.e., in two. The two-dimensional formula is

$$\Omega_1(A; U) = \frac{1}{8\pi^2}\,\mathrm{tr}\,\epsilon^{\alpha\beta}\partial_\alpha U U^{-1} A_\beta + \frac{d^{-1}}{24\pi^2}\,\mathrm{tr}\,\epsilon^{\alpha\beta\gamma}a_\alpha a_\beta a_\gamma \quad , \tag{6.50}$$

where the last term is a symbolic representation of a (complicated) quantity that cannot be given an explicit general expression for arbitrary groups, but can be worked out in specific cases, for example SU(2). From Exercises 3.5 and 3.6, we find that its form is

$$\frac{1}{4\pi^2}\,\mathrm{tr}\,\epsilon^{\alpha\beta}\ln U\,\partial_\alpha \ln U\,\partial_\beta \ln U\left(\frac{1}{|\theta|^2} - \frac{\sin|\theta|}{|\theta|^3}\right) ,$$

$$|\theta|^2 \equiv -2\,\mathrm{tr}(\ln U)^2 \quad .$$

Clearly, Ω_1 is a 1-cochain.

(6) We call the passage from $\mathscr{P}$ to Ω_0 and then to Ω_1 a dimensional descent, which may be continued by applying the coboundary operation to $\Omega_1(A;U)$. (There is no point in applying it to $\Delta\Omega_0$, since $\Delta^2 = 0$.)

$$\Delta\Omega_1 = \Omega_1(A^{U_1}; U_2) - \Omega_1(A; U_1 U_2) + \Omega_1(A; U_1) \quad . \tag{6.51}$$

One verifies that $\Delta\Omega_1$ is again a total divergence of an object which lives in one lower dimension — now one.

$$\Delta\Omega_1 = \partial_\mu \Omega_2^\mu(A; U_1, U_2) \quad . \tag{6.52}$$

For SU(2) there is an explicit expression for the 2-cochain Ω_2 [48].

$$\begin{aligned}
\Omega_2(A; U_1, U_2) = &-\frac{1}{4\pi^2} \operatorname{tr} U_1 \frac{d}{dx} \ln U_{12} \frac{1-\cos|\theta_{12}|}{|\theta_{12}|^2} \\
&-\frac{1}{4\pi^2} \operatorname{tr} U_1 \left[\frac{d}{dx} U_{12}, U_{12}\right] \frac{\sin|\theta_{12}| - |\theta_{12}|}{|\theta_{12}|^3} \\
&-\frac{1}{8\pi^2} \operatorname{tr} U_1 U_{12} \frac{d}{dx} \left\{ \frac{1-\cos|\theta_{12}|}{|\theta_{12}|^2} + \frac{\sin|\theta_{12}|}{|\theta_{12}|} \right. \\
&\left. + \int^{|\theta_{12}|} dt \frac{1-\cos t}{t} \right\} \\
|\theta_{12}|^2 \equiv &-2 \operatorname{tr}(\ln U_{12})^2 \quad .
\end{aligned} \tag{6.53}$$

For arbitrary groups Ω_2 may be simply given only for infinitesimal gauge transformations, $U = e^\theta = I + \theta + \ldots$. Then,

$$\begin{aligned}
&\Omega_2(A; I+\theta_1+\ldots, I+\theta_2+\ldots) \\
&= -\frac{1}{16\pi^2} \operatorname{tr} \left[\theta_1 \frac{d}{dx} \theta_2 - \theta_2 \frac{d}{dx} \theta_1\right] + \ldots \quad .
\end{aligned} \tag{6.54}$$

It follows that the integral of Ω_1 over two-space is a 1-cocycle,

$$\omega_1(A; U) \equiv \int d^2x\, \Omega_1(A; U) \quad , \tag{6.55a}$$

$$\Delta\omega_1 = \int d^2x\,\partial_\mu \Omega_2^\mu(A; U_1, U_2) = 0 \text{ (mod integer)} \quad , \tag{6.55b}$$

and similarly one can show that integrating Ω_2 over one-space yields a 2-cocycle.

$$\omega_2(A; U_1, U_2) \equiv \int dx\,\Omega_2(A; U_1, U_2) \quad , \tag{6.56a}$$

$$\Delta\omega_2 = 0 \text{ (mod integer)} \quad . \tag{6.56b}$$

With this, the dimensional descent terminates.

The sequence of formulas (6.44)-(6.54) may be compactly presented with the help of differential forms. In our notation, we depart from the conventional use of the wedge product; rather, the forms dx^μ are taken as anti-commuting variables $dx^\mu dx^\nu = -\,dx^\nu dx^\mu$ [122]. Thus, we have

$$A \equiv A_\mu dx^\mu \,, F \equiv \frac{1}{2} F_{\mu\nu} dx^\mu dx^\nu \,, \quad F = dA + A^2 \quad . \tag{6.57}$$

The ladder of descent, beginning with the four dimensional Chern-Pontryagin term is

$$\mathscr{P} = -\frac{1}{2!(2\pi)^2}\,\mathrm{tr}\,F^2 = d\Omega_0(A)$$

$$\Omega_0(A) = -\frac{1}{2!(2\pi)^2}\,\mathrm{tr}\,(dAA + \tfrac{2}{3}A^3)$$

$$\Delta\Omega_0 \equiv d\Omega_1(A; U)$$

$$\Omega_1(A; U) = \frac{1}{8\pi^2}\,\mathrm{tr}\,aA + \frac{d^{-1}}{24\pi^2}\,\mathrm{tr}\,a^3$$

$$\Delta\Omega_1 \equiv d\Omega_2(A; U_1, U_2)$$

$$\Omega_2(A; I + \theta_1 \ldots, I + \theta_2 + \ldots)$$

$$= -\frac{1}{16\pi^2}\,\mathrm{tr}\,(\theta_1 d\theta_2 - d\theta_1 \theta_2) + \ldots \tag{6.58}$$

Here, $a \equiv dUU^{-1}$. It is also useful to record the infinitesimal variation of the Chern-Simons 0-cochain [compare Eq. (3.34)],

$$\delta\Omega_0(A) = -\frac{1}{4\pi^2}\,\mathrm{tr}\,F\delta A + \frac{1}{8\pi^2}\,d\,\mathrm{tr}(A\delta A) \quad , \tag{6.59}$$

which for variations that are infinitesimal gauge transformations

$$\delta A = d\theta + [A, \theta] \equiv D\theta \quad , \tag{6.60}$$

reads

$$\delta\Omega_0(A)|_{\delta A = D\theta} = -\frac{1}{8\pi^2}\,\mathrm{tr}\,dA\,d\theta \quad . \tag{6.61}$$

Also, with an infinitesimal gauge functions we have

$$\Omega_1(A; I + \theta + \ldots) = \frac{1}{8\pi^2}\,\mathrm{tr}\,d\theta A + \ldots \tag{6.62}$$

One may consider similar chains of descent beginning in any even dimension. Thus, starting in two, we obtain a sequence of cochains [compare Eq. (3.51)].

$$\begin{aligned} \mathscr{P} &= -\frac{i}{2\pi}\,\mathrm{tr}\,F = d\Omega_0(A) \\ \Omega_0(A) &= -\frac{i}{2\pi}\,\mathrm{tr}\,A \qquad \Delta\Omega_0 = d\Omega_1(A; U) \\ \Omega_1(A; U) &= -\frac{i}{2\pi}\,\mathrm{tr}\ln U \quad . \end{aligned} \tag{6.63}$$

The six-dimensional dimensional ladder, which will play an important role below, reads [compare Eq. (3.52)]

$$\begin{aligned} \mathscr{P} &= \frac{i}{3!(2\pi)^3}\,\mathrm{tr}\,F^3 = d\Omega_0(A) \\ \Omega_0(A) &= \frac{i}{3!(2\pi)^3}\,\mathrm{tr}\,[(dA)^2A + \tfrac{3}{2}dA\,A^3 + \tfrac{3}{5}A^5] \\ \Delta\Omega_0 &= d\Omega_1(A; U) \end{aligned}$$

$$\Omega_1(A;U) = \frac{i}{12(2\pi)^3}\,\mathrm{tr}\Bigg[dA\,Aa + AdAa + A^3a$$

$$-\frac{1}{2}(Aa)^2 - Aa^3 + \frac{d^{-1}}{5}\,a^5\Bigg]$$

$$\Delta\Omega_1 = d\Omega_2(A;U_1,U_2)$$

$$\Omega_2(A;I+\theta_1+\ldots,I+\theta_2+\ldots)$$

$$= \frac{i}{12(2\pi)^3}\,\mathrm{tr}\,A(d\theta_1 d\theta_2 - d\theta_2 d\theta_1) + \ldots \qquad (6.64)$$

Again, we have recorded just the infinitesimal form of Ω_2; only that portion can be simply given and later the following further infinitesimal expression will be used.

$$\delta\Omega_0(A) = \frac{i}{16\pi^3}\,\mathrm{tr}\,F^2\delta A - \frac{i}{48\pi^3}\,d\,\mathrm{tr}(FA + AF - \tfrac{1}{2}A^3)\delta A \;. \qquad (6.65)$$

With a gauge transformation (6.60), this becomes

$$\delta\Omega_0(A)\Big|_{\delta A = D\theta}$$

$$= \frac{i}{48\pi^3}\,\mathrm{tr}\left((dA)^2 + \tfrac{1}{2}dAA^2 - \tfrac{1}{2}AdAA + \tfrac{1}{2}A^2dA\right)d\theta \;. \qquad (6.66)$$

Also,

$$\Omega_1(A;I+\theta+\ldots) = \frac{i}{12(2\pi)^3}\,\mathrm{tr}(dAA + AdA + A^3)d\theta + \ldots \qquad (6.67)$$

We recognize in (6.66) and (6.67), both of which describe the response of the five-dimensional Chern-Simons term to infinitesimal gauge transformations, structures that also occur in the anomalous divergence of gauge currents, see (4.26). Also, the first term in the right-hand side of (6.65) is seen in (4.27). Why this should be so will be explained presently. Finally, we note that integrals of the cochains Ω are cocycles ω, $\Delta\omega = 0$.

$$\omega_0(A) = \int d^5x\,\Omega_0(A) \quad . \tag{6.68}$$

$$\omega_1(A;U) = \int d^4x\,\Omega_1(A;U) \quad . \tag{6.69}$$

$$\omega_2(A;U_1,U_2) = \int d^3x\,\Omega_2(A;U_1,U_2) \quad . \tag{6.70}$$

6.4.1 0-Cocycle

The 0-cocycle, the integral of the Chern-Simons term, contributes to the action of an odd-dimensional gauge theory: the quantity in (6.47) is the three-dimensional topological mass term; the object in (6.68) plays an anologous role in a five-dimensional gauge theory. The cocycle $\omega_0(A)$ changes by an integer under gauge transformations; hence it enters the action multiplied by $2\pi\hbar n$ as in (5.26). [Since there is no one-dimensional, i.e., quantum mechanical gauge theory, no such role is assigned to the one-dimensional 0-cocycle arising in the sequence of descent (6.63). However, in quantum mechanics it can occur, see Exercise (5.8)].

6.4.2 1-Cocycle

The 1-cocycle (6.55) and (6.58) is a two-dimensional quantity and it measures the response of the wave functional occurring in the Hamiltonian description of a topologically massive gauge theory with Chern-Simons term in three-dimensional space-time, compare (5.23). A similar expression, involving the. four-dimensional 1-cocycle of (6.69) arises in the Hamiltonian description of a guage theory with Chern-Simons term in five-dimensional space time. Of course, we understand the presence of the 1-cocycle in the transformation law from the fact that the Lagrangian is not gauge invariant, but changes by a total time derivatives, see (5.24).

The 1-cocycle is an even dimensional object: it is defined in two dimensions when it descends from the four-dimensional Chern-Pontryagin density as in (6.50) and (6.58), while the descendent of the six-dimensional Chern-Pontrayain density exists in four dimensions, see (6.64) and (6.69). In the above mentioned application, it plays a role in the even-space dimensional Hamiltonian description of gauge theories in odd-dimensional space-time.

But the 1-cocycle has another, different role in even-dimensional space-time. Consider an anomalous gauge theory coupled to chiral fermions and ignore the dynamics of the gauge field, concentrating only on the dynamics of chiral fermions in an external, prescribed gauge field. The non-vanishing divergence of the current as in (4.26) indicates that the fermionic determinants $\Delta^{\pm}(A)$ are not gauge invariant. (The signs refer to the chirality of the coupling.) One may inquire how $\Delta^{\pm}(A)$ changes under a gauge transformation. The answer for the determinant continued to Euclidean space is inferred from (4.26), (6.66) and (6.67).

$$\Delta^{\pm}(A^U) = e^{\mp 2\pi i \omega_1(A;U)} \Delta^{\pm}(A) \quad . \tag{6.71}$$

That is, the 1-cocycle measures the anomalous response to gauge transformations of the determinant for chiral fermions. This explains why the variation of the Chern-Simons term under an infinitesimal gauge transformation produces the same expression as the anomalous divergence of a chiral gauge current, see (4.26), (6.66) and (6.67) [123].

In two dimensions it is not the chiral determinant that transforms with the 1-cocycle, rather the chiral determinant times the additional factor $\exp[(-ie/4\pi) \times \int d^2x \operatorname{tr} A^\mu A_\mu]$

$$\Delta'^{\pm}(A) \equiv \Delta^{\pm}(A) \exp[(-ie/4\pi) \int d^2x \operatorname{tr} A^\mu A_\mu] \quad . \tag{6.72a}$$

$$\Delta'^{\pm}(A^U) = e^{\mp 2\pi i \omega_1(A;U)} \Delta^{\pm}(A) \quad . \tag{6.72b}$$

That this is correct follows from the fact that the anomalous divergence of the gauge current in two-dimensional chiral model possesses the additional contribution involving $\partial_\mu A^\mu$, see (4.17), which is not a differential form.

It is easy to show from its definition that the 1-cocycle satisfies

$$\omega_1(A^U; U^{-1}V) = \omega_1(A;V) - \omega_1(A;U) \quad . \tag{6.73}$$

Hence $e^{\pm 2\pi i \omega_1(A;V)}$ transforms under a gauge transformation U, where A is transformed conventionally and V as $U^{-1}V$, in the same way as a chiral fermionic determinant. It may be therefore used as an effective, phenomenological description of the fermionic determinant, with V interpreted as a σ-model variable, i.e., $2\pi\hbar\,\omega_1(A;\sigma)$ is the Wess-Zumino action [75], described in Section 5.4 [123]. Note that the coefficient is fixed, and can be generalized only into an arbitrary integer, consistent with the quantization condition [115].

Exercise 6.1 Derive Eq. (6.73).

Exercise 6.2 Show that the Wess-Zumino condition (see Exercise 4.3) is the 1-cocycle condition $\Delta\omega_1 = 0$, taken in infinitesimal form.

We observe that the 1-cocycle is trivial, since according to (6.72) it satisfies

$$\omega_1(A;U) = \pm \frac{i}{2\pi}\left(\ln \Delta^{\pm}(A^U) - \ln \Delta^{\pm}(A)\right) \quad . \tag{6.74}$$

We may use this fact for the pure gauge theory in odd-dimensional space-time, without fermions but with a Chern-Simons term in the action, to remove from the wave functional the phase that arises when a gauge transformation is performed, see (5.23). That is, we view the Euclidean fermionic determinant as a given even-dimensional non-local functional of A and multiply it into the wave functional of the Chern-Simons gauge theory, which also resides in even dimensions, since it arises in the Hamiltonian description of the theory in odd-dimensional space-time. In the fashion described above, this will modify the Lagrangian by a total time derivative, and render it manifestly gauge invariant [117, 124].

The anomalous divergence of a gauge current (4.26) is related to the gauge variation of the Chern-Simons terms. Similarly, the anomalous divergence of a symmetry current (4.25) is related to the arbitrary variation of the integrated Chern-Simons term, i.e., of the 0-cocycle, as is seen from (6.65) [125]. [The last term in that equation disappears upon integration; when comparing the surviving term to (4.25), it should be remembered that the factor of two in the latter relative to the former comes from the fact that (5.25) refers to Dirac fermions, i.e., two Weyl fermions.]

6.4.3 2-Cocycle

When the composition law for operators implemented a representation is projective with a 2-cocycle as in (6.26) the commutator algebra is modified. In our application to local gauge transformations, considered at fixed time in a Hamiltonian formulation, the finite operator $\mathscr{G}_U$ is expanded for infinitesimal transformations (E.53) as

$$\mathscr{G}_{I+\theta+\ldots} = 1 + \frac{i}{\hbar}\int dr\,\theta^a(r) G_a(r) + \ldots \quad , \tag{6.75}$$

where $G_a(r)$ is the infinitesimal generator, and the 2-cocycle is expanded as

$$i2\pi\omega_2(A; I+\theta_1+\ldots, I+\theta_2+\ldots)$$

$$= \frac{1}{2}\int \mathrm{d}r\,\mathrm{d}r'\,\theta_1^a(r)\theta_2^b(r')S_{ab}(A;r,r')+\ldots \qquad (6.76)$$

The infinitesimal composition law departs from the Lie algebra (3.26) — S provides an extension [116].

$$\frac{i}{\hbar}[G_a(r), G_b(r')] = f_{abc}G_c(r)\delta(r-r') - i\hbar S_{ab}(A;r,r') \ . \qquad (6.77)$$

The generator consists of two contributions $(i\hbar/g)[D_i(\delta/\delta A_i)]_a$ which implements the gauge field variables, and $(1/g)J_a^0$ which does that job on chiral fermions, see (4.2) and (4.3).

It is known that whenever a dynamical current, i.e., a source current in a gauge theory, possess an anomalous divergence, then also the equal-time commutators for its components must contain anomalies [6, 126]. The reason is that a divergence may be represented by commutation with the translation generators P^μ and these commutators must be anomalous since the divergence is. Only $P^0 = H$ involves dynamics and since H contains J_a^μ, one can expect that the $[J_a^\mu, J_b^0]$ equal time commutator is anomalous, and indeed perturbative calculations have exhibited the anomaly [6, 126].

Thus, we may anticipate that the commutator (6.77) does indeed possess an anomaly, coming from the anomalous $[J_a^0, J_b^0]$ commutator and also from the commutator of $D_i(\delta/\delta A_i)_a$ with J_b^0. The latter induces infinitesimal gauge transformation on any explicit A_a dependence of J_b^0 — naively there is none, but the anomaly produces it. It is natural to suggest that the three-dimensional 2-cocycle that occurs in the descent from the six-dimensional Chern-Pontryagin density and the five-dimensional Chern-Simons terms, see (6.64) and (6.70), occurs in the three-dimensional, fixed time description of an anomalous gauge theory in four-dimensional space-time as an anomaly in the equal-time algebra (6.77). Similarly, the one-dimensional 2-cocycle in the descent from the four-dimensional Chern-Pontryagin term and the three-dimensional Chern-Simons term, see (6.56) and (6.58), is relevant to anomalous, two-dimensional gauge theories in a Hamiltonian description. This suggestion appears especially plausible when it is remembered that the higher-dimensional Chern-Simons term generates the anomalous divergence. Of course, such a conjecture should be verified with direct computations. The old calculations [6, 126] validate the result in Abelian theories and non-Abelian theories have also been investigated [127], [128] with results consistent with the conjecture.

The explicit forms for the anomalous extensions, given by the mathematical descent arguments, are in three dimensions [see (6.64)]

$$S_{ab}(A;r,r') = \frac{\mathrm{i}}{24\pi^2}\ \mathrm{tr}\, T^a\,\{T^b, T^c\}\ \epsilon^{ijk}\,\partial_i A^j_c \partial_k \delta(r-r')\ , \tag{6.78}$$

while in one dimension the answer is [see (6.58)]

$$S_{ab}(A;x,y) = \frac{\mathrm{i}}{4\pi}\,\delta_{ab}\,\delta'(x-y)\quad . \tag{6.79}$$

It should be recognized that the Abelian 2-cocycles are trivial. In three dimensions we have

$$\omega_2(A;\theta_1,\theta_2) = \int A d\theta_1 d\theta_2 = \Delta\alpha_1\quad ,$$

$$\alpha_1(A;\theta) = -\int A \mathrm{d} A\theta\quad . \tag{6.80}$$

while in one dimension

$$\omega_2(\theta_1;\theta_2) = \int(\mathrm{d}\theta_1\theta_2 - \theta_1\mathrm{d}\theta_2) = \Delta\alpha_1\quad ,$$

$$\alpha_1(A;\theta) = \frac{1}{2}\int A\theta\quad . \tag{6.81}$$

However, the non-Abelian 2-cocycles are non-trivial.

The problem with an anomalous gauge theory is now apparent from (6.77): one cannot require that states be annihilated by the Gauss' law generator, when the representation composition law is projective. Therefore, gauge invariance is lost.

Exercise 6.3. Find the form for the infinitesimal contribution S_{ab} to a 2-cocycle $\omega_2(A; V_1, V_2)$ when the cocycle is trivial. Show that in the trivial case extension in the commutator algebra may be removed by redefining the generator $G_a(r)$.

6.4.4 3-Cocycle

Field theoretic 3-cocycles are under present investigation, and a complete understanding is not yet at hand. But hints already exist that they may have a role in quantum field theory. In verifying the anomalous commutators (6.77), (6.78) and (6.79), it is necessary to compute commutators of $\boldsymbol{E}_a$, $\boldsymbol{A}_a$ and J_a^0 with each other, since all these objects occurs in G_a. When the calculation is performed in perturbation theory with the Bjorken-Johnson-Low (BJL) technique [129], it is found in the three-dimensional case that $\boldsymbol{E}_a$ cannot be represented by $\mathrm{i}/(\delta/\delta \boldsymbol{A}_a)$, because the perturbatively computed commutator does not vanish.

$$[E_a^i(r), E_b^j(r')] = \pm \frac{\hbar}{24\pi^2} \epsilon^{ijk} \operatorname{tr}\{T^a, T^b\} A^k(r)\delta(r-r') \quad . \tag{6.82}$$

The formula (6.82) does not satisfy the Jacobi identity. This result has also been understood in an analysis of the chiral anomaly for an anomalous gauge theory with the help of Berry's phase in the quantum adiabatic theorem [130].

Also a violation of the Jacobi identity appears in the quark model. When the Schwinger term in the commutator between time and space components of a current is a c-number, the Jacobi identity for triple commutators of spatial current components must fail [131]. Since deep-inelastic scattering data indicates that the Schwinger term is indeed a c-number [132], consistent with quark-model BJL calculations [133], the Jacobi identity for spatial current components should fail in the quark model, and this has been verified in perturbative calculations BJL [134]. The quark-model algebra of time and space components of vector and axial vector currents closes on U(6) × U(6) [135], and the above remarks indicate that a 3-cocycle occurs. However, a well-defined mathematical formulation is problematical, since the Schwinger term very likely is quadradically divergent [136].

6.5 Higher-Dimensional Gauge Dynamics Suggested by the Chern-Simons Term

In four-dimensional space-time, the Yang-Mills Hamiltonian density (2.67), defined on three-space, may be written as

$$\mathscr{E} = \frac{1}{2}\left(\boldsymbol{E}_a^2 + \frac{8\pi^2}{g^2}\left(\frac{\delta\,\omega_0(A)}{\delta \boldsymbol{A}_a}\right)^2\right) \quad . \tag{6.83}$$

This follows from (3.34). A similar formula holds in a degenerate way for the one-dimensional Hamiltonian density in two-dimensional space-time, since here the variation of the 0-cocycle is a constant. However, the energy density for conventional gauge dynamics in higher (even) dimensions does not possess the form (6.83) because the potential energy, which remains $\frac{1}{4} F^{ij}_a F_{aij}$ in arbitrary dimensions, is not related to variations of higher dimensional Chern-Simons terms.

An equivalent, covariant formulation is the following. The energy momentum tensor (2.33) may be presented in a manifestly self-dual form in four dimensions.

$$\theta^{\mu\nu} = \frac{1}{g^2} \operatorname{tr}(F^{\mu\alpha} F^{\nu}_{\alpha} + {}^*F^{\mu\alpha} {}^*F^{\nu}_{\alpha}) \quad ,$$

$${}^*F^{\mu\nu} \equiv \tfrac{1}{2} \epsilon^{\mu\nu\alpha\beta} F_{\alpha\beta} \quad . \tag{6.84}$$

This geometrically attractive formula is not immediately applicable in dimensionality other than four.

However, we suggest that the above expressions be generalized to higher dimensions, by insisting that the potential energy density always be given by the square of the variation of the Chern-Simons 0-cocycle [117].

Thus in five-space (relevant to a gauge theory in six-dimensional space-time) $(\delta\omega_0(A)/\delta A^i_a) \propto \epsilon^{ijklm} \operatorname{tr} F_{jk} F_{lm}$, compare (6.65), and the square of this would be the potential energy. Correspondingly, (6.84) would be generalized by retaining that expression, but defining the dual second-rank tensor by

$${}^*F^{\mu\nu}_a \propto \epsilon^{\mu\nu\alpha\beta\gamma\delta} \operatorname{tr} T^a F_{\alpha\beta} F_{\gamma\delta} \quad . \tag{6.85}$$

This provides new higher-dimensional gauge theories, with potentially interesting properties.

6.6 *The Dirac Sea and Chiral Anomalies*

While the physical origin of anomalies remains obscure and we must be content with formal, mathematical derivations, one can give the following discussion, based on the fermionic negative energy sea, which must be filled to define the vacuum of a second quantized theory of fermions [137].

To begin let me restate the fundamental puzzle of chiral anomalies. Consider a massless Dirac fermion [both chiralities] moving in a background gauge field. The dynamics is governed by a Lagrangian which splits into separate right and left parts.

$$\mathcal{L} = i\hbar\bar{\psi}(\not{\partial} + \not{A})\psi$$

$$= i\hbar\bar{\psi}_+(\not{\partial} + \not{A})\psi_+ + i\hbar\bar{\psi}_-(\not{\partial} + \not{A})\psi_- \ ,$$

$$\psi = \psi_+ + \psi_- \quad , \qquad \psi_\pm \equiv \tfrac{1}{2}(1 \pm i\gamma_5) \quad . \tag{6.86}$$

In the first-quantized theory, where ψ is a wavefunction and $\bar{\psi}\gamma^\mu\psi$ is a probability current, the separate right and left currents are conserved; and the separate probabilities $\int dr\, \psi_\pm\, \psi_\pm$ are time-independent. In the second quantized theory, where ψ becomes an operator, the anomaly phenomenon renders the separate right and left currents no longer conserved, and the right and left charges are not time-independent. Nevertheless, the sum of right and left — the vector current — is conserved, while the divergence of the difference between the right and left currents — the axial vector current — is beset by the anomaly. Our task then is to understand what causes the separate non-conservation of left and right currents even though there is no coupling between the two apparent in (6.86).

Evidently, the problem arises from the second quantization procedure, hence we review it. We set A^0 to zero, obtain the eigenmodes of the Hamiltonian in the background field $\boldsymbol{A}$, and define the second quantized vacuum by filling the negative energy modes and leaving the positive energy modes empty. The background $\boldsymbol{A}$ is chosen in a specific functional form so that the anomaly is non-vanishing. This requires a time-dependence, but we chose a static potential and model the time variation by an adiabatic change $\boldsymbol{A} \to \boldsymbol{A} + \delta\boldsymbol{A}$.

The simplest model to study is 2-dimensional and Abelian — the Schwinger model. As we shall see, this model also holds the key to the anomaly phenomenon in higher dimensions. The Lagrangian is as in (4.19) and the axial current possess the anomaly (4.15), which is proportional to $\partial_t A^1$. The eigenmodes to be second quantized satisfy a one-dimensional Dirac equation.

$$H\psi_E = \alpha\left(\frac{d}{idx} - eA\right)\psi_E = E\psi_E \quad ,$$

$$\alpha \equiv \gamma^0\gamma^1 \quad , \qquad A \equiv A^1 \quad . \tag{6.87}$$

With "Dirac" matrices (4.7), the modes are

$$\psi_+ = \begin{pmatrix} e^{ikx} \\ 0 \end{pmatrix} \quad \text{with} \quad E = -p + eA \quad ,$$

$$\psi_- = \begin{pmatrix} 0 \\ e^{ikx} \end{pmatrix} \quad \text{with} \quad E = p - eA \quad ,$$

$$p \equiv \hbar k \quad . \tag{6.88}$$

Second quantization is performed by filling the negative energy sea. For $A = 0$, the energy-momentum dispersion is given in Fig. 4, where the right-hand branch corresponds to fermions of one chirality, and the left-hand branch to those of the other chirality. The negative energy states are filled, as indicated by the filled circles, the positive energy states are empty, as indicated by the empty circles. As A increases from 0 to δA, empty states in the right-hand branch acquire negative energy, while filled states of the left-hand branch become positive energy states; i.e., there is a net production of right-handed antiparticles and left-handed particles; see Fig. 5. This is why the separate right and left charges are conserved, but their sum is. Put in another way, the separation between positive and negative energy states of definite chirality cannot be achieved gauge-invariantly, since changing A from 0 to a constant δA is a gauge transformation, yet particles are produced.

The same analysis is applicable to 4-dimensional massless electrodynamics where the anomaly involves $^*F^{\mu\nu}F_{\mu\nu} \propto \boldsymbol{E} \cdot \boldsymbol{B} \propto \partial_t \boldsymbol{A} \cdot \boldsymbol{B}$. Choosing for the background a potential with x and y components corresponding to a constant magnetic field in the z direction and a spatially homogeneous A_z, the Hamiltonian is

$$\begin{aligned} H &= \alpha \cdot (p - eA) \\ &= \alpha_\perp \cdot (p - eA)_\perp + \alpha_z (p_z - eA_z) \quad . \end{aligned} \tag{6.89}$$

Here, the perpendicular ($\perp$) directions are x, y, and $A_\perp$ gives rise to a constant B_z. It is known that a 2-dimensional Dirac Hamiltonian in a constant magnetic

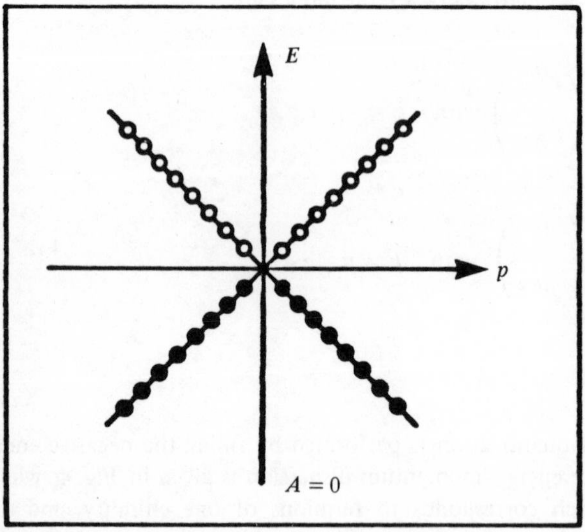

Fig. 4. Energy-momentum dispersion law at $A = 0$. Empty circles are empty states; filled circles are filled states.

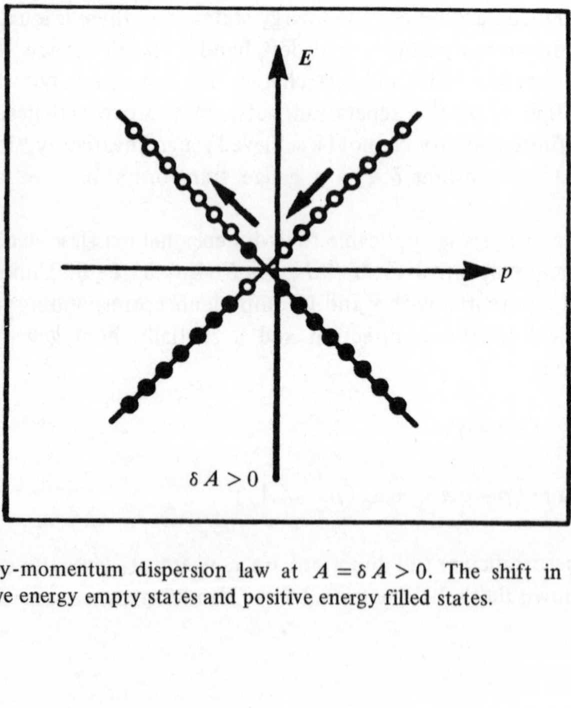

Fig. 5. Energy-momentum dispersion law at $A = \delta A > 0$. The shift in energy gives rise to negative energy empty states and positive energy filled states.

field in the [missing] third direction possesses a vanishing eigenvalue. Hence, the above Hamiltonian (6.89), restricted to that zero-mode subspace, reduces to the 1-dimensional problem (6.87).

In general, an anomaly in $d = 2n$ dimensions requires a Hamiltonian analysis in $2n - 1$ dimensions; that Hamiltonian has a zero eigenvalue in $2n - 2$ dimensions, leaving a 1-dimensional subspace which is analyzed as in (6.87). Moreover, the $2n - 2$ dimensional zero mode is insured by the Atiyah-Singer index theorem, i.e., by the anomaly in 2 dimensions lower.

We saw the above dimensional reduction operating from 4 to 2 dimensions. In 6 dimensions with a non-Abelian gauge theory, the requisite 4 dimensional zero modes are produced by the instanton configuration.

We thus see that the negative energy sea is responsible for the non-conservation of chirality even though the dynamics is chirally invariant. This effect was called "anomalous" because its discovery was an unexpected surprise. However, a better name might be quantum mechanical symmetry breaking — a symmetry breaking mechanism which like Heisenberg's spontaneous symmetry breaking, attributes physical asymmetry to the vacuum state and not to the dynamics. Here however, unlike in Heisenberg's case, it is not vacuum degeneracy, but the very definition of the vacuum that is responsible.

6.7 Discussion

The non-trivial topological effects in quantized gauge theories arise from the non-trivial properties of non-Abelian gauge groups. In the case of the Dirac monopole, even the Abelian U(1) group possesses a non-trivial structure. Moreover, possibilities for coupling fermions are strongly restricted by the interplay between the topology of the gauge group and the chiral anomaly.

Specific results for model building include the recognition of the vacuum angle in quantum chromodynamics, the emergence of quantization conditions on the parameters of the theory, and constraints on fermion content.

An intriguing question may be raised. Following usual procedures, we reject theories whose fermionic determinant is gauge non-invariant. However, in three dimensions there is another option: the gauge field action may be modified in a local, but gauge non-invariant fashion, so the complete theory is gauge invariant, but parity violating. Can gauge invariance problem with other "inconsistent" theories be cured, without altering fermion content [66], [113]?

Quantized parameters do not seem to arise in fundamental physical theories like quantum chromodynamics or the various unified models. But they do occur in Lagrangians which provide a phenomenological description for a limited dynamical range, like the high-temperature three-dimensional gauge theory, or the nonlinear σ model for low-energy, current-algebra controlled processes. In this respect the situation is analogous to condensed-matter theory, where non-trivial topological effects, like flux quantization and charge fractionization, operate on a phenomenological level, while the fundamental many-body Coulomb Hamiltonian is unconstrained.

Since the chiral anomaly associated with massless fermions, and with topological structure seems fundamental to the theory, one may speculate that further understanding in this area will shed light on the origin of mass and chirality — questions which thus far have no satisfactory answers.

In this context it is important to stress the fact that the chiral anomaly is not an obscure consequence of perturbative infinities, but reflects a true fact about quantized gauge theories with fermions. In two dimensions, this has been appreciated for a long time, since here the anomaly may be established non-perturbatively, see (4.9) and (4.10). The mathematical discussion in terms of cocycles gives an alternative non-perturbative derivation of non-Abelian anomalies, one that applies to four dimensions as well.

One may ask why structures in physical dimensions — the action in four space-time and the Hamiltonian in three-space — should be described by mathematical objects that descend from the six-dimensional Chern-Pontryagin density and the five-dimensional Chern-Simons density. While such questions often have no answer, in the present instance the reason can be given [138]. What is involved is a topological obstruction to defining gauge invariantly the chiral fermion determinant. To expose this obstruction, one embeds the gauge potential into a two-parameter homotopy family and the obstruction manifests itself in the vanishing of the determinant as the parameters are varied. This vanishing corresponds to zero eigenvalues of the Dirac equation in six dimensions (2 homotopy + 4 coordinate) and these are counted by the six-dimensional Atiyah-Singer index theorem, involving the Chern-Pontryagin 6-form.

While the connection between anomalies and topology is a beautiful example of conjunction of interests between the two sister sciences of physics and mathematics, there still remains something further to understand. The topological/mathematical arguments are relevant in topologically non-trivial situations, which arise in the non-Abelian theories. However, anomalies occur even for topologically trivial gauge theories, like those based on the Abelian U(1) group. Also, the topological analyses naturally apply to integrated quantities, while anomalies are intrinsically local densities. Thus, the full mathematical significance of anomalies remains still to be exposed.

Acknowledgement

I thank Bryce DeWitt and Raymond Stora and the organizers of the 1983 Les Houches summer school, devoted to relativity, groups and topology, for giving me the opportunity to lecture. I am also very grateful to Roberta Young for her assistance in preparing the text. Finally I thank all the participants in the school for listening to me. The research reported here was supported in part through funds provided by the US Department of Energy, under contract DE-AC02-76ERO3069.

Solutions to Selected Exercises*

Exercise 2.2.

$$\begin{aligned} D_\mu M &= \partial_\mu M + [A_\mu, M] \\ [D_\mu, D_\nu]M &= D_\mu D_\nu M - (\mu \leftrightarrow \nu) = D_\mu(\partial_\nu M + [A_\nu, M]) - (\mu \leftrightarrow \nu) \\ &= \partial_\mu(\partial_\nu M + [A_\nu, M]) + [A_\mu, \partial_\nu M + [A_\nu, M]] - (\mu \leftrightarrow \nu) \\ &= [\partial_\mu A_\nu - \partial_\nu A_\mu, M] + [A_\mu, [A_\nu, M]] - [A_\nu, [A_\mu, M]]. \end{aligned}$$

In the last equality,.the last two terms are rewritten with the help of the Jacobi identity as $[[A_\mu, A_\nu], M]$, which when combined with the first term gives the desired result.

Exercise 2.3.

$$\begin{aligned} F_{\mu\nu} &\equiv \partial_\mu A_\nu - \partial_\nu A_\mu + [A_\mu, A_\nu] \\ \delta F_{\mu\nu} &= \partial_\mu \delta A_\nu - \partial_\nu \delta A_\mu + [\delta A_\mu, A_\nu] + [A_\mu, \delta A_\nu] \\ &= D_\mu \delta A_\nu - D_\mu \delta A_\mu. \end{aligned}$$

Exercise 2.4. (a) $j^\mu_\Theta = (2/g^2)\,\mathrm{tr}\, F^{\mu\nu} D_\nu \Theta$. By (2.8b),

$$\partial_\mu j^\mu_\Theta = \frac{2}{g^2}\,\mathrm{tr}\, D_\mu F^{\mu\nu} D_\nu \Theta + \frac{2}{g^2}\,\mathrm{tr}\, F^{\mu\nu} D_\mu D_\nu \Theta.$$

$D_\mu F^{\mu\nu}$ vanishes by the equation of motion; in the remaining term, the anti-symmetry of $F^{\mu\nu}$ and Exercise (2.2) give

$$\partial_\mu j^\mu_\Theta = \frac{1}{g^2}\,\mathrm{tr}\, F^{\mu\nu}[F_{\mu\nu}, \Theta],$$

which vanishes by the cyclicity of the trace.

(b) Since j^μ_a may be obtained from j^μ_Θ by setting Θ to T^a, the result (a) above implies $\partial_\mu j^\mu_a = 0$.

Exercise 2.5.

$$\dot{Q}^a = \int d\mathbf{r}\, \partial_0 j^0_a = \int d\mathbf{r} (\partial_\mu j^\mu_a - \boldsymbol{\nabla} \cdot \mathbf{j}_a).$$

In the last integral, the first integrand vanishes by current conservation, and the second leads to a surface integral at infinity. Hence $\dot{Q}^a = 0$, provided $\mathbf{j}_a$ falls off sufficiently rapidly at $r = \infty$. The field equation gives

$$j^\mu_a = -f_{abc} F^{\mu\nu b} A^c_\nu = \frac{1}{g}\partial_\nu F^{\nu\mu a}.$$

---

* Prepared with the assistance of C. Burges and S.G. Jo.

Hence

$$j_a^0 = \frac{1}{g}\partial_i F^{i0a} = \frac{1}{g}\nabla\cdot\boldsymbol{E}_a,$$

where $E_a^i = F^{i0a}$. It follows that

$$Q = \frac{1}{g}\int \mathrm{d}\boldsymbol{r}\,\nabla\cdot\boldsymbol{E} = \frac{1}{g}\int \mathrm{d}\boldsymbol{S}\cdot\boldsymbol{E},$$

where $Q \equiv Q^a T^a$. Under a gauge transformation U,

$$Q \to Q^U, \qquad \boldsymbol{E} \to \boldsymbol{E}^U = U^{-1}\boldsymbol{E}U; \qquad Q^U = \frac{1}{g}\int \mathrm{d}\boldsymbol{S}\cdot U^{-1}\boldsymbol{E}U.$$

The surface is at $r = \infty$; also we assume U is independent of angles at $r = \infty : U \xrightarrow[r\to\infty]{} U_\infty$. Thus

$$Q^U = U_\infty^{-1}\left(\frac{1}{g}\int \mathrm{d}\boldsymbol{S}\cdot\boldsymbol{E}\right)U_\infty = U_\infty^{-1} Q U_\infty;$$

i.e. Q is gauge covariant.

Exercise 2.6.

$$j_\theta^0 = \frac{2}{g^2}\operatorname{tr} F^{0\nu}D_\nu\Theta = \frac{2}{g^2}\partial_\nu \operatorname{tr} F^{0\nu}\Theta - \frac{2}{g^2}\operatorname{tr} D_\nu F^{0\nu}\Theta.$$

The last term vanishes by the equation of motion. Thus

$$\int \mathrm{d}\boldsymbol{r}\, j_\theta^0 = \frac{-2}{g^2}\int \mathrm{d}\boldsymbol{r}\nabla\cdot\operatorname{tr}\boldsymbol{E}\Theta = \frac{-2}{g^2}\operatorname{tr}\int \mathrm{d}\boldsymbol{S}\cdot\boldsymbol{E}\Theta.$$

If Θ approaches an angle-independent-limit at $r = \infty$, where the surface integral is evaluated

$$\Theta \xrightarrow[r\to\infty]{} \Theta_\infty,$$

we then have

$$\int \mathrm{d}\boldsymbol{r}\, j_\theta^0 = -\frac{2}{g^2}\operatorname{tr}\Theta_\infty Q.$$

Exercise 2.7. By substituting (2.26) into (E2.4) one finds that

$$\delta_f F_{\mu\nu} = f^\alpha\partial_\alpha F_{\mu\nu} + (\partial_\mu f^\alpha)F_{\alpha\nu} + (\partial_\nu f^\alpha)F_{\mu\alpha},$$

which according to (2.27) is $L_f F_{\mu\nu}$.

Exercise 2.8.

$$\begin{aligned}[\delta_f, \delta_g]x^\mu &= \delta_f\delta_g x^\mu - (f \leftrightarrow g) = -\delta_f g^\mu(x) - (f \leftrightarrow g)\\ &= -\delta_f x^\alpha \partial_\alpha g^\mu - (f \leftrightarrow g) = f^\alpha \partial_\alpha g^\mu - g^\alpha \partial_\alpha f^\mu\\ &= h^\mu = -\delta_h x^\mu.\end{aligned}$$

$$\begin{aligned}[\delta_f, \delta_g]A_\mu &= \delta_f\delta_g A_\mu - (f \leftrightarrow g) = \delta_f(g^\alpha \partial_\alpha A_\mu + (\partial_\mu g^\alpha)A_\alpha) - (f \leftrightarrow g)\\ &= g^\alpha\partial_\alpha\delta_f A_\mu + (\partial_\mu g^\alpha)\delta_f A_\alpha - (f \leftrightarrow g)\\ &= g^\alpha \partial_\alpha(f^\beta \partial_\beta A_\mu + (\partial_\mu f^\beta)A_\beta) + \partial_\mu g^\alpha(f^\beta \partial_\beta A_\alpha + (\partial_\alpha f^\beta)A_\beta) - (f \leftrightarrow g)\\ &= (g^\alpha \partial_\alpha f^\beta - f^\alpha\partial_\alpha g^\beta)\partial_\beta A_\mu + \partial_\mu(g^\alpha \partial_\alpha f^\beta - f^\alpha \partial_\alpha g^\beta)A_\beta\\ &= -h^\beta \partial_\beta A_\mu - (\partial_\mu h^\beta)A_\beta = -\delta_h A_\mu.\end{aligned}$$

$$\begin{aligned}\partial_\mu h_\nu + \partial_\nu h_\mu &= \partial_\mu(f^\alpha \partial_\alpha g_\nu - g^\alpha \partial_\alpha f_\nu) + (\mu \leftrightarrow \nu)\\ &= f^\alpha \partial_\alpha(\partial_\mu g_\nu + \partial_\nu g_\mu) - g^\alpha\partial_\alpha(\partial_\mu f_\nu + \partial_\nu f_\mu)\\ &\quad + \partial_\mu f^\alpha \partial_\alpha g_\nu - \partial_\alpha f_\mu \partial_\nu g^\alpha + \partial_\nu f^\alpha \partial_\alpha g_\mu - \partial_\alpha f_\nu \partial_\mu g^\alpha.\end{aligned}$$

Because f^α and g^α are Killing vectors, the above is equal to $(\partial_\mu f_\alpha + \partial_\alpha f_\mu)\partial^\alpha g_\nu + (\partial_\nu f_\alpha + \partial_\alpha f_\nu)\partial^\alpha g_\mu$, which vanishes.

Exercise 2.9.

$$\begin{aligned}\partial_\mu \theta^{\mu\nu} &= \frac{2}{g^2}\partial_\mu \operatorname{tr}(F^{\mu\alpha}F^\nu{}_\alpha - \tfrac{1}{4}g^{\mu\nu}F^{\alpha\beta}F_{\alpha\beta})\\ &= \frac{2}{g^2}\operatorname{tr}(D_\mu F^{\mu\alpha}F^\nu{}_\alpha + F^{\mu\alpha}D_\mu F^\nu{}_\alpha - \tfrac{1}{2}F_{\alpha\beta}D^\nu F^{\alpha\beta}).\end{aligned}$$

The first covariant divergence vanishes by the equation of motion; the last is re-expressed by the Bianchi identity, leaving

$$\partial_\mu \theta^{\mu\nu} = \frac{2}{g^2}\operatorname{tr}(F_{\alpha\beta}D^\alpha F^{\nu\beta} + \tfrac{1}{2}F_{\alpha\beta}D^\alpha F^{\beta\nu} + \tfrac{1}{2}F_{\alpha\beta}D^\beta F^{\nu\alpha}),$$

which is seen to vanish.

Exercise 2.10. The Noether current is

$$\begin{aligned}j^\mu_f &= \frac{\delta\mathscr{L}_{\mathrm{YM}}}{\delta\partial_\mu A^a_\nu}\delta_f A^a_\nu - f^\mu\mathscr{L}\\ &= \frac{2}{g^2}\operatorname{tr} F^{\mu\nu}(f^\alpha\partial_\alpha A_\nu + (\partial_\nu f^\alpha)A_\alpha) - f^\mu\mathscr{L}_{\mathrm{YM}}\\ &= \frac{2}{g^2}\operatorname{tr}(F^{\mu\nu}F_{\alpha\nu} - \tfrac{1}{4}g^\mu{}_\alpha F^{\beta\nu}F_{\beta\nu})f^\alpha + \frac{2}{g^2}\operatorname{tr} F^{\mu\nu}(f^\alpha \partial_\nu A_\alpha + [A_\nu, f^\alpha A_\alpha] + (\partial_\nu f^\alpha)A_\alpha).\end{aligned}$$

The first term is recognized to be the current (2.33)–(2.34). The second term is collected into

$$\frac{2}{g^2}\operatorname{tr} F^{\mu\nu} D_\nu (f^\alpha A_\alpha) = \frac{2}{g^2}\partial_\nu \operatorname{tr} F^{\mu\nu} f^\alpha A_\alpha - \frac{2}{g^2}\operatorname{tr} D_\nu F^{\mu\nu} f^\alpha A_\alpha.$$

The last term vanishes by the equation of motion, while from the first we identify the super-potential as $-(2/g^2)\operatorname{tr} F^{\mu\nu} f^\alpha A_\alpha$.

Exercise 2.11.

$$\begin{aligned}\delta_f F_{\mu\nu} &= D_\mu \delta_f A_\nu - (\mu \leftrightarrow \nu) = D_\mu (f^\alpha F_{\alpha\nu}) - (\mu \leftrightarrow \nu)\\ &= (\partial_\mu f^\alpha) F_{\alpha\nu} - (\partial_\nu f^\alpha) F_{\alpha\mu} + f^\alpha (D_\mu F_{\alpha\nu} - D_\nu F_{\alpha\mu}).\end{aligned}$$

With the Bianchi identity, this may be rewritten as

$$\delta_f F_{\mu\nu} = f^\alpha D_\alpha F_{\mu\nu} + (\partial_\mu f^\alpha) F_{\alpha\nu} + (\partial_\nu f^\alpha) F_{\mu\alpha}.$$

Exercise 2.12.

$$\begin{aligned}[\delta_f, \delta_g] A_\mu &= \delta_f \delta_g A_\mu - (f \leftrightarrow g) = \delta_f g^\alpha F_{\alpha\mu} - (f \leftrightarrow g)\\ &= g^\alpha [f^\beta D_\beta F_{\alpha\mu} + (\partial_\alpha f^\beta) F_{\beta\mu} + (\partial_\mu f^\beta) F_{\alpha\beta}] - (f \leftrightarrow g)\\ &= (g^\alpha f^\beta - f^\alpha g^\beta) D_\beta F_{\alpha\mu} - h^\beta F_{\beta\mu} + (g^\alpha \partial_\mu f^\beta - f^\alpha \partial_\mu g^\beta) F_{\alpha\beta}\\ &= -\delta_h A_\mu + \tfrac{1}{2}(g^\alpha f^\beta - f^\alpha g^\beta)(D_\beta F_{\alpha\mu} - D_\alpha F_{\beta\mu}) + (g^\alpha \partial_\mu f^\beta + f^\beta \partial_\mu g^\alpha) F_{\alpha\beta}.\end{aligned}$$

With the help of the Bianchi identity, the last line becomes

$$-\delta_h A_\mu + g^\alpha f^\beta D_\mu F_{\alpha\beta} + \partial_\mu (g^\alpha f^\beta) F_{\alpha\beta},$$

which is equivalent to (E2.9).

Exercise 2.13. Operate on the conformal Killing equation (E2.10) with $\partial^\mu \partial^\nu$ to get (for $d > 1$) $\Box\, \partial_\alpha f^\alpha = 0$. Next operate on (E2.10) with ∂_α and use (E2.10) again to re-express $\partial_\alpha f_\nu$ in terms of $\partial_\nu f_\alpha$ and $g_{\nu\alpha} \partial_\beta f^\beta$. The result is

$$\partial_\nu (\partial_\alpha f_\mu - \partial_\mu f_\alpha) = \frac{2}{d}(g_{\nu\mu} \partial_\alpha \partial_\beta f^\beta - g_{\nu\alpha} \partial_\mu \partial_\beta f^\beta).$$

Finally differentiate with respect to x^α:

$$\Box \partial_\nu f_\mu = \partial_\mu \partial_\nu \partial_\alpha f^\alpha (1 - 2/d).$$

Since the right-hand side is symmetric, we may take the symmetric part of the left-hand side, which by the conformal Killing equation involves the vanishing quantity $\Box \partial_\alpha f^\alpha$. Hence for $d \neq 2$, we learn that $\partial_\mu \partial_\nu \partial_\alpha f^\alpha = 0$, whose solution is $\partial_\alpha f^\alpha = d(a + 2c_\alpha x^\alpha)$, where a

and c_α are constants and d is inserted for later convenience. Thus we learn that

$$\partial_\nu(\partial_\alpha f_\mu - \partial_\mu f_\alpha) = 4g_{\nu\mu}c_\alpha - 4g_{\nu\alpha}c_\mu,$$
$$\partial_\alpha f_\mu - \partial_\mu f_\alpha = 4x_\mu c_\alpha - 4x_\alpha c_\mu + 2\omega_{\mu\alpha}$$

with $\omega_{\mu\alpha}$ an antisymmetric constant, while the conformal Kiling equation becomes

$$\partial_\alpha f_\mu + \partial_\mu f_\alpha = g_{\alpha\mu}(2a + 4c_\beta x^\beta)$$

which implies

$$\partial_\alpha f_\mu = \omega_{\mu\alpha} + g_{\mu\alpha}a + 2g_{\mu\alpha}c_\beta x^\beta + 2x_\mu c_\alpha - 2x_\alpha c_\mu.$$

The most general solution is

$$f_\mu = a_\mu + \omega_{\mu\alpha}x^\alpha + ax_\mu + 2x_\mu c_\alpha x^\alpha - x^2 c_\mu,$$

which coincides with (2.28), (2.29), (2.40) and (2.41). For $d = 2$, we proceed differently. It is useful to introduce light-cone coordinates $x^\pm \equiv \frac{1}{2}\sqrt{2}(x^0 \pm x^1)$. With these components $g_{++} = g_{--} = g^{++} = g^{--} = 0$, $g_{+-} = g^{+-} = 1$. The $+\ -$ component of (E2.10) is vacuous, while the $+\ +$ and $-\ -$ components imply $\partial_\pm f_\pm = 0$. Thus, the $+$ [upper] component of a two-dimensional conformal Killing vector f^μ is an arbitrary function of x^+, and the $-$ [upper] component depends only on x^-.

Exercise 2.14.

$$\begin{aligned}\delta_f \mathscr{L}_{\mathrm{YM}} &= \frac{1}{g^2}\,\mathrm{tr}\,F^{\mu\nu}\delta_f F_{\mu\nu} \\ &= \frac{1}{g^2}\,\mathrm{tr}\,F^{\mu\nu}(f^\alpha D_\alpha F_{\mu\nu} + (\partial_\mu f^\alpha)F_{\alpha\nu} + (\partial_\nu f^\alpha)F_{\mu\alpha}) \\ &= \frac{1}{2g^2} f^\alpha \partial_\alpha\,\mathrm{tr}\,F^{\mu\nu}F_{\mu\nu} + \frac{1}{g^2}\,\mathrm{tr}\,F^{\mu\nu}F^\alpha{}_\nu(\partial_\mu f_\alpha + \partial_\alpha f_\mu).\end{aligned}$$

The conformal Killing equation satisfied by f_μ allows rewriting this as $\partial_\alpha(f^\alpha \mathscr{L}_{\mathrm{YM}})$. The current is

$$\frac{\delta \mathscr{L}_{\mathrm{YM}}}{\delta \partial_\mu A_\nu^a}\,\delta_f A_\nu^a - \Omega_f^\mu = \theta^{\mu\nu} f_\nu.$$

Conformal symmetry holds only for four-dimensional (classical) Yang–Mills theories, and a related fact is that for $d \neq 4$ the gauge coupling constant is not dimensionless (in units where $\hbar$ and the velocity of light are dimensionless). However, the non-interacting theory and the Abelian gauge theory are conformally invariant, with a modified field transformation law: one uses

$$\delta_f A_\mu = L_f A_\mu + \left(\frac{1}{2} - \frac{2}{d}\right)\partial_\alpha f^\alpha A_\mu$$

Exercise 2.15.

$$x'^{\mu} = \frac{x^{\mu}+c^{\mu}x^2}{1+2c\cdot x+c^2x^2} = \frac{c^{\mu}}{c^2}+\left(g^{\mu\nu}-\frac{2c^{\mu}c^{\nu}}{c^2}\right)\frac{1}{c^2}\left(\frac{x_{\nu}}{x^2}\right)+\mathrm{O}(c^{-3})$$

The $\mathrm{O}(c^{-1})$ term is a translation, $g^{\mu\nu}-2c^{\mu}c^{\nu}/c^2$ is an (improper) Lorentz transformation matrix, $1/c^2$ effects a dilatation, and the remainder is an inversion. Also if $x_1^{\mu} = x^{\mu}/x^2$, $x_2^{\mu} = x_1^{\mu}+c^{\mu}$, $x_3^{\mu} = x_2^{\mu}/x_2^2$, one finds that x_3^{μ} is a conformal transform of x^{μ}.

Exercise 2.16. If f^{α} and g^{β} are conformal Killing vector, their Lie bracket h_{μ} satisfies

$$\begin{aligned}
\partial_{\mu}h_{\nu}+\partial_{\nu}h_{\mu} &= \partial_{\mu}(f^{\alpha}\partial_{\alpha}g_{\nu})+\partial_{\nu}(f^{\alpha}\partial_{\alpha}g_{\mu})-(f\leftrightarrow g)\\
&= \partial_{\mu}f^{\alpha}\partial_{\alpha}g_{\nu}+f^{\alpha}\partial_{\mu}\partial_{\alpha}g_{\nu}+\partial_{\nu}f^{\alpha}\partial_{\alpha}g_{\mu}+f^{\alpha}\partial_{\nu}\partial_{\alpha}g_{\mu}-(f\leftrightarrow g)\\
&= \frac{2}{d}g_{\mu\nu}f^{\alpha}\partial_{\alpha}\partial_{\beta}g^{\beta}+(-\partial^{\alpha}f_{\mu}+\frac{2}{d}g^{\alpha}{}_{\mu}\partial_{\alpha}f^{\beta})\partial_{\alpha}g_{\nu}\\
&\quad +(-\partial^{\alpha}f_{\nu}+\frac{2}{d}g^{\alpha}{}_{\nu}\partial_{\beta}f^{\beta})\partial_{\alpha}g_{\mu}-(f\leftrightarrow g)\\
&= \frac{2}{d}g_{\mu\nu}(f^{\alpha}\partial_{\alpha}\partial_{\beta}g^{\beta}-g^{\alpha}\partial_{\alpha}\partial_{\beta}f^{\beta})\\
&\quad +\frac{2}{d}(\partial_{\beta}f^{\beta}\partial_{\mu}g_{\nu}-\partial_{\beta}g^{\beta}\partial_{\mu}f_{\nu})+\frac{2}{d}(\partial_{\beta}f^{\beta}\partial_{\nu}g_{\mu}-\partial_{\beta}g^{\beta}\partial_{\nu}f_{\mu})\\
&= \frac{2}{d}g_{\mu\nu}(f^{\alpha}\partial_{\alpha}\partial_{\beta}g^{\beta}-g^{\alpha}\partial_{\alpha}\partial_{\beta}f^{\beta})\\
&= \frac{2}{d}g_{\mu\nu}\partial_{\beta}(f^{\alpha}\partial_{\alpha}g^{\beta}-g^{\alpha}\partial_{\alpha}f^{\beta}) = \frac{2}{d}g_{\mu\nu}\partial_{\alpha}h^{\alpha}.
\end{aligned}$$

Exercise 2.18. The response of the matter action I_{M} to a gauge transformation is given by

$$\left(D_{\mu}\frac{\delta}{\delta A_{\mu}}\right)_a I_{\mathrm{M}}$$

since the field equations insure that the variation of I_{M} with respect to matter fields vanishes. But $\delta I_{\mathrm{M}}/\delta A_{\mu}$ is proportional to the current J^{μ}. Hence gauge invariance implies $D_{\mu}J^{\mu} = 0$.

Exercise 2.19.

$$\begin{aligned}
j_a^{\mu}T^a \equiv j^{\mu} &= J^{\mu}+[A_{\nu}, F^{\mu\nu}]\\
\partial_{\mu}j^{\mu} &= \partial_{\mu}J^{\mu}+[\partial_{\mu}A_{\nu}, F^{\mu\nu}]+[A_{\nu}, \partial_{\mu}F^{\mu\nu}]\\
&= -[A_{\nu}, J^{\nu}]+\tfrac{1}{2}[\partial_{\mu}A_{\nu}-\partial_{\nu}A_{\mu}, F^{\mu\nu}]+[A_{\nu}, J^{\nu}-[A_{\mu}, F^{\mu\nu}]]+D_{\mu}J^{\mu}\\
&= \tfrac{1}{2}[[A_{\nu}, A_{\mu}], F^{\mu\nu}]-[A_{\nu}, [A_{\mu}, F^{\mu\nu}]]+D_{\mu}J^{\mu} = D_{\mu}J^{\mu} = 0.
\end{aligned}$$

Also the field equation may be written as

$$\partial_\mu F^{\mu\nu} = J^\nu - [A_\mu, F^{\mu\nu}] = j^\nu.$$

Exercise 2.20. The gauge transformation which accompanies the coordinate transformation in the definition of δ_f is generated by $f^\alpha \mathscr{A}^\alpha$. Hence the ordinary derivatives in (2.63) and (2.64) are converted to covariant ones.

Exercise 2.23.

$$\frac{\mathrm{d}}{\mathrm{d}t}(\boldsymbol{r}\times m\dot{\boldsymbol{r}}) = \dot{\boldsymbol{r}}\times m\ddot{\boldsymbol{r}} = \boldsymbol{r}\times(e\dot{\boldsymbol{r}}\times g\hat{\boldsymbol{r}}/r^2) = \frac{\mathrm{d}}{\mathrm{d}t}(eg\hat{\boldsymbol{r}}).$$

$$f^i F_{ij} = \varepsilon^{ikl} r^k \omega^l(-\varepsilon^{ijm}B^m) = \boldsymbol{B}\cdot\boldsymbol{r}\omega^j - \boldsymbol{B}\cdot\boldsymbol{\omega} r^j$$
$$= \partial_j(g\hat{\boldsymbol{r}}\cdot\boldsymbol{\omega}).$$

Exercise 2.24. $\nabla\times\boldsymbol{A} = g\hat{\boldsymbol{r}}/r^2$. A decomposition of this equation in the spherical $\hat{r}$, $\hat{\theta}$ and $\hat{\phi}$ directions reads

$$\frac{1}{r^2\sin\theta}[\partial_\theta(r\sin\theta\, A^\phi) - \partial_\phi(rA^\theta)] = g/r^2,$$

$$\frac{1}{r}\partial_r(rA^\phi) - \frac{1}{r\sin\theta}\partial_\phi A^r = 0,$$

$$\frac{1}{r}\partial_r(rA^\theta) - \frac{1}{r}\partial_\theta A^r = 0.$$

Particular solutions are

$$A^\phi = g\frac{\sin\theta}{\cos\theta\pm 1}, \qquad A^r = 0, \qquad A^\theta = 0.$$

The general solution differs from the particular one by an arbitrary gauge transformation. This is gauge equivalent to (2.70a) with the gauge function $U = \exp(\sigma^a\theta^a/2\mathrm{i})$, $\theta^a = \theta\varepsilon^{ai3}\hat{r}^i/\sqrt{r^2-z^2}$, in the sense that

$$A_\mu\Big|_{(2.70\mathrm{b})} = U^{-1}A_\mu U\Big|_{(2.70\mathrm{a})} + U^{-1}\partial_\mu U.$$

Of course U may be followed by any gauge transformation that leaves the three-axis invariant, i.e., of the form $\exp \mathrm{i}\sigma^3\theta$. [This corresponds to an Abelian gauge transformation of (2.70b).] Spherical symmetry follows from Exercise (2.23).

Exercise 3.1. The transformation $\delta q_n = a(t)$, $\delta\dot{q}_n = \dot{a}(t)$ *leaves* L invariant provided A is transformed according to $\delta A = -\dot{a}(t)/e$. Equations of motion are obtained by varying q_n.

$$\dot{p}_1 = -V'(q_1-q_2), \qquad \dot{p}_2 = V'(q_1-q_2), \qquad p_n = m_n(\dot{q}_n + eA).$$

Varying A gives $p_1+p_2 = 0$. When $A = 0$, which may be achieved by a gauge transformation; $p_n = m_n\dot{q}_n$.

Exercise 3.2. The transformation $\delta r^i = \varepsilon^{ij} r^j \omega(t)$, $\delta \dot{r}^i = \varepsilon^{ij}\dot{r}^j\omega(t) + \varepsilon^{ij} r^j \dot{\omega}(t)$, $\delta A = -\dot{\omega}/e$ leaves L invariant. The equation of motion is

$$\dot{p}^i + e\varepsilon^{ij} A p^j = -\hat{r}^i V', \qquad p^i = m(\dot{r}^i + eA\varepsilon^{ij} r^j)$$

while the constraint requires $\varepsilon^{ij} r^i p^j = 0$. With $A = 0$, the constraint is easily solved in polar coordinates, where the Hamiltonian is the usual radial Hamiltonian, but with total angular momentum set to zero: $H = p_r^2/2m + V(r)$.

Exercise 3.3. The one-photon state is $A_{\mathrm{T}}^i(\boldsymbol{p})\Psi_0(\boldsymbol{A})$, where

$$A_{\mathrm{T}}^i(\boldsymbol{p}) = (\delta^{ij} - \hat{p}^i \hat{p}^j) \int \mathrm{d}\boldsymbol{r}\, e^{i\boldsymbol{p}\cdot\boldsymbol{r}} A^j(\boldsymbol{r}),$$

while the two-photon state is given by

$$(2A_{\mathrm{T}}^i(\boldsymbol{p}) A_{\mathrm{T}}^i(\boldsymbol{q}) - (\delta^{ij} - \hat{p}^i\hat{p}^j)\frac{\hbar}{p}(2\pi)^3 \delta(\boldsymbol{p}+\boldsymbol{q}))\Psi_0(\boldsymbol{A}).$$

Exercise 3.4. States satisfying the Gauss law

$$i\hbar \boldsymbol{\nabla} \cdot \frac{\delta}{\delta \boldsymbol{A}} \Psi(\boldsymbol{A}) = \rho \Psi(\boldsymbol{A})$$

are of the form

$$\Psi(\boldsymbol{A}) = \left(\exp \frac{i}{\hbar} \int \rho \frac{1}{\boldsymbol{\nabla}^2} \boldsymbol{\nabla} \cdot \boldsymbol{A}\right) \Phi(\boldsymbol{A}_{\mathrm{T}}),$$

where $\Phi(\boldsymbol{A}_{\mathrm{T}})$ is an arbitrary functional of the transverse potential. Substitution of this into $H\Psi = E\Psi$ yields $H_c \Phi = E_c \Phi$, where

$$H_c = \tfrac{1}{2} \int \mathrm{d}\boldsymbol{r} \left[-\hbar \frac{\delta^2}{\delta \boldsymbol{A}_{\mathrm{T}}^2} + \boldsymbol{B}^2 \right], \qquad E_c = E - \tfrac{1}{2} \int \rho \frac{1}{\boldsymbol{\nabla}^2} \rho.$$

Exercise 3.8. We shall use the modified transformation law (2.36). Together with the results of Exercise (2.14) we learn that $\bar{\delta}_f \mathscr{L} = \partial_\mu (f^\mu \mathscr{L})$, hence the Noether current

$$\frac{\delta \mathscr{L}}{\delta \partial_\mu A_\nu^a} \bar{\delta}_f A_\nu^a - f^\mu \mathscr{L}$$

is

$$\theta^{\mu\nu} f_\nu + \frac{\delta}{\delta \partial_\mu A_\nu^a}\left(-\frac{\hbar\theta}{16\pi^2} \operatorname{tr} {}^*F^{\alpha\beta} F_{\alpha\beta}\right) f^\gamma F_{\gamma\nu}^a - f^\mu \left(-\frac{\hbar\theta}{16\pi^2} \operatorname{tr} {}^*F^{\alpha\beta} F_{\alpha\beta}\right).$$

The additional θ-dependent terms cancel identically.

Exercise 3.9. The Hamiltonian follows from the energy–momentum tensor, and is therefore conventional. The canonical momentum is

$$\boldsymbol{\Pi}_a = -\boldsymbol{E}_a + \frac{\hbar\theta g^2}{8\pi^2} \boldsymbol{B}_a,$$

hence the Hamiltonian reads

$$H = \frac{1}{2}\int d\mathbf{r}\left\{\left(\boldsymbol{\Pi}_a - \frac{\hbar\theta g^2}{8\pi^2}\mathbf{B}_a\right)^2 + \mathbf{B}_a^2\right\}.$$

Exercise 4.2. The integrability condition on anomalies in gauge source currents reads

$$\left(\partial_\mu \frac{\delta}{\delta A_\mu^a(x)} + g f_{aa'a''} A_\mu^{a'}(x)\frac{\delta}{\delta A_\mu^{a''}(x)}\right)(D_\nu J^\nu(y))_b - (a \leftrightarrow b,\ x \leftrightarrow y)$$
$$= -g f_{abc}(D_\mu J^\mu(x))_c \delta(x-y).$$

Exercise 4.3. The singlet current divergence is

$$D_\mu J^\mu = \frac{\mathrm{i}}{4}\left(\frac{\hbar g^2}{48\pi^2}\right)\varepsilon^{\mu\alpha\beta\gamma}\partial_\mu[a_\alpha \partial_\beta a_\gamma + A_\alpha^a \partial_\beta A_\gamma^a + \tfrac{1}{4} g \varepsilon_{abc} A_\alpha^a A_\beta^b A_\gamma^c].$$

The above is U(1) but not SU(2) invariant. The triplet current divergence is neither U(1) nor SU(2) covariant.

$$D_\mu J_a^\mu = \frac{\mathrm{i}}{4}\left(\frac{\hbar g^2}{48\pi^2}\right)\varepsilon^{\mu\alpha\beta\gamma}\partial_\mu[A_\alpha^a \partial_\beta a_\gamma + a_\alpha \partial_\beta A_\gamma^a + \tfrac{1}{4} g \varepsilon_{abc} a_\alpha A_\beta^b A_\gamma^c].$$

If we add

$$-\mathrm{i}\hbar\left(\frac{g^2}{48\pi}\right)^2 \int dx\, \varepsilon^{\mu\alpha\beta\gamma}\varepsilon_{abc} a_\mu A_\alpha^a A_\beta^b A_\gamma^c$$

to the logarithm of the determinant then the singlet anomaly becomes SU(2) × U(1) invariant.

$$D_\mu J^\mu = \frac{\mathrm{i}}{4}\left(\frac{\hbar g^2}{48\pi^2}\right)\varepsilon^{\mu\alpha\beta\gamma}\partial_\mu[a_\alpha \partial_\beta a_\gamma + A_\alpha^a \partial_\beta A_\gamma^a + \tfrac{1}{3} g \varepsilon_{abc} A_\alpha^a A_\beta^b A_\gamma^c]$$
$$= \mathrm{i}\left(\frac{\hbar g^2}{48\pi^2}\right)\left({}^*f^{\mu\nu} f_{\mu\nu} + {}^*F^{\mu\nu a} F_{\mu\nu}^a\right),$$

$$f_{\mu\nu} \equiv \partial_\mu a_\nu - \partial_\nu a_\mu.$$

The triplet anomaly is as above, but with no trilinear term. The result is not SU(2) gauge covariant, though it is U(1) invariant. We see that this procedure makes the triplet anomaly U(1) invariant, but there is no modification which would make it SU(2) or SU(2) × U(1) covariant.

Exercise 4.4. The fermion zero mode is $\psi = \binom{u}{0}$, where u is a two-component spinor (labeled by indices n) and a two-component isospinor (labeled by indices a). The normalized formula is

$$u_{na} = \frac{1}{\pi}\frac{\lambda}{(\lambda^2 + x^2)^{3/2}}\begin{pmatrix} 0 & 1 \\ -1 & 0 \end{pmatrix}_{na}.$$

Exercise 5.1.

$$A_1 = \frac{g\sin\theta}{r(\cos\theta-1)}\hat{\phi}, \qquad A_2 = \frac{g\sin\theta}{g(\cos\theta+1)}\hat{\phi};$$

$$\dot{r}(t) = R\frac{2\pi}{t_2-t_1}\hat{\phi}.$$

The potential along the path ($\theta = \pi/2$) is

$$A_1 = -\frac{g}{R}\hat{\phi}, \qquad A_2 = \frac{g}{R}\hat{\phi};$$

$$e\int_{t_1}^{t_2} \mathrm{d}t\, \dot{r}(t)\cdot A_{1,2} = \mp g2\pi.$$

The difference between the two action integrals is $4\pi eg$.

Exercise 5.2. Since $F^{\mu\nu} = \varepsilon^{\mu\nu\alpha}\,{}^*F_\alpha$,

$$\begin{aligned}\varepsilon_{\alpha\beta\nu}D^\beta D_\mu F^{\mu\nu} &= \varepsilon_{\alpha\beta\nu}D^\beta D_\mu \varepsilon^{\mu\nu\gamma}\,{}^*F_\gamma\\ &= D^\mu D_\mu\,{}^*F_\alpha - D^\beta D_\alpha\,{}^*F_\beta\\ &= D^\mu D_\mu\,{}^*F_\alpha - [D^\beta, D_\alpha]\,{}^*F_\beta\\ &= D^\mu D_\mu\,{}^*F_\alpha - \varepsilon_{\alpha\beta\gamma}[{}^*F^\beta, {}^*F^\gamma].\end{aligned}$$

On the other hand, the field equation states that the left-hand side is equal to $-\mu\varepsilon_{\alpha\beta\nu}D^\beta\,{}^*F^\nu = -\mu D^\beta F_{\alpha\beta} = -\mu^2\,{}^*F_\alpha$.

Exercise 5.3. We rewrite (2.33) in terms of the dual tensor as

$$\theta^{\mu\nu} = -\frac{2}{g^2}\operatorname{tr}[{}^*F^\mu\,{}^*F^\nu - \tfrac{1}{2}g^{\mu\nu}\,{}^*F^2].$$

Hence

$$\begin{aligned}\partial_\mu\theta^{\mu\nu} &= -\frac{2}{g^2}\operatorname{tr}[{}^*F_\mu D^\mu\,{}^*F^\nu - {}^*F_\mu D^\nu\,{}^*F^\mu]\\ &= -\frac{2}{g^2}\operatorname{tr}\,{}^*F_\mu\varepsilon^{\mu\nu\alpha}\varepsilon_{\alpha\beta\gamma}D^\beta\,{}^*F^\gamma = \frac{2}{g^2}\operatorname{tr}\,{}^*F_\mu\varepsilon^{\mu\nu\alpha}D^\beta F_{\beta\alpha}.\end{aligned}$$

Thus far only the Bianchi identity has been used. But with the field equation we see that $\partial_\mu\theta^{\mu\nu} = (2\mu/g^2)\operatorname{tr}\varepsilon^{\nu\alpha\beta}\,{}^*F_\alpha\,{}^*F_\beta = 0$.

Exercise 5.6. The fourth component of the instanton in (3.60) is $A_4 = \mathrm{i}\boldsymbol{\sigma}\cdot\boldsymbol{r}/(\tau^2+r^2+\lambda^2)$. Hence the required gauge function U is determined by

$$U_\tau^{-1}A_4U + U_\tau^{-1}\frac{\mathrm{d}}{\mathrm{d}\tau}U = 0 \Leftrightarrow$$

$$U_\tau = \exp\frac{-\mathrm{i}\boldsymbol{\sigma}\cdot\boldsymbol{r}}{\sqrt{r^2+\lambda^2}}\left(\tan^{-1}\frac{\tau}{\sqrt{r^2+\lambda^2}} + \frac{\pi}{2}\right).$$

The transformed spatial components $U_\tau^{-1}A_iU_\tau + U_\tau^{-1}\partial_iU_\tau$ coincide with A_i at $\tau = -\infty$, since there $U = I$, hence they vanish becasue A_i does. At $\tau = +\infty$, A_i also vanishes and the transformed vector potential is the pure gauge $U_\infty^{-1}\partial_iU_\infty$, with

$$U_\infty = \exp\frac{-i\pi\boldsymbol{\sigma}\cdot\boldsymbol{r}}{\sqrt{r^2+\lambda^2}},$$

which is a non-trivial gauge of the type (E3.13).

Exercise 5.7.

$$L = \tfrac{1}{2}m(\dot{r}^i + eA\varepsilon^{ij}r^j)^2 - V(r) + cA.$$

The Chern–Simons action is $I_{CS} = c\int dt\, A(t)$. Under a gauge transformation,

$$I_{CS} \to I_{CS} - \frac{c}{e}\int dt\frac{d}{dt}\omega(t);$$

hence I_{CS} changes by $(c/e)\Delta\omega$, where $\Delta\omega$ is the difference between the end-point values of $\omega(t)$. We compactify the space (one-dimensional line), and demand that at the initial point ω vanishes, and at the endpoint it is an integral multiple of 2π, so that $e^{i\omega(t)}$ is single-valued at $t = 0, \beta$. This then requires that $c/e = n\hbar$, where n is an integer. The same quantization emerges when one examines the constraint, obtained by varying A:

$$\varepsilon^{ij}r^ip^j = c/e, \qquad p^i \equiv m(\dot{r}^i + eA\varepsilon^{ij}r^j).$$

In a Hamiltonian formulation of the theory this equation must be imposed as a condition on states:

$$\varepsilon^{ij}r^ip^j\psi = \frac{c}{e}\psi.$$

But the operator is recognized as the angular momentum whose eigenvalues are integral multiplets of $\hbar$.

Exercise 5.8.

$$L = i\hbar a^\dagger\left(\frac{d}{dt} + ieA\right)a.$$

This theory is formally gauge invariant:

$$\delta a = i\omega(t), \qquad \delta a^\dagger = -i\omega(t), \qquad \delta A = -\frac{1}{e}\dot{\omega}(t).$$

The functional determinant, obtained by integrating out the fermions, is

$$\Delta(A) = \det\left(i\frac{d}{dt} - eA\right),$$

and may be evaluated. We consider time to be compactified, running between 0 and β, and examine the eigenvalue equation

$$\left(\mathrm{i}\frac{\mathrm{d}}{\mathrm{d}t} - eA\right)\psi(t) = \lambda\psi(t).$$

The determinant is then defined as the regulated product of the eigenvalues. The differential equation is solved by

$$\psi(t) = c\exp -\mathrm{i}\left[\lambda t + e\int_0^t \mathrm{d}\tau\, A(\tau)\right],$$

where c is a constant. To obtain a well-defined eigenvalue problem, boundary conditions need to be specified. One may require either periodicity, $\psi(0) = \psi(\beta)$, or antiperiodicity, $\psi(0) = -\psi(\beta)$. In the former case the eigenvalues are $\lambda_+ = 2\pi(n-a)$, while in the latter they are $\lambda_- = 2\pi(n+\frac{1}{2}-a)$, where n is any integer and $a \equiv (e/2\pi)\int_0^\beta \mathrm{d}\tau A(\tau)$. The formal determinant in the periodic case is

$$\Delta_+(A) = -2\pi a \prod_{n=\pm 1,\pm 2,\ldots} 2\pi(n-a) = -2\pi a \prod_{n=1,2,\ldots} (a^2-n^2),$$

while the antiperiodic one leads to

$$\Delta_-(A) = \prod_{n=0,\pm 1,\pm 2,\ldots} 2\pi(n+\tfrac{1}{2}-a) = \prod_{n=1,2,\ldots} \pi^2(4a^2-(2n-1)^2).$$

To regulate these products, we divide them by the analogous expressions in the absence of A (without the zero mode in the periodic case). Thus

$$\Delta_+(A) = \pi a \prod_{n=1,2,\ldots}\left(1-\frac{a^2}{n^2}\right) = \sin\pi a,$$

$$\Delta_-(A) = \prod_{n=1,2,\ldots}\left(1-\frac{4a^2}{(2n-1)^2}\right) = \cos\pi a.$$

(Overall constant numerical factors have been dropped.) Under a gauge transformation, $a \to a+\text{integer}$, because we assume that the gauge function at $t = 0$ vanishes, and at $t = \beta$ is an integral multiple of 2π, so that $e^{\mathrm{i}\omega(t)}$ is single-valued at $t = 0, \beta$. When that integer is odd, it is seen that the determinant is not single-valued, but changes sign.

The choice of boundary conditions is a formal device. To bring physics into the problem, one thinks of the determinant as the vacuum-to-vacuum amplitude in the presence of the source $A(t)$. Thus one considers

$$W(A) = -\mathrm{i}\hbar \ln\det(I-GeA),$$

where G is the Green's function for $\mathrm{i}\,\mathrm{d}/\mathrm{d}t$. Which Green's function should be used is decided by considering the free propagator $\langle 0|Ta(t)a^\dagger(t)|0\rangle = \theta(t-t')$. (In this theory there are only two states: the no-fermion state, and the one-fermion state.) Hence $G(t,t') = -\mathrm{i}\theta(t-t')$. When $W(A)$ is expanded in powers of eA, one sees that only the first term contributes:

$$W(A) = \tfrac{1}{2}\hbar e\int_{-\infty}^{\infty} \mathrm{d}t\, A(t).$$

[We take $\theta(0) = \frac{1}{2}$]. Thus $\Delta(A) = \exp i\pi a$, where now

$$a \equiv \frac{e}{2\pi} \int_{-\infty}^{\infty} dt\, A(t).$$

Again this is not invariant against gauge transformations with gauge functions that change by an odd integer multiple of 2π as t passes from minus to plus infinity. Note that the relationship between the present, charge conjugation symmetric, evaluation and the earlier ones is $\Delta(A) = \Delta_- + i\Delta_+$.

References

[1] R. Jackiw, Rev. Mod. Phys. 49 (1977) 681; A. Neveu, Rep. Prog. Phys. 40 (1977) 709; L. Faddeev and V. Korepin, Phys. Rep. 42C (1978) 1.

[2] R. Rajaraman, Solitons and Instantons (North-Holland, Amsterdam, 1982).

[3] T. Eguchi, P. Gilkey and A. Hanson, Phys. Rep. 66 (1980) 213; Y. Choquet-Bruhat, C. DeWitt-Morette with M. Dillard-Bleick, Analysis, Manifolds and Physics, revised edition (North-Holland, Amsterdam, 1982).

[4] C. N. Yang and R. Mills, Phys. Rev. 96 (1954) 191. Precursors include O. Klein, in New Theories in Physics (International Institute of Intellectual Cooperation, Paris and Warsaw, 1939) and R. Shaw, Ph.D. Thesis (Cambridge Univ., 1955).

[5] T. T. Wu and C. N. Yang, Phys. Rev. D12 (1975) 3843.

[6] See R. Jackiw, "Field Theoretic Investigations in Current Algebra", which appears in this volume.

[7] E. Bessel-Hagen, Math. Ann. 84 (1921) 258.

[8] R. Jackiw, Phys. Rev. Lett. 41 (1978) 1635.

[9] The procedure for constructing the renormalized energy-momentum tensor for (an ordinary, not gauge) quantum field theory is explained in C. Callan, S. Coleman and R. Jackiw, Ann. Phys. (NY) 59 (1970) 42.

[10] S. Coleman and R. Jackiw, Ann. Phys. (NY) 67 (1971) 552.

[11] M. Gell-Mann and F. Low, Phys. Rev. 95 (1954) 1300. For further appreciations of the renormalization group see A. Guth, K. Huang and R. Jaffe, eds., Asymptotic Realms of Physics (MIT Press, Cambridge, MA, 1983).

[12] S. Adler, J. Collins and A. Duncan, Phys. Rev. D15 (1977) 1712; J. Collins, A. Duncan and S. Jogelkar, Phys. Rev. D16 (1977) 438; N. Nielsen, Nucl. Phys. B120 (1977) 212.

[13] For a review see N. Birrell and P. Davies, Quantum Fields in Curved Space (Cambridge Univ. Press, Cambridge, 1982).

[14] In supersymmetric theories the anomaly of the axial vector current belongs in the same supermultiplet as the anomaly in the trace of the energy-momentum tensor; S. Ferrara and B. Zumino, Nucl. Phys. B87 (1975) 207. This suggests that topological properties of the former should be reflected in the latter. However, the connection has not been developed as yet.

[15] For a summary see W. Marciano and H. Pagels, Phys. Rep. 36C (1978) 137; K. Huang, Quarks, Leptons and Gauge Fields (World Scientific, Singapore, 1982).

[16] For a summary see K. Huang, ref. 15; J. Taylor, Gauge Theories of Weak Interactions (Cambridge Univ. Press, Cambridge, 1976).

[17] S. Weinberg, Phys. Rev. Lett. 19 (1967) 1264; A. Salam, in Elementary Particle Theory, ed. N. Svartholm (Almqvist and Wiksell, Stockholm, 1968).
[18] For a summary of this theoretical work see E. Farhi and R. Jackiw, Dynamical Gauge Symmetry Breaking (World Scientific, Singapore, 1982).
[19] S. Glashow, A. Salam and S. Weinberg, Nobel lectures in physics, 1979, Rev. Mod. Phys. 52 (1980) 515.
[20] D. Eardley and V. Moncrief, Comm. Math. Phys. 83 (1982) 171; Y. Choquet-Bruhat and D. Christodoulou, Ann. Sci. Ec. Norm. Sup 14 (1981) 481.
[21] S. Coleman, Phys. Lett 70B (1977) 59.
[22] J. Goldstone and R. Jackiw, Phys. Rev. D11 (1975) 1486; review in refs. 1 and 2.
[23] S. Coleman, in New Phenomena in Subnuclear Physics, ed. A. Zichichi (Plenum, New York, 1977); S. Deser, Phys. Lett. 64B (1976) 463.
[24] V. DeAlfaro, S. Fubini and G. Furlan, Phys. Lett. 65B (1976) 163.
[25] B. Schechter, Phys. Rev. D16 (1977) 3015, M. Lüscher, Phys. Lett. 70B (1977) 321.
[26] For a review see A. Actor, Rev. Mod. Phys. 51 (1979) 461.
[27] For example the most general rotationally invariant SU(2) and SU(3) gauge potentials are constructed by R. Jackiw, Acta Phys. Austr., Suppl. XXII (1980) 383; Gu Chaohao and Hu Hesheng, Comm. Math. Phys. 79 (1981) 75. A more mathematical, global approach has been given by P. Forgàcs and N. Manton, Comm. Math. Phys. 72 (1980) 15; J. Harnard, S. Shnider and L. Vinet, J. Math. Phys. 21 (1980) 2719.
[28] N. Manton, Nucl. Phys. B158 (1979) 141; M. Meyer, Acta Phys. Austr. Suppl. XXIII (1981) 477.
[29] R. Jackiw and N. Manton, Ann. Phys. (NY) 127 (1980) 257.
[30] However, S. Adler has argued that this may be a useful framework for understanding confinement; see S. Adler and T. Piran, Rev. Mod. Phys. 56 (1984) 1.
[31] J. Mandula, Phys. Rev. D14 (1976) 3497.
[32] R. Jackiw, L. Jacobs and C. Rebbi, Phys. Rev. D20 (1979) 474.
[33] For a review see R. Jackiw, in Geometrical and Topological Methods in Gauge Theories, eds. J. Harnard and S. Shnider (Lecture Notes in Physics 129, Spinger, Berlin, 1980).
[34] R. Jackiw and P. Rossi, Phys. Rev. D21 (1979) 426.
[35] T. T. Wu and C. N. Yang, in Properties of Matter Under Unusual Conditions, eds. H. Mark and S. Fernbach (Interscience, New York, 1968).
[36] J. Goldstone and R. Jackiw, as reported by R. Jackiw, in Gauge Theories and Modern Field Theories, eds. R. Arnowitt and P. Nath (MIT Press, Cambridge, MA, 1976); E. Tomboulis and G. Woo, Nucl. Phys. B107 (1976) 221.
[37] G. 't Hooft, Nucl. Phys. B79 (1974) 276; A. Polyakov, Zh. Eksp. Teor. Fiz. Pis'ma Red. 20 (1974) 430 [JETP Lett. 20 (1974) 194].
[38] Early work is described by S. Coleman, ref. 23; more recent results are summarized by P. Rossi, Phys. Rep 86 (1983) 317; J. Burzlaff, Commun. Dublin Inst. Adv. Studies, Ser. A 27 (1983).
[39] B. Cabrera, Phys. Rev. Lett. 48 (1982) 1378.
[40] Y. Zeldovich and M. Khlopov, Phys. Lett. 79B (1978) 239; J. Preskill, Phys. Rev. Lett. 43 (1979) 1365.
[41] V. Rubakov, Zh. Eksp. Teor. Fiz. Pis'ma Red. 33 (1981) 658 [JETP Lett. 33 (1981) 644], Nucl. Phys. B203 (1982) 311; C. Callan, Phys. Rev. D25 (1982)

2141 and D26 (1982) 2058; V. Rubakov and M. Sevebryakov, Nucl. Phys. B218 (1983) 240.

[42] For a review, see R. Jackiw, in Quantum Structure of Space and Time, eds. M. Duff and C. Isham (Cambridge Univ. Press, Cambridge 1982).

[43] L. Faddeev, Teor. Mat. Fiz. 1 (1970) 1 [Theor. Math. Phys. 1 (1970) 1].

[44] H. Weyl, Theory of Groups and Quantum Mechanics (Dover, New York, 1950).

[45] J. Goldstone and R. Jackiw, Phys. Lett 74B (1978) 81; A. Izergin, V. Korepin, M. Semenov-Tian-Shansky and L. Faddeev, Teor. Mat. Fiz. 38 (1979) 3 [Theor. Math. Phys. 38 (1979) 1].

[46] R. Jackiw and C. Rebbi, Phys. Rev. Lett. 37 (1976) 172.

[47] C. Callan, R. Dashen and D. Gross, Phys. Lett. 63B (1976) 334.

[48] C. Cronström and J. Mickelsson, J. Math. Phys. 24 (1983) 2528; H. -Y. Guo, S. -K. Wang and Ke Wu, Stony Brook preprint ITP-SB-85-44 (1985); M. Laursen, G. Schierholz and U. Wiese, DESY preprint 85-062 (1985).

[49] H. Loos, Phys. Rev. 188 (1969) 2342. The Schrödinger picture approach to an interacting field theory is beset by divergences, just as is the covariant perturbation expansion. Indeed, even a free theory has vacuum energy divergences, as in (E3.5). Consequently actual calculations must be renormalized by a procedure which is more cumbersome than covariant renormalization theory; see K. Symanzik, Nucl. Phys. B190 [FS3] (1981) 1.

[50] The following problem in choosing a gauge, viz. in selecting χ_b, has been noted by V. Gribov, Nucl. Phys. B139 (1978) 1. Many common choices that are used in perturbation theory, e.g. the Coulomb or Lorentz conditions, do not fix the gauge uniquely, and correspondingly the gauge compensating determinant det $\{\delta_a \chi_b(A)\}$ vanishes for sufficiently large A_μ. This is a consequence of the non-trivial topological structure of non-Abelian gauge fields; see R. Jackiw, I. Muzinich and C. Rebbi, Phys. Rev. D17 (1978) 1576. Moreover the "Gribov ambiguity" has been shown to afflict any gauge condition that can be defined on a compact space; see I. Singer, Comm. Math. Phys. 60 (1978) 7. It does not appear that this ambiguity presents a true obstacle to the quantum theory. First of all, there are gauge choices, not definable on a compact space, but well defined in Minkowski space, which are not afflicted by this defect, e.g. axial gauges where one component of A_μ is set to zero. To be sure, perturbation theory is not conveniently carried out in these gauges, but then perturbative calculations may be safely performed in the Gribov-ambiguous Coulomb or Lorentz gauges, since perturbation theory intrinsically never reaches "large" potentials, being an expansion about $A_\mu = 0$. In a sense the entire Gribov problem is rather like the problem encountered when a plane is coordinatized by circular coordinates. A singularity is necessarily present and the functional measure $\mathscr{D}r$ becomes $(\det r)\ \mathscr{D}r\mathscr{D}\theta$, vanishing at $r = 0$. But no noteworthy dynamical significance is attached to the singularity. For more on this, see N. Christ and T. D. Lee, Phys. Rev. D22 (1980) 939; as well as the review by R. Jackiw, in New Frontiers in High-Energy Physics, eds. B. Kursunuglu, A. Perlmutter and L. Scott (Plenum, New York, 1978).

[51] This subject is further discussed in the mathematical works of ref. 3, and from a physicist's perspective by R. Jackiw, C. Nohl and C. Rebbi, in Particles and Fields, eds. D. Boal and A. Kamal (Plenum, New York, 1978).

[52] N. Christ and R. Jackiw, Phys. Lett. 91B (1980) 228.

[53] For further discussion of this approximation procedure in quantum mechanics, see D. McLaughlin, J. Math. Phys. 13 (1972) 1099.
[54] G. 't Hooft, Phys. Rev. D14 (1976) 3432, (E) 18 (1978) 2199; F. Ore, Phys. Rev. D16 (1977) 2577.
[55] A. Belavin, A. Polyakov, A. Schwartz and Y. Tyupkin, Phys. Lett. 59B (1975) 85.
[56] R. Jackiw and C. Rebbi, Phys. Rev. D14 (1976) 517.
[57] A. Schwartz, Phys. Lett. 67B (1977) 172; R. Jackiw and C. Rebbi, Phys. Lett. 67B (1977) 189.
[58] R. Jackiw, C. Nohl and C. Rebbi, Phys. Rev. D15 (1977) 1642.
[59] For a summary, see R. Jackiw, C. Nohl and C. Rebbi, ref. 51.
[60] M. Atiyah, V. Drinfeld, N. Hitchin and Y. Manin, Phys. Lett. 65A (1978) 185.
[61] For an extensive, but still incomplete list of applications, see B. Zumino, Y. -S. Wu and A. Zee, Nucl. Phys. B239 (1984) 477; W. Bardeen, Fermilab preprint Conf-85/111-T (1985).
[62] Chiral anomalies in the general coordinate transformation group have been found in theories on $(4d + 2)$-dimensional manifolds by L. Alvarez-Gaumé and E. Witten, Nucl. Phys. B234 (1984) 269, which appears in this volume. This work, together with that on higher-dimensional axial-vector current anomalies has reawakened interest in string models.
[63] K. Wilson, Nobel lecture in physics, 1982, Rev. Mod. Phys. 55 (1983) 583.
[64] Y. Frischman, D. Gepner and S. Yankielowicz, Phys. Lett. 130B (1983) 66. H. Nielsen and M. Ninomiya, Phys. Lett. 130B (1983) 389. W. Bardeen, S. Elitzur, Y. Frishman and E. Rabinovici, Nucl. Phys. B218 (1983) 445; A. Niemi and G. Semenoff, Phys. Rev. Lett. 51 (1983) 2077; R. Jackiw, Phys. Rev. D29 (1984) 2375, and Comm. Nucl. Part. Phys. 13 (1984) 15, 141; Y. Srivastava and A. Widom, Lett. Nuovo Cim. 17 (1984) 285; M. Friedman, J. Sokoloff, A. Widom and Y. Srivastava, Phys. Rev. Lett. 52 (1984) 1587; K. Ishikawa, Phys. Rev. Lett. 53 (1984) 1615; G. Semenoff, Phys. Rev. Lett. 53 (1984) 2449; R. Hughes, Phys. Lett. 148B (1984) 215; A. Widom, M. Friedman and Y. Srivastava, Phys. Rev. B31 (1985) 6588; K. Johnson, in The Santa Fe Meeting, eds. T. Goldman and M. Nieto (World Scientific, Singapore, 1985).
[65] Modern research on chiral anomalies began with the four-dimensional calculations of J. Bell and R. Jackiw, Nuovo Cim. 60A (1969) 47 and S. Adler, Phys. Rev. 177 (1969) 2426; for a review see ref. 6. The relevant graphs and their peculiar properties were previously noted by H. Fukuda and Y. Miyamoto, Prog. Theor. Phys. 4 (1949) 347; J. Steinberger, Phys. Rev. 76 (1949) 1180. J. Schwinger, Phys. Rev. 82 (1951) 664. The two-dimensional chiral anomaly is first discussed in K. Johnson, Phys. Lett. 5 (1963) 253.
[66] D. Gross and R. Jackiw, Phys. Rev. D6 (1972) 477; C. Bouchiat, J. Iliopoulos and Ph. Meyer, Phys. Lett. 38B (1972) 519. An attempt at using an anomalous gauge theory for (spontaneous?) breaking of gauge invariance is found in R. Jackiw and R. Rajaraman, Phys. Rev. Lett. 54 (1985) 1219, (E) 2060; R. Rajaraman, Phys. Lett. 154B (1985) 305 and CERN preprint TH 4227/85 (1985); J. Lott and R. Rajaraman (in preparation).
[67] K. Johnson, ref. 65.
[68] J. Schwinger, Phys. Rev. 128 (1962) 2425.
[69] The fact that chiral two-dimensional currents are not conserved suggests that the

conventional two-dimensional energy-momentum tensor also fails to be conserved; i.e., there is an anomaly in the coordinate reparametrization group, as shown by L. Alvarez-Gaumé and E. Witten, more generally in $4d + 2$ dimensions, ref. 62. The point is that the fermionic two-dimensional energy-momentum tensor may be written in terms of the fermionic currents, and when the latter are not conserved neither is the former: see S. Coleman, D. Gross and R. Jackiw, Phys. Rev. 180 (1969) 1359. But of course the theory is also afflicted by the loss of gauge invariance.

[70] Vector meson mass generation in this theory was established by J. Schwinger, ref. 68, who solved the model. The connection with the axial vector anomaly is given by R. Jackiw, in Laws of Hadronic Matter, ed. A. Zichichi (Academic New York, 1975) and in Asymptotic Realms of Physics, ref. 11.

[71] A. D'Adda, A. Davis and P. DiVecchia, Phys. Lett. 121B (1983) 335; A. Polyakov and P. Wiegman, Phys. Lett. 131B (1983) 121. R. Gamboa-Saravi, M. Muschietti, F. Schaposnik, F. Solonim, Ann. Phys. (NY) 157 (1984) 360.

[72] For a summary of the older work see R. Jackiw, ref. 6.

[73] I. Gerstein and R. Jackiw, Phys. Rev. 181 (1969) 1955.

[74] W. Bardeen, Phys. Rev. 184 (1969) 1848. In this work various forms for the anomaly, consistent with the algebraic structure of a non-Abelian group are given.

[75] J. Wess and B. Zumino, Phys. Lett. 37B (1971) 95.

[76] B. Zumino, Nucl. Phys. B253 (1985) 477.

[77] P. Frampton and T. Kephart, Phys. Rev. Lett. 50 (1983) 1343, 1347, Phys. Rev. D28 (1983) 1010; W. Bardeen and B. Zumino, Nucl. Phys. B244 (1984) 421, which appears in this volume; K. Fujikawa, Phys. Rev. D31 (1985) 341; L. Alvarez-Gaumé and P. Ginsparg, Ann. Phys. (NY) 161 (1985) 423.

[78] M. Paranjape, MIT PhD thesis (1984), Phys. Lett. B156 (1985) 376; L. Alvarez-Gaumé, W. Bardeen, P. Ginsparg, B. Zumino, ref. 77.

[79] H. Georgi and S. Glashow, Phys. Rev. D6 (1972) 429. D. Gross and R. Jackiw, ref. 66.

[80] D. Gross and R. Jackiw, ref. 66.

[81] S. Adler and W. Bardeen, Phys. Rev. 182 (1969) 1517. Absence of radiative corrections to axial anomalies in QCD is established in S. -Y. Pi and S. S. Shei, Phys. Rev. D11 (1975) 2946.

[82] This observation may provide a resolution of the following puzzle. In a supersymmetric theory the divergence of the axial vector current and of the scale current, i.e., the trace of the energy-momentum tensor, belong in the same super multiplet; see ref. 14. How, then, can one but not the other be renormalized? Indeed, recently a calculation has been announced claiming that a supersymmetric regularization of super-Yang-Mills theory produces identical renormalization of both, by the Gell-Mann-Low renormalization group function; M. Grisaru and P. West, Nucl. Phys. B254 (1985) 249.

[83] J. Bell and R. Jackiw, ref. 65; R. Jackiw and K. Johnson, Phys. Rev. 182 (1969) 1459.

[84] G. 't Hooft, Phys. Rev. Lett. 37 (1976) 8; and ref. 54.

[85] D. Sutherland, Nucl. Phys. B2 (1967) 433; M. Veltman, Proc. Roy. Soc. A301 (1967) 107.

[86] J. Bell and R. Jackiw, ref. 65.

[87] K. Fujikawa, Phys. Rev. Lett. 42 (1979) 1195; Phys. Rev. D21 (1980) 2848; (E) 22 (1980) 1499 and ref. 77.

[88] G. 't Hooft, ref. 84; R. Jackiw and C. Rebbi, ref. 56.

[89] L. Brown, R. Carlitz and C. Lee, Phys. Rev. D16 (1977) 417; R. Jackiw and C. Rebbi, Phys. Rev. D16 (1977) 1052. For a review see R. Jackiw, C. Nohl and C. Rebbi, ref. 51.

[90] R. Crewther, P. DiVecchia, G. Veneziano and E. Witten, Phys. Lett. 88B (1979) 123.

[91] S. Glashow, R. Jackiw and S. S. Shei, Phys. Rev. 187 (1969) 1916; S. Weinberg, Phys. Rev. D11 (1975) 3583.

[92] G 't Hooft, ref. 84.

[93] S. Deser, R. Jackiw and S. Templeton, Phys. Rev. Lett. 48 (1982) 975, and Ann. Phys. (NY) 140 (1982) 372.

[94] Another review of this material is by R. Jackiw, in Gauge Theories of the Eighties, eds. R. Raitio and J. Lindfors (Lecture Notes in Physics 181. Spinger, Berlin, 1983).

[95] R. Jackiw and S. Templeton, Phys. Rev. D23 (1981) 2291; J. Schonfeld, Nucl. Phys. B185 (1981) 157.

[96] H. Nielsen and P. Olesen, Nucl. Phys. B61 (1973) 45.

[97] J. Goldstone and E. Witten (unpublished). Their argument is presented in ref. 94.

[98] R. Jackiw and S. Templeton, ref. 95; S. Templeton, Phys. Lett 103B (1981) 134; Phys. Rev. D24 (1981) 3134; M. Bergère and F. David, Ann. Phys. (NY) 142 (1982) 416. E. Gundelman and Z. Radulovic, Phys. Rev. D27 (1983) 357, 30 (1984) 1338.

[99] R. Feynman, Nucl. Phys. B188 (1981) 479; I. Singer, Phys. Scripta 24 (1981) 817.

[100] R. Jackiw and S. Templeton, ref. 95.

[101] N. Redlich, Phys. Rev. Lett. 52 (1984) 18, Phys. Rev. D29 (1984) 2366. L. Alvarez-Gaumé and E. Witten, ref. 62. Alternative derivations and further discussion of these results are found in R. Jackiw, A. Niemi and G. Semenoff, ref. 64. J. Lott, Phys. Lett. 145B (1984) 179.

[102] The argument is analogous to one employed to the same end by E. Witten, Phys. Lett. 117B (1982) 324 in his analysis of a four-dimensional fermionic determinant which appears in this volume; see also Sec. 5.3.

[103] N. Redlich, ref. 101. The calculation method is that of J. Schwinger, ref. 65.

[104] For a review see D. Gross, R. Pisarski and L. Yaffe, Rev. Mod. Phys. 53 (1981) 43.

[105] Field theoretic finite-temperature calculations are explained in L. Dolan and R. Jackiw, Phys. Rev. D9 (1974) 3320.

[106] A temperature-induced Chern-Simons term is discussed in N. Redlich and L. Wijewardhana, Phys. Rev. Lett. 54 (1985) 970; A. Niemi and G. Semenoff; Phys. Rev. Lett. 54 (1985), 2166; K. Tsokos, Phys. Lett. 157B (1985) 413; H. -Y. Guo and W. -Y. Zhao, Comm. Theor. Phys. (Beijing) (in press).

[107] R. Jackiw and K. Johnson, Phys. Rev. D8 (1973) 2386; J. Cornwall and R. Norton, Phys. Rev. D8 (1973) 3338.

[108] Research on the quantum Hall effect from this point of view is presented by M. Friedman, R. Hughes, K. Ishikawa, R. Jackiw, K. Johnson, S. Sokoloff, Y. Srivastava and A. Widom, ref. 64.

[109] In the supersymmetry context, topological terms, related to the ones discussed here are analyzed in W. Siegel, Nucl. Phys. B156 (1979) 135; P. Townsend, K. Pilch and P. Van Nieuwenhuizen, Phys. Lett. 136B (1984) 38, 137B (1984) 443; S. Deser and R. Jackiw, Phys. Lett. 139B (1984) 371. M. Pernici, K. Pilch and P. Van Nieuwenhuizen, Phys. Lett. 143B (1984) 103.

[110] Three-dimensional gravity is further discussed in S. Deser, R. Jackiw and G. 't Hooft, Ann. Phys. (NY) 152 (1984) 220; S. Deser and R. Jackiw, Ann. Phys. (NY) 153 (1984) 405. In these papers the classical theory with sources, which is not trivial, is examined. This work is reviewed in R. Jackiw, Nucl. Phys. B252 (1985) 343.

[111] E. Witten, ref. 102.

[112] J. Goldstone (unpublished). Another version of this argument is given by E. D'Hoker and E. Farhi, Phys. Lett. 134B (1984) 86.

[113] E. D'Hoker and E. Farhi, ref. 112, have suggested that a soliton quantized as a fermion can compensate for the gauge non-invariance of the fermion sector.

[114] J. Bagger and E. Witten, Phys. Lett. 115B (1982) 202.

[115] E. Witten, Nucl. Phys. B223 (1983) 422, which appears in this volume.

[116] A. Polyakov and P. Wiegman, ref. 71. These authors cite S. Novikoc, Dokl. Akad. Nauk SSSR 260 (1981) 31 [Sov. Math. Dokl. 24 (1981) 222] as having independently discovered "multivalued" actions with quantized parameters.

[117] R. Jackiw, in E. Fradkin Festschrift, eds. I. Batalin. C. Isham and G. Vilkovisky (Adam Hilger, Bristol, 1985).

[118] M. Asorey and D. Mitter, Phys. Lett 153B (1985) 147; Y. -S. Wu and A. Zee, Nucl. Phys. B258 (1985) 157. In this paper the analogy with a magnetic monopole is sharpened through or a study of the topology of the Yang-Mills configuration space.

[119] L. Faddeev, Phys. Lett. 145B (1984) 81; L. Faddeev and S. Shatashvili, Teor. Mat. Fiz. 60 (1984) 206; J. Mickelsson, Comm. Math. Phys. 97 (1985) 361; I. Singer (to be published).

[120] R. Jackiw, Phys. Rev. Lett. 54 (1985) 159, 2380; Phys. Lett. 154B (1985) 303, 2380; B. Grossman, Phys. Lett. 152B (1985) 93; and B. -Y. Hou, Chinese Phys. Lett. (in press); Y. -S. Wu and A. Zee, Phys. Lett. 152B (1985) 98; J. Mickelsson, Phys. Rev. Lett. 54 (1985) 2379.

[121] See from example P. Carruthers and M. Nieto, Am. J. Phys. 33 (1965) 537.

[122] This convention for differential forms and for their multiplication is explained for example in R. Buck, Advanced Calculus, 2nd ed. (McGraw-Hill, New York, 1965).

[123] J. Goldstone (unpublished); R. Stora, in Progress in Gauge Field Theory, eds. G. 't Hooft, A. Jaffe, H. Lehmann, P. Mitter, I. Singer and R. Stora, (Plenum, New York, 1984); B. Zumino, Y. -S Wu and A. Zee, ref. 61, B. Zumino, ref. 76.

[124] The procedure of removing the trivial 1-cocycle has been carried out completely for the Abelian theory in S. Deser, R. Jackiw and S. Templeton, ref. 93. The non-Abelian case is being studied by D. Gonzales and N. Redlich (in preparation).

[125] A. Niemi and G. Semenoff, ref. 64.

[126] R. Jackiw and K. Johnson, ref. 83; D. Gross and R. Jackiw Nucl. Phys. B14 (1969) 269. For a review see ref. 6.

[127] L. Faddeev, Nuffield Workshop, Cambridge (1985); I. Frenkel and I. Singer (in preparation).

[128] S. -G. Jo, Nucl. Phys. B259 (1985) 616, Phys. Lett. B in press.
[129] For a description of this method, see ref. 6.
[130] A. Niemi and G. Semenoff, Phys. Rev. Lett. 55 (1985) 927.
[131] F. Bucella, G. Veneziano, R. Gatto and S. Okubo, Phys. Rev. 149 (1965) 1268.
[132] R. Jackiw, R. Van Royen and G. West, Phys. Rev. D2 (1970) 2473. For a summary, see ref. 6.
[133] T. Nagylaki, Phys. Rev. 158 (1967) 1534; D. Boulware and R. Jackiw, Phys. Rev. 186 (1969) 1442; D. Boulware and J. Herbert, Phys. Rev. D2 (1972) 1055.
[134] K. Johnson and F. Low, Prog. Theor. Phys. (Kyoto) Suppl. 37-38 (1966) 74.
[135] This has been established in QCD by M. A. B. Bég, Phys. Rev. D11 (1975) 1165.
[136] R. Jackiw and G. Preparata, Phys. Rev. Lett. 22 (1969) 975, (E) 1162; M. A. B. Bég, J. Bernstein, D. Gross, R. Jackiw and A. Sirlin, Phys. Rev. Lett. 25 (1970) 1232.
[137] This argument has been developed by many people. I learned it from R. Feynman.
[138] M. Atiyah and I. Singer, Proc. Natl. Acad. Sci. U.S.A. 81 (1984) 2597; L. Alvarez-Gaumé and P. Ginsparg, Nucl. Phys. B243 (1984) 449; J. Lott, Comm. Math. Phys. 97 (1985) 371.

CHIRAL ANOMALIES AND DIFFERENTIAL GEOMETRY[†] [*]

Bruno Zumino
Lawrence Berkeley Laboratory and Department of Physics
University of California
Berkeley, CA 94720, USA

[†]This contribution first appeared in *Relativity, Groups and Topology II*, Les Houches 1983, eds., B. S. DeWitt and R. Stora (North-Holland, Amsterdam, 1984).

[*]This work was supported in part by the Director, Office of Energy Research, Office of High Energy and Nuclear Physics, Division of High Energy Physics of the U.S. Department of Energy, under Contract DE-AC03-76SF00098, and in part by the National Science Foundation, under Research Grant No. PHY-81-18547.

1. Introduction

In these lectures I shall describe [0] a number of properties of chiral anomalies from a geometric point of view*. I follow mostly work done in collaboration with Raymond Stora [1]. Some of the results are contained in a recent paper written in collaboration with Wu Yong-Shi and Anthony Zee [2], to which I refer also for an extensive list of old and new references on chiral anomalies. It is possible that the methods and results described in these lectures are fully known in mathematics. On the other hand, several crucial formulas have not been given before (at any rate not explicitly) and their physical relevance is emphasized here.

As an introduction to the main subject let us consider some examples of the relevance of topology to physics:

(1) *The Dirac monopole.* The action integral for an electron in the field of a magnetic monopole is given by

$$I = \int_1^2 L_{\mathrm{kin}} \mathrm{d}t + e\int_1^2 \boldsymbol{A}(\boldsymbol{x}, t)\cdot\frac{\mathrm{d}\boldsymbol{x}}{\mathrm{d}t}\,\mathrm{d}t. \tag{1.1}$$

The integral is over a path joining the point 1 to the point 2. Let us consider the second term in the action. If we deform the path of integration, keeping the end points fixed, and then come back to the original path, the action returns to its original value, provided the deformation was not too large. However, if we swing the path about the position of the monopole and come back to the original path, we cut all flux lines of the Coulomb-like magnetic field. The action changes by an integer multiple of eg (g is the monopole charge),

$$I' = I + neg. \tag{1.2}$$

One can say that the space of paths is infinitely connected. Classically

*A very interesting (and rather mathematical) paper relevant to the subject of these lectures is: L. Bonora and P. Cotta-Ramusino, Commun. Math. Phys. 87 (1983) 589. See also L. Bonora, P. Cotta-Ramusino and C. Reina, Phys. Lett. 126B (1983) 305.

this fact is not very important, since I' gives the same equations of motion as I, but quantum mechanically it gives rise to a problem. For instance, the path integral

$$Z = \int \mathscr{D}(\text{path})\, e^{iI} \tag{1.3}$$

is not well defined, unless

$$neg = m2\pi, \quad m, n \text{ integers}, \tag{1.4}$$

which is the Dirac quantization condition. (If this quantization condition is not satisfied, the path integral could be defined as vanishing by destructive interference, when one integrates over the infinitely connected space of paths.) This is a well known example of quantization of classical parameters due to topology.

Other examples are:

(2) *Effective or phenomenological Lagrangians which arise as solutions of the anomalous Ward identities* (*see section* 4). Witten [3] and Balachandran, Nair and Trahern [4] have observed that a phenomenon similar to that occurring for the Dirac monopole occurs here, except that the one-dimensional path is replaced by a four-dimensional sphere.

(3) *Nonlinear σ-model coupled to supergravity* [5]. One finds that Newton's constant has to take quantized values, i.e. multiples of F_π^{-2}.

(4) *Three-dimensional Yang–Mills theory* [6]. The topological mass of the vector field has a quantized value.

(5) *In the Weinberg–Salam model there may exist heavy* (*unstable*) *soliton states* [7]. The Higgs sector of the model has a global $SU(2)_L \times SU(2)_R$ symmetry. It is a linear σ-model but for large Higgs mass it can be approximated by a nonlinear one, hence may have soliton solutions.

A common feature of all of these examples is that they make use of homotopy groups, so a list of the homotopy groups of the classical groups may be useful as a guide for a systematic search. Without going into the details of the definitions let us say roughly that the qth-homotopy group Π_q of a (topological) space X is the set of mappings of S^q (the q-dimensional sphere) into the space X, where two mappings are considered as equivalent when one can be continuously deformed into the other. We are interested in homotopy groups of groups, i.e., the space X is a classical compact Lie group G. Table 1 lists the homotopy groups of the classical groups [8].

The table exhibits the Bott periodicity theorem. Provided the group is

Table 1
Homotopy groups of the classical groups [8][a]

$\Pi_q : q$	U(N)	O(N)	Sp (N)
	$N > q/2$	$N > q+1$	$N > (q-2)/4$
0	0	$\mathbb{Z}_2$ [P]	0
1	$\mathbb{Z}$ [EM]	$\mathbb{Z}_2$ [spin]	0
2	0	0	0
3	$\mathbb{Z}$	$\mathbb{Z}$	$\mathbb{Z}$ [Instantons]
4	0	0	$\mathbb{Z}_2$ [Witten]
5	$\mathbb{Z}$ [Chir. Lag.]	0	$\mathbb{Z}_2$
6	0	0	0
7	$\mathbb{Z}$	$\mathbb{Z}$	$\mathbb{Z}$
8	0	$\mathbb{Z}_2$	0
period:	2	8	8

[a] $\mathbb{Z}$ is the integers, $\mathbb{Z}_2$ is the group of two elements.

sufficiently "large" (the inequalities are indicated) the homotopy groups follow a series of period 2 in the first column [U(N)] and of period 8 in the other two columns [O(N) and Sp(N)]. Observe also that the homotopy groups for O(N) and for Sp(N) follow the same pattern, only shifted by 4 (half the period). See Milnor's book [9], last chapter, for a proof of the Bott periodicity theorem.

Remarks:

(1) $\Pi_2 = 0$ for all three classes and also for the exceptional groups (E. Cartan).

(2) Π_0 refers to the connectedness of the group,
$\Pi_0(\mathrm{O}(N)) = \mathbb{Z}_2$ is related to parity,
$\Pi_1(\mathrm{O}(N \geqslant 3)) = \mathbb{Z}_2$ is related to spin.

(3) The Dirac monopole has to do with $\Pi_1(\mathrm{U}(1)) = \mathbb{Z}$. Note that for the 't Hooft–Polyakov monopole the relevant quantity is $\Pi_2(\mathrm{SU}(2)/\mathrm{U}(1)) = \Pi_1(\mathrm{U}(1))$. For homotopy groups of quotients of groups see Hilton [10].

(4) The instanton has to do with $\Pi_3(\mathrm{SU}(2)) = \mathbb{Z}$. Note that $\mathrm{Sp}(1) = \mathrm{SU}(2)$.

(5) Witten [11] has pointed out that an SU(2) gauge theory with an odd number of chiral fermion doublets is inconsistent. This is related to $\Pi_4(\mathrm{Sp}(1)) = \mathbb{Z}_2$.

(6) Chiral Lagrangians (cf. section 4) are related to $\Pi_5(\mathrm{U}(N \geqslant 3)) = \mathbb{Z}$.

(7) $\Pi_3 = \mathbb{Z}$. This fact is related to chiral solitons [12, 13].

2. Chiral Anomalies and Differential Forms

A simple way to introduce the subject of this section—the anomalies associated with chiral fermions—is to consider the Lagrangians

$$L = i\bar{\psi}\gamma^\mu(\partial_\mu - i\lambda_k A_\mu{}^k)\psi \tag{2.1}$$

in four-dimensional space–time, where the ψ are Dirac spinors, $A_\mu{}^k$ a set of external vector fields and λ^k the generators of a representation of an internal symmetry group [like SU(2), SU(3)]. Let us furthermore introduce an axial current operator

$$J_\mu{}^5 = \bar{\psi}\gamma_\mu\gamma_5\psi \tag{2.2}$$

which is a singlet under the internal group, and look at its classical conservation equations

$$\partial^\mu J_\mu{}^5 = 0. \tag{2.3}$$

It is well established [14] that in the one-loop approximation of perturbation theory the classical conservation equation breaks down. If one requires vector gauge invariance, the axial vector equation takes the form

$$\partial^\mu J_\mu{}^5 = -\frac{1}{16\pi^2}\varepsilon^{\mu\nu\rho\sigma}\,\mathrm{Tr}\,F_{\mu\nu}F_{\rho\sigma}. \tag{2.4}$$

Here

$$F_{\mu\nu} = -i\lambda_k F_{\mu\nu}{}^k,$$

and $F_{\mu\nu}{}^k$ is the usual Yang–Mills field strength associated with the fields $A_\mu{}^k$ ($\varepsilon_{0123} = 1$). In terms of the latter, the "singlet" or "Abelian" anomaly, as we shall call the rhs of eq. (2.4), can be written as

$$\partial^\mu J_\mu{}^5 = -\frac{1}{4\pi^2}\varepsilon^{\mu\nu\rho\sigma}\,\mathrm{Tr}\,\partial_\mu(A_\nu\partial_\rho A_\sigma + \tfrac{2}{3}A_\nu A_\rho A_\sigma). \tag{2.5}$$

Equally well one may consider two currents constructed analogously to the above: fermions are split into left/right ones,

$$\psi_{\substack{\mathrm{L}\\ \mathrm{R}}} = \frac{1\pm\gamma_5}{2}\psi, \tag{2.6}$$

and one starts from a Lagrangian in which they are coupled to corresponding left/right vector fields,

$$L = i\bar{\psi}_\mathrm{L}\gamma^\mu(\partial_\mu - iA_{\mu\mathrm{L}}^{\,k}\lambda_k)\psi_\mathrm{L} + i\bar{\psi}_\mathrm{R}\gamma^\mu(\partial_\mu - iA_{\mu\mathrm{R}}^{\,k}\lambda_k)\psi_\mathrm{R}. \tag{2.7}$$

Now, all currents

$$J_{\mu i}^{H} = \bar{\psi}_H \gamma_\mu \lambda_i \psi_H, \quad H = \mathrm{L, R}, \tag{2.8}$$

are covariantly conserved in the classical approximation but lead, upon proper definition [14], to anomalous equations

$$D_H^\mu J_{\mu i}^{H} = \eta_H \frac{1}{24\pi^2} \varepsilon^{\mu\nu\rho\sigma} \operatorname{Tr}\left(\lambda_i \partial_\mu \left(A_\nu^H \partial_\rho A_\sigma^H + \tfrac{1}{2} A_\nu^H A_\rho^H A_\sigma^H\right)\right)$$

$$H = \mathrm{L, R}, \qquad \eta_\mathrm{L} = -\eta_\mathrm{R} = -1, \tag{2.9}$$

in higher order. The rhs of eq. (2.9) will be called the non-Abelian anomaly. Comparing the factor 1/2 in front of the trilinear A-term with the corresponding factor 2/3 in eq. (2.5) it is clear that the non-Abelian anomaly cannot be rewritten in terms of Yang–Mills curls. Nevertheless there is an intricate relation between the two types of anomalies, which will be cleared up in the subsequent lectures. To point this out, differential geometric methods will be used, which are going to be introduced presently.

In terms of *differential forms* the Yang–Mills fields $A_\mu{}^k$ will be represented by

$$A = -\mathrm{i} A_\mu{}^k \lambda_k \mathrm{d}x^\mu, \tag{2.10}$$

a matrix of one-forms (i.e., having anti-commuting elements), and the field strength by

$$F = \mathrm{d}A + A^2, \tag{2.11}$$

a matrix of two-forms (wedge symbol suppressed, matrix multiplication understood, elements of F commuting). It is easy to check that the Bianchi identity

$$DF \equiv \mathrm{d}F + [A, F] = 0 \tag{2.12}$$

holds, by making use of the fact that

$$\mathrm{d}^2 = 0 \tag{2.13}$$

and that d anti-commutes with the one-form A. The operation D in eq. (2.12) is the covariant differential. In order to translate eqs. (2.5)–(2.9) into the language of forms we associate with currents $J_\mu{}^5$ one-forms $J_\mu{}^5 \mathrm{d}x^\mu$, go over to their duals

$$*J^5 = \frac{1}{3!} \varepsilon_{\nu\lambda\mu\rho} J^{5\nu} \mathrm{d}x^\lambda \mathrm{d}x^\mu \mathrm{d}x^\rho,$$

and then observe that evaluating the divergence (resp. the covariant divergence) is performed with the exterior derivation d (resp. the covariant D):

$$d{*}J^5 = (\partial^\lambda J_\lambda{}^5)\frac{1}{4!}\varepsilon_{\mu\nu\rho\sigma}dx^\mu dx^\nu dx^\rho dx^\sigma, \tag{2.14}$$

hence

$$d{*}J^5 \propto \mathrm{Tr}\, F^2 = d\,\mathrm{Tr}\,(AdA+\tfrac{2}{3}A^3), \tag{2.15}$$

$$(D{*}J^5)_i = -G_i(A) \propto d\,\mathrm{Tr}\,\lambda_i(AdA+\tfrac{1}{2}A^3). \tag{2.16}$$

As indicated in section 1, the fact that the anomalies are d-operating on something is crucial for their interrelation, so let us derive this fact. Consider, e.g., $\mathrm{Tr}\, F^2$. Observe first

$$d\,\mathrm{Tr}\, F^2 = \mathrm{Tr}\,(dFF+FdF) = 2\,\mathrm{Tr}\, dFF. \tag{2.17}$$

Adding zero in the form $2\,\mathrm{Tr}\,[A, F]F$ we obtain

$$d\,\mathrm{Tr}\, F^2 = 2\,\mathrm{Tr}\, DFF = 0, \tag{2.18}$$

due to the Bianchi identity (2.12). In order to find the form of which $\mathrm{Tr}\, F^2$ is the derivative we have to perform an integration, and therefore we look first of all at the variation of $\mathrm{Tr}\, F^2$ induced by varying A to $A+\delta A$:

$$F = dA+A^2,$$

$$\delta F = d\delta A+\delta AA+A\delta A = D(\delta A) \tag{2.19}$$

(the signs are correct, A being a one-form).

$$\delta\,\mathrm{Tr}\, F^2 = 2\,\mathrm{Tr}\,\delta FF = 2\,\mathrm{Tr}\, D(\delta A)F$$

$$= 2D\,\mathrm{Tr}\,\delta AF = 2d\,\mathrm{Tr}\,\delta AF. \tag{2.20}$$

We have used the Bianchi identity and the fact that $\mathrm{Tr}\,\delta AF$ is a scalar.

Let us now introduce the variation of A via a parameter t by

$$A_t = tA, \qquad F_t = tdA+t^2A^2 = tF+(t^2-t)A^2. \tag{2.21}$$

Equation (2.20) may now be written as

$$\delta\,\mathrm{Tr}\, F_t^2 = 2d\,\mathrm{Tr}\,\delta A_t F_t, \tag{2.22}$$

and with $\delta = \delta t\partial/\partial t$ we have by integration

$$\int_0^1 \delta t\frac{\partial}{\partial t}\mathrm{Tr}\, F_t^2 = 2d\int_0^1 \delta t\,\mathrm{Tr}\, AF_t, \tag{2.23}$$

hence

$$\mathrm{Tr}\, F^2 = 2\mathrm{d} \int_0^1 \delta t\, \mathrm{Tr}\, A(tF + (t^2 - t)A^2) = \mathrm{d}\, \mathrm{Tr}\, (AF - \tfrac{1}{3}A^3). \tag{2.24}$$

Calling the integral, which is a three-form, ω_3, we write

$$\begin{aligned}\omega_3 &= 2 \int_0^1 \delta t\, \mathrm{Tr}\, A(t\mathrm{d}A + t^2 A) \\ &= \mathrm{Tr}\, (A\mathrm{d}A + \tfrac{2}{3}A^3) = \mathrm{Tr}\, (AF - \tfrac{1}{3}A^3),\end{aligned} \tag{2.25}$$

and have thus verified that

$$\mathrm{Tr}\, F^2 = \mathrm{d}\omega_3. \tag{2.26}$$

Analogously one can proceed for higher powers $\mathrm{Tr}\, F^n$, since

$$\mathrm{d}\, \mathrm{Tr}\, F^n = n\, \mathrm{Tr}\, \mathrm{d}\, F F^{n-1} = n\, \mathrm{Tr}\, DFF^{n-1} = 0. \tag{2.27}$$

The result is

$$\mathrm{Tr}\, F^n = \mathrm{d}\omega_{2n-1}, \tag{2.28}$$

$$\omega_{2n-1} = n\, \mathrm{Tr} \int_0^1 \delta t A F_t^{n-1}. \tag{2.29}$$

Explicitly for $n = 3$:

$$\begin{aligned}\omega_5 &= 3\, \mathrm{Tr} \int_0^1 \delta t A(t\mathrm{d}A + t^2 A^2)^2 \\ &= \mathrm{Tr}\, (A(\mathrm{d}A)^2 + \tfrac{3}{5}A^5 + \tfrac{3}{2}A^3\mathrm{d}A),\end{aligned} \tag{2.30}$$

$$\mathrm{Tr}\, F^3 = \mathrm{d}\omega_5. \tag{2.31}$$

The above considerations have to be generalized to the case where $\mathrm{d}(\cdots) \neq 0$. The appropriate tool is the so-called *homotopy operator* k: it has the properties

$$\mathrm{d}k + k\mathrm{d} = 1, \tag{2.32}$$

$$k^2 = 0, \qquad \mathrm{d}^2 = 0. \tag{2.33}$$

Let us suppose for the moment that k exists and check on the known case above what k does. Apply eq. (2.32) to $\mathrm{Tr}\, F^2$:

$$(\mathrm{d}k + k\mathrm{d})\, \mathrm{Tr}\, F^2 = \mathrm{Tr}\, F^2; \tag{2.34}$$

since $\mathrm{d}\,\mathrm{Tr}\,F^2 = 0$ this is simply

$$\mathrm{d}(k\,\mathrm{Tr}\,F^2) = \mathrm{Tr}\,F^2. \tag{2.35}$$

Hence, if k is known, $\mathrm{Tr}\,F^2$ is readily expressed as a $\mathrm{d}(\cdots)$.

The construction of k proceeds algebraically. Build out of F and A all those formal polynomials that vanish at $F = 0$, $A = 0$. Define an operation d on them by

$$dA = F - A^2, \tag{2.36}$$

$$dF = FA - AF, \tag{2.37}$$

and the rule that it acts as anti-derivation (commutes with F, anti-commutes with A, and is linear on sums). Check that

$$d^2 = 0. \tag{2.38}$$

Indeed: $d^2A = d(F-A^2) = FA - AF - dAA + AdA = 0.$ Similarly $d^2F = d(FA - AF) = 0$ (work out!).

Define another operation l by

$$lA = 0, \tag{2.39}$$

$$lF = \delta A \tag{2.40}$$

and the antiderivation rule. Then verify that

$$ld + dl = \delta; \tag{2.41}$$

on A: $ldA + dlA = l(F - A^2) = \delta A$
on F: $ldF + dlF = l(FA - AF) + d\delta A = \delta AA + A\delta A + \delta dA = \delta F.$
Here we have assumed that δ commutes with d.

These definitions of d and l can thus be extended to all formal polynomials (vanishing at $F = 0$, $A = 0$) and, in fact, be applied to families A_t, F_t depending on a parameter t:

$$l_tA_t = 0, \quad l_tF_t = \delta A_t \equiv \delta t\frac{\partial A_t}{\partial t}, \tag{2.42}$$

with $A_0 = 0$, $F_0 = 0$.

The anti-commutation relation (2.41) becomes

$$l_td + dl_t = \delta = \delta t\frac{\partial}{\partial t}, \tag{2.43}$$

and integrating over t from 0 to 1 yields an explicit representation

$$k \equiv \int_0^1 l_t \tag{2.44}$$

with

$$kd + dk = 1. \tag{2.45}$$

Let us illustrate these abstract considerations by an example. Choose as polynomial AF. Then

$$d(AF) = F^2 - AFA, \tag{2.46}$$

$$ld(AF) = \delta AF + F\delta A + A\delta AA. \tag{2.47}$$

Choosing the t-family

$$\begin{aligned} A &\to A_t = tA, \\ \delta A &\to \delta A_t = \delta t \frac{\partial A_t}{\partial t} = \delta t A, \\ F &\to F_t = tF + (t^2 - t)A^2, \end{aligned} \tag{2.48}$$

we have

$$\begin{aligned} ld(AF) \to l_t d(A_t F_t) &= \delta A_t F_t + F_t \delta A_t + A_t \delta A_t A_t \\ &= \delta t (AF_t + F_t A + t^2 A^3). \end{aligned} \tag{2.49}$$

Hence, integrating over t from 0 to 1

$$\begin{aligned} kd(AF) &= \int_0^1 \delta t (AF_t + F_t A + t^2 A^3) \\ &= \tfrac{1}{2}(AF + FA). \end{aligned} \tag{2.50}$$

On the other hand

$$l(AF) = -A\delta A \to l_t(A_t F_t) = -(\delta t) t A^2, \tag{2.51}$$

$$\int_0^1 l_t(A_t F_t) = k(AF) = -\tfrac{1}{2}A^2, \tag{2.52}$$

$$dk(AF) = -\tfrac{1}{2}dA^2 = -\tfrac{1}{2}(FA - AF). \tag{2.53}$$

Adding eqs. (2.50) and (2.53),

$$(kd + dk)(AF) = AF, \tag{2.54}$$

as desired. The lesson we learn therefore is that one must perform first of all the l-operation term by term and then integrate. It is to be noted also that l_t depends on t since $l_t F_t = \delta A_t = \delta t \partial A_t/\partial t$ is a variation along the one-parameter family at the point t. On the contrary d is t-independent. Observe that, in the example discussed above, one can verify by direct computation that the square of the operator k vanishes,

$$k^2 = 0.$$

Actually, it is not difficult to show that this is a general fact, when k is defined by means of the family (2.48). We leave the proof as an exercise to the reader.

A word of caution. Equations such as (2.24) and (2.26) are really valid only locally, in some finite neighborhood in x-space. It is however well known that they can be given a global meaning by using a connection on a principal fibre bundle, rather than a vector potential on the base (x-space).

Observe that the forms ω_{2n-1} are local expressions, constructed with the gauge potential and its derivatives up to some finite order, all calculated at a given point [see eq. (2.29)].

Finally we emphasize that in defining the operators d, l and k and in studying their properties [from formula (2.36) on] we have treated A and F as purely algebraic objects from which one can freely form polynomials. No special relations (such as particular commutation relations) have been used and the polynomials were not restricted by any symmetry or invariance property.

3. Transformation Properties of the Anomalies

The key question whose answer eventually leads to the characterization of the anomalies is: how do they transform under a gauge transformation?

We have seen in the last section that

$$\operatorname{Tr} F^n = d\omega_{2n-1}{}^0, \tag{3.1}$$

with

$$\omega_{2n-1}{}^0 = k \operatorname{Tr} F^n = n \int_0^1 \delta t \operatorname{Tr} A F_t^{n-1}, \tag{3.2}$$

(the additional superscript ^{0} is introduced for later convenience), where k was the homotopy operator. Under a finite gauge transformation $g(x)$

the field A transforms into

$$A_g = g^{-1}Ag + g^{-1}\mathrm{d}g, \tag{3.3}$$

hence $F = \mathrm{d}A + A^2$ into

$$F_g = g^{-1}Fg. \tag{3.4}$$

Under this transformation $\mathrm{Tr}\, F^n$ is clearly invariant, but how does

$$\omega_{2n-1}{}^0 = \omega_{2n-1}{}^0(A, F),$$

understood as function of A and F, change? Certainly $\omega_{2n-1}{}^0$ may (and will, in general) change by a term $\mathrm{d}\alpha$, α being a $(2n-2)$-form, since this contribution is annihilated by applying the d operator yielding $\mathrm{Tr}\, F^n$. But it turns out, that

$$\omega_{2n-1}{}^0(A_g, F_g) = \omega_{2n-1}{}^0(A, F) + \mathrm{d}\alpha_{2n-2} + \omega_{2n-1}{}^0(g^{-1}\mathrm{d}g, 0), \tag{3.5}$$

i.e. the transformed $\omega_{2n-1}{}^0$ contains besides $\mathrm{d}\alpha$ a term which globally cannot be written as $\mathrm{d}(\cdots)$ and nevertheless is annihilated by d, the form $\omega_{2n-1}{}^0(g^{-1}\mathrm{d}g, 0)$ is closed. Let us now derive this result.

Dropping for the moment the indices we write the gauge transformed $\omega_{2n-1}{}^0(A_g, F_g)$:

$$\begin{aligned}\omega(A_g, F_g) &= \omega(g^{-1}Ag + g^{-1}\mathrm{d}g, g^{-1}Fg)\\ &= \omega(A+V, F),\end{aligned} \tag{3.6}$$

$$V \equiv \mathrm{d}gg^{-1}, \qquad \mathrm{d}V = V^2, \tag{3.7}$$

since ω is given by a trace. We want to use now the homotopy operator k for obtaining information about $\omega(A+V, F)$, but $\omega(A+V, F) \neq 0$ at $A = 0$, $F = 0$, so we have to subtract $\omega(V, 0)$. It is convenient to subtract one more term: $\omega(A, F)$. Hence consider

$$\Omega \equiv \omega(A+V, F) - \omega(V, 0) - \omega(A, F); \tag{3.8}$$

observe that

$$\mathrm{d}\Omega = 0. \tag{3.9}$$

Indeed:

$$\begin{aligned}\mathrm{d}\omega(A+V, F) &= \mathrm{Tr}\, F^n,\\ -\mathrm{d}\omega(V, F) &= 0,\\ -\mathrm{d}\omega(A, F) &= -\mathrm{Tr}\, F^n.\end{aligned}$$

Recall eq. (2.32),

$$(\mathrm{d}k+k\mathrm{d})\Omega = \Omega \tag{3.10}$$

$$\text{i.e.,}\quad \mathrm{d}(k\Omega) = \Omega.$$

We have thus identified the $(2n-2)$-form α:

$$\alpha_{2n-2} = k(\omega_{2n-1}{}^0(A+V,F)-\omega_{2n-1}{}^0(A,F)-\omega_{2n-1}{}^0(V,0)). \tag{3.11}$$

This completes the proof of eq. (3.5) if we also use

$$\omega(\mathrm{d}gg^{-1},0) = \omega(g^{-1}\mathrm{d}g,0). \tag{3.12}$$

Actually, $k^2 = 0$ and $\omega_{2n-1}{}^0(A,F) = k\,\mathrm{Tr}\,F^n$ eliminate the second term in eq. (311). Also, $k\omega_{2n-1}{}^0(V,0) = 0$, so that

$$\alpha_{2n-2} = k(\omega_{2n-1}{}^0(A+V,F)). \tag{3.13}$$

Exercise. Calculate α_{2n-2} for $n = 2, 3$. (Note: in actual calculations it may be simpler to carry along the term $\omega_{2n-1}{}^0(A,F)$ in Ω.)

The result of this calculation is:

For $n = 2$:

$$\alpha_2 = -\mathrm{Tr}\,(VA), \qquad V = \mathrm{d}gg^{-1}. \tag{3.14}$$

For $n = 3$:

$$\begin{aligned}\alpha_4 &= \mathrm{Tr}\,(-\tfrac{1}{2}V(FA+AF)+\tfrac{1}{2}VA^3+\tfrac{1}{4}VAVA+\tfrac{1}{2}V^3A)\\ &= \mathrm{Tr}\,(-\tfrac{1}{2}V(A\mathrm{d}A+\mathrm{d}AA)-\tfrac{1}{2}VA^3+\tfrac{1}{4}VAVA+\tfrac{1}{2}V^3A).\end{aligned} \tag{3.15}$$

So we have:

$$\omega_3{}^0(A_g,F_g) = \omega_3{}^0(A,F)+\mathrm{d}\alpha_2-\tfrac{1}{3}\mathrm{Tr}\,(g^{-1}\mathrm{d}g)^3, \tag{3.16}$$

$$\omega_5{}^0(A_g,F_g) = \omega_5{}^0(A,F)+\mathrm{d}\alpha_4+\tfrac{1}{10}\mathrm{Tr}\,(g^{-1}\mathrm{d}g)^5. \tag{3.17}$$

Equation (3.16) has a well known application to instantons. Equation (3.17) will be used in section 4 and could in principle serve as a definition for the anomaly as well, but a slightly more sophisticated derivation yields the anomaly in a more convenient form, so let us proceed to this one.

We shall distinguish the differentiation in the direction of x from that in the direction of the group, and denote the former by d, the latter by δ:

$$\mathrm{d} = \mathrm{d}x^\mu \frac{\partial}{\partial x^\mu}, \tag{3.18}$$

$$\delta = \mathrm{d}t^r \frac{\partial}{\partial t^r} \tag{3.19}$$

(x^μ are coordinates in space–time; t^r any parameters on which the group elements may depend). In gauge transformations too, we shall separate these variations:

$$A \to g^{-1}Ag + g^{-1}\mathrm{d}g + g^{-1}\delta g; \tag{3.20}$$

g depends on both x and t, while A is a form in x alone. Clearly, for

$$\Delta = d + \delta, \qquad \Delta^2 = 0 \tag{3.21}$$

since

$$\mathrm{d}^2 = \delta^2 = \mathrm{d}\delta + \delta\mathrm{d} = 0. \tag{3.22}$$

For

$$\mathscr{A} \equiv g^{-1}Ag + g^{-1}\mathrm{d}g, \tag{3.23}$$

$$v \equiv g^{-1}\delta g, \tag{3.24}$$

one verifies

$$\delta\mathscr{A} = -\mathrm{d}v - v\mathscr{A} - \mathscr{A}v = -\mathscr{D}v, \tag{3.25}$$

$$\delta v = -v^2. \tag{3.26}$$

Now $F = \mathrm{d}A + A^2$, therefore

$$\mathscr{F} = \mathrm{d}\mathscr{A} + \mathscr{A}^2 = g^{-1}Fg. \tag{3.27}$$

Notice that also

$$\mathscr{F} = \Delta(\mathscr{A} + v) + (\mathscr{A} + v)^2, \tag{3.28}$$

as easily verified. This implies that

$$\Delta\omega_{2n-1}{}^0(\mathscr{A} + v, \mathscr{F}) = \mathrm{d}\omega_{2n-1}{}^0(\mathscr{A}, \mathscr{F}), \tag{3.29}$$

and both sides equal

$$\mathrm{Tr}\,\mathscr{F}^n = \mathrm{Tr}\,F^n. \tag{3.30}$$

Let us expand $\omega_{2n-2}{}^0(\mathscr{A} + v, \mathscr{F})$ in powers of v:

$$\omega_{2n-1}{}^0(\mathscr{A} + v, \mathscr{F}) = \omega_{2n-1}{}^0(\mathscr{A}, \mathscr{F}) + \omega_{2n-2}{}^1 + \cdots + \omega_0{}^{2n-1}, \tag{3.31}$$

where the superscript indicates the power of v. Equation (3.29) implies a set of relations

$$\begin{aligned}
&\delta\omega_{2n-1}{}^{0}+\mathrm{d}\omega_{2n-2}{}^{1}=0,\\
&\delta\omega_{2n-2}{}^{1}+\mathrm{d}\omega_{2n-3}{}^{2}=0,\\
&\vdots\\
&\delta\omega_{1}{}^{2n-2}+\mathrm{d}\omega_{0}{}^{2n-1}=0,\\
&\delta\omega_{0}{}^{2n-1}=0.
\end{aligned}\tag{3.32}$$

We shall see later that $\omega_{2n-2}{}^{1}$ is to be identified with the anomaly. Let us calculate it explicitly. Now, from eqs. (2.29) and (2.21), we see that

$$\omega_{2n-1}{}^{0}(\mathscr{A}+v,\mathscr{F})=n\int_0^1\delta t\,\mathrm{Tr}\,((\mathscr{A}+v)\hat{\mathscr{F}}_t^{n-1}),\tag{3.33}$$

where

$$\begin{aligned}
\hat{\mathscr{F}}_t&=t\mathscr{F}+(t^2-t)(\mathscr{A}+v)^2\\
&=\mathscr{F}_t+(t^2-t)\{\mathscr{A},v\}+(t^2-t)v^2.
\end{aligned}\tag{3.34}$$

It is convenient to replace the trace by the symmetrized trace,

$$\mathrm{Str}\,(B_1,B_2,\ldots,B_n)=\sum_{\mathrm{Perm.}}\frac{1}{n!}\mathrm{Tr}\,(B_{P(1)}\cdots B_{P(n)}).\tag{3.35}$$

To first order in v, eq. (3.33) gives

$$\begin{aligned}
&n\int_0^1\delta t\,\mathrm{Str}\,(v\mathscr{F}_t^{n-1}+(t^2-t)\mathscr{A}[\mathscr{F}_t^{n-2}\{\mathscr{A},v\}+\mathscr{F}_t^{n-3}\{\mathscr{A},v\}\mathscr{F}_t+\cdots])\\
&\quad=n\int_0^1\delta t\,\mathrm{Str}\,(v\mathscr{F}_t^{n-1}+(t^2-t)(n-1)\mathscr{A}\{\mathscr{A},v\}\mathscr{F}_t^{n-2}).
\end{aligned}$$

Using the invariance of Str, one can rewrite this as

$$\begin{aligned}
&n\int_0^1\delta t\,\mathrm{Str}\,(v\mathscr{F}_t^{n-1}+(t^2-t)(n-1)(\{\mathscr{A},\mathscr{A}\}v\mathscr{F}_t^{n-2}+\mathscr{A}v[\mathscr{A},\mathscr{F}_t^{n-2}]))\\
&\quad=n\int_0^1\delta t\,\mathrm{Str}(v[\mathscr{F}_t^{n-1}+(t-1)(n-1)(t\{\mathscr{A},\mathscr{A}\}\mathscr{F}_t^{n-2}-\mathscr{A}[\mathscr{A}_t,\mathscr{F}_t^{n-2}])]).
\end{aligned}$$

Now observe that

$$\mathrm{d}\mathscr{F}_t^{n-2}=-[\mathscr{A}_t,\mathscr{F}_t^{n-2}]$$

and

$$\frac{\partial \mathscr{F}_t}{\partial t} = \mathrm{d}\mathscr{A} + t\{\mathscr{A}, \mathscr{A}\}.$$

The above expression becomes

$$n \int_0^1 \delta t \,\mathrm{Str}\left(v \left[\mathscr{F}_t^{n-1} + (t-1)(n-1) \times \left(\left(\frac{\partial \mathscr{F}_t}{\partial t} - \mathrm{d}\mathscr{A} \right) \mathscr{F}_t^{n-2} + \mathscr{A}\mathrm{d}\mathscr{F}_t^{n-2} \right) \right] \right),$$

and finally, integrating by parts with respect to t, we find the result for the anomaly,

$$\omega_{2n-2}{}^1 = n(n-1) \int_0^1 \delta t(1-t)\,\mathrm{Str}\,(v\mathrm{d}\,(\mathscr{A}\mathscr{F}_t^{n-2})). \tag{3.36}$$

Let us give the explicit expressions for $n = 2, 3$:

$$\omega_2{}^1 = \mathrm{Tr}\,(v\mathrm{d}\mathscr{A}), \qquad n = 2, \tag{3.37}$$

$$\omega_4{}^1 = \mathrm{Tr}\,(v\mathrm{d}(\mathscr{A}\mathrm{d}\mathscr{A} + \tfrac{1}{2}\mathscr{A}^3)), \quad n = 3. \tag{3.38}$$

Equation (3.36) is very convenient because it exhibits the anomaly in the canonical form in which the differential operates on a function of $\mathscr{A}$ and $\mathscr{F}$, while v is not differentiated. Equation (3.38) agrees with (2.16).

Equation (3.37) gives a two-form in x-space, which is the non-Abelian anomaly in two dimensions. Similarly $\omega_4{}^1$ gives the non-Abelian anomaly in four dimensions and generally $\omega_{2n-2}{}^1$ in $2n-2$ dimensions. One may wonder whether the other forms, $\omega_{2n-k}{}^{k-1}$ $(2 \leqslant k \leqslant 2n)$, are also relevant to physics. If one is interested in four-dimensional space–time, one must take

$$2n-k = 4, \qquad k-1 = 2n-5. \tag{3.39}$$

For any $n \geqslant 3$ this gives a four-form in x-space, $\omega_4{}^{2n-5}$, in which the infinitesimal gauge transformation v occurs an odd number of times, $2n-5$. There is an infinite number of such forms, as n varies. According to unpublished work by I. Singer, these generalized anomalies can be identified as obstructions to a definition of the Dirac propagator in an external potential, globally in the space of all potentials.

4. Identification and Use of the Anomalies

In section 3 we have defined the form $\omega_{2n-2}{}^1$ to be the non-Abelian anomaly. We now wish to justify this definition. To this end we go back to section 2, generalize appropriately the Lagrangian (2.7) to arbitrary (even) space–time dimensions, renormalize in the one-loop approximation, and consider the functional of one-particle-irreducible Green's functions $W[A]$ to this order. Gauge transformations are now represented by functional differential operators

$$X_i(x) = -\partial_\mu \frac{\delta}{\delta A_\mu{}^i} - \left(A_\mu \times \frac{\delta}{\delta A_\mu}\right)_i. \tag{4.1}$$

The cross-product is constructed with the structure constants of the respective simple, compact Lie group under consideration. One may convince oneself that the X_i form the algebra

$$[X_i(x), X_j(y)] = f_{ij}{}^k X_k(x)\delta(x-y), \tag{4.2}$$

and also that their action on $W[A]$ just yields the current (non)-conservation equation

$$X_i(x)W[A] = G_i[A](x). \tag{4.3}$$

($G_i = 0$ would correspond to the conservation of the respective currents.) Now the mere existence of the functional $W[A]$, which we suppose to be ensured by appropriate renormalization, implies a consistency condition for the possible G_i. Acting twice on $W[A]$ and using eq. (4.2) we derive [15]

$$X_i(x)G_j(y) - X_j(y)G_i(x) = f_{ij}{}^k G_k(x)\delta(x-y). \tag{4.4}$$

Trivial solutions of these equations are, of course, readily found:

$$G_i(x) = X_i(x)\hat{G}[A] \tag{4.5}$$

($\hat{G}[A]$ e.g. local) is a solution. But eq. (4.4) has not yet been solved in all generality. The anomalies (2.9) arise as solutions of (4.4) which are not variations of a *local* functional in the basic fields of the theory: This feature we take as definition for the general case: we regard as anomaly any solution of (4.4) which is not a variation of a *local* functional in the basic fields (ψ, A_μ).

Before showing that $\omega_{2n-2}{}^1$ does solve just (4.4) we have to reformulate the problem somewhat.

Let us introduce anti-commuting scalar fields (Faddeev–Popov fields)

$v_i(x)$ and the notation

$$v \cdot X \equiv \sum_i \int \mathrm{d}x\, v_i(x) X_i(x), \qquad v \cdot G \equiv \sum_i \int \mathrm{d}x\, v_i(x) G_i(x). \tag{4.6}$$

Then the consistency conditions (4.4) turn into

$$v \cdot X\, v \cdot G - \tfrac{1}{2}(v \times v) \cdot G = 0. \tag{4.7}$$

Similarly the gauge transformation on ${A_\mu}^i$ may be reformulated:

$$\delta A_\mu = v \cdot X A_\mu. \tag{4.8}$$

Interpreting (4.7) as an invariance property of $v \cdot G$ suggests to transform v_i also:

$$\delta v_i = -\tfrac{1}{2}(v \times v)_i, \tag{4.9}$$

i.e., eq. (4.7) becomes

$$\delta(v \cdot G) = 0. \tag{4.10}$$

It should be clear that eqs. (4.8) and (4.9) are nothing but the BRS-transformations and eq. (4.10) the Slavnov identity for the special case of currents. This statement is confirmed by showing

$$\delta^2 = 0, \tag{4.11}$$

i.e., the transformations (4.8), (4.9) are nilpotent. (*Exercise.* check eq. (4.11).)

Let us now go over to forms

$$A = -\mathrm{i}{A_\mu}^k \lambda_k \mathrm{d}x^\mu, \tag{4.12}$$

$$v = -\mathrm{i}v^k \lambda_k, \tag{4.13}$$

where v is a zero-form with values in the Lie-algebra, and re-express eqs. (4.8) and (4.9) as

$$\delta A = -\mathrm{d}v - vA - Av \equiv -Dv. \tag{4.14}$$

$$\delta v = -v^2. \tag{4.15}$$

The consistency equation (4.10) becomes

$$\delta \int \mathrm{Tr}\, vG[A] = 0 \tag{4.16}$$

or, equivalently,

$$\delta\, \mathrm{Tr}\, vG[A](x) = \mathrm{d}\chi, \tag{4.17}$$

(d: exterior x-derivative; χ: some quantity). Hence the δ defined in eqs. (4.14) and (4.15) and the d in eq. (4.17) fulfill the algebra (3.25), (3.26), and will be identified with those operators. What remains to be shown is thus only that (4.17) can be identified with

$$\delta\omega_{2n-2}{}^{1} = \mathrm{d}(-\omega_{2n-3}{}^{2}), \tag{4.18}$$

i.e., $\omega_{2n-2}{}^{1}$ with $\mathrm{Tr}\, vG[A](x)$ and χ with $-\omega_{2n-2}{}^{2}$. Indeed, let us look at the system (3.32):

$$\begin{aligned}
&\mathrm{Tr}\, F^n - \mathrm{d}\omega_{2n-1}{}^{0} = 0,\\
&\delta\, \mathrm{Tr}\, F^n = 0,\\
&\delta\omega_{2n-1}{}^{0} + \mathrm{d}\omega_{2n-2}{}^{1} = 0,\\
&\delta\omega_{2n-2}{}^{1} + \mathrm{d}\omega_{2n-3}{}^{2} = 0,\\
&\vdots
\end{aligned}$$

We see that $\omega_{2n-2}{}^{1}$ is linear in v and satisfies the consistency condition. The problem of finding the most general solution of the consistency condition will not be discussed here [16].

In order to derive physical consequences from the presence of the anomalies we use the approach of phenomenological Lagrangians [15]. We permit the presence of another multiplet of fields ξ_i (Lorentz scalars) and try to adjust its transformation law under the gauge group so that the anomaly can be derived as variation of a local functional of the gauge fields plus the fields ξ_i.

It turns out [15] that the law of nonlinear realization,

$$\xi \to \xi'(\alpha, \xi): \mathrm{e}^{\alpha}\mathrm{e}^{\xi} = \mathrm{e}^{\xi'}, \tag{4.19}$$

is the correct one and that

$$W[A, \xi] = \int_0^1 \delta t\, \mathrm{e}^{-t\xi\cdot X}\xi \cdot G[A] \tag{4.20}$$

satisfies

$$\alpha \cdot (X+Z)W[A, \xi] = \alpha \cdot G[A], \tag{4.21}$$

i.e., fulfills the anomalous Ward identity. Here

$$Z_i = H_{ij}\frac{\delta}{\delta\xi_j}, \qquad H_{ij} = \left.\frac{\partial\xi'_j}{\partial\alpha_i}\right|_{\alpha=0} \tag{4.22}$$

generates the transformation of ξ [X is given in eq. (4.1)]. The identification $\xi_i = 1/F_\pi \pi_i$ in the local action

$$W[A,\xi]+F_\pi{}^2/2\ \mathrm{Tr}\int \mathrm{d}x\ \partial^\mu e^\xi \partial_\mu e^{-\xi}+\text{normal solution}$$

shows that the anomaly contributes additional pion–pion and pion–vector interactions in the σ-model-type phenomenological action. Examples where these arguments have been successfully applied are the processes $\pi^0 \to 2\gamma$, $\eta \to \gamma\pi\pi$, etc. [15].

One can show directly [15] that eq. (4.20) gives a solution of the anomalous Ward identity. Here instead we first rewrite it in a more geometric form from which this fact will follow. The factor $e^{-t\xi\cdot X}$ transforms A into $A_{g(t)}$, where

$$g(t) = e^{-t\xi}, \tag{4.23}$$

which we may understand as a family of group elements parametrized by t. Hence varying this parameter is a variation δ in group space

$$g^{-1}(t)\,\delta g(t) = -\xi\delta t, \tag{4.24}$$

$$v = -\xi\delta t. \tag{4.25}$$

Thus

$$\begin{aligned} W[A,\xi] &= \int \mathrm{d}x \int_0^1 \delta t\, e^{-t\xi\cdot X}\xi_i G_i[A](x) \\ &= \int \mathrm{d}x \int_0^1 \delta t \xi_i G_i[A_{g(t)}](x) \\ &= -\int \mathrm{d}x \int_0^1 \mathrm{Tr}\, v\cdot G[A_{g(t)}](x). \end{aligned} \tag{4.26}$$

Interchanging the order of integration we can interpret the integral in group space: for any fixed t the x-integral is in fact one over the corresponding configuration $g_t(x)$ in group space, $t = 0$ parametrizing the identity e and $t = 1$ the element $g(x)$. We therefore write (up to a sign)

$$W[A;g(x)] = \int_e^{g(x)} \omega_4{}^1(\mathscr{A},v). \tag{4.27}$$

Using the expansion

$$\omega_5(\mathscr{A}+v) = \omega_5{}^0(\mathscr{A})+\omega_4{}^1(\mathscr{A},v)+\cdots,$$

we first note that

$$\omega_5{}^0(\mathscr{A}) = 0, \tag{4.28}$$

since $\omega_5{}^0(\mathscr{A})$ is a five-form purely in x, but space–time here is four-dimensional; next we see that for the special parametrization (4.23)

$$v^2 = v^3 = \cdots = 0, \tag{4.29}$$

since there is only one independent differential δt. So we can write

$$W[A; g(x)] = \int_e^{g(x)} \omega_5(\mathscr{A}+v). \tag{4.30}$$

In higher dimensions we would write similarly

$$W[A, g(x)] = \int_e^{g(x)} \omega_{2n-1}(\mathscr{A}+v) \tag{4.31}$$

($2n-2$ dimensional space–time). Observe that

$$(\mathrm{d}+\delta)\omega_5(\mathscr{A}+v) = \mathrm{Tr}\,(g^{-1}Fg) = \mathrm{Tr}\,F^3, \tag{4.32}$$

which is a six-form purely in x and therefore vanishes in four dimensions. Therefore eq. (4.30) is invariant under deformations of the integration manifold, provided the limits of integration are kept fixed [similarly for eq. (4.31)].

We now show that W satisfies the anomalous Ward identity. This is a consequence of the second of eqs. (3.32), slightly re-interpreted. Let us perform a gauge transformation

$$A \to h^{-1}Ah + h^{-1}\mathrm{d}h = A_h, \qquad g(x) \to h^{-1}g(x). \tag{4.33}$$

Observe that

$$\mathscr{A}+v = g^{-1}Ag + g^{-1}(\mathrm{d}+\delta)g = a(A, g)$$

satisfies

$$a(A_h, g) = a(A, hg). \tag{4.34}$$

Therefore

$$W[A_h, h^{-1}g] = \int_e^{h^{-1}g} \omega_5(a(A_h, g')) = \int_e^{h^{-1}g} \omega_5(a(A, hg')). \tag{4.35}$$

Change the integration variable from g' to $g'' = kg'$, where $k = h$ at the upper limit and $k = e$ at the lower limit. Then

$$W[A_h, h^{-1}g] = \int_e^g \omega_5(\mathscr{a}(A, hk^{-1}g'')). \tag{4.36}$$

If h is infinitesimal,

$$h(x) = e + m(x), \tag{4.37}$$

then

$$k = e + s, \tag{4.38}$$

where $s = m(x)$ at the upper limit and $s = 0$ at the lower limit, and

$$hk^{-1} = e + m(x) - s = e + n. \tag{4.39}$$

So we must make an infinitesimal transformation (drop the double-primes)

$$\delta_m g = ng, \tag{4.40}$$

and correspondingly

$$\delta_m \mathscr{a} = \mathscr{a}(A, g+ng) - \mathscr{a}(A, g) = -\mathrm{d}\mathscr{V} - \mathscr{a}\mathscr{V} - \mathscr{V}\mathscr{a}, \tag{4.41}$$

where

$$\mathscr{V} = g^{-1}\delta_m g = g^{-1}ng. \tag{4.42}$$

Now, in analogy with eq. (3.32),

$$\delta_m \omega_5(\mathscr{a}(A, g)) = -(\mathrm{d}+\delta)\omega_4{}^1(g^{-1}ng, \mathscr{a}) \tag{4.43}$$

(here δ_m is an even variation), therefore

$$\delta_m W = -\int_e^g (\mathrm{d}+\delta)\omega_4{}^1. \tag{4.44}$$

The right-hand side can be evaluated by Stokes' theorem. Since n vanishes at the upper limit, while $n = m(x)$ at the lower limit, we finally obtain the desired equation

$$\delta_m W = \int_x \omega_4{}^1(m(x), A). \tag{4.45}$$

The expression (4.30) for the effective Lagrangian W can be simplified if one makes use of eqs. (3.5) and (3.17), which implies

$$\omega_5(\mathscr{A} + v) = \omega_5(\mathscr{A}) + (\mathrm{d}+\delta)\alpha_4 + \omega_5(g^{-1}(\mathrm{d}+\delta)g). \tag{4.46}$$

Now, the first term $\omega_5(A)$ in the rhs vanishes because it is a five-form purely in x. The second term can be integrated by Stokes' theorem. Therefore eq. (4.10), using eq. (3.17), gives

$$W[A, g(x)] = \int_x \alpha_4(V, A) + \tfrac{1}{10} \int_{(5)}^{g(x)} \operatorname{Tr}(g^{-1}(\mathrm{d}+\delta)g)^5. \tag{4.47}$$

Here α_4 and V are as given in eqs. (3.14), (3.15). The last integral is extended over a five-dimensional manifold in group space having the sphere $g(x)$ as boundary (as x varies over S_4, $g(x)$ describes a sphere in group space). The integral is invariant under deformations of the five-manifold because the integrand is a closed form (in general

$$\mathrm{d}\operatorname{Tr} V^{2n-1} = (2n-1)\operatorname{Tr} V^{2n} = 0 \tag{4.48}$$

for $V = \mathrm{d}gg^{-1}$, $\mathrm{d}V = V^2$). In (4.47) the dependence on the vector fields is explicit, since α_4 is explicitly known. It is polynomial only. One could use the simplified form (4.47) to show that W satisfies the anomalous Ward identity (4.45) (*exercise* for the reader).

The last term in eq. (4.47) is an integral in group space. For a group [like SU(3)] with a nontrivial Π_5, there exist five-cycles C_5 such that the integral (write simply d for $\mathrm{d}+\delta$)

$$\int_{C_5} \operatorname{Tr}(g^{-1}\mathrm{d}g)^5 \neq 0 \tag{4.49}$$

does not vanish; suitably normalized (see below) it equals an integer. This means that although the integral in eq. (4.47) is unchanged if one performs small deformations of the five-manifold, it is ambiguous for large deformations. This fact leads to a quantization of the effective action, as mentioned in the Introduction. The normalization to be chosen is 2π times that which gives an integer for the integral (4.49).

5. Normalization of the Anomalies

The normalization of the form $\operatorname{Tr} F^n$, which enters in the Abelian anomaly (2.5), (2.15), can be related to the formula for the index of the Dirac operator. This gives the correctly normalized Abelian (or singlet) anomaly in $2n$ dimensions. The connection between this and the non-Abelian anomaly in $2n-2$ dimensions permits then to find that normalization also. So both normalizations can be determined completely from purely geometric arguments.

First the singlet anomaly. In (compactified) Euclidean space–time, one writes*

$$\partial^\mu J_{\mu\ \mathrm{reg}}^{\ 5} = C(x) - 2 \sum_{\text{zero modes}} \phi_a^\dagger \gamma_5 \phi_a, \tag{5.1}$$

where $C(x)$ is the anomaly, ϕ_a are normalized zero modes of the Dirac operator with a given external potential, and γ_5 means the analogue of γ_5 in any number of dimensions. $J_{\mu\ \mathrm{reg}}^{\ 5}$ is the (suitably regularized) axial vector current. The factor of 2 comes from carrying out the divergence, which gives the Dirac operator once on the spinor on the right and once on that on the left.

Integrating (5.1) over all space–time, the left-hand side gives zero. Therefore

$$\int C(x)\mathrm{d}x = 2 \int \Sigma \phi_a^\dagger \gamma_5 \phi_a = 2(n_+ - n_-), \tag{5.2}$$

where n_+ (n_-) is the number of zero modes of positive (negative) chirality. Their difference is the index. Now it is known (see, e.g., ref. [17], eq. (7.22)) that the index is given by the integral of the Chern character

$$\mathrm{ch}\,(V) = \mathrm{Tr}\exp\left(\frac{\mathrm{i}}{2\pi} F\right). \tag{5.3}$$

More precisely, in $2n$ dimensions:

$$n_+ - n_- = \int \frac{1}{n!}\left(\frac{\mathrm{i}}{2\pi}\right)^n \mathrm{Tr}\,F^n. \tag{5.4}$$

Therefore ($F = \frac{1}{2} F_{\mu\nu} \mathrm{d}x^\mu \mathrm{d}x^\nu$)

$$C(x) = \frac{2}{n!}\frac{\mathrm{i}^n}{(2\pi)^n 2^n}\,\mathrm{Tr}\,F_{\mu_1\mu_2}F_{\mu_3\mu_4}\cdots F_{\mu_{2n-1}\mu_{2n}}\varepsilon^{\mu_1\mu_2\cdots\mu_{2n}}. \tag{5.5}$$

(This is real because $F_{\mu\nu} = -\mathrm{i}F_{\mu\nu}^{\ \ i}\lambda_i$). This requires, of course, that one knows somehow the correct formula for the index. A nice derivation (for physicists) based on quantum mechanical supersymmetry, has been given recently by Alvarez–Gaumé [18] and by Friedan and Windey [19].

In order to determine the normalization of the non-Abelian anomaly, we shall proceed as follows. Since the non-Abelian anomaly determines

*For a derivation of eq. (5.1) see: N.K. Nielsen and B. Schroer, Nucl. Phys. B127 (1977) 493.

the phenomenological Lagrangian (see section 4), we shall require that it be normalized so that the Lagrangian satisfies the quantization condition. As we shall see, the normalization of the non-Abelian anomaly is then related directly to that of the index formula, without the extra factor of 2 necessary for the singlet anomaly. Of course this procedure is, in a sense, like going backwards, and the normalization of the non-Abelian anomaly can be computed directly in perturbation theory. What we are saying is that the result of perturbation theory agrees with the correct normalization for the phenomenological Lagrangian, as required by geometric considerations.

Remember that

$$\operatorname{Tr} F^n = d\omega_{2n-1}(A, F). \tag{5.6}$$

where

$$\omega_{n-1}(A, F) = n\int_0^1 \delta t \operatorname{Tr} AF_t^{n-1}, \tag{5.7}$$

$$F_t = tF + (t^2 - t)A^2. \tag{5.8}$$

One finds

$$\omega_{2n-1}(V, 0) = n\int_0^1 \delta t (t^2 - t)^{n-1} \operatorname{Tr} V^{2n-1}. \tag{5.9}$$

The integral is easily carried out (successive integrations by parts) with the result

$$n\int_0^1 \delta t (t^2 - t)^{n-1} = (-1)^{n-1}\frac{(n-1)!n!}{(2n-1)!}. \tag{5.10}$$

Multiplying (5.10) by the factor in front of the index formula, $(1/n!)(i/2\pi)^n$, (without the extra factor of 2) gives

$$\frac{1}{n!}\left(\frac{i}{2\pi}\right)^n(-1)^{n-1}\frac{(n-1)!n!}{(2n-1)!} = (-1)^{n-1}\left(\frac{i}{2\pi}\right)^n\frac{(n-1)!}{(2n-1)!}. \tag{5.11}$$

From the fact that the index is an integer one can then deduce that the form

$$\left(\frac{i}{2\pi}\right)^n\frac{(n-1)!}{(2n-1)!}\operatorname{Tr} V^{2n-1}, \qquad V = dgg^{-1}, \tag{5.12}$$

also integrates to a integer, the integral being performed over a $(2n-1)$-sphere in group space (see [20]). Now we know that the phenomenological Lagrangian must be normalized with an additional factor of 2π. This means that the non-Abelian anomaly in $2n-2$ dimensions is given by (up to a sign)

$$\frac{1}{n!}\frac{\mathrm{i}^n}{(2\pi)^{n-1}}\,\omega_{2n-2}{}^1, \tag{5.13}$$

with $\omega_{2n-2}{}^1$ given by the expansion (3.31) or by (3.36). For $n = 3$, formula (5.13) with (3.38) agrees with (2.9).

The normalization argument of this section can be made more precise by considering, instead of the phenomenological Lagrangian, the connected vacuum functional

$$W[A] = \frac{1}{\mathrm{i}}\log Z[A],$$

where $Z[A]$ is the vacuum functional obtained by integrating over the fermion fields but keeping the external gauge fields A_μ fixed. By definition $Z[A]$ is a single-valued function of its argument A_μ. Now, the change in $W[A]$ under an infinitesimal gauge transformation is given by eq. (4.3). The change in $W[A]$ under a finite gauge transformation can be easily worked out by iterating eq. (4.3). Using arguments very similar to those of section 4, the change in $W[A]$ can be written as an integral of the differential form $\omega_{2n-1}{}^0(A_g, F_g)$ given in eq. (3.5). Now, it is possible to find a family of gauge transformations which, starting from the identity, moves in group space and comes back to the identity, but such that $W[A]$ changes by an amount proportional to

$$\int_{C_5}\omega_{2n-1}(g^{-1}\mathrm{d}g, 0) \propto \int_{C_5}\mathrm{Tr}\,(g^{-1}\mathrm{d}g)^{2n-1}.$$

For suitable cycles C_5 this is different from zero. On the other hand, $Z = \exp(\mathrm{i}W[A])$ must be single-valued, by definition. Therefore, $W[A]$ must change by a multiple of 2π. This gives the normalization of the anomaly in agreement with section 5.

Appendix I. A Simple Formula for α_{2n-2}

It is often useful to have a simple explicit formula for the differential form α_{2n-2} occurring in eqs. (3.5) and (3.14)–(3.17). The formulas (3.11) or (3.13) are sufficient, but they still require some work to evaluate α_{2n-2}. In this appendix we derive such a simple, explicit formula, (A.16).

We define a connection depending upon two parameters λ and μ,

$$\mathscr{A}_{\lambda,\mu} = \lambda A - \mu V, \tag{A.1}$$

where, as in the text,

$$V = \mathrm{d}gg^{-1}, \qquad \mathrm{d}V = V^2. \tag{A.2}$$

The corresponding field strength is

$$\mathscr{F}_{\lambda,\mu} = \mathrm{d}\mathscr{A}_{\lambda,\mu} + \mathscr{A}^2_{\lambda,\mu} \tag{A.3}$$

and it satisfies the Bianchi identities

$$\mathrm{d}\mathscr{F}_{\lambda,\mu} = -[\mathscr{A}_{\lambda,\mu}, \mathscr{F}_{\lambda,\mu}]. \tag{A.4}$$

Differentiating (A.3) one finds

$$\frac{\partial \mathscr{F}_{\lambda,\mu}}{\partial \lambda} = \mathrm{d}A + \{\mathscr{A}_{\lambda,\mu}, A\} \tag{A.5}$$

and

$$\frac{\partial \mathscr{F}_{\lambda,\mu}}{\partial \mu} = -\mathrm{d}V - \{\mathscr{A}_{\lambda,\mu}, V\}. \tag{A.6}$$

We consider the integral

$$n \int \mathrm{Tr}\,((\delta\lambda A - \delta\mu V)\mathscr{F}_{\lambda,\mu}{}^{n-1}) \tag{A.7}$$

over a one-dimensional path, which is a clockwise triangle in the λ, μ plane going from the origin to the point (0, 1), to (1, 0), and back to the origin. On the segment from (1, 0) to (0, 0),

$$\mu = 0, \qquad \mathscr{A}_{\lambda,\mu} = \lambda A, \qquad \mathscr{F}_{\lambda,\mu} = \lambda \mathrm{d}A + \lambda^2 A^2 = F_\lambda$$

[as defined in eq. (2.48)], and therefore eq. (A.7) equals, by (2.29)

$$n \int_1^0 \mathrm{Tr}\,\delta\lambda\, A F_\lambda^{n-1} = -\omega_{2n-1}(A, F). \tag{A.8}$$

On the segment from (0, 0) to (0, 1),

$$\lambda = 0, \qquad \mathscr{A}_{\lambda,\mu} = -\mu V, \qquad \mathscr{F}_{\lambda,\mu} = -\mu \mathrm{d}V + \mu^2 V^2 = (\mu^2 - \mu)V^2.$$

Therefore eq. (A.7) equals

$$-n\int_0^1 \mathrm{Tr}\,(\delta\mu V((\mu^2-\mu)V^2)^{n-1}) = -\omega_{2n-1}(V,0). \tag{A.9}$$

On the segment from (0, 1) to (1, 0),

$$\lambda+\mu = 1, \qquad \mathscr{A}_{\lambda,\mu} = \lambda A + (\lambda-1)V,$$

$$\mathscr{F}_{\lambda,\mu} = F_\lambda + (\lambda^2-\lambda)(V^2+\{A, V\}).$$

Therefore eq. (A.7) equals

$$n\int_0^1 \delta\lambda\,\mathrm{Tr}\,(A+V)(F_\lambda+(\lambda^2-\lambda)(V^2+\{A, V\}))^{n-1}$$
$$= \omega_{2n-1}(A+V, F). \tag{A.10}$$

Finally, eq. (A.7) integrated over the clockwise triangle equals

$$\omega_{2n-1}(A+V, F) - \omega_{2n-1}(A, F) - \omega_{2n-1}(V, 0), \tag{A.11}$$

which is the expression we would like to equate to $\mathrm{d}\alpha_{2n-2}$. If we consider $\mathrm{Tr}\,(A\mathscr{F}_{\lambda,\mu}{}^{n-1})$ and $-\mathrm{Tr}\,(V\mathscr{F}_{\lambda,\mu}{}^{n-1})$ as the two components of a two-vector in the plane, we can apply Stokes' theorem to eq. (A.7) and transform it into an integral over the *inside* of the triangle,

$$n\iint \mathrm{Tr}\left(\left(A\frac{\partial}{\partial\mu} + V\frac{\partial}{\partial\lambda}\right)\mathscr{F}_{\lambda,\mu}{}^{n-1}\right). \tag{A.12}$$

Using eqs. (A.5) and (A.6), expression (A.12) becomes

$$n(n-1)\iint \mathrm{Str}\,((-A\mathrm{d}V + V\mathrm{d}A)\mathscr{F}_{\lambda,\mu}{}^{n-2}$$
$$-A\{\mathscr{A}_{\lambda,\mu}, V\}\mathscr{F}_{\lambda,\mu}{}^{n-2} + V\{\mathscr{A}_{\lambda,\mu}, A\}\mathscr{F}_{\lambda,\mu}{}^{n-2}). \tag{A.13}$$

Using the invariance of Str, the last two terms can be rewritten

$$\mathrm{Str}\,(-\{\mathscr{A}_{\lambda,\mu}, V\}A\mathscr{F}_{\lambda,\mu}{}^{n-2} + V\{\mathscr{A}_{\lambda,\mu}, A\}\mathscr{F}_{\lambda,\mu}{}^{n-2})$$
$$= \mathrm{Str}\,(VA[\mathscr{A}_{\lambda,\mu}, \mathscr{F}_{\lambda,\mu}{}^{n-2}])$$
$$= -\mathrm{Str}\,(VA\mathrm{d}\mathscr{F}_{\lambda,\mu}{}^{n-2}), \tag{A.14}$$

where we have also used eq. (A.4). Therefore expression (A.13) becomes

$$-n(n-1)\mathrm{d}\iint \mathrm{Str}\,(VA\mathscr{F}_{\lambda,\mu}{}^{n-2}), \tag{A.15}$$

which finally gives the desired formula

$$\alpha_{2n-2} = -n(n-1)\iint \mathrm{Str}\,(VA\mathscr{F}_{\lambda,\mu}{}^{n-2}), \tag{A.16}$$

as a two-dimensional integral over the interior of the triangle: in eq. (A.16)

$$\iint = \int_0^1 \delta\lambda \int_0^{1-\lambda} \delta\mu. \tag{A.17}$$

One may have preferred a one-dimensional integral formula for α_{2n-2}, like that for ω_{2n-1}, but eq. (A.16) is just as easy to evaluate. In the expansion of $\mathscr{F}_{\lambda,\mu}{}^{n-2}$ one encounters only the integrals

$$\int_0^1 \delta\lambda \int_0^{1-\lambda} \delta\mu\, \lambda^h \mu^k = \frac{h!k!}{(h+k+2)!}. \tag{A.18}$$

As an exercise, the reader may check that eq. (A.16) agrees with eqs. (3.14) (obvious) and (3.15) and then go on to the next case, $n = 4$.

References

[0] Standard mathematical references for the subject of these lectures are:
M. Flanders, Differential Forms with Applications to Physical Sciences (Academic, New York, 1963).
S.S. Chern, Complex Manifolds without Potential Theory, especially the Appendix: Geometry of Characteristic Classes (Springer, New York, 1979).
S. Kobayashi and K. Nomizu, Foundations of Differential Geometry (Interscience, New York, 1969).
Y. Choquet-Bruhat and C. DeWitt-Morette, with M. Dillard-Bleick, Analysis, Manifolds and Physics, 2nd, Rev. Ed. (North-Holland, Amsterdam, 1979).

[1] R. Stora and B. Zumino, in preparation.

[2] B. Zumino, Wu Yong-Shi and A. Zee, Washington preprint 40048-17 P3 (Univ. of Washington, 1983), Nuclear Physics, to be published.

[3] E. Witten, Nucl. Phys. B223 (1983) 422.

[4] A.P. Balachandran, V.P. Nair and C.G. Trahern, Phys. Rev. D27 (1983) 1369.

[5] J. Bagger and E. Witten, Phys. Lett. 115B (1982) 202.

[6] S. Deser, R. Jackiw and S. Templeton, Ann. Phys. 140 (1982) 372.

[7] J.M. Gipson, preprint VPI-HEP-83/1 (Virginia Polytechnic Inst., Blacksburg, VA, 1983).

[8] M.F. Atiyah, Proc. Solvay conference, Austin, TX, Nov. 1982.
[9] J. Milnor, Morse Theory (Princeton Univ. Press, 1973).
[10] P.J. Hilton, An Introduction to Homotopy Theory (Cambridge Univ. Press, 1966).
[11] E. Witten, Phys. Lett. 117B (1982) 324.
[12] T.H.R. Skyrme, Proc. Roy. Soc. A260 (1961) 12; Nucl. Phys. 31 (1962) 556.
D. Finkelstein and J. Rubinstein, J. Math. Phys. 9 (1968) 1762.
L.D. Faddeev, Lett. Math. Phys. 1 (1976) 289.
[13] E. Witten, Nucl. Phys. B223 (1983) 433.
G.S. Adkins, C.R. Nappi and E. Witten, IAS preprint 83, to appear in Nucl. Phys. B.
G.S. Adkins and C.R. Nappi, Princeton preprint (Princeton Univ., Aug. 1983).
[14] J.S. Bell and R. Jackiw, Nuovo Cim. 60A (1969) 47.
Ş. Adler, Phys. Rev. 177 (1969) 2426.
W.A. Bardeen, Phys. Rev. 184 (1969) 1848.
D.J. Gross and R. Jackiw, Phys. Rev. D6 (1972) 477.
C. Bouchiat, J. Iliopoulos and Ph. Meyer, Phys. Lett. 38B (1972) 519.
C. Becchi, Commun. Math. Phys. 39 (1975) 329.
[15] J. Wess and B. Zumino, Phys. Lett. 37B (1971) 95.
[16] H.P.W. Gottlieb and S. Mărculescu, Nucl. Phys. B49 (1972) 633.
[17] T. Eguchi, P.B. Gilkey and A.J. Hanson, Phys. Rep. 66 (1980) p. 213.
[18] L. Alvarez-Gaumé, Harvard preprint TP-83/A029 (Harvard Univ., Cambridge, MA, 1983).
[19] D. Friedan and P. Windey, in preparation.
[20] R. Bott and R. Seeley, Commun. Math. Phys. 62 (1978) 235.

CONSISTENT AND COVARIANT ANOMALIES IN GAUGE AND GRAVITATIONAL THEORIES[†] [*]

William A. Bardeen

Fermi National Laboratory
P. O. Box 500, Batavia IL 60510, USA

Bruno Zumino

Lawrence Berkeley Laboratory and Department of Physics
University of California
Berkeley, CA 94720, USA

[†] Reprinted with permission from *Nuclear Physics* **B244** (1984) 421-453.

[*] This work was supported in part by the Director, Office of Energy Research, Office of High Energy and Nuclear Physics, Division of High Energy Physics of the US Department of Energy under contracts DE-AC03-76SF00098 and DE-AC02-76-CHO-3000 and in part by the National Science Foundation under research grant PHY-81-18547.

The gauge structure of anomalies and the related currents is analyzed in detail. We construct the covariant forms for both the currents and the anomalies for general gauge theories in even-dimensional space-time. The results are then extended to determine the structure of gravitational anomalies. These can always be interpreted as anomalies for local Lorentz transformations.

1. Introduction

The gauge principle is used as the fundamental basis for present theories of all known forces, from electromagnetism to gravitation. Anomalies [1–5] result when gauge invariance cannot be maintained in the quantum theory. A complete understanding of anomalies is essential for the full application of these theories to physical problems.

The anomaly is usually defined as the gauge variation of the connected vacuum functional in the presence of external gauge fields. When an anomaly occurs, this variation does not vanish and the vacuum functional is not gauge invariant. The gauge currents are no longer covariantly conserved but have the anomalies as their divergence. As a consequence of its definition the anomaly satisfies certain consistency conditions [6] which restricts its functional form. For the non-singlet, non-abelian, chiral anomaly, the consistency conditions imply that the anomaly cannot have a covariant expression. Similarly, the anomaly implies that the non-singlet gauge currents cannot have covariant transformation properties.

However, a number of authors [7–9] have recently presented explicit calculations of the non-singlet anomaly and have obtained covariant results. The same situation occurs for the case of gravitational anomalies. In the work by Alvarez-Gaumé and Witten [10] they are presented in covariant form, while the gravitational consistency conditions would imply that they should have a non-covariant form.

In this paper, we clarify the situation by showing that both the covariant and the non-covariant anomalies can be correct forms for the covariant divergence of different currents. For the gravitational anomalies, the two forms correspond to different energy momentum tensors. We shall use the term "consistent" anomaly to refer to the covariant divergence of the current J_μ obtained by varying the vacuum functional with respect to the external gauge potential. The "covariant" anomaly is obtained by modifying the current by adding to it a local function of the gauge potential. The resulting current $\tilde{J}_\mu$ is determined so as to be covariant under local gauge transformations, which implies that its covariant divergence is also covariant. The consistent anomaly has fundamental significance, since it reflects directly the gauge dependence of the vacuum functional. The related covariant anomaly, on the other hand, is distinguished by its simple gauge transformation properties and the covariant current may have significance when used to construct gauge invariant couplings to other fields. As shown in this paper, it is always possible to construct the covariant forms of the current and of the anomaly from the knowledge of the consistent anomaly. Hence the anomaly cancellation conditions are the same for either form. We note that our ability to modify the form of the anomaly by changing the definition of the local currents is different from the ambiguity in the form of the anomaly arising from the addition of local functions of the gauge potential to the vacuum functional [3].

Let us illustrate the situation by the case of non-abelian gauge anomalies in two space-time dimensions. The consistent anomaly is known to be*

$$D_\mu J^\mu = c\,\partial_\mu A_\lambda \varepsilon^{\lambda\mu} \tag{1.1}$$

* Actually, in two-dimensional Minkowski space-time, the anomaly can be written in one of the forms

$$c\,\partial_\mu A_\lambda\left(\varepsilon^{\lambda\mu} \pm \eta^{\lambda\mu}\right),$$

which differ respectively from (1.1) by the gauge variation of the local functionals

$$\pm \tfrac{1}{2}c\int A_\mu A_\lambda \eta^{\mu\lambda} = \pm\tfrac{1}{2}c\int A_\mu A^\mu .$$

Roman Jackiw has emphasized that these forms are more natural than (1.1) since for chiral (antichiral) spinors in two dimensions the Dirac lagrangian depends only on $A_0 + A_1$, $(A_0 - A_1)$ and therefore the anomaly should also depend only on those combinations. We prefer to ignore this peculiarity of the two-dimensional Minkowski case and illustrate our point using the form (1.1) which is perfectly analogous to the abnormal parity expressions valid in four and higher dimensions.

where c is a certain constant and a matrix notation has been used for both the current and the gauge potential. The right-hand side satisfies the consistency condition [6] but is non-covariant. The current J^μ also transforms non-covariantly. We now define a new current

$$\tilde{J}^\mu = J^\mu + cA_\lambda \varepsilon^{\lambda\mu}. \tag{1.2}$$

Its covariant divergence is

$$D_\mu \tilde{J}^\mu = c\partial_\mu A_\lambda \varepsilon^{\lambda\mu} + \partial_\mu\left(cA_\lambda \varepsilon^{\lambda\mu}\right) + c\left[A_\mu, A_\lambda\right]\varepsilon^{\lambda\mu} = cF_{\mu\lambda}\varepsilon^{\lambda\mu}, \tag{1.3}$$

where

$$F_{\mu\lambda} = \partial_\mu A_\lambda - \partial_\lambda A_\mu + \left[A_\mu, A_\lambda\right] \tag{1.4}$$

is the Yang-Mills field strength. The right-hand side of (1.3) is now covariant. The current $\tilde{J}^\mu$ may also be shown to be covariant, but it cannot be obtained from the variation of a vacuum functional with respect to the gauge field A_μ, since the covariant anomaly does not satisfy the consistency condition. Observe that the linearized right-hand side of (1.3) is twice the right-hand side of (1.1) (this factor becomes $1 + \frac{1}{2}\nu$ in ν dimensions and may be considered as a Bose symmetry factor for the linearized anomaly). We emphasize the care which is needed in interpreting the linearized calculations.

In this paper we discuss various aspects of the gauge structure of anomalies and their currents. In sect. 2 we study the gauge dependence of the currents and their anomalies and apply conventional methods to construct the covariant currents and anomalies for four-dimensional gauge theories. In sect. 3 we discuss the structure of the consistent anomaly in arbitrary even space-time dimensions and give also the explicit expressions for the covariant currents and the covariant anomalies. This is done by using the compact notation of exterior differential forms and the techniques described in refs. [11–16]. The needed results are collected and, in part, rederived in appendix A.

Our results are generalized to include gravitational anomalies in sect. 4. A theory with spinor fields in curved space must be formulated so that it is covariant under general coordinate transformations (which we shall call Einstein transformations) as well as under local Lorentz transformations. Local Lorentz invariance of the connected vacuum functional is usually assumed and the gravitational anomalies are taken to be anomalies of the Einstein transformations. In sect. 5 we shall formulate the consistency conditions for the combined Einstein and Lorentz anomalies [13] and we shall find the form of these anomalies. We also show that the Einstein anomalies can always be transformed into Lorentz anomalies (and vice versa) by adding local corrections to the vacuum functional. Hence it is always possible to define the vacuum functional so that all gravitational anomalies are indeed violations of local Lorentz invariance alone. This appears to us a preferred canonical form for the gravitational anomalies. The treatment of gravitational anomalies in sects. 4 and 5 relates their structure to that of gauge anomalies.

Throughout this paper the anomalies will be expressed in terms of symmetric invariant polynomials which shall not be further specified. The particular polynomial appropriate to each situation depends on the spin of the particles propagating in the loops of the vacuum functional and can be determined by an explicit perturbation calculation, as done in the paper by Alvarez-Gaumé and Witten [10] for the gravitational anomaly. The correct polynomial can also be determined directly from the appropriate index theorem. This approach is discussed in a paper by Alvarez, Singer and Zumino [17].

2. Gauge structure of currents and anomalies in four dimensions

The consistent anomaly is determined by the gauge dependence of the connected vacuum functional defined in the presence of external gauge fields $A_\mu{}^a(x)$. The vacuum functional $W[A]$ may be considered as a non-local function of these gauge fields. Under infinitesimal gauge transformations the gauge potentials transform according to

$$T_\Lambda A_\mu{}^a = (D_\mu \Lambda)^a = (\partial_\mu \Lambda + [A_\mu, \Lambda])^a,$$

$$T_\Lambda F_{\mu\nu}{}^a = ([F_{\mu\nu}, \Lambda])^a, \tag{2.1}$$

where Λ^a is the infinitesimal gauge parameter. The gauge dependence of the vacuum functional defines the anomaly

$$\begin{aligned} T_\Lambda W[A] &= \int \mathrm{d}x \frac{\delta W}{\delta A_\mu{}^a} T_\Lambda A_\mu{}^a \\ &= \int \mathrm{d}x J^\mu{}_a(x)(D_\mu \Lambda)^a \\ &= -\int \mathrm{d}x D_\mu J^\mu{}_a(x) \Lambda^a(x) \\ &= \int \mathrm{d}x \Lambda^a(x) G_a(A), \end{aligned} \tag{2.2}$$

where $G_a(A)$ is the anomaly and the current $J^\mu{}_a(x)$ is defined as the functional derivative of the vacuum functional.

The consistency condition follows from considering the commutator of two gauge transformations on the vacuum functional

$$(T_\Lambda T_{\Lambda'} - T_{\Lambda'} T_\Lambda) W[A] = T_{[\Lambda, \Lambda']} W[A]. \tag{2.3}$$

Using (2.2) this implies

$$\int dx\, (\Lambda'^a T_\Lambda G_a - \Lambda^a T_{\Lambda'} G_a) = \int dx\, [\Lambda, \Lambda']^a G_a. \tag{2.4}$$

The consistent anomaly must obey this consistency condition (2.4).

The consistent anomaly also determines the gauge dependence of the basic non-abelian current $J^\mu{}_a$. Naively, this current would be expected to transform covariantly under gauge transformations. The effect of the anomalies can be determined by evaluating in two ways the commutator of a gauge variation and the variation which defines the current

$$(\delta_B T_\Lambda - T_\Lambda \delta_B) W[A], \tag{2.5}$$

where δ_B is defined by

$$\delta_B A_\mu{}^a = B_\mu{}^a, \tag{2.6}$$

$$\delta_B W[A] = \int dx \frac{\delta W}{\delta A_\mu{}^a} \delta_B A_\mu{}^a = \int dx J^\mu{}_a(x) B_\mu{}^a(x). \tag{2.7}$$

The commutator may be evaluated directly

$$\delta_B T_\Lambda - T_\Lambda \delta_B = \delta_{[B, \Lambda]}. \tag{2.8}$$

Applying this operator to the vacuum functional we obtain

$$\begin{aligned} (\delta_B T_\Lambda - T_\Lambda \delta_B) W[A] &= \int dx\, \{(\delta_B G_a)\Lambda^a - (T_\Lambda J^\mu{}_a) B_\mu{}^a\} \\ &= \int dx J^\mu{}_a ([B, \Lambda])^a. \end{aligned} \tag{2.9}$$

This gives immediately the gauge transformation properties of the non-abelian current

$$\int dx\, (T_\Lambda J^\mu{}_a) B_\mu{}^a = -\int dx\, ([\Lambda, J^\mu])_a B_\mu{}^a + \int dx\, (\delta_B G_a) \Lambda^a. \tag{2.10}$$

The first term on the right-hand side of (2.10) gives the usual transformation property of the current while the second term is dictated by the consistent anomaly. The basic current $J^\mu{}_a$ will only be covariant if the anomaly vanishes.

We shall now demonstrate the existence of a covariant non-abelian current $\tilde{J}_\mu{}^a$ and compute its covariant divergence. This result was obtained independently by Paranjape and Goldstone [18]* and can also be inferred from some work by Niemi and Semenoff [19]. In subsequent chapters we shall generalize these results to gauge and gravitational anomalies in higher dimensional space-times.

To construct the covariant non-abelian current we must find a local polynomial in the gauge potential, $X^\mu{}_a(A)$, with an anomalous gauge transformation property opposite to that of the basic current

$$\int \mathrm{d}x\,(T_\Lambda X^\mu{}_a)B_\mu{}^a = -\int \mathrm{d}x\,([\Lambda, X^\mu])_a B_\mu{}^a - \int \mathrm{d}x\,(\delta_B G_a)\Lambda^a . \tag{2.11}$$

The covariant current is then given by

$$\tilde{J}^\mu{}_a = J^\mu{}_a + X^\mu{}_a(A), \tag{2.12}$$

since (2.10) and (2.11) imply

$$T_\Lambda \tilde{J}^\mu{}_a = -([\Lambda, \tilde{J}^\mu])_a . \tag{2.13}$$

It is not obvious that an appropriate local expression $X^\mu{}_a(A)$ can always be found.

In four dimensions, the consistent non-abelian anomaly for spin one-half fermions is well known [3–4]:

$$G^a(x) = -\frac{1}{24\pi^2}\varepsilon^{\mu\nu\rho\sigma}\mathrm{Tr}\{\lambda^a\partial_\mu(A_\nu\partial_\rho A_\sigma + \tfrac{1}{2}A_\nu A_\rho A_\sigma)\}, \tag{2.14}$$

where Tr is the trace over Fermi multiplets and λ^a is the gauge coupling matrix. Eq. (2.11) for $X^\mu{}_a$ becomes

$$\begin{aligned}\int \mathrm{d}x\,\{T_\Lambda X^\mu{}_a + ([\Lambda, X^\mu])_a\}B_\mu{}^a &= -\int \mathrm{d}x\,(\delta_B G_a)\Lambda^a \\ &= \frac{i}{48\pi^2}\int \mathrm{d}x\,\varepsilon^{\mu\nu\rho\sigma}\partial_\mu\Lambda^a B_\nu{}^b \\ &\quad\times\mathrm{Tr}\{(\lambda_a\lambda_b + \lambda_b\lambda_a)F_{\rho\sigma} - \lambda_a\lambda_b A_\rho A_\sigma \\ &\quad -\lambda_b\lambda_a A_\rho A_\sigma - \lambda_a A_\rho \lambda_b A_\sigma\} .\end{aligned} \tag{2.15}$$

From the three possible terms for the polynomial $X^\mu{}_a$ we find the unique result

$$X^\mu{}_a = \frac{1}{48\pi^2}\varepsilon^{\mu\nu\rho\sigma}\,\mathrm{Tr}\{\lambda_a(A_\nu F_{\rho\sigma} + F_{\rho\sigma}A_\nu - A_\nu A_\rho A_\sigma)\} . \tag{2.16}$$

By applying a group transformation to (2.16) we can reproduce (2.15).

* We thank R. Jackiw for informing us of this work.

We may now compute the covariant anomaly $\tilde{G}_a$, by a direct evaluation of the covariant divergence of the current

$$\tilde{G}_a = -D_\mu \tilde{J}^\mu{}_a = G_a - D_\mu X^\mu{}_a$$
$$= -\frac{1}{32\pi^2}\varepsilon^{\mu\nu\rho\sigma}\mathrm{Tr}\{\lambda_a F_{\mu\nu}F_{\rho\sigma}\}. \tag{2.17}$$

We observe that the covariant anomaly may be expressed solely in terms of a product of field strengths as expected by covariance. The linearized form of the consistent anomaly (2.14) and of the covariant anomaly (2.17) are the same except that the covariant anomaly is three times larger.

We emphasize the need for a complete specification of the structure of the anomalous currents before the gauge anomalies can be properly interpreted. The consistent anomaly is directly related to the gauge dependence of the vacuum functional. It is appropriate for the study of anomaly cancellation between fermion multiplets but also for the derivation of physical consequences of anomalous non-dynamical currents such as the flavor chiral currents in QCD [6, 20, 21]. The covariant current, on the other hand, has a simple gauge structure and may have physical significance when coupled to other external non-gauge fields. Since the covariant anomaly is directly related to the consistent anomaly, it may also be used to study anomaly cancellation. In the above discussion we have focussed on the ambiguities in defining appropriate non-abelian currents. There is also the ambiguity in defining the vacuum functional, as one is always free to modify the vacuum functional by adding local polynomials in the gauge fields. This freedom is exploited when we use the functional for gauging dynamically different anomaly free subgroups.

3. Chiral anomalies in higher dimensions

Following the notation of refs. [11, 12] we now describe the Yang-Mills field strength by means of the Lie-algebra valued 2-form

$$F = \mathrm{d}A + A^2, \tag{3.1}$$

where d denotes exterior differentiation, and

$$A = A_\mu \mathrm{d}x^\mu \tag{3.2}$$

is the gauge potential 1-form. Explicitly

$$F = \tfrac{1}{2}F_{\mu\nu}\mathrm{d}x^\mu \mathrm{d}x^\nu, \qquad F_{\mu\nu} = \partial_\mu A_\nu - \partial_\nu A_\mu + [A_\mu, A_\nu], \tag{3.3}$$

(the differentials dx^μ anticommute). Let

$$P(F_1, F_2, \dots F_n) \tag{3.4}$$

be a symmetric invariant polynomial of degree n in the Lie-algebra valued variables $F_1, \dots, F_n$. For compact notation, if some of the F_i's are equal, say $F_4 = F_5 = \dots F_n = F$, we shall write (3.4) as

$$P(F_1, F_2, F_3, F^{n-3}). \tag{3.5}$$

Using the Bianchi identities for the field strength,

$$dF = FA - AF, \tag{3.6}$$

one shows easily that

$$dP(F^n) = 0. \tag{3.7}$$

Actually one can write

$$P(F^n) = d\omega_{2n-1}(A, F), \tag{3.8}$$

where the $(2n-1)$-form is given by

$$\omega_{2n-1}(A, F) = n\int_0^1 dt\, P(A, F_t^{n-1}), \tag{3.9}$$

with

$$F_t = t\,dA + t^2A^2 = tF + (t^2 - t)A^2. \tag{3.10}$$

The consistent non-singlet anomaly is obtained as follows. Introduce an odd (anticommuting) Lie-algebra valued element v and an infinitesimal gauge transformation $\mathcal{g}$

$$\mathcal{g}A = -Dv = -dv - \{A, v\},$$

$$\mathcal{g}F = Fv - vF,$$

$$\mathcal{g}v = -v^2, \tag{3.11}$$

which satisfies

$$\mathcal{g}^2 = d\mathcal{g} + \mathcal{g}d = d^2 = 0. \tag{3.12}$$

$\mathcal{J}$ is the generator of a Becchi-Rouet-Stora transformation [22]. If we introduce

$$\mathcal{A} = A + v, \tag{3.13}$$

and a corresponding field strength

$$\mathcal{F} = (\mathrm{d} + \mathcal{J})\mathcal{A} + \mathcal{A}^2 \tag{3.14}$$

we find easily that

$$\mathcal{F} = F. \tag{3.15}$$

Therefore

$$(\mathrm{d} + \mathcal{J})\omega_{2n-1}(A + v, F) = P(F^n) = \mathrm{d}\omega_{2n-1}(A, F). \tag{3.16}$$

Let us expand in powers of v

$$\omega_{2n-1}(A + v, F) = \omega^0_{2n-1} + \omega^1_{2n-2} + \cdots + \omega^{2n-1}_0, \tag{3.17}$$

where the superscript indicates the power of v and the subscript the degree of the form. Eq. (3.16) implies a set of relations

$$\begin{aligned} &\mathcal{J}\omega^0_{2n-1} + \mathrm{d}\omega^1_{2n-2} = 0, \\ &\mathcal{J}\omega^1_{2n-2} + \mathrm{d}\omega^2_{2n-3} = 0, \\ &\qquad \cdots \\ &\mathcal{J}\omega^{2n-2}_1 + \mathrm{d}\omega^{2n-1}_0 = 0, \\ &\qquad \mathcal{J}\omega^{2n-1}_0 = 0. \end{aligned} \tag{3.18}$$

The consistent anomaly is given by the integral* of ω^1_{2n-2}. The consistency condition, which can be written as

$$\mathcal{J}\int \omega^1_{2n-2} = 0, \tag{3.19}$$

follows from the second equation of (3.18), the second terms integrate to zero. One can derive a convenient explicit formula for the anomaly [11]

$$\int \omega^1_{2n-2} = n(n-1)\int_0^1 \mathrm{d}t\,(1-t)\int P(\mathrm{d}v, A, F_t^{n-2}). \tag{3.20}$$

* Actually, if the polynomial P is normalized as in the index formula, so that it integrates to an integer, the correct normalization of the anomaly requires an extra factor 2π (see e.g. ref. [12]), or $2\pi i$ in euclidean space.

In the gauge transformation (3.11) the infinitesimal parameter is odd (anticommuting) and transforms like a Faddeev-Popov ghost. If one prefers, one can rewrite the consistency condition in terms of gauge transformations T_Λ with even (commuting) infinitesimal parameter Λ

$$T_\Lambda A = D\Lambda = \mathrm{d}\Lambda + [A, \Lambda],$$

$$T_\Lambda F = [F, \Lambda], \tag{3.21}$$

and T_Λ does not operate on the parameter itself. The anomaly, with Λ replaced for v, is a linear functional of Λ. Denoting it with $\Lambda \cdot G[A, F]$ (the dot indicates integration as well as summation over internal symmetry indices), it satisfies the consistency condition

$$T_\Lambda(\Lambda' \cdot G) - T_{\Lambda'}(\Lambda \cdot G) = [\Lambda, \Lambda'] \cdot G, \tag{3.22}$$

which is equivalent to (3.19). This is the form used in sect. 2; it follows from the definition

$$T_\Lambda W[A] = \Lambda \cdot G, \tag{3.23}$$

and justifies the above construction. If we define the current $(n-1)$-form (which is dual to the usual current vector J^μ)

$$J = \frac{\delta W}{\delta A}, \tag{3.24}$$

(3.23) can be written as

$$\Lambda \cdot DJ \equiv \Lambda \cdot (\mathrm{d}J + \{A, J\}) = \Lambda \cdot G \tag{3.25}$$

(remember that J is odd).

How does J transform under gauge transformations? As explained in sect. 2 we evaluate in two ways the commutator

$$(\delta T_\Lambda - T_\Lambda \delta) W[A], \tag{3.26}$$

where δ is defined by

$$\delta A = B, \qquad \delta F = DB = \mathrm{d}B + \{A, B\}. \tag{3.27}$$

Here, the increment B is odd and the operation δ is even. Since

$$T_\Lambda = D\Lambda \cdot \frac{\delta}{\delta A} = (\mathrm{d}\Lambda + [A, \Lambda]) \cdot \frac{\delta}{\delta A}, \tag{3.28}$$

the commutator equals

$$[B,\Lambda]\cdot\frac{\delta W}{\delta A}=[B,\Lambda]\cdot J. \tag{3.29}$$

On the other hand, using (3.23), and again (3.24), the commutator equals

$$\delta(\Lambda\cdot G)-T_\Lambda(B\cdot J). \tag{3.30}$$

Equating (3.29) with (3.30) we obtain

$$\begin{aligned}T_\Lambda(B\cdot J)&=-[B,\Lambda]\cdot J+\delta(\Lambda\cdot G)\\&=-B\cdot[\Lambda,J]+\delta(\Lambda\cdot G).\end{aligned} \tag{3.31}$$

The first term on the right-hand side would be the covariant transformation law appropriate to the adjoint representation. When there is an anomaly the second term shows that J does not transform covariantly. In (3.31) B is taken not to change under the gauge transformation generated by T_Λ. If instead we stipulate that B transforms according to the adjoint representation

$$T_\Lambda B=[B,\Lambda], \tag{3.32}$$

(3.31) becomes simply

$$T_\Lambda(B\cdot J)=\delta(\Lambda\cdot G). \tag{3.33}$$

Together with (3.32) and

$$\delta\Lambda=0, \tag{3.34}$$

(3.33) completely specifies the transformation law of the current J in terms of the anomaly $\Lambda\cdot G[A,F]$.

Is it possible to find a *local* $(\nu-1)$-form X such that the new current

$$\tilde{J}=J+X \tag{3.35}$$

transforms covariantly? This means

$$T_\Lambda(B\cdot\tilde{J})=0 \tag{3.36}$$

and therefore we must require

$$T_\Lambda(B\cdot X)=-\delta(\Lambda\cdot G). \tag{3.37}$$

This equation for X can be rewritten in terms of the anticommuting parameter v, instead of Λ,

$$\mathscr{G}(B\cdot X)=\delta\int\omega^1_{2n-2}(v,A,F), \tag{3.38}$$

with

$$\delta v=0. \tag{3.39}$$

Here we have used the fact that the anomaly is given by

$$v\cdot G[A,F]=\int\omega^1_{2n-2}(v,A,F). \tag{3.40}$$

Now, it is very easy to solve (3.38) in general. We use the relation, explained in appendix A,

$$\delta=\mathrm{d}l+l\mathrm{d}, \tag{3.41}$$

where δ is defined by (3.27), (3.39), d is the exterior differentiation, and the odd operation l is given by

$$lA=0,\qquad lF=B,\qquad lv=0. \tag{3.42}$$

Applying (3.41) to ω^1_{2n-2} we find

$$\begin{aligned}\delta\omega^1_{2n-2}&=\mathrm{d}(l\omega^1_{2n-2})+l\,\mathrm{d}\omega^1_{2n-2}\\&=\mathrm{d}(l\omega^1_{2n-2})-l\mathscr{G}\omega^0_{2n-1},\end{aligned} \tag{3.43}$$

where we have used (3.18). Now, the operators l and $\mathscr{G}$ anticommute

$$l\mathscr{G}+\mathscr{G}l=0. \tag{3.44}$$

Upon integration over (compactified) space-time, the first term on the right-hand side of (3.43) vanishes and we obtain finally

$$\delta\int\omega^1_{2n-2}=\mathscr{G}\int l\omega^0_{2n-1}. \tag{3.45}$$

Clearly we can drop the superscript zero on the right-hand side. Comparing with the equation for X, (3.38), we see that it is solved by

$$B\cdot X=\int l\omega_{2n-1}. \tag{3.46}$$

The explicit formula (3.9) for ω_{2n-1} can be used to find an explicit formula for X, since the operator l is easy to apply and, from (3.10),

$$lF_t = tB. \tag{3.47}$$

One finds easily

$$B \cdot X = n(n-1)\int_0^1 \mathrm{d}t\, t \int P(B, A, F_t^{n-2}). \tag{3.48}$$

For instance, for an internal symmetry such as SU(N) in four dimensions, we start from

$$P(F^3) = c\,\mathrm{Tr}\,F^3 \tag{3.49}$$

(where c is a known constant). Then

$$\omega_5 = c\,\mathrm{Tr}\left(F^2 A - \tfrac{1}{2}FA^3 + \tfrac{1}{10}A^5\right), \tag{3.50}$$

and applying l directly,

$$\begin{aligned} l\omega_5 &= c\,\mathrm{Tr}\left[B\left(FA + AF - \tfrac{1}{2}A^3\right)\right] \\ &= c\,\mathrm{Tr}\left[B\left(\mathrm{d}A\,A + A\,\mathrm{d}A + \tfrac{3}{2}A^3\right)\right], \end{aligned} \tag{3.51}$$

which gives

$$X = c\left(\mathrm{d}A\,A + A\,\mathrm{d}A + \tfrac{3}{2}A^3\right). \tag{3.52}$$

Alternatively one can use (3.48) with exactly the same result:

$$\mathrm{Tr}(BX) = 6c\,\mathrm{STr}\left[BA\left(\tfrac{1}{3}F - \tfrac{1}{12}A^2\right)\right], \tag{3.53}$$

where STr is the totally symmetrized trace (see ref. [11]).

Since the current $\tilde{J} = J + X$ transforms covariantly, its covariant divergence must also be covariant. In order to compute it we need $DX = \mathrm{d}X + \{A, X\}$. The simplest way to obtain this is to observe that, integrating by parts,

$$v \cdot DX = Dv \cdot X \tag{3.54}$$

(remember that v is odd), and it would seem that the right-hand side can be obtained directly from (3.48) just by making the substitution

$$B = Dv. \tag{3.55}$$

Strictly speaking this is not allowed since both B and v are odd and so is the operation D. To be precise we must first rewrite (3.48) as

$$C\cdot X=n(n-1)\int_0^1 \mathrm{d}t\,t\int P\left(C,A,F_t^{\,n-2}\right), \tag{3.56}$$

where C is even. Actually (3.56) follows from (3.48), and vice versa. Now we can set correctly, in (3.56),

$$C=Dv\,, \tag{3.57}$$

which gives

$$Dv\cdot X=n(n-1)\int_0^1 \mathrm{d}t\,t\int P\left(\mathrm{d}v+\{A,v\},A,F_t^{\,n-2}\right). \tag{3.58}$$

We also know that

$$v\cdot DJ=Dv\cdot J=n(n-1)\int_0^1 \mathrm{d}t\,(1-t)\int P\left(\mathrm{d}v,A,F_t^{\,n-2}\right), \tag{3.59}$$

where we have used the explicit form (3.20) for the anomaly. Adding (3.58) and (3.59) we obtain

$$\begin{aligned} v\cdot D\tilde{J} &= n(n-1)\int_0^1 \mathrm{d}t\int P\left(\mathrm{d}v+t\{A,v\},A,F_t^{\,n-2}\right) \\ &= n(n-1)\int_0^1 \mathrm{d}t\int P\left(v,\mathrm{d}A+t\{A,A\},F_t^{\,n-2}\right) \\ &= n\int_0^1 \mathrm{d}t\,\frac{\mathrm{d}}{\mathrm{d}t}\int P\left(v,F_t^{\,n-1}\right) \\ &= n\int P\left(v,F^{n-1}\right). \end{aligned} \tag{3.60}$$

In going from the first to the second form of this expression we have used the invariance of the symmetric polynomial P. The last expression (3.60) shows the covariant form of the anomaly. The result should be compared with (3.59) or equivalently

$$v\cdot DJ=\int \omega_{2n-2}^{1}(v,A,F). \tag{3.61}$$

Now, it is clearly

$$\omega_{2n-1}(A,F)=P\left(A,F^{n-1}\right)+\cdots \tag{3.62}$$

where the dots denote higher non-linear terms. This implies that

$$\omega^1_{2n-2}(v, A, F) = P(v, F^{n-1}) + \cdots . \tag{3.63}$$

Therefore the leading (least non-linear) term in (3.60) is n times larger than the leading term in (3.61). The relation between n and the dimension ν of space-time is

$$\nu + 2 = 2n . \tag{3.64}$$

As mentioned in the introduction, this factor can be understood diagrammatically as a result of Bose symmetrization.

4. Purely gravitational anomalies

Infinitesimal Einstein transformations are specified in terms of infinitesimal parameters $\xi^\mu(x)$ and operate on tensors as Lie derivatives. For instance, on a scalar field $A(x)$

$$E_\xi A = \xi^\mu \partial_\mu A , \qquad \partial_\mu = \frac{\partial}{\partial x^\mu} , \tag{4.1}$$

while on the metric tensor $g_{\mu\nu}(x)$

$$\begin{aligned} E_\xi g_{\mu\nu} &= \xi^\lambda \partial_\lambda g_{\mu\nu} + \partial_\mu \xi^\lambda g_{\lambda\nu} + \partial_\nu \xi^\lambda g_{\mu\lambda} \\ &= D_\mu \xi_\nu + D_\nu \xi_\mu . \end{aligned} \tag{4.2}$$

They satisfy the commutation relations

$$[E_{\xi_1}, E_{\xi_2}] = E_{[\xi_2, \xi_1]} , \tag{4.3}$$

where

$$([\xi_1, \xi_2])^\mu = \xi_1^\lambda \partial_\lambda \xi_2^\mu - \xi_2^\lambda \partial_\lambda \xi_1^\mu . \tag{4.4}$$

If the connected vacuum functional in an external gravitational field $W[g_{\mu\nu}]$ is not Einstein invariant

$$E_\xi W = H_\xi , \tag{4.5}$$

the anomaly H_ξ must satisfy the consistency condition

$$E_{\xi_1} H_{\xi_2} - E_{\xi_2} H_{\xi_1} = H_{[\xi_2, \xi_1]} , \tag{4.6}$$

which is the analogue of (2.4).

It is not difficult to find a solution of the consistency condition (4.6) in terms of the form $\omega^1_{2n-2}(v, A, F)$ which gives the anomaly in the case of gauge theories. In differential geometry the Levi-Civita connection $\Gamma_{\lambda\mu}{}^{\rho}$ plays the role of gauge potential and the Riemann tensor $R_{\nu\lambda\mu}{}^{\rho}$ the role of field strength. If we introduce the 1-forms

$$(\Gamma)_{\mu}{}^{\rho} = \Gamma_{\lambda\mu}{}^{\rho} \mathrm{d}x^{\lambda}, \tag{4.7}$$

the Riemann tensor is given by the 2-forms

$$(R)_{\mu}{}^{\rho} = (\mathrm{d}\Gamma + \Gamma^2)_{\mu}{}^{\rho} = \tfrac{1}{2} R_{\nu\lambda\mu}{}^{\rho} \mathrm{d}x^{\nu} \mathrm{d}x^{\lambda}, \tag{4.8}$$

with

$$R_{\nu\lambda\mu}{}^{\rho} = \partial_{\nu}\Gamma_{\lambda\mu}{}^{\rho} - \partial_{\lambda}\Gamma_{\nu\mu}{}^{\rho} + \Gamma_{\nu\mu}{}^{\sigma}\Gamma_{\lambda\sigma}{}^{\rho} - \Gamma_{\lambda\mu}{}^{\sigma}\Gamma_{\nu\sigma}{}^{\rho}. \tag{4.9}$$

Under an infinitesimal Einstein transformation the connection transforms as

$$E_{\xi}\Gamma_{\lambda\mu}{}^{\rho} = \xi^{\sigma}\partial_{\sigma}\Gamma_{\lambda\mu}{}^{\rho} + \partial_{\lambda}\xi^{\sigma}\Gamma_{\sigma\mu}{}^{\rho} + \partial_{\mu}\xi^{\sigma}\Gamma_{\lambda\sigma}{}^{\rho} - \Gamma_{\lambda\mu}{}^{\sigma}\partial_{\sigma}\xi^{\rho} - \partial_{\lambda}\partial_{\mu}\xi^{\rho}. \tag{4.10}$$

The last three terms have exactly the form of a gauge transformation with infinitesimal gauge parameter

$$\Lambda_{\mu}{}^{\rho} = -\partial_{\mu}\xi^{\rho}, \tag{4.11}$$

while the first two terms have the form of a Lie derivative of $\Gamma_{\lambda\mu}{}^{\rho}$, treated as a vector with lower index λ and ignoring the other two indices μ and ρ. In terms of the 1-form (4.7) we can write

$$E_{\xi}\Gamma = \mathfrak{L}_{\xi}\Gamma + T_{\Lambda}\Gamma, \tag{4.12}$$

where

$$T_{\Lambda}\Gamma = D\Lambda = \mathrm{d}\Lambda + [\Gamma, \Lambda] \tag{4.13}$$

is a gauge transformation with infinitesimal parameter (4.11). In (4.12) the Lie derivative is defined as usual on forms, to operate only on those indices which are saturated with differentials, so that it corresponds to only the first two terms in (4.10). The well-known formula applies

$$\mathfrak{L}_{\xi} = \mathrm{d}i_{\xi} + i_{\xi}\mathrm{d}, \tag{4.14}$$

where i_{ξ} is the (odd) inner product operator, for instance

$$i_{\xi}\left(\Gamma_{\lambda\mu}{}^{\rho}\mathrm{d}x^{\lambda}\right) = \Gamma_{\lambda\mu}{}^{\rho}\xi^{\lambda}. \tag{4.15}$$

In general, for a form of higher degree, i_{ξ} substitutes the vector ξ^{μ} for each differential (one after the other), for instance

$$i_\xi R_{\nu\lambda}\,dx^\nu dx^\lambda = R_{\nu\lambda}\xi^\nu dx^\lambda - R_{\nu\lambda}\,dx^\nu \xi^\lambda = 2R_{\nu\lambda}\xi^\nu dx^\lambda,$$

$$R_{\nu\lambda} = -R_{\lambda\nu}. \tag{4.16}$$

The effect of an Einstein transformation on the Riemann curvature 2-form is given by

$$E_\xi R = \mathfrak{L}_\xi R + T_\Lambda R, \tag{4.17}$$

where now

$$T_\Lambda R = [R, \Lambda]. \tag{4.18}$$

In a space-time of ν dimensions a ν-form has maximal degree and its differential vanishes. Therefore (4.14) becomes

$$\mathfrak{L}_\xi \omega_\nu = d(i_\xi \omega_\nu) \tag{4.19}$$

and the integral vanishes (with suitable boundary conditions)

$$\mathfrak{L}_\xi \int \omega_\nu = \int d(i_\xi \omega_\nu) = 0. \tag{4.20}$$

In the dual description, more familiar to physicists, a ν-form corresponds to a density $\mathfrak{D}$, (4.19) corresponds to

$$\mathfrak{L}_\xi \mathfrak{D} = \partial_\mu(\xi^\mu \mathfrak{D}), \tag{4.21}$$

and (4.20) corresponds to

$$\mathfrak{L}_\xi \int \mathfrak{D}\, dx = \int \partial_\mu(\xi^\mu \mathfrak{D})\, dx = 0. \tag{4.22}$$

The relation between Einstein and gauge transformations expressed by (4.12) and (4.17) shows that one can reduce the problem of finding consistent Einstein anomalies to that of finding consistent gauge anomalies. Indeed the gauge anomaly, in the form (3.40), immediately gives a consistent Einstein anomaly in the form

$$H_\xi = \Lambda \cdot G[\Gamma, R] = -\int \partial_\rho \xi^\mu G_\mu{}^\rho(\Gamma, R), \tag{4.23}$$

with the same function $G[\Gamma, R]$. Indeed

$$E_{\xi_1} H_{\xi_2} = (\mathfrak{L}_{\xi_1} + T_{\Lambda_1})\Lambda_2 \cdot G = -\int \partial_\rho \xi_2^\nu \partial_\lambda(\xi_1^\lambda G_\nu^\rho) + T_{\Lambda_1}\Lambda_2 \cdot G, \tag{4.24}$$

so that

$$E_{\xi_1}H_{\xi_2} - E_{\xi_2}H_{\xi_1} = \int \left(\xi_1^\lambda \partial_\lambda \partial_\rho \xi_2^\nu - \xi_2^\lambda \partial_\lambda \partial_\rho \xi_1^\nu\right) G_\nu{}^\rho + [\Lambda_1, \Lambda_2] \cdot G. \tag{4.25}$$

Finally, using (4.11), the right-hand side of (4.25) becomes equal to

$$-\int \partial_\rho \left(\xi_2^\lambda \partial_\lambda \xi_1^\nu - \xi_1^\lambda \partial_\lambda \xi_2^\nu\right) G_\nu{}^\rho = H_{\xi_2 \cdot \partial \xi_1 - \xi_1 \cdot \partial \xi_2}. \tag{4.26}$$

Observe that the consistent gravitational anomaly given by (4.23) does not depend explicitly on the metric, but only on the connection (and through it on the metric) even though the connected vacuum functional $W[g_{\mu\nu}]$ cannot be expressed in terms of the connection alone. For instance, in two dimensions, up to a known numerical factor, the consistent gravitational anomaly is

$$H_\xi \propto -\int \partial_\rho \xi^\mu \partial_\nu \Gamma_{\lambda\mu}{}^\rho \varepsilon^{\nu\lambda} \mathrm{d}^2 x, \tag{4.27}$$

which corresponds to the non-abelian anomaly [see (1.1)]

$$\Lambda \cdot G \propto \int \mathrm{Tr}(\Lambda \partial_\nu A_\lambda) \varepsilon^{\nu\lambda} \mathrm{d}^2 x. \tag{4.28}$$

In higher dimensions the consistent gravitational anomaly is just as easily written, once the appropriate invariant polynomial (3.4) is known. In a Riemann space the Riemann tensor (4.8) is antisymmetric

$$R_{\mu\rho} = -R_{\rho\mu}. \tag{4.29}$$

As a consequence the invariant symmetric polynomial (2.4) vanishes except for even n

$$n = 2m, \tag{4.30}$$

which corresponds to a space-time dimension

$$\nu = 2n - 2 = 4m - 2 \tag{4.31}$$

(see the analogous argument below, leading to (6.12) and (6.13)).

In terms of the energy momentum "tensor"

$$\theta^{\mu\nu} = 2\frac{\delta W}{\delta g_{\mu\nu}}, \tag{4.32}$$

(4.5) can be written

$$\int \xi_\nu D_\mu \theta^{\mu\nu} \mathrm{d}x = -H_\xi = -\Lambda \cdot G, \tag{4.33}$$

or

$$D_\mu \theta^\mu{}_\nu = \partial_\mu G_\nu{}^\mu(\Gamma, R). \tag{4.34}$$

Here the covariant derivative is that appropriate to a symmetric tensor density

$$D_\mu \theta^{\mu\nu} = \partial_\mu \theta^{\mu\nu} - \theta^{\mu\rho} \Gamma_{\mu\rho}{}^\nu, \tag{4.35}$$

but $\theta^{\mu\nu}$ is not a tensor density when the anomaly does not vanish. How does it transform under Einstein transformations? We follow an argument similar to that given in sect. 2. Evaluate in two different ways the commutator

$$(E_\xi \delta_\varphi - \delta_\varphi E_\xi) W[g_{\mu\nu}], \tag{4.36}$$

where we define

$$\delta_\varphi = \int \varphi_{\mu\nu} \frac{\delta}{\delta g_{\mu\nu}} \mathrm{d}x, \tag{4.37}$$

an operation which gives $g_{\mu\nu}$ an arbitrary symmetric increment $\varphi_{\mu\nu}$. Since

$$E_\xi = \int (E_\xi g_{\mu\nu}) \frac{\delta}{\delta g_{\mu\nu}} \mathrm{d}x, \tag{4.38}$$

the commutator equals

$$-\int (E_\xi \varphi_{\mu\nu}) \frac{\delta W}{\delta g_{\mu\nu}} \mathrm{d}x. \tag{4.39}$$

On the other hand, using (4.5), the commutator equals

$$\int \varphi_{\mu\nu} E_\xi \frac{\delta W}{\delta g_{\mu\nu}} \mathrm{d}x - \delta_\varphi H_\xi. \tag{4.40}$$

Equating (4.39) and (4.40), and using (4.32), we obtain

$$\int \{ \varphi_{\mu\nu} E_\xi \theta^{\mu\nu} + (E_\xi \varphi_{\mu\nu}) \theta^{\mu\nu} \} \mathrm{d}x = 2\delta_\varphi H_\xi, \tag{4.41}$$

or

$$E_\xi \int \varphi_{\mu\nu} \theta^{\mu\nu} \, \mathrm{d}x = 2\delta_\varphi H_\xi . \tag{4.42}$$

In this equation E_ξ transforms $\varphi_{\mu\nu}$ like a tensor, while $\theta^{\mu\nu}[g_{\rho\sigma}]$ transforms as it follows from the transformation law of $g_{\rho\sigma}$ [eq. (4.2)]. If the anomaly on the right-hand side were zero, the left-hand side would be Einstein invariant, i.e. $\theta^{\mu\nu}$ would transform like a tensor density.

5. The covariant energy momentum tensor and its covariant anomaly

Is it possible to find a symmetric local $Y^{\mu\nu}$ such that the new energy momentum tensor

$$\tilde{\theta}^{\mu\nu} = \theta^{\mu\nu} + Y^{\mu\nu} \tag{5.1}$$

transforms like a tensor density? This means

$$E_\xi \int \varphi_{\mu\nu} \tilde{\theta}^{\mu\nu} \, \mathrm{d}x = 0 , \tag{5.2}$$

and therefore we must require

$$E_\xi \int \varphi_{\mu\nu} Y^{\mu\nu} \mathrm{d}x = -2\delta_\varphi H_\xi . \tag{5.3}$$

A solution $Y^{\mu\nu}$ of this equation can be easily found in terms of the solution of the analogous problem discussed in sect. 3. There we found a $(\nu - 1)$-form X which satisfied (3.37)

$$T_\Lambda B \cdot X = -\delta\Lambda \cdot G , \tag{3.37}$$

where

$$\delta = \int B \cdot \frac{\delta}{\delta A} . \tag{5.4}$$

In view of the relation (4.23) between G and H_ξ, it is clear that we can use (3.37) with the substitutions $A \to \Gamma$, $F \to R$ and taking also (4.11) and

$$B = B[\varphi] = \delta_\varphi \Gamma , \tag{5.5}$$

since then $\delta \to \delta_\varphi$ in the right-hand side of (3.37). The result is that $Y^{\mu\nu}$ is given by

$$2B[\varphi] \cdot X = \int \varphi_{\mu\nu} Y^{\mu\nu} \mathrm{d}x . \tag{5.6}$$

In order to make this expression more explicit we observe that the standard expression for the Christoffel connection

$$\Gamma_{\lambda\mu}{}^{\rho} = \tfrac{1}{2} g^{\rho\sigma} (\partial_\sigma g_{\lambda\mu} - \partial_\lambda g_{\mu\sigma} - \partial_\mu g_{\lambda\sigma}) , \tag{5.7}$$

implies

$$B_{\lambda\mu}{}^{\rho}(\varphi) = \tfrac{1}{2} g^{\rho\sigma} (D_\sigma \varphi_{\lambda\mu} - D_\lambda \varphi_{\mu\sigma} - D_\mu \varphi_{\lambda\sigma}) ,$$

$$(B[\varphi])_\mu{}^{\rho} = B_{\lambda\mu}{}^{\rho}(\varphi)\, \mathrm{d}x^\lambda . \tag{5.8}$$

Substituting into (5.6) and integrating by parts the covariant derivatives one finds easily the explicit form of $Y^{\mu\nu}$ in terms of that of X, but we shall not carry it out. We point out that the argument which leads to (5.6) is based on the identification

$$E_\xi = \mathfrak{L}_\xi + T_\Lambda , \tag{5.9}$$

and on the fact that $\mathfrak{L}_\xi$ gives zero when applied to the quantities we are interested in.

Since the new energy momentum tensor is really a tensor density, its covariant divergence will also be covariant. We can work it out without unnecessary computations if we observe that

$$\int (D_\mu \xi_\nu + D_\nu \xi_\mu) Y^{\mu\nu} \mathrm{d}x = -2 \int \xi_\nu D_\mu Y^{\mu\nu} \mathrm{d}x . \tag{5.10}$$

Therefore we can use (5.6) with the substitution

$$\varphi_{\mu\nu} = D_\mu \xi_\nu + D_\nu \xi_\mu = E_\xi g_{\mu\nu} , \tag{5.11}$$

which gives

$$\delta_\varphi = E_\xi , \tag{5.12}$$

and from (5.5), (4.12) and (4.13),

$$B[\varphi_{\mu\nu}] = E_\xi \Gamma = \mathfrak{L}_\xi \Gamma + D\Lambda . \tag{5.13}$$

We obtain in this way

$$\int \xi_\nu D_\mu Y^{\mu\nu} \mathrm{d}x = -\mathfrak{L}_\xi \Gamma \cdot X + \Lambda \cdot DX . \tag{5.14}$$

We shall now use the result (3.60) which represents the solution of the analogous problem for the non-abelian current. Eq. (3.60) can be written as (use Λ instead of v)

$$\Lambda \cdot DX - \Lambda \cdot G = n \int P(\Lambda, F^{n-1}). \tag{5.15}$$

Combining (4.33), (5.14) and (5.15) with the definition (5.1) we obtain

$$\int \xi_\nu D_\mu \tilde{\theta}^{\mu\nu} \mathrm{d}x = n \int P(\Lambda, R^{n-1}) - \mathcal{L}_\xi \Gamma \cdot X. \tag{5.16}$$

The right-hand side is still not obviously covariant but the two terms can be combined because, as we shall show below,

$$\mathcal{L}_\xi \Gamma \cdot X = -nP\left(i_\xi \Gamma, R^{n-1}\right). \tag{5.17}$$

Since

$$\begin{aligned} \Lambda_\mu{}^\nu + \left(i_\xi \Gamma\right)_\mu{}^\nu &= -\partial_\mu \xi^\nu + \xi^\lambda \Gamma_{\lambda\mu}{}^\nu \\ &= -D_\mu \xi^\nu \equiv M_\mu{}^\nu, \end{aligned} \tag{5.18}$$

we finally obtain the fully covariant result

$$\int \xi_\nu D_\mu \tilde{\theta}^{\mu\nu} \mathrm{d}x = n \int P(M, R^{n-1}). \tag{5.19}$$

Note that, again, the leading (least non-linear) term in (5.19) is $n = \frac{1}{2}(\nu + 2)$ times larger than the corresponding term on the right-hand side of (4.33). The covariant form of the anomaly given by the right-hand side of (5.19) is also expressed in terms of the connection alone. The metric does not occur explicitly, just as it does not in the consistent form.

It remains for us to prove (5.17). In ν space-time dimensions, $(\nu + 1)$ forms vanish, therefore

$$\Gamma^2 \cdot X = \mathrm{d}\Gamma \cdot X = 0. \tag{5.20}$$

If we apply to these forms the operator i_ξ, where the vector ξ^μ is tangent to the ν-dimensional space-time manifold, we still get zero

$$\begin{aligned} 0 &= i_\xi(\Gamma^2 \cdot X) = (i_\xi \Gamma \Gamma) \cdot X - (\Gamma i_\xi \Gamma) \cdot X + \Gamma^2 \cdot i_\xi X \\ &= i_\xi \Gamma \cdot (\Gamma X + X \Gamma) + \Gamma^2 \cdot i_\xi X, \end{aligned} \tag{5.21}$$

$$0 = i_\xi(\mathrm{d}\Gamma \cdot X) = (i_\xi \mathrm{d}\Gamma) \cdot X + \mathrm{d}\Gamma \cdot i_\xi X. \tag{5.22}$$

Now

$$\mathfrak{L}_\xi \Gamma \cdot X = i_\xi \mathrm{d}\Gamma \cdot X + \mathrm{d} i_\xi \Gamma \cdot X$$

$$= -\mathrm{d}\Gamma \cdot i_\xi X - i_\xi \Gamma \cdot \mathrm{d}X. \tag{5.23}$$

Subtracting (5.21) from (5.23) we obtain

$$\mathfrak{L}_\xi \Gamma \cdot X = -R \cdot i_\xi X - i_\xi \Gamma \cdot DX. \tag{5.24}$$

Now, again in $\nu = 2n-2$ dimensions, the $(2n-1)$-form

$$\omega_{2n-1}(\Gamma, R) = 0 \tag{5.25}$$

vanishes. Apply again i_ξ

$$0 = i_\xi \int \omega_{2n-1} = i_\xi \Gamma \cdot \frac{\delta}{\delta \Gamma} \int \omega_{2n-1} + i_\xi R \cdot \frac{\delta}{\delta R} \int \omega_{2n-1}$$

$$= i_\xi \Gamma \cdot G + i_\xi R \cdot X, \tag{5.26}$$

where we have used equations analogous to (3.17) and (3.46). This gives

$$R \cdot i_\xi X = i_\xi \Gamma \cdot G. \tag{5.27}$$

Combining (5.27) with (5.24) we obtain

$$\mathfrak{L}_\xi \Gamma \cdot X = -i_\xi \Gamma \cdot (DX + G), \tag{5.28}$$

and finally using (3.60), we prove (5.17). Observe that occasionally, in our derivations, we use results proven earlier for odd quantities or operations and apply them to even quantities or operations and vice versa. This is permissible if proper care is exercised and we leave it to the reader to be properly careful so as not to make sign mistakes.

6. Einstein anomalies and Lorentz anomalies are equivalent

As explained in the introduction, local Lorentz invariance could also be spoiled by anomalies. In this case the connected vacuum functional must be considered as a functional of the vielbein field $e_{\mu a}$ and cannot be assumed to depend on the metric tensor. Let us work in the euclidean. Under local rotations of infinitesimal parameter $\theta_{ab} = -\theta_{ba}$ the vielbein field transforms as

$$L_\theta e_{\mu a} = e_{\mu b} \theta_{ba}, \tag{6.1}$$

while under Einstein transformations we have

$$E_\xi e_{\mu a} = \xi^\lambda \partial_\lambda e_{\mu a} + \partial_\mu \xi^\lambda e_{\lambda a}. \tag{6.2}$$

It is easy to see that the full Lie algebra consists of (4.3), (4.4) together with

$$\left[L_{\theta_1}, L_{\theta_2}\right] = L_{[\theta_1, \theta_2]}, \tag{6.3}$$

$$\left[L_\theta, E_\xi\right] = L_{\xi \cdot \partial\theta}. \tag{6.4}$$

If there are Lorentz anomalies

$$L_\theta W = K_\theta, \tag{6.5}$$

they must satisfy the consistency conditions

$$L_{\theta_1} K_{\theta_2} - L_{\theta_2} K_{\theta_1} = K_{[\theta_1, \theta_2]}, \tag{6.6}$$

$$L_\theta H_\xi - E_\xi K_\theta = K_{\xi \cdot \partial\theta}. \tag{6.7}$$

It is consistent to assume that there are only Einstein anomalies ($K_\theta = 0$) and we have discussed this case in sect. 4. It is also consistent to assume that there are only rotational anomalies ($H_\xi = 0$). A consistent form for the rotational anomaly is easy to find, the orthogonal rotation group can be treated like an internal symmetry. The gauge potential is the Cartan-Weyl connection

$$\alpha_{ab} = -\alpha_{ba} = \alpha_{\mu ab} \mathrm{d}x^\mu, \tag{6.8}$$

and the field strength is the Riemann tensor referred to local orthonormal frames

$$R_{ab} = -R_{ba} = (\mathrm{d}\alpha + \alpha^2)_{ab} = \tfrac{1}{2} R_{\mu\nu ab} \mathrm{d}x^\mu \mathrm{d}x^\nu \tag{6.9}$$

(customarily the connection (6.8) is denoted by the letter ω; here we depart from the usual notation in order to avoid confusion with the forms ω of sect. 3). The solution of the consistency condition (6.6) can be written immediately in terms of (3.9) and (3.17)

$$K_\theta = \int \omega^1_{2n-2}(\theta, \alpha, R) = \theta \cdot G[\alpha, R]. \tag{6.10}$$

We note again that, because of the antisymmetry of the matrix R_{ab},

$$P(R^n) = (-1)^n P(R^n). \tag{6.11}$$

Therefore, in the case of the orthogonal group, the symmetric polynomial P will vanish unless n is even

$$n = 2m, \tag{6.12}$$

which corresponds to a space-time dimension

$$\nu = 2n - 2 = 4m - 2. \tag{6.13}$$

Only in these dimensions can there be rotational (Lorentz) anomalies.

Just as the Einstein anomaly (4.23) does not depend explicitly on the metric, but only through the connection Γ, so the rotational anomaly (6.10) does not depend explicitly on the vielbein, but only through the connection α. Indeed, the functional forms of the two anomalies are directly related. However, *there is* a vielbein field, and in this the theory of gravitation is not like other gauge theories, a fact which cannot be sufficiently stressed. Let us use matrix notation and denote by E the vielbein matrix $e_{\mu a}$. The field H defined by

$$E = \mathrm{e}^{H} \tag{6.14}$$

behaves, in a certain sense, like a Goldstone field for both Einstein transformations and local rotations. Under an infinitesimal Einstein transformation

$$E_{\xi} H = \xi^{\lambda} \partial_{\lambda} H + T_{\Lambda} H, \tag{6.15}$$

where $T_{\Lambda} H$ is defined by

$$T_{\Lambda} \mathrm{e}^{H} = -\Lambda \mathrm{e}^{H}, \tag{6.16}$$

and Λ is given by (4.11). The finite version of (6.16) is

$$\mathrm{e}^{T_{\Lambda}} \mathrm{e}^{H} = \mathrm{e}^{-\Lambda} \mathrm{e}^{H}. \tag{6.17}$$

Similarly, under a finite rotation

$$\mathrm{e}^{L_{\theta}} \mathrm{e}^{H} = \mathrm{e}^{H} \mathrm{e}^{\theta}. \tag{6.18}$$

This suggests that, using the vielbein field, one should be able to construct a local functional whose Einstein variation gives the Einstein anomaly and whose Lorentz variation gives the rotational anomaly. This is indeed possible by mimicking the solution of the anomalous chiral Ward identities obtained using a Goldstone field [6]. Define the functional

$$S[E, \Gamma] = \int_0^1 \mathrm{d}t \int_x \mathrm{Tr}(H G[\Gamma_t]), \tag{6.19}$$

where

$$\Gamma_t = e^{-tH}\Gamma e^{tH} + e^{-tH} d\, e^{tH}. \tag{6.20}$$

In appendix B we verify that

$$E_\xi S \doteq \int_x \partial_\mu \xi^\nu G_\nu{}^\mu[\Gamma] = -H_\xi. \tag{6.21}$$

On the other hand, one can express S in terms of E and α, instead of E and Γ. We recall the relation between the Christoffel and the Cartan connection

$$\Gamma = E\alpha E^{-1} + E\, d E^{-1}. \tag{6.22}$$

This implies that

$$\Gamma_t = e^{(1-t)H}\alpha\, e^{-(1-t)H} + e^{(1-t)H} d\, e^{-(1-t)H}. \tag{6.23}$$

Changing the integration variable from t to

$$\tau = 1 - t, \tag{6.24}$$

we see that

$$S[E,\Gamma] = \int_0^1 d\tau \int_x \mathrm{Tr}(HG[\alpha_\tau]) \equiv S'[E,\alpha], \tag{6.25}$$

where

$$\alpha_\tau = e^{\tau H}\alpha\, e^{-\tau H} + e^{\tau H} d\, e^{-\tau H}. \tag{6.26}$$

Using perfectly analogous arguments as for (6.21) one shows that

$$L_\theta S = L_\theta S' = K_\theta. \tag{6.27}$$

The functional S (or S') is local, in the sense that it is the integral of an expression constructed with derivatives of the vielbein and of the connection up to a finite order. It is highly non-linear and uniquely defined only for relatively weak fields. Nevertheless, it can be used to redefine the connected vacuum functional so as to eliminate either the Einstein anomalies (by changing W into $W+S$) or the Lorentz anomalies (by changing W into $W-S'$). In this sense Einstein and Lorentz anomalies are different aspects of the same thing. It seems convenient to choose the pure Lorentz anomaly (vanishing Einstein anomaly) as the canonical form of the gravitational anomaly: the formalism is then more directly related to the case of internal gauge symmetries and the absence of Einstein anomalies gives a more satisfactory geometrical picture*.

Finally, we remark that formulas (6.19) and (6.25) for the functional S' can be written in a more intrinsic form (see ref. [12]). We have preferred to use here the special choice of local coordinates of (6.19) and (6.25) in order to render manifest the locality in x of the functional.

7. Conclusion

An anomaly is a local expression which satisfies the consistency condition, but which is not the gauge variation of a local functional of the gauge potential, or the metric tensor, in the gravitational case. Here a local functional of certain fields means an integral over x of an expression constructed with the fields and their x derivatives up to some finite order. The consistent anomalies discussed in this paper satisfy both the above conditions. In order to show that they cannot be obtained from a local functional one has to enumerate all possible candidate expressions of the correct dimension and with the correct power of the fields and check that there is no combination which reproduces the anomaly when one performs a gauge variation. In general, the proof is rather cumbersome, but it can be considerably simplified by going over to the covariant form of the anomalies. Since the covariant current $\tilde{J}_\mu$ is obtained from the original current J_μ by adding to it a local expression, one can reduce the problem to that of finding a covariant current which is a local expression in the gauge potential and whose covariant divergence gives the covariant anomaly. Similarly, in the gravitational case, one can ask whether there exists a covariant energy momentum tensor (which means that it is really a tensor) constructed locally from the metric tensor and satisfying the anomalous equation. The number of possible candidates is greatly restricted by the condition that these quantities be tensors. A further restriction comes from the fact that the covariant anomaly has a known form possessing "abnormal parity", i.e. it is constructed with epsilon tensors (corresponding to it being an exterior form). This would require the

* Tom Banks has observed that it is possible (for instance in six space-time dimensions) to regularize the lagrangian for chiral spinors with a mass term which is Einstein covariant. Such a regularization, however, cannot be Lorentz covariant. Evaluated in this way the anomaly would naturally appear as a pure Lorentz anomaly.

In the case of an internal gauge symmetry based on the compact Lie group $\mathcal{G}$, one knows [12] that the existence of anomalies requires that the homotopy group $\Pi_{\nu+1}[\mathcal{G}]$ contain the group Z of the integers (here ν is the dimensions of space-time). For the orthogonal group of local rotations $O(\nu)$, one is then led to consider $\Pi_{\nu+1}[O(\nu)]$. Now, it is known that this homotopy group contains Z only for $\nu = 4m - 2$ (m an integer > 1) and otherwise is finite (see e.g. ref. [23]). So, one expects that only for these space-time dimensions can there be a topological Lorentz anomaly. This condition is the same as given in (6.13). For $\nu = 2$ there is no topological anomaly, but there still is a Lorentz anomaly, in the local sense discussed in this paper. The connection between the local and the topological meaning of the anomalies is discussed in ref. [17].

current and the energy momentum tensor also to have abnormal parity, since no epsilon tensors can be generated by taking the covariant divergence. With these restrictions, it is not difficult to show that no such local quantities exist [24] (for the energy momentum tensor one must also use the fact that it is symmetric).

The authors wish to thank O. Alvarez and I. Singer for numerous illuminating discussions and L. Alvarez-Gaumé, E. Witten and J. Thierry-Mieg for useful correspondence. Discussions and correspondence with R. Stora have been very helpful and stimulating.

Some additional recent papers on anomalies are listed as refs. [25].

Appendix A

ALGEBRAIC STRUCTURE

In this appendix we collect and rederive, in part, the results and techniques described in refs. [11–13] and used in the main text of this paper. The reader will notice that the structure described below is completely algebraic. In the text we use the resulting formulas for the gauge potential 1-form A and the field strength 2-form F, and for forms which are functions of them. However, the arguments given in this appendix apply to any expression which is a polynomial in two free variables A and F (free means not restricted by algebraic relations), say with complex coefficients. In particular we do not assume that the polynomials are symmetric or invariant, nor do we assume that A and F commute or that they satisfy specific commutation relations. In addition to A and F we shall also use two more variables v and B. A, v and B are odd (anticommuting), F is even. On these variables we define the (odd) antiderivatives d, $\mathfrak{s}$ and l and the (even) derivative δ with the properties

$$\mathrm{d}A = F - A^2, \qquad \mathrm{d}F = FA - AF, \tag{A.1}$$

$$\mathfrak{s}F = Fv - vF, \qquad \mathfrak{s}v = -v^2, \qquad \mathfrak{s}B = -vB - Bv, \tag{A.2}$$

$$lA = 0, \qquad lF = B, \qquad lv = 0, \qquad lB = 0, \tag{A.3}$$

$$\delta A = B, \qquad \delta v = 0, \qquad \delta B = 0, \tag{A.4}$$

$$\mathfrak{s}A + \mathrm{d}v = -vA - Av, \tag{A.5}$$

$$\delta F - \mathrm{d}B = AB + BA. \tag{A.6}$$

The differentiation operators satisfy

$$\mathrm{d}^2 = l^2 = \mathcal{J}^2 = 0\,, \tag{A.7}$$

$$\mathrm{d}\mathcal{J} + \mathcal{J}d = l\mathcal{J} + \mathcal{J}l = 0\,, \tag{A.8}$$

$$l\mathrm{d} + \mathrm{d}l = \delta\,. \tag{A.9}$$

The algebraic consistency of all these relations (A.1) to (A.9) is not hard to verify. For instance, to see that $\mathrm{d}^2 = 0$, apply d^2 on A

$$\begin{aligned} \mathrm{d}(\mathrm{d}A) &= \mathrm{d}(F - A^2) = \mathrm{d}F - \mathrm{d}A\,A + A\,\mathrm{d}A \\ &= FA - AF - (F - A^2)A + A(F - A^2) = 0\,. \end{aligned} \tag{A.10}$$

Similarly, on F

$$\begin{aligned} \mathrm{d}(\mathrm{d}F) &= \mathrm{d}(FA - AF) = \mathrm{d}F\,A + F\,\mathrm{d}A - \mathrm{d}A\,F + A\,\mathrm{d}F \\ &= (FA - AF)A + F(F - A^2) - (F - A^2)F + A(FA - AF) \\ &= 0\,. \end{aligned} \tag{A.11}$$

Now apply d on (A.5). After a little algebra we find, using (A.1) and (A.5),

$$\mathrm{d}\mathcal{J}A + \mathrm{d}^2 v = -\mathcal{J}\,\mathrm{d}A\,. \tag{A.12}$$

This shows that $\mathrm{d}^2 = 0$ on v as well, provided d and $\mathcal{J}$ anticommute.

Let us also verify, in few cases, the important relation (A.9). On A

$$l(\mathrm{d}A) = l(F - A^2) = lF = B\,, \tag{A.13}$$

$$\mathrm{d}(lA) = 0\,. \tag{A.14}$$

Comparing the sum of (A.13) and (A.14) with (A.4), we see that (A.9) is valid on A. Let us check it on F:

$$l(\mathrm{d}F) = l(FA - AF) = BA + AB\,, \tag{A.15}$$

$$\mathrm{d}(lF) = \mathrm{d}B\,. \tag{A.16}$$

Therefore

$$(l\mathrm{d} + \mathrm{d}l)F = \mathrm{d}B + BA + AB = \delta F\,, \tag{A.17}$$

using (A.6).

Because of the properties of derivatives and antiderivatives all this extends immediately to polynomials in the variables A, F, v and B.

Appendix B

SOLUTION OF THE ANOMALY EQUATION

In this appendix we verify that there exists a local functional whose Einstein variation gives the Einstein anomaly and whose Lorentz variation gives the Lorentz anomaly. The functional is given by (6.19)

$$S[E,\Gamma]=\int_0^1 \mathrm{d}t \int_x \mathrm{Tr}(HG[\Gamma_t]), \tag{B.1}$$

where we use matrix notation and denote by E the vielbein matrix $e_{\mu\alpha}$. Then

$$E=\mathrm{e}^H, \tag{B.2}$$

$$\Gamma_t=\mathrm{e}^{-tH}\Gamma\mathrm{e}^{tH}+\mathrm{e}^{-tH}\mathrm{d}\,\mathrm{e}^{tH}. \tag{B.3}$$

An Einstein transformation is given by (5.9)

$$E_\xi=\mathcal{L}_\xi+T_\Lambda, \tag{B.4}$$

where

$$\Lambda_\mu{}^\nu=-\partial_\mu\xi^\nu, \tag{B.5}$$

and the effect of T_Λ on Γ is given by (4.13)

$$T_\Lambda\Gamma=\mathrm{d}\Lambda+[\Gamma,\Lambda]. \tag{B.6}$$

We see that (B.4) is valid also on E, if we take

$$T_\Lambda E=-\Lambda E, \tag{B.7}$$

$$\mathcal{L}_\xi E=\xi^\lambda\partial_\lambda E, \tag{B.8}$$

which agrees with (4.14)

$$\mathcal{L}_\xi=i_\xi\mathrm{d}+\mathrm{d}i_\xi, \tag{B.9}$$

if we treat E as a zero-form.

Now, since S is the integral of a form of maximum degree, $\mathfrak{L}_\xi$ applied to S gives zero. Therefore, we need only evaluate the effect of T_Λ. It is easy to verify that

$$T_\Lambda \Gamma_t = \mathrm{d}\Lambda_t + [\Gamma_t, \Lambda_t], \tag{B.10}$$

where

$$\Lambda_t \doteq \mathrm{e}^{-tH} \Lambda \,\mathrm{e}^{tH} + \mathrm{e}^{-tH} T_\Lambda \mathrm{e}^{tH}. \tag{B.11}$$

Observe that, from (B.2) and (B.7),

$$T_\Lambda \mathrm{e}^H = -\Lambda \,\mathrm{e}^H. \tag{B.12}$$

Therefore

$$\Lambda_0 = \Lambda, \qquad \Lambda_1 = 0, \tag{B.13}$$

$$\begin{aligned}\frac{\partial \Lambda_t}{\partial t} &= -H\mathrm{e}^{-tH}\Lambda\,\mathrm{e}^{tH} + \mathrm{e}^{-tH}\Lambda\,\mathrm{e}^{tH}H - H\mathrm{e}^{-tH}T_\Lambda \mathrm{e}^{tH} + \mathrm{e}^{-tH}T_\Lambda(\mathrm{e}^{tH}H) \\ &= \left[\mathrm{e}^{-tH}\Lambda\,\mathrm{e}^{tH} + \mathrm{e}^{-tH}T_\Lambda \mathrm{e}^{tH}, H\right] + T_\Lambda H \\ &= [\Lambda_t, H] + T_\Lambda H. \end{aligned}\tag{B.14}$$

It is not necessary to know explicitly $T_\Lambda H$ and $T_\Lambda \mathrm{e}^{tH}$. Let us compute

$$T_\Lambda S = \int_0^1 \mathrm{d}t \int_x \mathrm{Tr}(T_\Lambda H G[\Gamma_t] + H T_\Lambda G[\Gamma_t]). \tag{B.15}$$

Now, according to (B.10),

$$T_\Lambda G[\Gamma_t] = T_{\Lambda_t} G[\Gamma_t], \tag{B.16}$$

if we define

$$T_{\Lambda_t} \Gamma_t = \mathrm{d}\Lambda_t + [\Gamma_t, \Lambda_t]. \tag{B.17}$$

The consistency condition gives

$$\int_x \mathrm{Tr}\big(H T_{\Lambda_t} G[\Gamma_t]\big) = \int_x \mathrm{Tr}(\Lambda_t T_H G[\Gamma_t]) + \int_x \mathrm{Tr}([\Lambda_t, H] G[\Gamma_t]), \tag{B.18}$$

where we have defined

$$T_H \Gamma_t = \mathrm{d}H + [\Gamma_t, H] = \frac{\partial \Gamma_t}{\partial t}. \tag{B.19}$$

So, (B.15) gives

$$T_\Lambda S = \int_0^1 \mathrm{d}t \int_x \mathrm{Tr}\left\{ (T_\Lambda H + [\Lambda_t, H]) G[\Gamma_t] + \Lambda_t \frac{\partial G[\Gamma_t]}{\partial t} \right\}$$

$$= \int_0^1 \mathrm{d}t \int_x \mathrm{Tr}\left\{ \frac{\partial \Lambda_t}{\partial t} G[\Gamma_t] + \Lambda_t \frac{\partial G[\Gamma_t]}{\partial t} \right\}$$

$$= \int_0^1 \mathrm{d}t \frac{\partial}{\partial t} \int_x \mathrm{Tr}(\Lambda_t G[\Gamma_t]) = - \int_x \mathrm{Tr}(\Lambda G[\Gamma]) . \qquad \text{(B.20)}$$

In conclusion

$$E_\xi S = \mathfrak{L}_\xi S + T_\Lambda S = T_\Lambda S = -H_\xi . \qquad \text{(B.21)}$$

The effect of local rotations can be evaluated in an analogous manner, using the expression (6.25) in terms of E and α. Here there is not even the Lie derivative term, since local rotations are exactly like ordinary gauge transformations. The result is eq. (6.27).

References

[1] H. Fukuda and Y. Miyamoto, Prog. Theor. Phys. 4 (1949) 347;
J. Steinberger, Phys. Rev. 76 (1949) 1180;
J. Schwinger, Phys. Rev. 82 (1951) 664;
L. Rosenberg, Phys. Rev. 129 (1963) 2786;
J. Bell and R. Jackiw, Nuovo Cim. 60A (1969) 47;
S. Adler, Phys. Rev. 177 (1969) 2426
[2] I.S. Gerstein and R. Jackiw, Phys. Rev. 181 (1969) 1955;
R. Jackiw and K. Johnson, Phys. Rev. 182 (1969) 1459;
S. Adler and W. Bardeen, Phys. Rev. 182 (1969) 1517;
S. Adler and D.G. Boulware, Phys. Rev. 184 (1969) 1740;
R.W. Brown, C.-C. Shi and B.-L. Young, Phys. Rev. 186 (1969) 1491
[3] W.A. Bardeen, Phys. Rev. 184 (1969) 1848
[4] D.J. Gross and R. Jackiw, Phys. Rev. D6 (1972) 477
[5] C. Bouchiat, J. Iliopoulos and Ph. Meyer, Phys. Lett. 38B (1972) 519
[6] J. Wess and B. Zumino, Phys. Lett. B37 (1971) 95
[7] K. Fujikawa, Phys. Rev. Lett. 42 (1979) 1195; 44 (1980) 1733; Phys. Rev. D21 (1980) 2848; D22 (1980) 1499 (E); D23 (1981) 2262
[8] P.H. Frampton and T.W. Kephart, Phys. Rev. Lett. 50 (1983) 1343, 1347; Phys. Rev. D28 (1983) 1010
[9] P.K. Townsend and G. Sierra, Nucl. Phys. B222 (1983) 493
[10] L. Alvarez-Gaumé and E. Witten, Nucl. Phys. B234 (1984) 269
[11] B. Zumino, Wu Yong-Shi and A. Zee, Nucl. Phys. B239 (1984) 477
[12] B. Zumino, Les Houches lectures 1983, notes by K. Sibold, ed. R. Stora and B. DeWitt (North-Holland) to be published
[13] R. Stora, Cargèse lectures 1983
[14] R. Stora and B. Zumino, in preparation
[15] C. Gomez, Salamanca preprint DFTUS 06/83 (1983)

[16] L. Alvarez-Gaumé and P. Ginsparg, Harvard preprint HUTP-83/A081 (1983)
[17] O. Alvarez, I.M. Singer and B. Zumino, Comm. Math. Phys., to be published
[18] M. Paranjape, MIT thesis (1984);
M. Paranjape and J. Goldstone, in preparation
[19] A.J. Niemi and G.W. Semenoff, Phys. Rev. Lett. 51 (1983) 2077
[20] E. Witten, Nucl. Phys. B223 (1983) 422; B223 (1983) 433
[21] A.P. Balachandran, V.P. Nair and C.G. Trahern, Phys. Rev. D27 (1983) 1369
[22] C. Becchi, A. Rouet and R. Stora, Comm. Math. Phys. 42 (1975) 127
[23] Encyclopedic Dictionary of Mathematics, ed. S. Iyanaga and Y. Kawada, translation reviewed by K.O. May (MIT, 1980)
[24] O. Alvarez and B. Zumino, in preparation
[25] A.P. Balachandran, G. Marmo, V.P. Nair and C.G. Trahern, Phys. Rev. D25 (1982) 2713;
A. Adrianov, L. Bonora and R. Gamba-Saravi, Phys. Rev. D26 (1982) 2821;
T. Matsuki Phys. Rev. D28 (1983) 2107;
T. Matsuki and A. Hill, Ohio preprint (1983);
Chou Kuang-Chao, Guo Han-Ying, Wu Ke and Song Xing-Chang, Phys. Lett. 134B (1984) 67;
Chou Kuang-Chao, Gou Han-Ying and Wu Ke, Peking University preprint A.S-TIP-83-033;
J.M. Gipson, Virginia Polytechnic preprint VPI-HEP-83/8;
S. Elitzur and V.P. Nair, Inst. for Adv. Study preprint, Princeton (1984);
M.F. Atiyah and I.M. Singer, Proc. Nat. Acad. Sci. 81 (April 1984)

AN SU(2) ANOMALY[†] [*]

Edward Witten
Joseph Henry Laboratories
Princeton University
Princeton, New Jersey 08544, USA

[†]Reprinted with permission from *Physics Letters* **117B** (1982) 324-328.
[*]This work was supported in part by the National Science Foundation under Grant No. PHY80-19754.

A new restriction on fermion quantum numbers in gauge theories is derived. For instance, it is shown that an SU(2) gauge theory with an odd number of left-handed fermion doublets (and no other representations) is mathematically inconsistent.

It has been a long-standing puzzle to elucidate the properties of an SU(2) gauge theory with a single doublet of left-handed (Weyl) fermions. This theory defies simple phenomenological descriptions. There is no obvious attractive channel in which a fermion condensate could form, consistent with Fermi statistics and Lorentz invariance. But it is hard to believe that the fermions could remain massless in the presence of strong SU(2) gauge forces at long distances.

This puzzle persists (in the absence of other representations) whenever the number of elementary fermion doublets is odd. An even number of doublets, even if they have zero bare mass, could pair up and become massive Dirac fermions through spontaneous chiral symmetry breaking. With an odd number of elementary doublets, however, there would always be one massless doublet left over after any assumed chiral symmetry breaking, as long as the SU(2) gauge symmetry remains unbroken.

Of course, there is no real paradox here. Perhaps our heuristic pictures of strongly interacting gauge theories are inadequate. However, the facts noted above do suggest that something is strange about an SU(2) gauge theory with an odd number of elementary doublets. The purpose of this paper is to determine precisely what is strange about these theories; we will see that they are mathematically inconsistent! The inconsistency arises from a problem somewhat analogous to the Adler–Bell–Jackiw anomaly.

Although a hamiltonian approach exists, let us first look at this problem from the point of view of euclidean functional integrals. The starting point is the fact [1] that the fourth homotopy group of SU(2) is nontrivial,

$$\pi^4(\mathrm{SU}(2)) = \mathrm{Z}_2. \tag{1}$$

[Note that we are dealing with the *fourth* homotopy group, while the *third* homotopy group, $\pi^3(\mathrm{SU}(2)) = \mathrm{Z}$, has entered in instanton studies [2]. The analogue of π^4 has entered in some recent studies [3] of 2 + 1 dimensional models.] Eq. (1) means that in four-dimensional euclidean space, there is a gauge transformation $U(x)$ such that $U(x) \to 1$ as $|x| \to \infty$, and $U(x)$ "wraps" around the gauge group in such a way that it cannot be continuously deformed to the identity. The fact that the homotopy group is Z_2 means that a gauge transformation that wraps twice around SU(2) in this way can be deformed to the identity. We will not need the detailed form of $U(x)$.

The existence of the topologically non-trivial mapping $U = U(x)$ means that when we carry out the euclidean path integral

$$\int (\mathrm{d}A_\mu) \exp\left(-\frac{1}{2g^2}\int \mathrm{d}^4x \,\mathrm{tr}\, F_{\mu\nu}F^{\mu\nu}\right), \tag{2}$$

we are actually double counting. For every gauge field A_μ, there is a conjugate gauge field

$$A_\mu^{\mathrm{U}} = U^{-1}A_\mu U - \mathrm{i}U^{-1}\partial_\mu U,$$

which makes exactly the same contribution to the functional integral. There is no way to eliminate this double counting because A_μ and A_μ^U lie in the same sector of field space; A_μ^U can be reached continuously from A_μ without passing through singularities or infinite action barriers. But, in the absence of fermions, the double counting is harmless and cancels out when one calculates vacuum expectation values.

Now let us include fermions. Introducing, say, a single doublet of left-handed fermions, we now must deal with

$$Z = \int d\psi\, d\bar{\psi} \int dA_\mu$$

$$\times \exp\left(-\int d^4x [(1/2g^2)\, \mathrm{tr}\, F_{\mu\nu}^2 + \bar{\psi} i \not{D} \psi]\right). \quad (3)$$

We would like to integrate out the fermions and discuss the effective theory with the fermions eliminated.

As is well known, for a theory with a doublet of *Dirac* fermions, the basic integral is

$$\int (d\psi\, d\bar{\psi})_{\mathrm{Dirac}} \exp(\bar{\psi} i \not{D} \psi) = \det i\not{D}. \quad (4)$$

Here the right-hand side is, formally, the infinite product of all eigenvalues of the hermitian operator $i\not{D} = i\gamma_\mu D^\mu$. Certain theories – those that are afflicted with Adler–Bell–Jackiw anomalies – are ill-defined because it is impossible to renormalize this formal product so as to get a gauge invariant answer. However, a doublet of Dirac fermions could have a gauge invariant bare mass; this means that Pauli–Villars regularization is available, and hence that the determinant in (4) can be defined satisfactorily. This determinant is completely gauge invariant – invariant both under infinitessimal gauge transformations and under the topologically non-trivial gauge transformation U discussed earlier.

Now, with the gauge group SU(2), a doublet of Dirac fermions is exactly the same as two left-handed or Weyl doublets. Hence the fermion integration with a single Weyl doublet would give precisely the square root of (4):

$$\int (d\psi\, d\bar{\psi})_{\mathrm{Weyl}} \exp(\bar{\psi} i \not{D} \psi) = (\det i\not{D})^{1/2}. \quad (5)$$

But an ambiguity arises here; the square root has two signs. As we will see, this ambiguity leads to trouble.

Picking a particular gauge field A_μ, we are free to define in an arbitrary way the sign of $(\det i\not{D})^{1/2}$ for this field. Once this is done, there is no further freedom; to satisfy the Schwinger–Dyson equations we must define the fermion integral $(\det i\not{D})^{1/2}$ to vary smoothly as A_μ is varied.

Defined in this way $(\det i\not{D})^{1/2}$ is certainly invariant under infinitessimal gauge transformations – since the sign does not change abruptly. But nothing guarantees that $(\det i\not{D})^{1/2}$ is invariant under the topologically non-trivial gauge transformation U. In fact, as we will see, $(\det i\not{D})^{1/2}$ is odd under U. We will see that for any gauge field A_μ,

$$[\det i\not{D}(A_\mu)]^{1/2} = -[\det i\not{D}(A_\mu^U)]^{1/2}. \quad (6)$$

In other words, if one continuously varies the gauge field from A_μ to A_μ^U, one ends up with the opposite sign of the square root.

Before explaining why eq. (6) is valid, let us first discuss why it results in the mathematical inconsistency of the SU(2) theory with a single left-handed doublet. The partition function would be

$$Z = \int dA_\mu (\det i\not{D})^{1/2} \exp\left(-\frac{1}{2g^2}\int d^4x\, \mathrm{tr}\, F_{\mu\nu}^2\right). \quad (7)$$

But this vanishes identically, because the contribution of any gauge field A_μ is exactly cancelled by the equal and opposite contribution of A_μ^U! Likewise the path integral Z_X with insertion of any gauge invariant operator X is identically zero. So expectation values are indeterminate, $\langle X \rangle = Z_X/Z = 0/0$. For this reason, the theory is ill-defined.

One cannot avoid this problem by taking the absolute value of $(\det i\not{D})^{1/2}$; the resulting theory would not obey the Schwinger–Dyson equations. Nor can one consistently integrate over only "half" of field space, since A_μ and A_μ^U are continuously connected.

It remains to explain eq. (6). For convenience, take space–time to be a sphere of large volume so that the spectrum of $i\not{D}$ is discrete. We may as well assume there are no zero eigenvalues since otherwise $\det i\not{D}(A_\mu)$ vanishes and (6) is certainly true. The eigenvalues of $i\not{D}$ are real and (fig. 1) for every eigenvalue λ there is an eigenvalue $-\lambda$, since if $i\not{D}\psi = \lambda\psi$, then $i\not{D}(\gamma_5\psi) = -\lambda(\gamma_5\psi)$.

Taking the square root of $\det i\not{D}$ means that we want the product of only half of the eigenvalues, not all of them. We may suppose that for every pair of eigenvalues $(\lambda, -\lambda)$ we pick one or the other, but not both. For instance, for a particular gauge field A_μ we

(1)

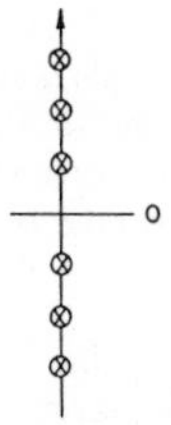

Fig. 1. The spectrum of the Dirac operator for a particular gauge field A_μ. The square root of the determinant may be defined – for this particular gauge field – as the product of the positive eigenvalues.

may define $(\det i\not{D})^{1/2}$ to be the product of the positive eigenvalues (fig. 1).

Now imagine varying the gauge field along a continuous path in field space from A_μ to A_μ^{U}. For instance, one may consider the gauge field $A_\mu^t = (1-t)A_\mu + tA_\mu^{\mathrm{U}}$, with t varied smoothly from zero to one. Let us follow the flow of the eigenvalues as a function of t. The spectrum of $i\not{D}$ is precisely the same at $t = 1$ as it is at $t = 0$. However, the individual eigenvalues may rearrange themselves as t is varied from zero to one.

As will be explained, the Atiyah–Singer index theorem predicts that such a rearrangement occurs. The simplest rearrangement allowed by the theorem is indicated in fig. 2. A single pair of eigenvalues $\{\lambda(t), -\lambda(t)\}$ cross at zero and change places as t is varied form zero to one.

In particular, one of the eigenvalues which was positive at $t = 0$ is negative by the time $t = 1$. If at $t = 0$, $(\det i\not{D})^{1/2}$ was defined as the product of the positive eigenvalues, then, following the eigenvalues continuously, by the time we reach $t = 1$ $(\det i\not{D})^{1/2}$ is the product of many positive eigenvalues and a single negative one. This means that $(\det i\not{D})^{1/2}$ has the opposite sign at $t = 1$ from its value at $t = 0$. The square root vanishes when the eigenvalue pair passes through zero ($t = t_0$ in fig. 2) and is negative for $t > t_0$.

(2)

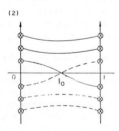

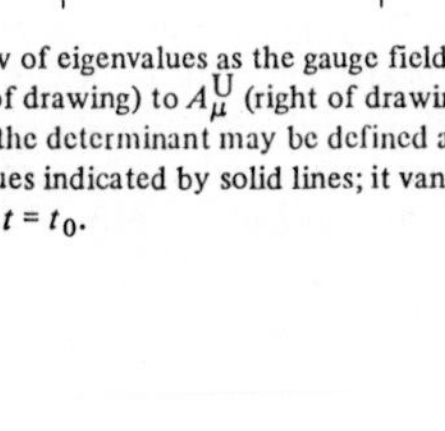

Fig. 2. The flow of eigenvalues as the gauge field is varied from A_μ (left of drawing) to A_μ^{U} (right of drawing). The square root of the determinant may be defined as the product of the eigenvalues indicated by solid lines; it vanishes and changes sign at $t = t_0$.

The Atiyah–Singer theorem permits more complicated rearrangements of eigenvalues, but the number of positive eigenvalues that become negative as t is varied from 0 to 1 is always *odd*. This is the basis for eq. (6).

The connection between the index theorem and the flow of eigenvalues is well known in mathematics [4] and has been discussed in the physics literature [5]. What is relevant for our problem is a slightly exotic form of the index theorem, namely the mod two index theorem for a certain *five*-dimensional Dirac operator [6].

Consider the five-dimensional Dirac equation for an SU(2) doublet of fermions,

$$\not{D}^{(5)}\Psi = \sum_{i=1}^{5} \gamma^i \left(\partial_i + \sum_{a=1}^{3} A_i^a T^a\right) \Psi = 0. \tag{8}$$

The spinor Ψ has eight components because the spinor representation of O(5) is four dimensional while an SU(2) doublet has two components.

The spinor representation of O(5) is pseudo-real, rather than real, and the doublet of SU(2) is likewise pseudo-real. But the tensor product of the spinor representation of O(5) with the doublet of SU(2) is a *real* representation of O(5) × SU(2). This means that in (8), we can take the gamma matrices γ^i to be real, symmetric 8 × 8 matrices while the anti-hermitian generators T^a of SU(2) are real, anti-symmetric matrices [‡1].

The five-dimensional Dirac operator $\not{D}^5$ for an SU(2) doublet is therefore a real, antisymmetric opera-

‡1 In fact, one can arrange Ψ as a two-component column vector of quaternions $\Psi = \binom{\psi_1}{\psi_2}$ – which would have eight real components. Such a column vector can be acted on from the left by a 2 × 2 unitary matrix of quaternions [making the group Sp(2) or O(5)] and on the right by a unitary quaternion [the group Sp(1) or SU(2)]. This is the desired eight-dimensional real representation of O(5) × SU(2).

(3)

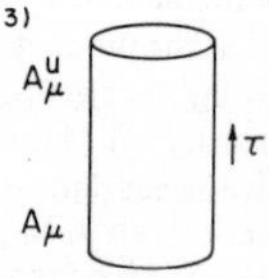

Fig. 3. A five-dimensional cylinder, $S^4 \times R$, on which an instanton-like gauge field is defined.

tor, acting on an infinite dimensional space. The eigenvalues of such a real, antisymmetric operator either vanish or are imaginary and occur in complex conjugate pairs. When the gauge field A_μ is varied, the number of zero eigenvalues of $\not{D}^5$ can change only if a complex conjugate pair of eigenvalues moves to – or away from – the origin. The number of zero eigenvalues of $\not{D}^5$ modulo two is therefore a topological invariant. It is known as the mod two index of the Dirac operator.

Now, consider a five dimensional cylinder $S^4 \times R$ (fig. 3). Let $x^\mu, \mu = 1, ..., 4$, be coordinates for S^4 while the position in the "time" direction (position in R) is called τ. Consider the following five-dimensional instanton-like SU(2) gauge field. For all x^μ and τ, $A_\tau = 0$. But $A_\mu(X^\sigma, \tau)$, $\mu = 1, ..., 4$, is – for each τ – a four-dimensional gauge field described as follows. For $\tau \to -\infty$ $A_\mu(x^\sigma, \tau)$ approaches the four-dimensional gauge field A_μ of our previous discussion ($t = 0$ in fig. 2). For $\tau \to +\infty$ $A_\mu(x^\sigma, \tau)$ approaches what we previously called A^U_μ ($t = 1$ in fig. 2). As τ varies from $-\infty$ to $+\infty$, $A_\mu(X^\sigma, \tau)$ varies *adiabatically* from A_μ to A^U_μ, along the same path in field space considered in fig. 2.

The mod two Atiyah–Singer index theorem predicts that the number of zero modes in this five-dimensional gauge field is odd – equal to one modulo two [‡2].

On the other hand, the number of zero modes of $\not{D}^5$, modulo two, can be calculated in terms of the eigenvalue flow of fig. 2. The Dirac equation $\not{D}^5 \Psi = 0$ can be written

$$\mathrm{d}\Psi/\mathrm{d}\tau = -\gamma^\tau \not{D}^4 \Psi, \tag{9}$$

where $\not{D}^4 = \Sigma_{i=1}^4 \gamma^\mu D_\mu$ is – at each τ – a four-dimensional Dirac operator.

Since $A_\mu(x^\sigma, \tau)$ evolves adiabatically in τ, (9) can be solved in the adiabatic approximation. We write $\Psi(x^\mu, \tau) = F(\tau)\, \phi^\tau(x^\mu)$, where $\phi^\tau(x^\mu)$ is a smoothly evolving solution of the eigenvalue equation

$$\gamma^\tau \not{D}^4 \phi^\tau(x^\mu) = \lambda(\tau) \phi^\tau(x^\mu). \tag{10}$$

The eigenvalues $\lambda(\tau)$ evolve on the curves of fig. 2 ($\mathrm{i}\not{D}^4$ and $\gamma^\tau \not{D}^4$ have the same spectrum). In the adiabatic limit, eq. (9) now reduces to $\mathrm{d}F/\mathrm{d}\tau = -\lambda(\tau) F(\tau)$, and the solution is

$$F(\tau) = F(0) \exp\left(- \int_0^\tau \mathrm{d}\tau' \, \lambda(\tau') \right). \tag{11}$$

This is normalizable only if λ is positive for $\tau \to +\infty$, and negative for $\tau \to -\infty$.

In the adiabatic approximation, the number of zero eigenvalues of $\not{D}^5$ is therefore equal to the number of eigenvalue curves in fig. 2 which pass from negative to positive values (or from positive to negative values) between $t = 0$ and $t = 1$. When corrections to the adiabatic approximation are considered, this gives the number of zero eigenvalues modulo two.

The Atiyah–Singer theorem, which requires that $\not{D}^5$ has an odd number of zero eigenvalues, therefore implies that the number of eigenvalue curves that pass from positive to negative values in fig. 2 is odd. This is precisely what we needed to show that $(\det \mathrm{i}\not{D})^{1/2}$ is odd under the topologically non-trivial gauge transformation U.

Now let us consider some generalizations. With n left-handed fermion doublets, the fermion integration would give $(\det \mathrm{i}\not{D})^{n/2}$. If n is even, the sign of the square root does not matter, but if n is odd, the theory suffers from the same inconsistency as before.

This persists even if additional gauge or Yukawa couplings are added to an SU(2) gauge theory. Since the fermion integration is necessarily either even or odd under U, if it is odd in a pure SU(2) gauge theory, it remains odd if additional gauge or Yukawa couplings are smoothly switched on. In particular, the standard $SU(3) \times SU(2) \times U(1)$ model of strong, weak, and electromagnetic interaction would be inconsistent if the number of left-handed fermion doublets were odd.

If one considers theories with SU(2) representations of isospin bigger than one half, the Atiyah–Singer

[‡2] Actually, in a special case one can easily find the zero mode. If one conformally compactifies $S^4 \times R$ to the five-sphere S^5, then on S^5 one can choose the instanton field to be invariant (up to a gauge transformation) under an SU(3) subgroup of the symmetry group O(6) of the five-sphere. The fermion zero mode is then the unique SU(3) invariant spinor field that can be defined.

theorem gives the following result. If one normalizes the SU(2) generators conventionally so that tr T_3^2 = 1/2 in the doublet representation, then the fermion integration is even or odd depending on whether tr T_3^2, evaluated among all the left-handed fermions, is an integer or half-integer. The inconsistent theories are those where tr T_3^2 is a half-integer. (In an ordinary instanton field, the number of fermion zero modes is 2 tr T_3^2, so the inconsistent theories are precisely those with an odd number of fermion zero modes in an instanton field.)

Considering gauge groups other than SU(2), we have

$$\pi^4(\mathrm{SU}(N)) = 0, \quad N > 2,$$

$$\pi^4(\mathrm{O}(N)) = 0, \quad N > 5,$$

$$\pi^4(\mathrm{Sp}(N)) = Z_2, \quad \text{any } N. \tag{12}$$

Thus non-trivial conditions arise only for Sp(N) groups.

Finally, let us note how this appears in a hamiltonian framework. Space permits only a brief statement of results.

From a hamiltonian viewpoint, one introduces the group G consisting of all gauge transformations $U(x, y, z)$ defined in *three*-dimensional space such that $U(\boldsymbol{x}) \to 1$ as $|\boldsymbol{x}| \to \infty$.

The fact that $\pi^4(\mathrm{SU}(2)) = Z_2$ means that $\pi^1(\mathrm{G}) = Z_2$. For the topologically non-trivial gauge transformation $U = U(x, y, z, t)$ that we have discussed is – at each t – an element of G. At $t \to -\infty$ or $t \to +\infty$ it is the identity in G; varying t from $-\infty$ to $+\infty$, U describes a loop in G which cannot be deformed away.

In canonical quantization, one encounters operators

$$Q^a(x) = g^{-2} D_i F^a_{0i}(x) - \bar{\psi}\gamma^0 T^a \psi, \tag{13}$$

which are generators of the Lie algebra of G. However, when a group – in this case G – is not simply connected, a representation of the Lie algebra does not necessarily provide a representation of the group. In general one gets a representation only of the simply connected covering group $\bar{\mathrm{G}}$. Since $\pi^1(\mathrm{G}) = Z_2$, the center of $\bar{\mathrm{G}}$ has a single non-trivial element P.

In quantum field theory, P – being in the center of $\bar{\mathrm{G}}$ – commutes with all fields and therefore is a c-number. Since $P^2 = 1$, we must have $P = +1$ or $P = -1$ (as an operator statement) in any given field theory. The theories in which the fermion integration is odd under U are the theories in which $P = -1$.

Theories with $P = -1$ are inconsistent for the following reason. According to Gauss's law, physical states $|\psi\rangle$ must be gauge invariant, obeying $Q^a|\psi\rangle = 0$ and hence $P|\psi\rangle = |\psi\rangle$. If $P = -1$, there are no states in the entire Hilbert space that obey Gauss's law.

Similar behavior can be seen in the models of ref. [3] by means of canonical quantization. This was one motivation for the present work.

I would like to thank S. Coleman and H. Georgi for discussions of the SU(2) theory with an odd number of doublets, I. Affleck and J. Harvey for discussions of fermion integration, and R. Jackiw for discussions about the models of ref. [3]. I also wish to thank W. Browder and J. Milnor for discussions about topology.

References

[1] S.T. Hu, Homotopy theory (Academic Press, New York, 1959) Ch. 11;
H. Toda, Composition methods in homotopy groups of spheres (Princeton U.P., Princeton, 1962).
[2] A.A. Belavin, A.M. Polyakov, A. Schwarz and Y. Tyupkin, Phys. Lett. 59B (1975) 85;
C. Callan R. Dashen and D.J. Gross, Phys. Lett. 63B (1976) 334;
R. Jackiw and C. Rebbi, Phys. Rev. Lett. 37 (1976) 334.
[3] S. Deser, R. Jackiw and S. Templeton, MIT preprint (1982).
[4] M.F. Atiyah, V. Patodi and I. Singer, Math. Proc. Camb. Philos. Soc. 79 (1976) 71.
[5] C.G. Callan, R. Dashen and D.J. Gross, Phys. Rev. D17 (1978) 2717;
J. Kiskis, Phys. Rev. D18 (1978) 3061.
[6] M.F. Atiyah and I.M. Singer, Ann. of Math. 93 (1971) 119.

GLOBAL ASPECTS OF CURRENT ALGEBRA[†][*]

Edward Witten
Joseph Henry Laboratories
Princeton University
Princeton, New Jersey 08544, USA

[†] Reprinted with permission from *Nuclear Physics* **B223** (1983) 422-432.
[*] This work was supported in part by NSF Grant PHY80-19754.

A new mathematical framework for the Wess-Zumino chiral effective action is described. It is shown that this action obeys an a priori quantization law, analogous to Dirac's quantization of magnetic change. It incorporates in current algebra both perturbative and non-perturbative anomalies.

The purpose of this paper is to clarify an old but relatively obscure aspect of current algebra: the Wess-Zumino effective lagrangian [1] which summarizes the effects of anomalies in current algebra. As we will see, this effective lagrangian has unexpected analogies to some 2 + 1 dimensional models discussed recently by Deser et al. [2] and to a recently noted SU(2) anomaly [3]. There also are connections with work of Balachandran et al. [4].

For definiteness we will consider a theory with $SU(3)_L \times SU(3)_R$ symmetry spontaneously broken down to the diagonal SU(3). We will ignore explicit symmetry-breaking perturbations, such as quark bare masses. With $SU(3)_L \times SU(3)_R$ broken to diagonal SU(3), the vacuum states of the theory are in one to one correspondence with points in the SU(3) manifold. Correspondingly, the low-energy dynamics can be conveniently described by introducing a field $U(x^\alpha)$ that transforms in a so-called non-linear realization of $SU(3)_L \times SU(3)_R$. For each space-time point x^α, $U(x^\alpha)$ is an element of SU(3): a 3×3 unitary matrix of determinant one. Under an $SU(3)_L \times SU(3)_R$ transformation by unitary matrices (A, B), U transforms as $U \to AUB^{-1}$.

The effective lagrangian for U must have $SU(3)_L \times SU(3)_R$ symmetry, and, to describe correctly the low-energy limit, it must have the smallest possible number of derivatives. The unique choice with only two derivatives is

$$\mathcal{L} = \tfrac{1}{16} F_\pi^2 \int d^4x \, \mathrm{Tr}\, \partial_\mu U \partial_\mu U^{-1}, \tag{1}$$

where experiment indicates $F_\pi \simeq 190$ MeV. The perturbative expansion of U is

$$U = 1 + \frac{2i}{F_\pi} \sum_{a=1}^{8} \lambda^a \pi^a + \cdots , \tag{2}$$

where λ^a (normalized so $\mathrm{Tr}\, \lambda^a \lambda^b = 2\delta^{ab}$) are the SU(3) generators and π^a are the Goldstone boson fields.

This effective lagrangian is known to incorporate all relevant symmetries of QCD. All current algebra theorems governing the extreme low-energy limit of Goldstone boson S-matrix elements can be recovered from the tree approximation to it. What is less well known, perhaps, is that (1) possesses an extra discrete symmetry that is *not* a symmetry of QCD.

The lagrangian (1) is invariant under $U \leftrightarrow U^{\mathrm{T}}$. In terms of pions this is $\pi^0 \leftrightarrow \pi^0$, $\pi^+ \leftrightarrow \pi^-$; it is ordinary charge conjugation. (1) is also invariant under the naive parity operation $\boldsymbol{x} \leftrightarrow -\boldsymbol{x}$, $t \leftrightarrow t$, $U \leftrightarrow U$. We will call this P_0. And finally, (1) is invariant under $U \leftrightarrow U^{-1}$. Comparing with eq. (2), we see that this latter operation is equivalent to $\pi^a \leftrightarrow -\pi^a$, $a = 1,\dots,8$. This is the operation that counts modulo two the number of bosons, N_{B}, so we will call it $(-1)^{N_{\mathrm{B}}}$.

Certainly, $(-1)^{N_{\mathrm{B}}}$ is not a symmetry of QCD. The problem is the following. QCD is parity invariant only if the Goldstone bosons are treated as pseudoscalars. The parity operation in QCD corresponds to $\boldsymbol{x} \leftrightarrow -\boldsymbol{x}$, $t \leftrightarrow t$, $U \leftrightarrow U^{-1}$. This is $P = P_0(-1)^{N_{\mathrm{B}}}$. QCD is invariant under P but not under P_0 or $(-1)^{N_{\mathrm{B}}}$ separately. The simplest process that respects all bona fide symmetries of QCD but violates P_0 and $(-1)^{N_{\mathrm{B}}}$ is $\mathrm{K}^+\mathrm{K}^- \to \pi^+\pi^0\pi^-$ (note that the ϕ meson decays to both $\mathrm{K}^+\mathrm{K}^-$ and $\pi^+\pi^0\pi^-$). It is natural to ask whether there is a simple way to add a higher-order term to (1) to obtain a lagrangian that obeys *only* the appropriate symmetries.

The Euler-Lagrangian equation derived from (1) can be written

$$\partial_\mu \left(\tfrac{1}{8} F_\pi^2 U^{-1} \partial_\mu U \right) = 0 . \tag{3}$$

Let us try to add a suitable extra term to this equation. A Lorentz-invariant term that violates P_0 must contain the Levi-Civita symbol $\varepsilon_{\mu\nu\alpha\beta}$. In the spirit of current algebra, we wish a term with the smallest possible number of derivatives, since, in the low-energy limit, the derivatives of U are small. There is a unique P_0-violating term with only four derivatives. We can generalize (3) to

$$\partial_\mu \left(\tfrac{1}{8} F_\pi^2 U^{-1} \partial_\mu U \right) + \lambda \varepsilon^{\mu\nu\alpha\beta} U^{-1} (\partial_\mu U) U^{-1} (\partial_\nu U) U^{-1} (\partial_\alpha U) U^{-1} (\partial_\beta U) = 0 , \tag{4}$$

λ being a constant. Although it violates P_0, (4) can be seen to respect $P = P_0(-1)^{N_{\mathrm{B}}}$.

Can eq. (4) be derived from a lagrangian? Here we find trouble. The only pseudoscalar of dimension four would seem to be $\varepsilon^{\mu\nu\alpha\beta} \mathrm{Tr}\, U^{-1}(\partial_\mu U) \cdot U^{-1}(\partial_\nu U) U^{-1} (\partial_\alpha U) U^{-1} (\partial_\beta U)$, but this vanishes, by antisymmetry of $\varepsilon^{\mu\nu\alpha\beta}$ and cyclic symmetry of the trace. Nevertheless, as we will see, there is a lagrangian.

Let us consider a simple problem of the same sort. Consider a particle of mass m constrained to move on an ordinary two-dimensional sphere of radius one. The lagrangian is $\mathfrak{L} = \frac{1}{2}m\int \mathrm{d}t\, \dot{x}_i^2$ and the equation of motion is $m\ddot{x}_i + mx_i(\Sigma_k \dot{x}_k^2) = 0$; the constraint is $\Sigma x_i^2 = 1$. This system respects the symmetries $t \leftrightarrow -t$ and separately $x_i \leftrightarrow -x_i$. If we want an equation that is only invariant under the combined operation $t \leftrightarrow -t$, $x_i \leftrightarrow x_i$, the simplest choice is

$$m\ddot{x}_i + mx_i\left(\sum_k \dot{x}_k^2\right) = \alpha\varepsilon_{ijk}x_j\dot{x}_k, \tag{5}$$

where α is a constant. To derive this equation from a lagrangian is again troublesome. There is no obvious term whose variation equals the right-hand side (since $\varepsilon_{ijk}x_ix_j\dot{x}_k = 0$).

However, this problem has a well-known solution. The right-hand side of (5) can be understood as the Lorentz force for an electric charge interacting with a magnetic monopole located at the center of the sphere. Introducing a vector potential $\boldsymbol{A}$ such that $\nabla \times \boldsymbol{A} = \boldsymbol{x}/|x|^3$, the action for our problem is

$$I = \int \left(\tfrac{1}{2}m\dot{x}_i^2 + \alpha A_i\dot{x}_i\right)\mathrm{d}t. \tag{6}$$

This lagrangian is problematical because A_i contains a Dirac string and certainly does not respect the symmetries of our problem. To explore this quantum mechanically let us consider the simplest form of the Feynman path integral, $\mathrm{Tr}\,\mathrm{e}^{-\beta H} = \int \mathrm{d}x_i(t)\mathrm{e}^{-I}$. In e^{-I} the troublesome term is

$$\exp\left(i\alpha\int_\gamma A_i\,\mathrm{d}x^i\right), \tag{7}$$

where the integration goes over the particle orbit γ: a closed orbit if we discuss the simplest object $\mathrm{Tr}\,\mathrm{e}^{-\beta H}$.

By Gauss's law we can eliminate the vector potential from (7) in favor of the magnetic field. In fact, the closed orbit γ of fig. 1a is the boundary of a disc D, and by Gauss's law we can write (7) in terms of the magnetic flux through D:

$$\exp\left(i\alpha\int_\gamma A_i\,\mathrm{d}x^i\right) = \exp\left(i\alpha\int_{\mathrm{D}} F_{ij}\,\mathrm{d}\Sigma^{ij}\right). \tag{8}$$

The precise mathematical statement here is that since $\pi_1(\mathrm{S}^2) = 0$, the circle γ in S^2 is the boundary of a disc D (or more exactly, a mapping γ of a circle into S^2 can be extended to a mapping of a disc into S^2).

The right-hand side of (8) is manifestly well defined, unlike the left-hand side, which suffers from a Dirac string. We could try to use the right-hand side of (8) in a Feynman path integral. There is only one problem: D isn't unique. The curve γ also bounds the disc D′ (fig. 1c). There is no consistent way to decide whether to choose

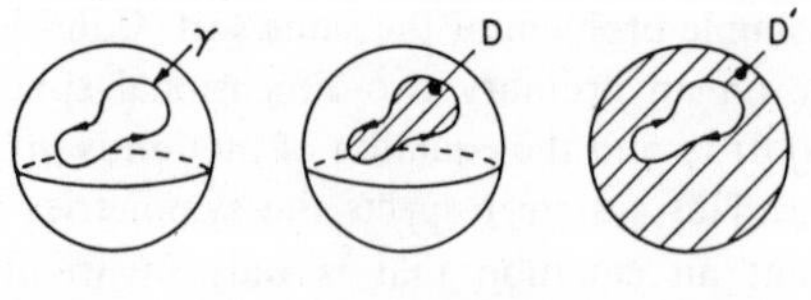

Fig. 1. A particle orbit γ on the two-sphere (part (a)) bounds the discs D (part (b)) and D′ (part (c)).

D or D′ (the curve γ could continuously be looped around the sphere or turned inside out). Working with D′ we would get

$$\exp\left(i\alpha\int_\gamma A_i\mathrm{d}x^i\right)=\exp\left(-i\alpha\int_{\mathrm{D}'}F_{ij}\mathrm{d}\Sigma^{ij}\right),\tag{9}$$

where a crucial minus sign on the right-hand side of (9) appears because γ bounds D in a right-hand sense, but bounds D′ in a left-hand sense. If we are to introduce the right-hand side of (8) or (9) in a Feynman path integral, we must require that they be equal. This is equivalent to

$$1=\exp\left(i\alpha\int_{\mathrm{D}+\mathrm{D}'}F_{ij}\mathrm{d}\Sigma^{ij}\right).\tag{10}$$

Since D + D′ is the whole two sphere S^2, and $\int_{S^2}F_{ij}\mathrm{d}\Sigma^{ij}=4\pi$, (10) is obeyed if and only if α is an integer or half-integer. This is Dirac's quantization condition for the product of electric and magnetic charges.

Now let us return to our original problem. We imagine space-time to be a very large four-dimensional sphere M. A given non-linear sigma model field U is a mapping of M into the SU(3) manifold (fig. 2a). Since $\pi_4(\mathrm{SU(3)})=0$, the four-sphere in SU(3) defined by U(x) is the boundary of a five-dimensional disc Q.

By analogy with the previous problem, let us try to find some object that can be integrated over Q to define an action functional. On the SU(3) manifold there is a unique fifth rank antisymmetric tensor ω_{ijklm} that is invariant under $\mathrm{SU(3)_L}\times\mathrm{SU(3)_R}$*. Analogous to the right-hand side of eq. (8), we define

$$\Gamma=\int_{\mathrm{Q}}\omega_{ijklm}\mathrm{d}\Sigma^{ijklm}.\tag{11}$$

* Let us first try to define ω at $U=1$; it can then be extended to the whole SU(3) manifold by an $\mathrm{SU(3)_L}\times\mathrm{SU(3)_R}$ transformation. At $U=1$, ω must be invariant under the diagonal subgroup of $\mathrm{SU(3)_L}\times\mathrm{SU(3)_R}$ that leaves fixed $U=1$. The tangent space to the SU(3) manifold at $U=1$ can be identified with the Lie algebra of SU(3). So ω, at $U=1$, defines a fifth-order antisymmetric invariant in the SU(3) Lie algebra. There is only one such invariant. Given five SU(3) generators A, B, C, D and E, the one such invariant is $\mathrm{Tr}\,ABCDE-\mathrm{Tr}\,BACDE\pm$ permutations. The $\mathrm{SU(3)_L}\times\mathrm{SU(3)_R}$ invariant ω so defined has zero curl ($\partial_i\omega_{jklmn}\pm$ permutations $=0$) and for this reason (11) is invariant under infinitesimal variations of Q; there arises only the topological problem discussed in the text.

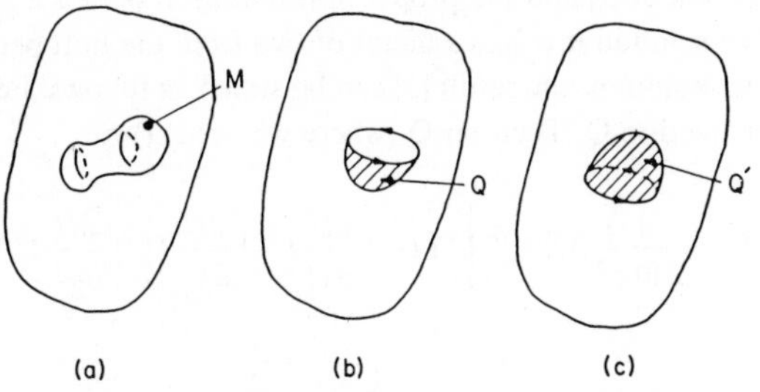

Fig. 2. Space-time, a four-sphere, is mapped into the SU(3) manifold. In part (a), space-time is symbolically denoted as a two sphere. In parts (b) and (c), space-time is reduced to a circle that bounds the discs Q and Q′. The SU(3) manifold is symbolized in these sketches by the interior of the oblong.

As before, we hope to include $\exp(i\Gamma)$ in a Feynman path integral. Again, the problem is that Q is not unique. Our four-sphere M is also the boundary of another five-disc Q′ (fig. 2c). If we let

$$\Gamma' = -\int_{Q'} \omega_{ijklm}\, d\Sigma^{ijklm}, \tag{12}$$

(with, again, a minus sign because M bounds Q′ with opposite orientation) then we must require $\exp(i\Gamma) = \exp(i\Gamma')$ or equivalently $\int_{Q+Q'} \omega_{ijklm}\, d\Sigma^{ijklm} = 2\pi \cdot \text{integer}$. Since Q + Q′ is a closed five-dimensional sphere, our requirement is

$$\int_S \omega_{ijklm}\, d\Sigma^{ijklm} = 2\pi \cdot \text{integer},$$

for any five-sphere S in the SU(3) manifold.

We thus need the topological classification of mappings of the five-sphere into SU(3). Since $\pi_5(\mathrm{SU}(3)) = Z$, every five sphere in SU(3) is topologically a multiple of a basic five sphere S_0. We normalize ω so that

$$\int_{S_0} \omega_{ijklm}\, d\Sigma^{ijklm} = 2\pi, \tag{13}$$

and then (with Γ in eq. (11)) we may work with the action

$$I = \tfrac{1}{16} F_\pi^2 \int d^4x\, \mathrm{Tr}\, \partial_\mu U\, \partial_\mu U^{-1} + n\Gamma, \tag{14}$$

where n is an arbitrary integer. Γ is, in fact, the Wess-Zumino lagrangian. Only the a priori quantization of n is a new result.

The identification of S_0 and the proper normalization of ω is a subtle mathematical problem. The solution involves a factor of two from the Bott periodicity theorem. Without abstract notation, the result [5] can be stated as follows. Let y^i, $i = 1\ldots5$ be coordinates for the disc Q. Then on Q (where we need it)

$$\mathrm{d}\Sigma^{ijklm}\,\omega_{ijklm} = -\frac{i}{240\pi^2}\mathrm{d}\Sigma^{ijklm}\left[\mathrm{Tr}\,U^{-1}\frac{\partial U}{\partial y^i}U^{-1}\frac{\partial U}{\partial y^j}U^{-1}\frac{\partial U}{\partial y^k}U^{-1}\frac{\partial U}{\partial y^l}U^{-1}\frac{\partial U}{\partial y^m}\right]. \tag{15}$$

The physical consequences of this can be made more transparent as follows. From eq. (2),

$$U^{-1}\partial_i U = \frac{2i}{F_\pi}\partial_i A + \mathrm{O}(A^2), \qquad \text{where } A = \Sigma\lambda^a\pi^a. \tag{16}$$

So

$$\omega_{ijklm}\mathrm{d}\Sigma^{ijklm} = \frac{2}{15\pi^2F_\pi^5}\mathrm{d}\Sigma^{ijklm}\,\mathrm{Tr}\,\partial_iA\,\partial_jA\,\partial_kA\,\partial_lA\,\partial_mA + \mathrm{O}(A^6)$$

$$= \frac{2}{15\pi^2F_\pi^5}\mathrm{d}\Sigma^{ijklm}\,\partial_i(\mathrm{Tr}\,A\,\partial_jA\,\partial_kA\,\partial_lA\,\partial_mA) + \mathrm{O}(A^6).$$

So $\int_Q\omega_{ijklm}\mathrm{d}\Sigma^{ijklm}$ is (to order A^5 and in fact also in higher orders) the integral of a total divergence which can be expressed by Stokes' theorem as an integral over the boundary of Q. By construction, this boundary is precisely space-time. We have, then,

$$n\Gamma = n\frac{2}{15\pi^2F_\pi^5}\int\mathrm{d}^4x\,\varepsilon^{\mu\nu\alpha\beta}\,\mathrm{Tr}\,A\,\partial_\mu A\,\partial_\nu A\,\partial_\alpha A\partial_\beta A + \text{higher order terms}. \tag{17}$$

In a hypothetical world of massless kaons and pions, this effective lagrangian rigorously describes the low-energy limit of $K^+K^-\to\pi^+\pi^0\pi^-$*. We reach the remarkable conclusion that in any theory with SU(3)×SU(3) broken to diagonal SU(3), the low-energy limit of the amplitude for this reaction must be (in units given in (17)) an integer.

What is the value of this integer in QCD? Were n to vanish, the practical interest of our discussion would be greatly reduced. It turns out that if N_c is the number of colors (three in the real world) then $n = N_c$. The simplest way to deduce this is a

* Our formula should agree for $n = 1$ with formulas of ref. [1], as later equations make clear. There appears to be a numerical error on p. 97 of ref. [1] ($\frac{1}{6}$ instead of $\frac{2}{15}$).

procedure that is of interest anyway, viz. coupling to electromagnetism, so as to describe the low-energy dynamics of Goldstone bosons and photons.

Let

$$Q = \begin{pmatrix} \frac{2}{3} & & \\ & -\frac{1}{3} & \\ & & -\frac{1}{3} \end{pmatrix}$$

be the usual electric charge matrix of quarks. The functional Γ is invariant under global charge rotations, $U \to U + i\varepsilon[Q, U]$, where ε is a constant. We wish to promote this to a local symmetry, $U \to U + i\varepsilon(x)[Q, U]$, where $\varepsilon(x)$ is an arbitrary function of x. It is necessary, of course, to introduce the photon field A_μ which transforms as $A_\mu \to A_\mu - (1/e)\partial_\mu \varepsilon$; e is the charge of the proton.

Usually a global symmetry can straightforwardly be gauged by replacing derivatives by covariant derivatives, $\partial_\mu \to D_\mu = \partial_\mu + ieA_\mu$. In the case at hand, Γ is not given as the integral of a manifestly $\mathrm{SU(3)_L} \times \mathrm{SU(3)_R}$ invariant expression, so the standard road to gauging global symmetries of Γ is not available. One can still resort to the trial and error Noether method, widely used in supergravity. Under a local charge rotation, one finds $\Gamma \to \Gamma - \int d^4x\, \partial_\mu \varepsilon J^\mu$ where

$$J^\mu = \frac{1}{48\pi^2} \varepsilon^{\mu\nu\alpha\beta} \mathrm{Tr}\big[Q(\partial_\nu U U^{-1})(\partial_\alpha U U^{-1})(\partial_\beta U U^{-1}) + Q(U^{-1}\partial_\nu U)(U^{-1}\partial_\alpha U)(U^{-1}\partial_\beta U)\big], \quad (18)$$

is the extra term in the electromagnetic current required (from Noether's theorem) due to the addition of Γ to the lagrangian. The first step in the construction of an invariant lagrangian is to add the Noether coupling, $\Gamma \to \Gamma' = \Gamma - e\int d^4x\, A_\mu J^\mu(x)$. This expression is still not gauge invariant, because J^μ is not, but by trial and error one finds that by adding an extra term one can form a gauge invariant functional

$$\tilde{\Gamma}(U, A_\mu) = \Gamma(U) - e\int d^4x\, A_\mu J^\mu + \frac{ie^2}{24\pi^2}\int d^4x\, \varepsilon^{\mu\nu\alpha\beta}(\partial_\mu A_\nu)A_\alpha \times \mathrm{Tr}\big[Q^2(\partial_\beta U)U^{-1} + Q^2 U^{-1}(\partial_\beta U) + QUQU^{-1}(\partial_\beta U)U^{-1}\big]. \quad (19)$$

Our gauge invariant lagrangian will then be

$$\mathcal{L} = \tfrac{1}{16}F_\pi^2 \int d^4x\, \mathrm{Tr}\, D_\mu U D_\mu U^{-1} + n\tilde{\Gamma}. \quad (20)$$

What value of the integer n will reproduce QCD results?

Here we find a surprise. The last term in (18) has a piece that describes $\pi^0 \to \gamma\gamma$. Expanding U and integrating by parts, (18) has a piece

$$A = \frac{ne^2}{48\pi^2 F_\pi} \pi^0 \varepsilon^{\mu\nu\alpha\beta} F_{\mu\nu} F_{\alpha\beta}. \tag{21}$$

This agrees with the result from QCD triangle diagrams [6] if $n = N_c$, the number of colors. The Noether coupling $-eA_\mu J^\mu$ describes, among other things, a $\gamma\pi^+\pi^0\pi^-$ vertex

$$B = -\tfrac{2}{3} ie \frac{n}{\pi^2 F_\pi^3} \varepsilon^{\mu\nu\alpha\beta} A_\mu \,\partial_\nu \pi^+ \,\partial_\alpha \pi^- \,\partial_\beta \pi^0. \tag{22}$$

Again this agrees with calculations [7] based on the QCD VAAA anomaly if $n = N_c$. The effective action $N_c \Gamma$ (first constructed in another way by Wess and Zumino) precisely describes all effects of QCD anomalies in low-energy processes with photons and Goldstone bosons.

It is interesting to try to gauge subgroups of $SU(3)_L \times SU(3)_R$ other than electromagnetism. One may have in mind, for instance, applications to the standard weak interaction model. In general, one may try to gauge an arbitrary subgroup H of $SU(3)_L \times SU(3)_R$, with generators K^σ, $\sigma = 1 \dots r$. Each K^σ is a linear combination of generators T_L^σ and T_R^σ of $SU(3)_L$ and $SU(3)_R$, $K^\sigma = T_L^\sigma + T_R^\sigma$. (Either T_L^σ or T_R^σ may vanish for some values of σ.) For any space-time dependent functions $\varepsilon^\sigma(x)$, let $\varepsilon_L = \Sigma_\sigma T_L^\sigma \varepsilon^\sigma(x)$, $\varepsilon_R = \Sigma_\sigma T_R^\sigma \varepsilon^\sigma(x)$. We want an action with local invariance under $U \to U + i(\varepsilon_L(x)U - U\varepsilon_R(x))$.

Naturally, it is necessary to introduce gauge fields $A_\mu^\sigma(x)$, transforming as $A_\mu^\sigma(x) \to A_\mu^\sigma(x) - (1/e_\sigma)\,\partial_\mu \varepsilon^\sigma + f^{\sigma\tau\rho}\varepsilon^\tau A_\mu^\rho$ where e_σ is the coupling constant corresponding to the generator K^σ, and $f^{\sigma\tau\rho}$ are the structure constants of H. It is useful to define $A_{\mu L} = \Sigma_\sigma e_\sigma A_\mu^\sigma T_L^\sigma$, $A_\mu^R = \Sigma_\sigma e_\sigma A_\mu^\sigma T_R^\sigma$.

We have already seen that Γ incorporates the effects of anomalies, so it is not very surprising that a generalization of Γ that is gauge invariant under H exists only if H is a so-called anomaly-free subgroup of $SU(3)_L \times SU(3)_R$. Specifically, one finds that H can be gauged only if for each σ,

$$\mathrm{Tr}(T_L^\sigma)^3 = \mathrm{Tr}(T_R^\sigma)^3, \tag{23}$$

which is the usual condition for cancellation of anomalies at the quark level.

If (23) is obeyed, a gauge invariant generalization of Γ can be constructed somewhat tediously by trial and error. It is useful to define $U_{\nu L} = (\partial_\nu U)U^{-1}$ and $U_{\nu R} = U^{-1}\partial_\nu U$. The gauge invariant functional then turns out to be

$$\tilde{\Gamma}(A_\mu, U) = \Gamma(U) + \frac{1}{48\pi^2}\int d^4x\, \varepsilon^{\mu\nu\alpha\beta} Z_{\mu\nu\alpha\beta},$$

where

$$
\begin{aligned}
Z_{\mu\nu\alpha\beta} = & -\mathrm{Tr}\left[A_{\mu L}U_{\nu L}U_{\alpha L}U_{\beta L} + (\mathrm{L}\to\mathrm{R})\right] \\
& + i\,\mathrm{Tr}\left[\left[(\partial_\mu A_{\nu L})A_{\alpha L} + A_{\mu L}(\partial_\nu A_{\alpha L})\right]U_{\beta L} + (\mathrm{L}\to\mathrm{R})\right] \\
& + i\,\mathrm{Tr}\left[(\partial_\mu A_{\nu R})U^{-1}A_{\alpha L}\,\partial_\beta U + A_{\mu L}U^{-1}(\partial_\nu A_{\alpha R})\,\partial_\beta U\right] \\
& -\tfrac{1}{2}i\,\mathrm{Tr}\left(A_{\mu L}U_{\nu L}A_{\alpha L}U_{\beta L} - (\mathrm{L}\to\mathrm{R})\right) \\
& + i\,\mathrm{Tr}\left[A_{\mu L}UA_{\nu R}U^{-1}U_{\alpha L}U_{\beta L} - A_{\mu R}U^{-1}A_{\nu L}UU_{\alpha R}U_{\beta R}\right] \\
& -\mathrm{Tr}\left[\left[(\partial_\mu A_{\nu R})A_{\alpha R} + A_{\mu R}(\partial_\nu A_{\alpha R})\right]U^{-1}A_{\beta L}U\right. \\
& \qquad \left. - \left[(\partial_\mu A_{\nu L})A_{\alpha L} + A_{\mu L}(\partial_\nu A_{\alpha L})\right]UA_{\beta R}U^{-1}\right] \\
& \qquad -\mathrm{Tr}\left[A_{\mu R}U^{-1}A_{\nu L}UA_{\alpha R}U_{\beta R} + A_{\mu L}UA_{\nu R}U^{-1}A_{\alpha L}U_{\beta L}\right] \\
& \qquad -\mathrm{Tr}\left[A_{\mu L}A_{\nu L}U(\partial_\alpha A_{\beta R})U^{-1} + A_{\mu R}A_{\nu R}U^{-1}(\partial_\alpha A_{\beta L})U\right] \\
& \qquad -i\,\mathrm{Tr}\left[A_{\mu R}A_{\nu R}A_{\alpha R}U^{-1}A_{\beta L}U - A_{\mu L}A_{\nu L}A_{\alpha L}UA_{\beta R}U^{-1}\right. \\
& \qquad\qquad \left. + \tfrac{1}{2}A_{\mu L}A_{\nu L}UA_{\alpha R}A_{\beta R}U^{-1} + \tfrac{1}{2}A_{\mu R}U^{-1}A_{\nu L}UA_{\alpha R}U^{-1}A_{\beta L}U\right]. \qquad (24)
\end{aligned}
$$

If eq. (22) for cancellation of anomalies is not obeyed, then the variation of $\tilde{\Gamma}$ under a gauge transformation does not vanish but is

$$
\begin{aligned}
\delta\tilde{\Gamma} = & -\frac{1}{24\pi^2}\int \mathrm{d}^4x\,\varepsilon^{\mu\nu\alpha\beta}\,\mathrm{Tr}\,\varepsilon_{\mathrm{L}}\left[(\partial_\mu A_{\nu L})(\partial_\alpha A_{\beta L}) - \tfrac{1}{2}i\partial_\mu(A_{\nu L}A_{\alpha L}A_{\beta L})\right] \\
& -(\mathrm{L}\to\mathrm{R}), \qquad (25)
\end{aligned}
$$

in agreement with computations at the quark level [8] of the anomalous variation of the effective action under a gauge transformation.

Thus, Γ incorporates all information usually associated with triangle anomalies, including the restriction on what subgroups H of $\mathrm{SU(3)_L}\times\mathrm{SU(3)_R}$ can be gauged. However, there is another potential obstruction to the ability to gauge a subgroup of $\mathrm{SU(3)_L}\times\mathrm{SU(3)_R}$. This is the non-perturbative anomaly [3] associated with $\pi_4(\mathrm{H})$. Is this anomaly, as well, implicit in Γ? In fact, it is.

Let H be an SU(2) subgroup of $\mathrm{SU(3)_L}$, chosen so that an SU(2) matrix W is embedded in $\mathrm{SU(3)_L}$ as

$$
\hat{W} = \left(\begin{array}{cc|c} \multicolumn{2}{c|}{\multirow{2}{*}{W}} & 0 \\ & & 0 \\ \hline 0 & 0 & 1 \end{array}\right).
$$

This subgroup is free of triangle anomalies, so the functional $\tilde{\Gamma}$ of eq. (23) is invariant under infinitesimal local H transformations.

However, is $\tilde{\Gamma}$ invariant under H transformations that cannot be reached continuously? Since $\pi_4(\mathrm{SU}(2)) = \mathrm{Z}_2$, there is one non-trivial homotopy class of SU(2) gauge transformations. Let W be an SU(2) gauge transformation in this non-trivial class. Under $\hat{W}$, $\tilde{\Gamma}$ may at most be shifted by a constant, independent of U and A_μ, because $\delta\tilde{\Gamma}/\delta U$ and $\delta\tilde{\Gamma}/\delta A_\mu$ are gauge-covariant local functionals of U and A_μ. Also $\tilde{\Gamma}$ is invariant under $\hat{W}^2$, since $\hat{W}^2$ is equivalent to the identity in $\pi_4(\mathrm{SU}(2))$, and we know $\tilde{\Gamma}$ is invariant under topologically trivial gauge transformations. This does not quite mean that $\tilde{\Gamma}$ is invariant under W. Since $\tilde{\Gamma}$ is only defined modulo 2π, the fact that $\tilde{\Gamma}$ is invariant under W^2 leaves two possibilities for how $\tilde{\Gamma}$ behaves under W. It may be invariant, or it may be shifted by π.

To choose between these alternatives, it is enough to consider a special case. For instance, it suffices to evaluate $\Delta = \tilde{\Gamma}(U=1,\ A_\mu = 0) - \tilde{\Gamma}(U = \hat{W},\ A_\mu = ie^{-1}(\partial_\mu \hat{W})\hat{W}^{-1})$. It is not difficult to see that in this case the complicated terms involving $\varepsilon^{\mu\nu\alpha\beta}Z_{\mu\nu\alpha\beta}$ vanish, so in fact $\Delta = \Gamma(U=1) - \Gamma(U=\hat{W})$. A detailed calculation shows that

$$\Gamma(U=1) - \Gamma(U=\hat{W}) = \pi . \tag{26}$$

This calculation has some other interesting applications and will be described elsewhere [9].

The Feynman path integral, which contains a factor $\exp(iN_c\tilde{\Gamma})$, hence picks up under W a factor $\exp(iN_c\pi) = (-1)^{N_c}$. It is gauge invariant if N_c is even, but not if N_c is odd. This agrees with the determination of the SU(2) anomaly at the quark level [3]. For under H, the right-handed quarks are singlets. The left-handed quarks consist of one singlet and one doublet per color, so the number of doublets equals N_c. The argument of ref. [3] shows at the quark level that the effective action transforms under W as $(-1)^{N_c}$.

Finally, let us make the following remark, which apart from its intrinsic interest will be useful elsewhere [9]. Consider $\mathrm{SU}(3)_\mathrm{L} \times \mathrm{SU}(3)_\mathrm{R}$ currents defined at the quark level as

$$J^a_{\mu\mathrm{L}} = \bar{q}\lambda^a\gamma_\mu\tfrac{1}{2}(1-\gamma_5)q , \qquad J^a_{\mu\mathrm{R}} = \bar{q}\lambda^a\gamma_\mu\tfrac{1}{2}(1+\gamma_5)q . \tag{27}$$

By analogy with eq. (17), the proper sigma model description of these currents contains pieces

$$J^{\mu a}_\mathrm{L} = \frac{N_c}{48\pi^2}\varepsilon^{\mu\nu\alpha\beta}\mathrm{Tr}\,\lambda^a U_{\nu\mathrm{L}}U_{\alpha\mathrm{L}}U_{\beta\mathrm{L}} ,$$

$$J^{\mu a}_\mathrm{R} = \frac{N_c}{48\pi^2}\varepsilon^{\mu\nu\alpha\beta}\mathrm{Tr}\,\lambda^a U_{\nu\mathrm{R}}U_{\alpha\mathrm{R}}U_{\beta\mathrm{R}} , \tag{28}$$

corresponding (via Noether's theorem) to the addition to the lagrangian of $N_c\Gamma$. In this discussion, the λ^a should be traceless SU(3) generators. However, let us try to construct an anomalous baryon number current in the same way. We define the baryon number of a quark (whether left-handed or right-handed) to be $1/N_c$, so that an ordinary baryon made from N_c quarks has baryon number one. Replacing λ^a by $1/N_c$, but including contributions of both left-handed and right-handed quarks, the anomalous baryon-number current would be

$$J^\mu = \frac{1}{24\pi^2}\varepsilon^{\mu\nu\alpha\beta}\operatorname{Tr} U^{-1}\partial_\nu U\, U^{-1}\partial_\alpha U\, U^{-1}\partial_\beta U. \tag{29}$$

One way to see that this is the proper, and properly normalized, formula is to consider gauging an arbitrary subgroup not of $SU(3)_L \times SU(3)_R$ but of $SU(3)_L \times SU(3)_R \times U(1)$, U(1) being baryon number. The gauging of U(1) is accomplished by adding a Noether coupling $-eJ^\mu B_\mu$ plus whatever higher-order terms may be required by gauge invariance. (B_μ is a U(1) gauge field which may be coupled as well to some $SU(3)_L \times SU(3)_R$ generator.) With J^μ defined in (29), this leads to a generalization of $\tilde{\Gamma}$ that properly reflects anomalous diagrams involving the baryon-number current (for instance, it properly incorporates the anomaly in the baryon number $SU(2)_L - SU(2)_L$ triangle that leads to baryon non-conservation by instantons in the standard weak interaction model). Eq. (29) may also be extracted from QCD by methods of Goldstone and Wilczek [10].

References

[1] J. Wess and B. Zumino, Phys. Lett. 37B (1971) 95
[2] S. Deser, R. Jackiw and S. Templeton, Phys. Rev. Lett. 48 (1982) 975; Ann. of Phys. 140 (1982) 372
[3] E. Witten, Phys. Lett. 117B (1982) 324
[4] A.P. Balachandran, V.P. Nair and C.G. Trahern, Syracuse University preprint SU-4217-205 (1981)
[5] R. Bott and R. Seeley, Comm. Math. Phys. 62 (1978) 235
[6] S.L. Adler, Phys. Rev. 177 (1969) 2426;
J.S. Bell and R. Jackiw, Nuovo Cim. 60 (1969) 147;
W.A. Bardeen, Phys. Rev. 184 (1969) 1848
[7] S.L. Adler and W.A. Bardeen, Phys. Rev. 182 (1969) 1517;
R. Aviv and A. Zee, Phys. Rev. D5 (1972) 2372
S.L. Adler, B.W. Lee, S.B. Treiman and A. Zee, Phys. Rev. D4 (1971) 3497
[8] D.J. Gross and R. Jackiw, Phys. Rev. D6 (1972) 477
[9] E. Witten, Nucl. Phys. B223 (1983) 433
[10] J. Goldstone and F. Wilczek, Phys. Rev. Lett. 47 (1981) 986

GRAVITATIONAL ANOMALIES†

Luis Alvarez-Gaumé[1]
Lyman Laboratory of Physics
Harvard University
Cambridge, MA 02138, USA

Edward Witten[2]
Joseph Henry Laboratories
Princeton University
Princeton, New Jersey 08544, USA

†Reprinted with permission from *Nuclear Physics* **B234** (1983) 269-330.

[1] Research supported in part by the National Science Foundation under grant no. PHY-82-15249.

[2] Research supported in part by the National Science Foundation under grant no. PHY-80-19754.

It is shown that in certain parity-violating theories in $4k+2$ dimensions, general covariance is spoiled by anomalies at the one-loop level. This occurs when Weyl fermions of spin-$\frac{1}{2}$ or -$\frac{3}{2}$ or self-dual antisymmetric tensor fields are coupled to gravity. (For Dirac fermions there is no trouble.) The conditions for anomaly cancellation between fields of different spin is investigated. In six dimensions this occurs in certain theories with a fairly elaborate field content. In ten dimensions there is a unique theory with anomaly cancellation between fields of different spin. It is the chiral $n=2$ supergravity theory, which is the low-energy limit of one of the superstring theories. Beyond ten dimensions there is no way to cancel anomalies between fields of different spin.

1. Introduction

The fermion anomaly in $(3+1)$-dimensional quantum field theory has a remarkable number of important applications. In the original version [1], one considers a massless fermion triangle diagram with one axial current and two vector currents. Requiring conservation of the vector currents, one finds, even for massless fermions, that the axial current is not conserved (fig. 1). This results in a breakdown of chiral symmetry in the presence of gauge fields that are coupled to the conserved vector currents. This breakdown is known to lead to an understanding of π^0 decay and to a resolution of the U(1) problem in QCD [2].

Another, equally significant facet of the anomaly arises if gauge fields are coupled not to vector currents but to linear combinations of vector and axial vector currents, as in the standard $SU(2)\times U(1)$ model of weak interactions. For instance, in a gauge theory with $V-A$ gauge couplings, one must consider (fig. 2) a fermion triangle with a $V-A$ current at each vertex. This diagram is again anomalous. Unless it cancels when summing over the fermion species running around the loop, the anomaly spoils conservation of the $V-A$ currents. But gauge theories with gauge fields coupled to non-conserved currents are inconsistent. So the $SU(2)\times U(1)$ model (or any gauge theory with non-vectorlike gauge couplings) is inconsistent unless the

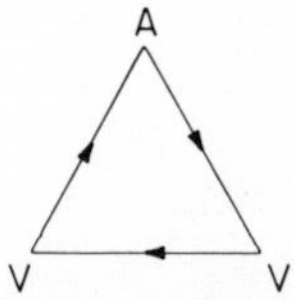

Fig. 1. The fermion triangle with one axial current and two vector currents.

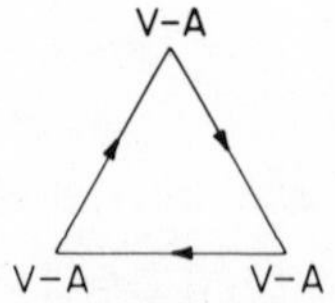

Fig. 2. The triangle diagram with a V − A insertion at each vertex.

anomalies cancel [3]. The classic triangle anomaly has other important applications as well [4].

These remarks can be generalized in various directions. First of all, in addition to the fermion triangle diagram with three currents, the triangle with one current and two energy-momentum tensors is anomalous (fig. 3) [5]. This has been widely discussed in connection with the breakdown of chirality conservation in a gravitational field [6].

Also, several authors have recently discussed the question of anomalies in space–times of higher dimension. In any even number of dimensions, there is an anomaly analogous to the triangle anomaly. In N dimensions, a fermion one-loop diagram with $\frac{1}{2}N+1$ external gluons is potentially anomalous. The anomaly can be evaluated [7] by analogy with the usual evaluation of the triangle in four dimensions. Cancellation of anomalies in higher dimensions is potentially of interest as a restriction on Kaluza–Klein theories. For instance, as noted by some of the authors of ref. [7], the supersymmetric Yang–Mills theory in ten dimensions is inconsistent because of hexagon anomalies. This problem is even more serious than the lack of renormalizability (which has been explicitly demonstrated [8]) because it represents a breakdown of gauge invariance which almost certainly cannot be cured by a short distance cut-off. This point will be developed in sect. 3; it is related to 't Hooft's observation [4] that anomalies can be understood in terms of low-energy physics.

The thrust of the present paper is the following. The need to cancel anomalies places restrictions on which theories can be coupled to gravity. Such a restriction arises in four dimensions as a straightforward consequence of the triangle anomaly of fig. 3. Specifically, the $SU(2) \times U(1)$ theory cannot be coupled to gravity unless the sum of the hypercharges of the left-handed fermions vanishes. This appears to

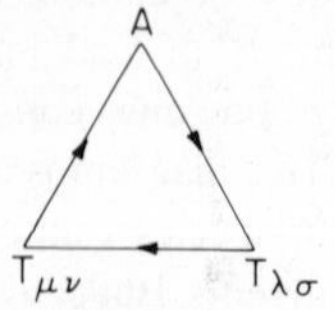

Fig. 3. The anomalous triangle with one axial current A and two insertions of the energy-momentum tensor.

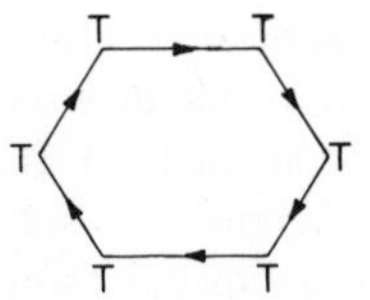

Fig. 4. The hexagon anomaly in ten dimensions; a diagram with six insertions of the energy-momentum tensor T.

be the case for known fermions (it is true for each generation of quarks and leptons). This will be further discussed in sect. 3.

Our main effort, however, will be devoted to unraveling the structure of anomalies in more than four dimensions. As we will see, in certain theories there are one-loop anomalies in the purely gravitational interactions. This sort of anomaly does not occur in four dimensions, and this is probably why it has not been previously noted*. Purely gravitational anomalies occur for Weyl fermions (of spin $\frac{1}{2}$ or spin $\frac{3}{2}$) in $N=4k+2$ dimensions, $k=0,1,2,\ldots$. For instance, in ten dimensions (fig. 4) the hexagon diagram with six external gravitons coupled to a Weyl fermion of definite chirality is anomalous. It is impossible to maintain Bose symmetry and gauge invariance for the external graviton lines. This anomaly arises for Weyl fermions of spin $\frac{1}{2}$ or spin $\frac{3}{2}$. We will find a similar anomaly in the gravitational coupling of self-dual antisymmetric tensor Bose fields. Because of these anomalies, Weyl fermions or self-dual tensors cannot be coupled to gravity in $4k+2$ dimensions, unless one arranges for anomaly cancellation between the different spins.

A number of interesting theories are affected by this discussion. For instance, in ten dimensions, there is an $n=1$ supergravity theory with one real Weyl gravitino (and one real spin-$\frac{1}{2}$ field of opposite chirality). This theory is anomalous.

In ten dimensions there are two theories with $n=2$ supergravity. One can be obtained by dimensional reduction from eleven-dimensional supergravity [10]. It is parity conserving, and like all parity conserving theories it is anomaly free. The other theory [11] has two gravitinos of the same chirality. It has one-loop anomalies in the gravitino, spin-$\frac{1}{2}$, and antisymmetric tensor couplings, but as we will see the anomalies of the fields of different spin cancel. This is, in fact, apparently the only ten-dimensional theory with anomaly cancellation between fields of different spin.

Since anomalies can be understood in terms of the *low*-energy behavior of a theory [4], the same considerations apparently apply to the supersymmetric string theories [12] which are of much interest as an approach to quantum gravity. The "type (I)" superstring theories reduce at low energy to $n=1$ supergravity in ten

* There is a well-known trace anomaly in four dimensions [9]. This concerns the anomalous trace of the energy-momentum tensor and is related to the Callan–Symanzik β function. It does not spoil the *conservation* of the energy-momentum tensor and does not ruin the mathematical consistency of coupling to gravity.

dimensions. They are presumably anomalous. (Introduction of a gauge group cannot remove the gravitational anomaly, as one may see from formulae of sect. 12. On the contrary [7] it introduces new anomalies.) The "type (II)" theories of closed superstrings only reduce at low energies to $n=2$ supergravity in ten dimensions. They appear to be free of hexagon anomalies. (Unfortunately, it is difficult to check these statements directly in the string context, because the simple light-cone formalism [12] is restricted to external graviton lines of $p^+ = \varepsilon^{+\mu} = 0$, for which the kinematical factor R defined in sect. 6 vanishes. Of course, a string theory might have anomalies that disappear in the field theoretic limit and survive when $p^+ = \varepsilon^{+\mu} = 0$, but this is outside the scope of our investigation.)

In addition to the purely gravitational anomalies, there are (as in four dimensions) anomalies in one-loop diagrams with both external gluons and external gravitons. In general, in n dimensions, the one-loop diagrams with $2r$ external gluons and $\frac{1}{2}n+1-2r$ external gravitons is anomalous for $0 \leq r \leq \frac{1}{4}(n+2)$. For $r=\frac{1}{4}(n+2)$ this has been discussed previously [7]. The cancellation of all these anomalies is a very severe restriction on the allowed fermion quantum numbers in Kaluza–Klein theory. The phenomenology of some of the anomaly-free theories will be discussed elsewhere [13].

The construction of effective actions in curved space–time has been reviewed in ref. [14]. Our results answer in the negative the question of whether the induced energy-momentum tensor for matter fields in a background geometry is always conserved.

Our calculations will reveal a connection between the gravitational anomalies in $4k+2$ dimensions and index theorems on curved manifolds in $4k+4$ dimensions. This connection was suggested by M.F. Atiyah (private communication) on the basis of properties of diffeomorphism groups. Anticipating this connection was of considerable help in finding ways to calculate gravitational anomalies.

2. Some generalities about anomalies

Before considering particular theories in detail, let us begin with some generalities about anomalies. Most of the remarks in this section are well known.

First of all, as noted in the original literature, the anomaly constitutes a breakdown of gauge invariance. Theories with anomalies are theories in which the effective action is not gauge invariant at the one-loop level. To be specific, consider in four-dimensional euclidean space a theory with gauge fields coupled to fermion fields in a complex representation of the gauge group. Let us denote the left- and right-handed fermions as ψ and $\bar{\chi}$ respectively (in euclidean space they are independent variables, not complex conjugates of each other; and we use different names to emphasize the fact they they are independent). The lagrangian for ψ and $\bar{\chi}$

interacting with gauge fields is

$$\mathcal{L}=\int d^4x\,\bar{\chi}(i\partial\!\!\!/-\gamma^\mu\Sigma A^a_\mu\lambda^a)\left(\frac{1-\gamma_5}{2}\right)\psi, \tag{1}$$

where the λ^a, normalized so that $\mathrm{Tr}\,\lambda^a\lambda^b=2\delta^{ab}$, are the gauge group generators acting on the left-handed fermions. If we define $A_\mu=\frac{1}{2}\Sigma\lambda^a A^a_\mu$, then the variation of A_μ under an infinitesimal gauge transformation is $A_\mu\to A_\mu-D_\mu\varepsilon$. Consequently, any functional $\Gamma(A)$ changes under an infinitesimal gauge transformation as

$$\begin{aligned}\Gamma(A_\mu)&\to\Gamma(A_\mu-D_\mu\varepsilon)\\&=\Gamma(A_\mu)-\int d^4x\,\mathrm{Tr}\,D_\mu\varepsilon(x)\frac{\delta\Gamma}{\delta A_\mu(x)}\\&=\Gamma(A_\mu)+\int d^4x\,\mathrm{Tr}\,\varepsilon(x)D_\mu\frac{\delta\Gamma}{\delta A_\mu(x)},\end{aligned}$$

so the generator of gauge transformations is $D_\mu\,\delta/\delta A^a_\mu$. Now, let Γ be the one-loop fermion effective action; thus, formally,

$$\exp[-\Gamma(A_\mu)]=\int d\bar{\chi}\,d\psi\,\exp\left\{-\int d^4x\left[\bar{\chi}iD\!\!\!\!/\left(\frac{1-\gamma_5}{2}\right)\psi\right]\right\}. \tag{2}$$

One often writes Γ as the logarithm of the determinant of the Dirac operator [15]; we will discuss shortly the limits of this formulation.

Although Γ is naively gauge invariant, the statement of the anomaly is that the variation of Γ under a gauge transformation does not vanish; rather

$$D_\mu\frac{\delta\Gamma}{\delta A^a_\mu}=-\frac{i}{48\pi^2}\varepsilon^{\mu\nu\alpha\beta}\,\mathrm{Tr}\,\lambda^a[2\partial_\mu A_\nu\partial_\alpha A_\beta-i\partial_\mu(A_\nu A_\alpha A_\beta)]. \tag{3}$$

Thus, the anomaly represents a failure of gauge invariance.

The connection of this loss of gauge invariance with the failure of current conservation due to the anomaly is as follows. The fermion current induced by an applied gauge field is

$$J^a_\sigma=\left\langle\bar{\chi}\gamma_\sigma\lambda^a\left(\frac{1-\gamma_5}{2}\right)\psi\right\rangle^{A_\mu},$$

where the expectation value is to be taken in the background field A_μ. However, from eq. (2) it follows that $J^a_\mu=\delta\Gamma/\delta A^a_\mu$ so the (covariant) divergence of the current is

$$D_\mu J^a_\mu=D_\mu\frac{\delta}{\delta A^a_\mu}\Gamma, \tag{4}$$

and the loss of gauge invariance is equivalent to a failure of current conservation.

In gauge theories, unitarity and (in the case of renormalizable theories) renormalizability depend upon gauge invariance. Once gauge invariance is lost, these properties are also lost.

Now let us return to eq. (3). The factor of i on the right-hand side of eq. (3) is of fundamental importance. (This factor has nothing to do with thresholds or unitarity; it is required by CPT in euclidean space for a parity violating amplitude proportional to $\varepsilon_{\mu\nu\alpha\beta}$.) Because of this factor of i, only the imaginary part of Γ is not gauge invariant. The real part of Γ is perfectly gauge invariant. In terms of the fermion integral, which equals $e^{-\Gamma}$, this means that the *modulus* of the fermion integral is gauge invariant; it is the *phase* of the fermion integral that is not gauge invariant.

There is a simple reason for this. First of all, consider the case in which the fermions are in a real representation of the gauge group. Such fermions are real (anticommuting) variables. For such real variables, the fermion integral $e^{-\Gamma}$ is real. Moreover, theories in which the fermions are in a real representation of the gauge group can be regularized by the Pauli–Villars method. The existence of this gauge invariant regularization ensures that the fermion effective action is gauge invariant. For fermions in a real representation of the gauge group, the effective action is real and gauge invariant.

Now consider fermions in a complex representation Q. Such fermions are complex objects and the fermion integral is not necessarily real. On the contrary, as we have noted, the parity-violating amplitudes are imaginary. Gauge invariance forbids bare masses for fermions in a complex representation Q, and for this reason there is no evident way to regularize such theories. So the effective action $\Gamma(\mathrm{Q})$ for fermions in the representation Q is complex and not necessarily gauge invariant.

Now consider fermions in the complex conjugate representation $\bar{\mathrm{Q}}$. Such variables are complex conjugates of the ones we have just considered, and the Dirac action for fermions in the $\bar{\mathrm{Q}}$ representation is the complex conjugate of the action for fermions in the Q representation. So the effective action $\Gamma(\bar{\mathrm{Q}})$ is the complex conjugate of $\Gamma(\mathrm{Q})$. (Differently put, in passing from Q to $\bar{\mathrm{Q}}$, the parity-violating amplitudes, which are imaginary, change sign, so the action is complex conjugated.) Hence $\Gamma(\mathrm{Q})+\Gamma(\bar{\mathrm{Q}})=2\,\mathrm{Re}\,\Gamma(\mathrm{Q})$.

But $\Gamma(\mathrm{Q})+\Gamma(\bar{\mathrm{Q}})$ is the effective action for fermions in the real representation $\mathrm{Q}+\bar{\mathrm{Q}}$. Because of our previous remarks, this effective action is gauge invariant. So $\mathrm{Re}\,\Gamma(\mathrm{Q})$ is gauge invariant, and only $\mathrm{Im}\,\Gamma(\mathrm{Q})$ may suffer from anomalies. Anomalies in $\mathrm{Im}\,\Gamma(\mathrm{Q})$ do in fact show up in triangle diagrams.

A variant of this can occur if Q is a pseudoreal but not real representation of the gauge group. In this case, bare masses and Pauli–Villars regularization are again not possible so there is no guarantee that $e^{-\Gamma}$ is gauge invariant. However, if Q is pseudoreal, then $\mathrm{Q}\oplus\mathrm{Q}$ is a real representation, so the corresponding fermion integral, which is $e^{-2\Gamma}$, can be regularized and is gauge invariant. The fact that $e^{-2\Gamma}$ is gauge invariant means that $e^{-\Gamma}$ is gauge invariant in absolute value but perhaps not in sign. In certain cases [16], it is indeed impossible to define the sign in a gauge invariant way.

There is another point of view about this which may seem cumbersome but does help clarify why the phase of the fermion integral is potentially anomalous.

We are accustomed to thinking of the fermion integral as the determinant of the Dirac operator. For fermions in a real representation, we consider the hermitian eigenvalue problem $i\not{D}\psi_i = \lambda_i\psi_i$ and define det $i\not{D}$ as the product of the λ_i. Since the λ_i are real, their product is real; this shows again that the fermion integral is real for fermions in a real representation. Of course, the product of the λ_i needs to be regularized. But this is easily accomplished. One may define det $i\not{D} = \prod F(\lambda_i)$ where $F(\lambda_i)$ is a suitably chosen function such that $F(\lambda) = \lambda$ for small λ and $F(\lambda) \to 1$ as $\lambda \to \infty$.

For fermions in a complex representation, life is more subtle. Our action is of the general form

$$\mathcal{L} = \bar{\chi} i\not{D}\left(\frac{1-\gamma_5}{2}\right)\psi, \tag{5}$$

where the ψ are left-handed fermions in a complex representation Q, and the $\bar{\chi}$ are right-handed fermions in the complex conjugate representation $\bar{\mathrm{Q}}$.

Let V be the vector space of left-handed fermion fields in the Q representation. Given a vector space V and an operator M (not necessarily hermitian) mapping V into V, it is possible to define the determinant of M. In the usual way, one solves the eigenvalue problem $M\psi_i = \lambda_i\psi_i$ (or one defines the λ_i to be the diagonal elements when M is put in Jordan canonical form) and one defines det $M = \prod \lambda_i$. In general, this determinant is a complex number.

We would like to define a determinant of the operator $D = \frac{1}{2}i\not{D}(1-\gamma_5)$. The problem is that D does not map V into itself; it maps V into $\tilde{\mathrm{V}}$, the vector space of *right-handed* fermions in the Q representation. Without further information, there is no way to define a determinant of an operator M that maps one vector space V into another space $\tilde{\mathrm{V}}$.

In the case at hand, V and $\tilde{\mathrm{V}}$ are Hilbert spaces; for ψ in V or $\tilde{\mathrm{V}}$, we define the norm of ψ as $\langle\psi|\psi\rangle = \int d^4x \sum_{\alpha i} |\psi_{\alpha i}(x)|^2$. Given an operator M from one Hilbert space V to another Hilbert space $\tilde{\mathrm{V}}$, a determinant can be defined *up to phase* as follows. Let $|\psi_i\rangle$ be an orthonormal basis for V. Let $|x_i\rangle = M|\psi_i\rangle$. Choose the $|\psi_i\rangle$ so $\langle x_i|x_j\rangle = 0$ for $i \neq j$, and define $\lambda_i = \sqrt{\langle x_i|x_i\rangle}$. The product of the λ_i is automatically real and is not a sensible definition of det M, since it does not reduce to the standard definition (which can be complex) when $\mathrm{V} = \tilde{\mathrm{V}}$. But it makes sense to adopt this definition of the *modulus* of the determinant: $|\det M| = \prod \lambda_i$.

What about the *phase* of det M? Given a single operator M from V to $\tilde{\mathrm{V}}$, there is no way to define this phase. But suppose we have two operators M and N from V to $\tilde{\mathrm{V}}$. Then $A = N^{-1}M$ maps V into V, so the determinant of A is well defined as a complex number. The definition ln det M − ln det N = ln det A defines the phase *difference* between det M and det N.

In our physical problem, we do not care about the overall phase of the determinant of $D=\frac{1}{2}i\not{D}(1-\gamma_5)$, since a constant can be absorbed in normalizing the partition function. We do care about *relative* phases. We arbitrarily pick a convenient gauge field, say $A^a_\mu=0$, and define det D to be, say, positive for this gauge field. For any other gauge field A^a_μ one attempts to define the phase of the determinant by saying $\ln\det D(A^a_\mu)-\ln\det D(A^a_\mu=0)=\ln\det(D^{-1}(A^a_\mu=0)D(A^a_\mu))$. When dealing with differential operators in infinite-dimensional function spaces, it is difficult to define the determinant of an operator such as $D^{-1}(A^a_\mu=0)D(A^a_\mu)$. This difficulty is one way of understanding the origin of anomalies.

Let us now return to the Feynman diagrams in which anomalies appear. In four dimensions, the simplest fermion diagram which (kinematically) can violate parity is the diagram with three external gluons. The diagram is indeed anomalous, as we have already discussed. A few general remarks about diagrammatic evaluation of anomalies will be useful later.

There is no good way to regularize the anomalous diagrams; if there were, there would be no anomaly. Because there is no good way to regularize these diagrams, they are potentially ambiguous. The potential ambiguity consists of the ability to add a polynomial in the external momenta. The reason for this is as follows. Any acceptable definition of the triangle must obey unitarity in each channel. Using unitarity, the triangle amplitude (or any one-loop amplitude) can be uniquely reconstructed from tree diagrams up to a polynomial in the external momenta. (See recent discussions by Coleman and Grossman and by Frishman et al. [4].) Therefore, the triangle amplitude is well defined modulo the ability to add such a polynomial. When one claims that a diagram is anomalous, one means that it is impossible to add a polynomial in the momenta so as to eliminate the anomaly and obtain an amplitude that obeys all physical principles.

It automatically follows from this that, regardless of the superficial degree of divergence of a diagram, anomalies are always finite. After all, the infinite part of a diagram is always a polynomial in the external momenta. Our freedom to redefine an amplitude by adding a polynomial includes the freedom to throw away all infinite pieces. Hence, relevant anomalies, if any, are always finite.

Even in unrenormalizable theories, anomalies that ruin gauge invariance occur only at the one-loop level. Multi-loop diagrams can be regularized in a gauge covariant way by Pauli–Villars regularization of the internal bose lines these diagrams necessarily contain.

When an anomaly occurs, it is impossible to define an amplitude to obey all physical principles. In this situation, different attempts at defining the amplitude may give different answers. For instance, consider the triangle diagram. It should obey Bose symmetry and current conservation in each of three external lines. There are two standard ways to define the triangle. One may insist on Bose symmetry in

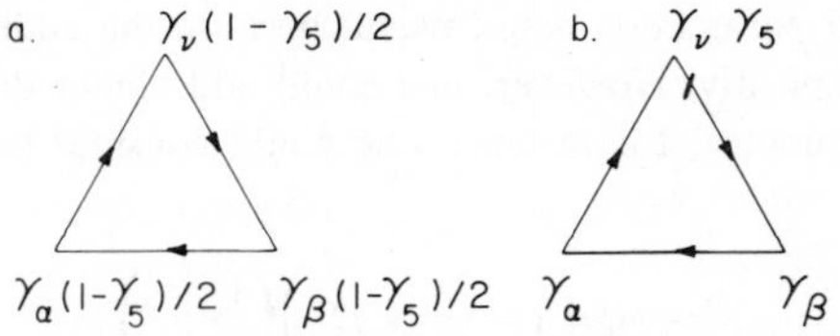

Fig. 5. Alternative forms of the fermion triangle.

each of the three lines and check for current conservation. Or one may insist on Bose symmetry and current conservation in two of the three lines and check for current conservation in the third one.

The difference is exemplified by the two diagrams of fig. 5. In fig. 5a there is a factor of $\frac{1}{2}\gamma_\mu(1-\gamma_5)$ at each vertex. In fig. 5b there are two insertions of γ_μ and one insertion of $\gamma_\nu\gamma_5$. Formally, for massless fermions, since $(\gamma_5)^2=1$ and γ_5 anticommutes with gamma matrices, fig. 5b is equal to just -2 times the parity-violating part of fig. 5a.

Suppose, though, that one defines these diagrams by Pauli–Villars regularization – by subtracting the contribution of a massive regulator field. Then figs. 5a and 5b are not equivalent in the presence of the regulator, and because of the anomaly they do not become equivalent even as the regulator mass goes to infinity. Fig. 5a, as the regulator mass goes to infinity, obeys Bose symmetry but violates current conservation in each channel, while fig. 5b respects Bose symmetry and current conservation in the vector channels but violates them in the axial vector channel.

If the amplitude of 5a obeyed current conservation its parity-violating part would serve as an acceptable definition of the amplitude of fig. 5b. If fig. 5b were conserved in the axial vector channel, then by Bose symmetrizing it one could get an acceptable definition of the amplitude of fig. 5a.

This observation is indispensable because in fact (at least if one uses Pauli–Villars regularization) it is much easier to calculate the amplitude of fig. 5b. When we calculate gravitational anomalies, the simplification from considering a diagram of type 5b will be essential.

As we have discussed, the existence of anomalies depends upon the fact that the one-loop diagrams cannot be regulated in a way that preserves chiral symmetry. Let us therefore examine this point briefly. To regularize the lagrangian

$$\mathscr{L}=\bar{\psi}i\not{D}\left(\frac{1-\gamma_5}{2}\right)\psi$$

by adding, say,

$$\frac{1}{\Lambda}\bar{\psi}D_\mu D^\mu\left(\frac{1-\gamma_5}{2}\right)\psi,$$

(with Λ as a cutoff parameter) is not useful because the additional term violates chiral symmetry explicitly. However, one could add higher-dimension terms that preserve chiral symmetry; for instance one could consider the chirally invariant lagrangian

$$\mathscr{L}' = \bar{\psi} i \not{D}\left(1 - \frac{1}{\Lambda^2} D_\mu D^\mu\right)\left(\frac{1-\gamma_5}{2}\right)\psi .$$

The reason that this regularization fails to eliminate anomalies is slightly subtle: Although $\mathscr{L}'$ conserves chiral symmetry, the naive current

$$J^a_\mu = \bar{\psi}\gamma_\mu \lambda^a \left(\frac{1-\gamma_5}{2}\right)\psi$$

is not conserved in the theory described by $\mathscr{L}'$. Rather one must find the appropriate conserved current by applying Noether's theorem to $\mathscr{L}'$. It includes additional pieces such as

$$\Delta J^a_\mu = \bar{\psi} i \gamma^\mu \lambda^a \left(-\frac{1}{\Lambda^2} D_\alpha D^\alpha\right)\psi .$$

In one-loop diagrams with gluons (or gravitons) coupled to the conserved currents derived from $\mathscr{L}'$, there are extra factors of momentum in the vertices which just cancel the improvement of the propagators. Therefore, the passage from $\mathscr{L}$ to $\mathscr{L}'$ as a regularization does not eliminate anomalies.

More generally, we can show that under broad assumptions the one-loop anomalies depend only on the quantum numbers of the elementary fields, and not on the specific lagrangian chosen. Let A and B be two appropriate differential operators; in the spin-$\frac{1}{2}$ case we may take $A = i\not{D}$ and $B = i\not{D}(1 - D_\mu D^\mu/\Lambda^2)$, for example. Assume that A and B conserve parity. Let us prove that the two theories with lagrangians

$$\mathscr{L}_1 = \bar{\psi} A \left(\frac{1-\gamma_5}{2}\right)\psi , \qquad \mathscr{L}_2 = \bar{\psi} B \left(\frac{1-\gamma_5}{2}\right)\psi$$

have the same one-loop anomalies. It is equivalent to prove that the theory with lagrangian

$$\mathscr{L} = \bar{\psi} A \left(\frac{1-\gamma_5}{2}\right)\psi + \bar{\psi} B \left(\frac{1+\gamma_5}{2}\right)\psi ,$$

is free of anomalies. But

$$\mathscr{L} = \bar{\psi}\tfrac{1}{2}(A+B)\psi - \bar{\psi}\tfrac{1}{2}(A-B)\gamma_5\psi ,$$

and a suitable regularization is simply to pass from $\mathscr{L}$ to

$$\mathscr{L}' = \bar{\psi}\tfrac{1}{2}(A+B)(1 + (-D_\mu D^\mu/\Lambda^2)^n)\psi - \bar{\psi}\tfrac{1}{2}(A-B)\gamma_5\psi ,$$

for some suitable integer *n*. This regularization improves the propagators in Feynman diagrams and it does not add extra powers of momentum in *parity-violating* vertices. Since all parity-violating amplitudes computed from $\mathscr{L}'$ are highly convergent, $\mathscr{L}'$ is free of anomalies, proving that the original theories with lagrangian

$$\bar{\psi}A\left(\frac{1-\gamma_5}{2}\right)\psi, \quad \text{or} \quad \bar{\psi}B\left(\frac{1-\gamma_5}{2}\right)\psi$$

had exactly the same anomalies. This observation will be quite useful when we calculate the gravitational anomaly of a spin-$\frac{3}{2}$ field in sect. 7.

3. Gravitational anomalies in four dimensions

Before attempting a systematic discussion of gravitational anomalies in *n* dimensions, let us discuss some implications of known facts in four dimensions.

Soon after the discovery of the standard triangle anomaly, it was pointed out that the fermion triangle with one axial current and two energy momentum tensors (fig. 3) has a similar anomaly [5]. This anomaly has been much discussed in connection with gravitational instantons. The axial vector current J^5_μ of massless fermions is not conserved, but obeys

$$D_\mu J^5_\mu = -\frac{1}{384\pi^2} R\tilde{R}\,, \tag{6}$$

where $R\tilde{R} = \frac{1}{2}\varepsilon^{\mu\nu\alpha\beta} R_{\mu\nu\sigma\tau} R_{\alpha\beta}{}^{\sigma\tau}$, $R_{\mu\nu\sigma\tau}$ being the usual Riemann curvature tensor. One of the results of our discussion in sect. 6 will be a relatively simple way to obtain eq. (6) from Feynman diagrams.

Violation of a global chirality conservation law in the presence of gravity is only one aspect of eq. (6). Another consequence arises if the anomalous axial current J^5_μ is coupled to a gauge field A_μ. In this case eq. (6) represents a breakdown of gauge invariance. It corresponds to an effective action Γ that is not gauge invariant but changes under the transformation $A_\mu \to A_\mu - \partial_\mu \varepsilon$ by an amount

$$\delta\Gamma = -\frac{1}{384\pi^2} \int \mathrm{d}^4x \sqrt{g}\,\varepsilon(x) R\tilde{R}(x)\,. \tag{7}$$

Actually, the anomaly takes this form if one defines the triangle diagram in a way that maintains general covariance and sacrifices current conservation. One may instead define a triangle amplitude that obeys current conservation and violates general covariance. This can be done as follows. The topological density $R\tilde{R}$ can be written as a total divergence $R\tilde{R} = D_\mu K^\mu$ where K^μ is a functional of the metric that is not generally covariant. One can replace the amplitude Γ by $\Gamma' = \Gamma + (1/384\pi^2)\int \mathrm{d}^4x \sqrt{g} A_\mu K^\mu$ which differs from Γ' by a local functional of the fields. One may readily see that Γ' is invariant under the gauge transformation $A_\mu \to A_\mu - \partial_\mu \varepsilon$. However, Γ' is not generally covariant because K^μ is not. It is not possible to define the triangle in a way that respects both gauge invariance and general covariance.

Therefore in four dimensions gauge theories cannot be consistently coupled to gravity unless the triangle anomaly of fig. 3 cancels when summing over all of the elementary Fermi fields.

Precisely this problem can potentially arise in the standard $SU(2)\times U(1)$ model of weak interactions. Let us take A_μ to be the U(1) gauge field that is coupled to hypercharge. Let Y be the hypercharge operator regarded as a matrix acting on the left-handed Fermi fields. The triangle diagram with one external hypercharge gauge boson and two external gravitons is proportional to Tr Y. A necessary condition for consistency of the $SU(2)\times U(1)$ model coupled to gravity is therefore

$$\mathrm{Tr}\, Y = 0\,. \tag{8}$$

If this condition does not hold, then either gauge invariance or general covariance is lost at the one-loop level.

In fact, eq. (8) does hold in nature for the fermions of each observed generation. However, this condition is usually interpreted as evidence for grand unification; in grand unified theories, the hypercharge operator is a generator of a simple group and must be traceless. We see that since gravity does exist in nature, eq. (8) is needed for simple mathematical consistency.

One might be sceptical of this claim on the grounds that general relativity is unrenormalizable. How do we know that eq. (8) for cancellation of the anomaly will not be modified by whatever cures the short distance behavior of quantum gravity?

The point is that it is possible to understand anomalies strictly on the basis of *long wavelength* physics. For instance, consider trying to study the low-energy limit of the gauge boson-graviton-graviton coupling on the basis of unitarity. (See Coleman and Grossman, and Frishman et al., ref. [4].) By imposing unitarity in each channel, one could reconstruct the amplitude from the possible zero mass intermediate states, up to a term that is analytic in the momenta at $p=q=r=0$. But the whole idea of the anomaly is that one cannot eliminate it by adding to the amplitude a term analytic in the momenta at zero momentum. Otherwise, on dimensional grounds, it would be a suitable polynomial of dimension four that could compensate for the anomaly, and one would simply add this polynomial to obtain an anomaly-free triangle.

Theories free of the usual triangle anomalies but with Tr $Y \neq 0$ are easily constructed. For instance, one may add to the standard model left-handed $SU(3)\times SU(2)$ singlets of hypercharges y_i. If $\sum y_i^3 = 0$, but $\sum y_i \neq 0$, this preserves the cancellation of the usual anomalies but introduces an anomaly in the coupling to gravity.

What specific consequences would this have? The usual electric charge operator is $Q = T_3 + \frac{1}{2}Y$. Since T_3 is traceless in any representation of SU(2), if Tr $Y \neq 0$ then Tr $Q \neq 0$. Such a world would therefore have massless electrically charged fermions

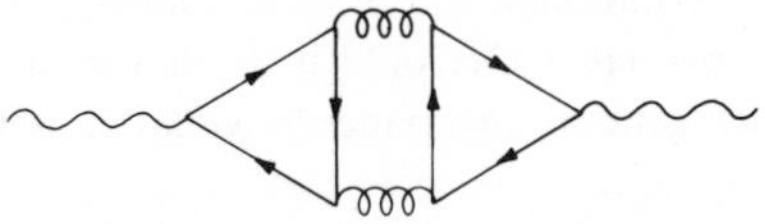

Fig. 6. A two-loop diagram that gives the photon a mass if Tr $Y \neq 0$. External wavy lines are photons; internal loopy lines are gravitons.

with parity-violating electromagnetic couplings. The anomaly would show up in diagrams with external photons and, by analogy with a similar discussion in the case of gauge theories [3], the photon would get a mass from a three-loop diagram with two internal gravitons (fig. 6). Despite the smallness of the gravitational constant, this photon mass is not necessarily negligible. It would be of order

$$m_\gamma^2 = \alpha G_N^2 \Lambda^6 , \tag{9}$$

where m_γ is the photon mass, G_N is Newton's constant, and Λ is an ultraviolet cut-off needed to make sense out of the divergent diagram. For $\Lambda = 500$ TeV, this gives a photon Compton wavelength of about 10^4 km, roughly the observational lower bound*. Thus the gravitational interactions would need to be cut off at rather "low" energies.

The anomaly of fig. 3 has one other interesting application. 't Hooft pointed out some years ago [4] that anomalies can serve as a restriction on the quantum numbers of composite massless particles. Lt J_μ be a conserved current in some quantum field theory, and assume that the corresponding conservation law is not spontaneously broken. Then the $\langle J_\mu J_\alpha J_\beta \rangle$ anomaly computed in terms of the elementary quanta must be precisely equal to the same anomaly computed in terms of the massless particles of the exact physical spectrum. Just the same condition must hold for the $\langle J_\mu T_{\alpha\beta} T_{\sigma\nu} \rangle$ anomaly. Therefore, the trace of any conserved charge Y evaluated among the elementary left-handed fermions must equal the same trace evaluated in the physical spectrum – unless the conservation of Y is spontaneously broken. This requirement should be imposed as a constraint in preon models.

4. Purely gravitational anomalies

In sect. 3, we discussed some implications of the anomaly that arises when gauge fields are coupled to gravity. It is natural to ask whether anomalies occur in theories with gravitational couplings only.

As we have discussed, the classical results about triangle anomalies concern the question of whether the fermion effective action is gauge invariant. Let us ask the analogous question for gravity. Let Γ be the one-loop effective action for matter

* We have assumed here that the triangle is defined in a way that preserves general coordinate invariance and sacrifices gauge invariance. Otherwise, the graviton would gain a mass.

fields propagating in a gravitational field. Γ is, of course, a functional of the metric tensor. Is Γ generally covariant? Or could it be that in some theories of matter fields coupled to gravity general covariance is violated at the one-loop level by anomalies?

Under the infinitesimal coordinate transformation $x^\mu \to x^\mu + \varepsilon^\mu$, the variation in the metric tensor is $\delta g_{\mu\nu} = -D_\mu \varepsilon_\nu - D_\nu \varepsilon_\mu$. The variation of Γ is

$$\delta\Gamma = -\int d^4x \sqrt{g}\, \delta\Gamma/\delta g_{\mu\nu}(D_\mu\varepsilon_\nu + D_\nu\varepsilon_\mu).$$

But $\delta\Gamma/\delta g_{\mu\nu}$ is $\frac{1}{2}\langle T_{\mu\nu}\rangle$, where $\langle T_{\mu\nu}\rangle$ is the expectation value of the energy-momentum tensor of the matter fields. Integrating by parts and using the symmetry of $T_{\mu\nu}$, we have then that $\delta\Gamma = \int d^4x \sqrt{g}\,\varepsilon_\nu D_\mu \langle T^{\mu\nu}\rangle$. So the question of whether the effective action is invariant under infinitesimal general coordinate transformations is equivalent to the question of whether the induced energy-momentum tensor of the matter fields is conserved. This is just analogous to the fact that in the case of gauge fields coupled to charged fermions, gauge invariance of the effective action is equivalent to conservation of the induced current.

Now, by reasoning analogous to the discussion in sect. 2, the real part of Γ is always generally covariant. But we will see that the imaginary part of Γ can suffer from anomalies.

But under what circumstances does Γ have an imaginary part? In euclidean space of n dimensions, the holonomy group of a riemannian manifold is $O(n)$ (or a subgroup thereof). Consider matter fields in some representation Q of $O(n)$. Their coupling to gravity involves the metric and connection – which are real – and the $O(n)$ matrices in the Q representation, which may be complex. Only in case Q in a complex representation does Γ have an imaginary part. But $O(n)$ has complex representations only if $n = 4k+2$ for some integer k, so it is only in $4k+2$ dimensions that the effective action may violate general covariance. In particular, for matter fields coupled to gravity in four dimensions, the one-loop action always respects general covariance.

Which complex representations of $O(n)$ are relevant? For fermions, we may consider spin-$\frac{1}{2}$ fields of definite chirality or spin-$\frac{3}{2}$ fields of definite chirality. As we will see, each of these fields gives rise to one-loop anomalies. In addition, it seems that there is one type of Bose field that suffers from an anomaly. The simplest complex bosonic representation of $O(4k+2)$ is an antisymmetric tensor $F_{\mu_1\cdots\mu_{2k+1}}$ with $2k+1$ indices that obeys a duality condition $F_{\mu_1\cdots\mu_{2k+1}} = \pm i/(2k+1)! \times \varepsilon_{\mu_1\cdots\mu_{2k+1}\nu_1\cdots\nu_{2k+1}} F^{\nu_1\cdots\nu_{k+1}}$. Certain very interesting theories [11, 17] contain an antisymmetric tensor field $A_{\mu_1\cdots\mu_{2k}}$ of $2k$ indices whose curl is constrained to obey such a condition. We will see that also for such a field, there is a one-loop anomaly. Most of the rest of this paper will consist of a detailed evaluation of anomalies for the spin-$\frac{1}{2}$, spin-$\frac{3}{2}$, and antisymmetric tensor fields.

First, however, let us discuss in more detail what is special about $4k+2$ dimensions. (See also ref. [18], where many of the points that follow have been made.)

In an odd number of dimensions the Lorentz group $O(1, n-1)$ has only one type of spinor representation. Its couplings to gravity conserve parity, leading to a real effective action that is free of perturbative anomalies.

In an even number of dimensions, the group $O(1, n-1)$ has two spinor representations related to each other by parity. One might hope to make a theory with parity-violating gravitational couplings by including fermions of one chirality only. This is only possible in $4k+2$ dimensions, for the following reason.

Let $\gamma_0, \gamma_1, \ldots \gamma_{n-1}$ be the Dirac gamma matrices, obeying $\{\gamma_\mu, \gamma_\nu\} = 2\eta_{\mu\nu}$. (Our signature is $(+---\cdots-)$.) The operator that distinguishes the two spinor representations is $\gamma_5 = \gamma_0\gamma_1 \ldots \gamma_{n-1}$. It commutes with all generators of $O(1, n-1)$.

Now in $4k$ dimensions, one may readily see that $(\gamma_5)^2 = -1$. Hence the eigenvalues of γ_5 are $\pm i$, and are exchanged by complex conjugation. This means that in $4k$ dimensions the CPT operation reverses the chirality of fermions. Acting on fermions of $\gamma_5 = +i$, CPT gives fermions of $\gamma_5 = -i$, and vice versa. Hence in $4k$ dimensions, CPT conservation requires the existence of an equal number of fermions of positive and negative chirality. The couplings of fermions to gravity in $4k$ dimensions conserve parity (or violate parity only by irrelevant non-minimal terms) and the effective action is real.

In $4k+2$ dimensions, the story is different. In this case $\gamma_5^2 = +1$, so γ_5 has eigenvalues ± 1. A CPT transformation maps particles of given helicity into particles of the same helicity. In this case, it is possible to consider a theory in which the gravitational couplings are chirally asymmetric. One may have more fermion multiplets of one chirality than of the other. For such theories, the gravitational couplings violate parity, the euclidean space effective action is complex and, as we will see, there are anomalies.

One more side to this story deserves a mention. Specializing to fields of spin $\frac{1}{2}$, in four dimensions the basic object is the four-component Majorana field. General covariance permits this field to have a mass, and hence its couplings to gravity can be regularized in a generally covariant way by the Pauli–Villars method. This being so, the effective action to which it gives rise is generally covariant.

However, in $4k+2$ dimensions the basic object is the fermion field of definite chirality. General covariance (or even global Lorentz invariance) forbids such a field to have a bare mass; a fermion mass term in $4k+2$ dimensions is not allowed for Fermi fields all of the same chirality. Because the mass is forbidden, Pauli–Villars regularization cannot be performed, and anomalies may occur.

Our discussion so far has concerned the question of whether the effective action is invariant under *infinitesimal* general coordinate transformations. If so, one must still address the question of whether the effective action is invariant under *non-infinitesimal* general coordinate transformations – transformations that cannot be reached by exponentiating an infinitesimal transformation. This question which is analogous to certain considerations in gauge theories [16], will be our subject in sect. 10. Here let us simply note a few kinematical facts that are relevant.

In dimensions other than $4k+2$, the basic Majorana spinor representation* of $O(1, n-1)$ is either real or pseudoreal; that is, it admits a second-order invariant tensor that is either symmetric or antisymmetric. For instance, in four dimensions the four-component Majorana spinor representation admits an antisymmetric invariant tensor $\varepsilon_{\alpha\beta}$. Corresponding to this the mass term is $i\varepsilon_{\alpha\beta}\psi^\alpha\psi^\beta$ is possible (it is usually written $\bar{\psi}\psi$).

In eight or nine dimensions, or more generally in $8k$ or $8k+1$ dimensions, the situation is different. In $8k$ or $8k+1$ dimensions, the Majorana spinor representation admits only a *symmetric* invariant $c_{\alpha\beta}$. One cannot in $8k$ or $8k+1$ dimensions write a mass term for a single Majorana fermion, since $c_{\alpha\beta}\psi^\alpha\psi^\beta=0$ by Fermi statistics.

Given several Majorana fermions $\psi^\alpha{}_i$, $i=1\cdots k$, one can write the mass term $c_{\alpha\beta}\psi^\alpha{}_i\psi^\beta{}_j M_{ij}$ where M_{ij} is an antisymmetric mass matrix. If k is even, all fermions can obtain mass this way, but if k is odd, the antisymmetric matrix M necessarily has a zero eigenvalue. This means that with an odd number of Majorana spinors in $8k$ or $8k+1$ dimensions Pauli–Villars regularization is not possible – the regulator field would always have had an even number of components. This suggests that non-perturbatively there may be difficulties with an odd number of Majorana fermions in $8k$ or $8k+1$ dimensions, and we will see in sect. 10 that this is the case. As in the case of the Z_2 anomaly in gauge theories, the difficulty has to do with the sign of the fermion integral.

5. The spin-$\frac{1}{2}$ anomaly in two dimensions

In this section we will illustrate the preceding discussion with a detailed calculation of the gravitational anomaly for a spin-$\frac{1}{2}$ Weyl fermion in two dimensions. We will carry out this discussion with a Minkowski signature. The advantage of two dimensions is that because of the simplicity of the relevant diagram we can calculate explicitly the whole relevant amplitude and then study its behavior under coordinate transformations. In our subsequent study of loop diagrams in more than two dimensions, we will only be able to study the anomalous behavior of the loop diagrams.

In two dimensions the Dirac lagrangian for a fermion in a gravitational field can be written

$$\mathcal{L}=\int \mathrm{d}^2x \det e\, \mathrm{e}^{\nu a}(\tfrac{1}{2}\bar{\psi}i\gamma_a \overleftrightarrow{\partial}_\nu\psi)\,. \tag{10}$$

Here $e_{\nu a}$ is the tetrad; the spin connection drops out of the Dirac lagrangian in two dimensions because of Fermi statistics. We will study the propagation of fermions

* By the Majorana spinor representation we mean the minimum-dimensional representation of the Clifford algebra by gamma matrices that are all real or all imaginary. In $8k$ or $8k+1$ dimensions they are real; for other values of the dimension they are imaginary.

in the weak gravitational field $g_{\mu\nu}=\eta_{\mu\nu}+h_{\mu\nu}$. We will investigate the behavior of the effective action under coordinate transformations, not local Lorentz transformations, so it will be adequate to make a simple gauge choice for the tetrad: $e_{\mu a}=\eta_{\mu a}+\frac{1}{2}h_{\mu a}$. At the linearized level, the interaction lagrangian is simply $\Delta\mathcal{L}=-\frac{1}{2}h^{\mu\nu}T_{\mu\nu}$ where $T_{\mu\nu}=\frac{1}{4}i\bar{\psi}(\gamma_\mu\overleftrightarrow{\partial}_\nu+\gamma_\nu\overleftrightarrow{\partial}_\mu)\psi$ is the fermion energy-momentum tensor.

We wish to calculate the effective action for fermions that obey a Weyl condition $\gamma_5\psi=-\psi$, where $\gamma^5=\gamma^0\gamma^1$. (We will discuss complex fermions that obey this condition, although in two dimensions one could impose a Majorana–Weyl condition; this involves dividing the effective action by two.) It is convenient to introduce light-cone coordinates $x^\pm=\sqrt{\frac{1}{2}}(x^0\pm x^1)$. The corresponding gamma matrices $\gamma^\pm=\sqrt{\frac{1}{2}}(\gamma^0\pm\gamma^1)$ obey $(\gamma^+)^2=(\gamma^-)^2=0$, $\gamma^+\gamma^-+\gamma^-\gamma^+=2$. Indices are raised and lowered as follows: $V^+=V_-$, $V^-=V_+$, $V^\mu W_\mu=V^+W_++V^-W_-=V_+W_-+V_-W_+$.

A fermion obeying $\gamma_5\psi=-\psi$ also obeys $0=\gamma^-\psi=\gamma_+\psi$. For this reason, the free equation of motion $0=(\gamma_+\partial_-+\gamma_-\partial_+)\psi$ reduces to $0=\partial_-\psi=\sqrt{\frac{1}{2}}(\partial/\partial t-\partial/\partial x)\psi$. So in two dimensions a fermion of negative chirality is simply an object that travels constantly in the $+x$ direction at the speed of light. Such an object, of course, cannot have a mass – it cannot be brought to rest – and this is ultimately why anomalies are possible, as we have previously discussed.

With $\gamma_-\psi=\partial_-\psi=0$, the only non-vanishing component of the energy-momentum tensor is $T_{++}=\frac{1}{2}i\bar{\psi}\gamma_+\overleftrightarrow{\partial}_+\psi$, and the linearized interaction of fermions with the gravitational field is $\Delta\mathcal{L}=-h_{--}\frac{1}{4}i\bar{\psi}\gamma_+\overleftrightarrow{\partial}_+\psi$. We will study the effective action to second order in the metric perturbation h, by studying the two-point function

$$U(p)=\int \mathrm{d}^2x\,\mathrm{e}^{ip\cdot x}\langle\Omega|T(T_{++}(x)T_{++}(0))|\Omega\rangle\,. \tag{11}$$

Now, it is possible to see without any computation that there must be an anomaly. The naive conservation law for T_{++} is $\partial_-T_{++}=0$; it leads to the naive Ward identity $p_-U=0$*. If true, this would imply $U=0$ for all non-zero p_-, and hence (by analyticity) for all p_-. But U, as the two-point function of the hermitian operator T_{++}, cannot vanish. So there must be an anomaly.

In fact, the anomaly is easily computed, using tricks introduced in [19]. After performing the Dirac algebra in fig. 7, one finds

$$\begin{aligned}U(p)&=-\tfrac{1}{4}\int\frac{\mathrm{d}k_+\mathrm{d}k_-}{(2\pi)^2}(2k+p)_+{}^2\frac{k_+}{k_+k_-+i\varepsilon}\,\frac{(k+p)_+}{(k+p)_+(k+p)_-+i\varepsilon}\\&=-\tfrac{1}{4}\int\frac{\mathrm{d}k_+\,\mathrm{d}k_-}{(2\pi)^2}(2k+p)_+{}^2\frac{1}{k_-+i\varepsilon/k_+}\,\frac{1}{(k+p)_-+i\varepsilon/(k+p)_+}\,.\end{aligned} \tag{12}$$

* Naively there is no equal time commutator term in this Ward identity. If one looks at (11) as a two-point function in flat space, the anomaly we will find can be regarded as an anomalous commutation relation $[T_{++}(x),T_{++}(y)]=(i/48\pi)\delta'''(x-y)+$tree level terms. It is closely related to the anomaly in the Virasoro algebra in string theories. But we will see that upon coupling T_{++} to the gravitational field, the anomaly is a breakdown of general covariance.

(7)

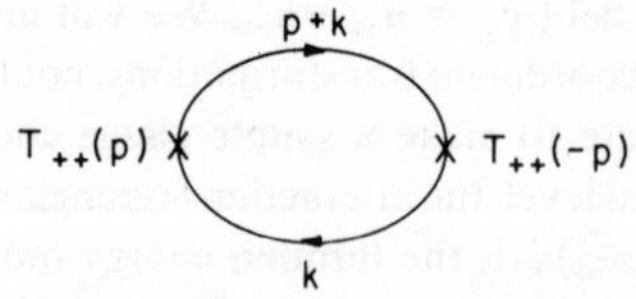

Fig. 7. The gravitational anomaly in two dimensions.

One now performs first the k_- integral by contour integration. It vanishes unless the poles at $k_-=-i\varepsilon/k_+$ and $k_-=-p_--i\varepsilon/(k+p)_+$ are on opposite sides of the real axis. So if, say, $p_+>0$, the k_+ integral can be restricted to $0>k_+>-p_+$. We have then

$$U(p)=\frac{i}{8\pi}\int_{-p_+}^{0}\mathrm{d}k_+\,\frac{(2k+p_+)^2}{p_-}$$

$$=\frac{i}{24\pi}\frac{p_+{}^3}{p_-}\,. \tag{13}$$

So $U(p)\neq 0$. What is more, as expected the anomaly is finite and is a polynomial in the momenta:

$$p_-U(p)=\frac{i}{24\pi}p_+{}^3\,. \tag{14}$$

Now, let us discuss the question of the covariance of the effective action. If we couple $-\frac{1}{2}h_{--}$ at each vertex in fig. 7, then this diagram represents $i\mathcal{L}_{\mathrm{eff}}(h_{--})$, $\mathcal{L}_{\mathrm{eff}}$ being the effective action for the gravitational field. Remembering to include a factor of $\frac{1}{2}$ for Bose statistics, (13) corresponds to the effective action

$$\mathcal{L}_{\mathrm{eff}}(h_{\mu\nu})=-\frac{1}{192\pi}\int\mathrm{d}^2p\,\frac{p_+{}^3}{p_-}\,h_{--}(p)h_{--}(-p)\,, \tag{15}$$

where, of course, $h_{--}(p)$ is the Fourier transform of the metric perturbation. We wish to discuss the behavior of (15) under coordinate transformations.

Under an infinitesimal general coordinate transformation $\delta x^\mu=\varepsilon^\mu$, $h_{\mu\nu}$ transforms as $\delta h_{\mu\nu}(x)=-\partial_\mu\varepsilon_\nu(x)-\partial_\nu\varepsilon_\mu(x)$. A coordinate transformation therefore corresponds in momentum space to

$$\delta h_{++}(p)=-2ip_+\varepsilon_+\,,$$

$$\delta h_{+-}(p)=-ip_-\varepsilon_+-ip_+\varepsilon_-\,,$$

$$\delta h_{--}(p)=-2ip_-\varepsilon_-\,. \tag{16}$$

Now, it is obvious that (15) is not invariant under this transformation; but that does not quite prove the existence of an anomaly. We must try to use our freedom of adding to the effective action a local functional of the fields so as to obtain a gauge invariant effective action. So we try $\mathcal{L}_{\text{eff}} \to \mathcal{L}_{\text{eff}} + \Delta\mathcal{L}$, where

$$\Delta\mathcal{L} = \int d^2p[Ap_+{}^2h_{--}(p)h_{+-}(-p) + Bp_+p_-h_{+-}(p)h_{+-}(-p) + Cp_+p_-h_{++}(p)h_{--}(-p) + Dp_-{}^2h_{++}(p)h_{+-}(-p)], \tag{17}$$

this being the most general Lorentz invariant polynomial of appropriate dimension. It is easy to see that regardless of the choice of A, B, C, and D the modified action $\mathcal{L}_{\text{eff}} + \Delta\mathcal{L}$ is not invariant under (16). This concludes the demonstration that the effective action for a Weyl fermion in two dimensions is not generally covariant.

To complete the picture, suppose we consider a massless Dirac fermion in $1+1$ dimensions, with both chiralities present. The effective action obtained from Feynman diagrams can be found by adding (15) to its parity conjugate:

$$\mathcal{L}_{\text{eff}}^{\text{Dirac}} = -\frac{1}{192\pi}\int d^2p\left(\frac{p_+{}^3}{p_-}h_{--}(p)h_{--}(-p) + \frac{p_-{}^3}{p_+}h_{++}(p)h_{++}(-p)\right). \tag{18}$$

Again, this is not invariant under coordinate transformations. Now, however, we can add a local counterterm to form an action

$$\bar{\mathcal{L}} = -\frac{1}{192\pi}\int d^2p\left(\frac{p_+{}^3}{p_-}h_{--}(p)h_{--}(-p) + \frac{p_-{}^3}{p_+}h_{++}(p)h_{++}(-p) + 2p_+p_-h_{++}(p)h_{--}(-p) - 4p_+{}^2h_{--}(p)h_{+-}(-p) - 4p_-{}^2h_{++}(p)h_{+-}(-p) + 4p_+p_-h_{+-}(p)h_{+-}(-p)\right), \tag{19}$$

which is easily seen to be invariant under coordinate transformations. It can be written more succinctly as

$$\bar{\mathcal{L}} = -\frac{1}{192\pi}\int d^2p\frac{R(p)R(-p)}{p_+p_-}, \tag{20}$$

where $R = p_+{}^2h_{--} + p_-{}^2h_{++} - 2p_+p_-h_{+-}$ is the linearized form of the curvature scalar.

As a by-product of this investigation we can extract a formula for the well-known trace anomaly. Classically, the energy momentum tensor for the massless Dirac field is traceless; it obeys $T_{+-} = 0$. As a consequence, at the linearized level h_{+-} does not couple to fermions and should not appear in the linearized effective action. We see, indeed, that h_{+-} is absent in the effective action (18) obtained from diagrams, but a dependence on h_{+-} is needed in (19) for general covariance.

The trace of the induced energy momentum tensor for a Dirac field in curved space is

$$\langle 2T_{+-}(p)\rangle = -2\frac{\delta\bar{\mathcal{L}}}{\delta h_{+-}(-p)} = -\frac{1}{24\pi}R(p);$$

this is the equation of the trace anomaly.

The evaluation of Feynman diagrams in more than two dimensions is considerably more difficult. It is useful to note that a simple argument ensures that if there is an anomaly in two dimensions, there is also an anomaly in $4k+2$ dimensions, for any k.

Consider a theory of spin-$\frac{1}{2}$ Weyl fermions in $4k+2$ dimensions. Suppose that the $4k+2$ dimensional space is of the form $M^2 \times B$ where M^2 is an asymptotically Minkowskian two-dimensional world and B is a compact space of $4k$ dimensions on which the Dirac field has a non-zero index. Massless fermions in the effective two-dimensional world are zero modes of the Dirac operator on B. The usual relation $\gamma_1\gamma_2 \cdots \gamma_{4k+2} = (\gamma_1\gamma_2).\ (\gamma_3\gamma_4 \cdots \gamma_{4k+2})$ shows that for chiral fermions in $4k+2$ dimensions (say $\gamma_1\gamma_2 \cdots \gamma_{4k+2} = +1$) the two-dimensional chirality (eigenvalue of $\gamma_1\gamma_2$) equals the $4k$-dimensional chirality (eigenvalue of $\gamma_3\gamma_4 \cdots \gamma_{4k+2}$). Hence if there is a non-zero index of the Dirac operator on B, so that the zero modes on B have preferentially one chirality, then the chiral theory in $4k+2$ dimensions reduces macroscopically to a chiral theory in two dimensions.

An anomaly free theory in n dimensions always remains anomaly free after any process of compactification. After all, absence of anomalies means that the Dirac determinant in n dimensions is well defined and generally covariant; if this is so the Dirac determinant must remain generally covariant after any valid approximation, such as an approximate reduction to a two-dimensional determinant. Since the Weyl theory in $4k+2$ dimensions can reduce – in the manner just described – to a Weyl theory in two dimensions, which we know to have an anomaly, the theory of Weyl fermions in $4k+2$ dimensions must have an anomaly for any k.

The restriction to $4k+2$ dimensions emerges, in this context, because the index of the Dirac operator always vanishes except in $4k$ dimensions.

As we will see in detail in subsequent sections, the anomaly in $4k+2$ dimensions involves several tensor structures. Not all of them survive reduction to two dimensions, so the trick of dimensional reduction is not a full substitute for a computation in $4k+2$ dimensions. But it does show that there is an anomaly in $4k+2$ dimensions for any k.

6. The spin-$\frac{1}{2}$ anomaly in $4k+2$ dimensions

In this section we will perform a diagrammatic evaluation of the gravitational anomaly in $4k+2$ dimensions.

We consider a diagram with n external graviton lines. The amplitude will depend on the n momenta $p_\mu^{(i)}$, $i = 1 \cdots n$ of the external gravitons and on their n symmetric polarization tensors $\varepsilon_{\mu\nu}^{(i)}$, $i = 1 \cdots n$. The momenta are not independent but obey one constraint, $\sum_{i=1}^n p_\mu^{(i)} = 0$.

By our general considerations, only the parity-violating amplitudes are anomalous. A parity-violating amplitude is necessarily proportional to the Levi–Civita symbol $\varepsilon_{\mu_1\mu_2\cdots\mu_{4k+2}}$, which must be contracted with the external momenta and polarization vectors. The epsilon symbol, being antisymmetric, can be contracted with at most one index from each symmetric tensor $\varepsilon_{\mu\nu}$, and with at most $(n-1)$ linearly independent momentum vectors p_μ. To saturate the epsilon symbol it must be, therefore, that $n+(n-1)\geqslant 4k+2$ or $n\geqslant 2k+2$. We see, then, that in $4k+2$ dimensions, diagrams with less than $2k+2$ external gravitons are free from anomalies. We will evaluate the anomaly in one-loop diagrams with precisely $2k+2$ external gravitons. Diagrams with more than $2k+2$ gravitons also have anomalies which probably can be determined in terms of the anomalous diagrams with $2k+2$ gravitons by consistency conditions of the Wess–Zumino type.

The lagrangian for a Weyl fermion in $4k+2$ dimensions is

$$S=\int \mathrm{d}x \det e\, \mathrm{e}^{\mu\alpha}\tfrac{1}{2}\bar{\psi}i\gamma_a\overleftrightarrow{D}_\mu\left(\frac{1-\gamma_5}{2}\right)\psi. \tag{21}$$

The covariant derivative of a spinor is $D_\mu\psi=\partial_\mu\psi+\frac{1}{2}\omega_{\mu cd}\sigma^{cd}\psi$, where $\omega_\mu{}^{cd}$ is the spin connection and $\sigma^{cd}=\frac{1}{4}[\gamma^c,\gamma^d]$. So $S=S_1+S_2$, with

$$S_1=\tfrac{1}{2}\int \mathrm{d}x \det e\, e^{\mu a}\bar{\psi}i\gamma_a\overleftrightarrow{\partial}_\mu\left(\frac{1-\gamma_5}{2}\right)\psi,$$

$$S_2=\tfrac{1}{2}\int \mathrm{d}x \det e\, e^{\mu a}\omega_\mu^{\ cd}i\bar{\psi}\{\gamma_a,\tfrac{1}{2}\sigma_{cd}\}\left(\frac{1-\gamma_5}{2}\right)\psi,$$

$$=\tfrac{1}{4}\int \mathrm{d}x \det e\, e^{\mu a}\omega_\mu^{\ cd}i\bar{\psi}\Gamma_{acd}\left(\frac{1-\gamma_5}{2}\right)\psi, \tag{22}$$

where Γ_{acd} is the antisymmetrized product of three gamma matrices, $\Gamma_{acd}=\frac{1}{6}(\gamma_a\gamma_c\gamma_d\pm\text{permutations})$. We will take ψ to be a complex spinor restricted only by the condition $\gamma_5\psi=-\psi$. In $8k+2$ dimensions (but not in $8k+6$ dimensions), it would be possible to restrict ψ by a Majorana condition, and in this case our subsequent formulae must be divided by two.

As in sect. 5 we study a metric $g_{\mu\nu}=\eta_{\mu\nu}+h_{\mu\nu}$, and work in the gauge $e_{\mu a}=\eta_{\mu a}+\frac{1}{2}h_{\mu a}$. The restriction to this gauge means that we will study anomalies in general coordinate transformations but not in local Lorentz transformations (though these could be studied by similar methods).

Vertices originating in S_1 have one or more external gravitons. Vertices originating in S_2 have at least two external gravitons, because, although $\mathrm{e}^{\mu a}\omega_\mu^{\ cd}$ contains a piece linear in $h_{\mu\nu}$, this vanishes when antisymmetrized in a, c, and d.

Parity violation only appears because of the factor of $\frac{1}{2}(1-\gamma_5)$ in the lagrangian. Of course, a trace containing γ_5 will vanish unless at least $4k+2$ gamma matrices are present. Actually, as we will see, in contracting a graviton line with its momentum to test for conservation, we always lose two gamma matrices. Hence, only diagrams with at least $4k+4$ gamma matrices in the numerator will be anomalous.

Consider a one-loop diagram with n_1 vertices originating from S_1 and n_2 vertices originating from S_2. Such a diagram has n_1+n_2 internal propagators, each with one gamma matrix in the numerator. It also has one gamma matrix at each vertex originating in S_1 and three gamma matrices at each vertex originating in S_2. The number of gamma matrices in the numerator of such a diagram is therefore $(n_1 + n_2) + n_1 + 3n_2 = 2(n_1 + 2n_2)$. We need, therefore, $2(n_1 + 2n_2) \geq 4k + 4$ if the diagram is to be anomalous, so $n_1 + 2n_2 \geq 2k + 2$.

We can get an inequality that runs in the opposite direction by counting external graviton lines. We want to look at diagrams with $2k+2$ external gravitons. There will be at least one external graviton for each vertex originating in S_1, and at least two for each vertex originating in S_2 so $2k+2 \geq n_1 + 2n_2$ for the diagrams of interest.

Combining these inequalities, we see that the anomalous diagrams with $2k+2$ gravitons have $n_1 + 2n_2 = 2k+2$, with precisely one graviton line attached to each vertex that originates in S_1 and precisely two attached to each vertex that originates in S_2. The interaction lagrangian therefore simplifies drastically; we may take

$$\mathscr{L}_1 = -\tfrac{1}{4} i h^{\mu\nu} \bar{\psi} \gamma_\mu \overleftrightarrow{\partial}_\nu \left(\frac{1-\gamma_5}{2} \right) \psi ,$$

$$\mathscr{L}_2 = -\tfrac{1}{16} i (h_{\lambda\alpha} \partial_\mu h_{\nu\alpha}) \bar{\psi} \Gamma^{\mu\lambda\nu} \left(\frac{1-\gamma_5}{2} \right) \psi . \tag{23}$$

The Feynman rules for these vertices are given in fig. 8.

Now we must discuss how we will regularize the one-loop diagram. We will use a procedure discussed at the end of sect. 2 and originally due to Adler. Although the one-loop diagrams have Bose symmetry in the external lines, the simplest method for extracting the anomaly does not preserve this symmetry. Instead of placing a factor $\frac{1}{2}(1-\gamma_5)$ at each vertex, we place such a factor at one vertex only. Then we introduce a Pauli–Villars regulator, subtracting from our diagram a similar diagram with a massive fermion of mass M (or we add and subtract suitable diagrams with regulator fields of suitable masses; but it is not necessary to do this explicitly). The amplitude constructed in this way has an anomaly only in the channel where $\frac{1}{2}(1-\gamma_5)$ is inserted. By Bose symmetrization, one can construct the anomaly that

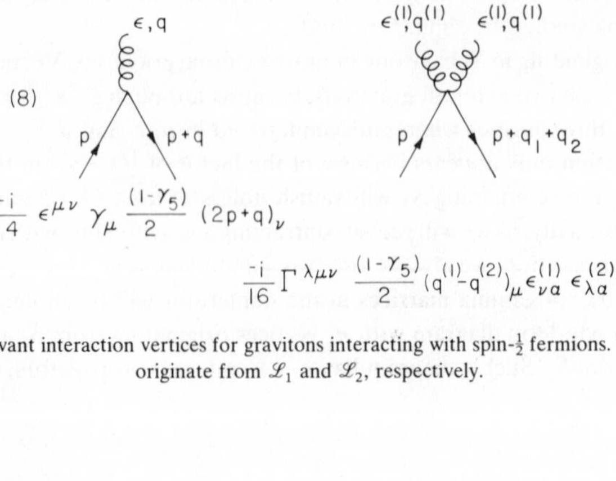

Fig. 8. The relevant interaction vertices for gravitons interacting with spin-$\frac{1}{2}$ fermions. The two vertices originate from $\mathscr{L}_1$ and $\mathscr{L}_2$, respectively.

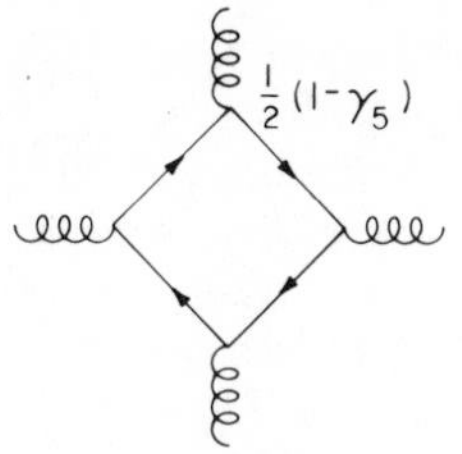

Fig. 9. An anomalous diagram with vertices originating from $\mathscr{L}_1$ only. Shown is the anomalous box diagram of six dimensions. A factor $\frac{1}{2}(1-\gamma_5)$ is inserted at one vertex only. From this amplitude is subtracted a like amplitude for a regulator fermion of mass M.

the Bose symmetric amplitude would have; it is equal, in each channel, to $1/(2k+2)$ times the anomaly we will calculate in one channel.

To indicate how the calculation goes we consider first a diagram (fig. 9) with vertices coming from $\mathscr{L}_1$ only. In the dangerous channel with an insertion of $\frac{1}{2}(1-\gamma_5)$ we take the graviton momentum to be p_μ and its polarization tensor to be $i(p_\mu\varepsilon_\nu + p_\nu\varepsilon_\mu)$; the amplitude with this polarization tensor should vanish because of invariance under the general coordinate transformation $x^\mu \to x^\mu + \varepsilon^\mu$. The other gravitons have momenta $p_\mu^{(i)}$, and to keep the algebra so simple as possible we assume a factorized form $\varepsilon_{\mu\nu}^{(i)} = \varepsilon_\mu^{(i)}\varepsilon_\nu^{(i)}$ for their polarization tensors; the general result can easily be reconstructed from this case.

As pointed out originally by Adler, in the regularized amplitude naive manipulations can be carried out. The anomaly appears only because for the massive regulator fermion, which we will call λ, the energy momentum tensor with insertion of $\frac{1}{2}(1-\gamma_5)$ is not conserved even formally. And it is only the regulator diagram that contributes to the anomaly.

In fact, if the regulator λ obeys $(i\not{D}-M)\lambda=0$, then $D_\nu(\bar\lambda\gamma_\mu \overleftrightarrow{D}_\nu \frac{1}{2}(1-\gamma_5)\lambda)=0$ while $D_\mu(\bar\lambda\gamma_\mu \overleftrightarrow{D}_\nu \frac{1}{2}(1-\gamma_5)\lambda)\neq 0$. Hence in the polarization tensor $i(p_\mu\varepsilon_\nu + p_\nu\varepsilon_\mu)$ of the dangerous channel, the term $p_\nu\varepsilon_\mu$ can be dropped; the other term causes trouble.

In fact, for a fermion of mass M, $D_\mu(\bar\lambda\gamma^\mu \overleftrightarrow{D}_\nu \frac{1}{2}(1-\gamma_5)\lambda) = -iM\bar\lambda \overleftrightarrow{D}_\nu\gamma_5\lambda$. Consequently, in the dangerous channel we may replace $i(p_\mu\varepsilon_\nu + p_\nu\varepsilon_\mu)\cdot(-\frac{1}{4}i\bar\lambda\gamma_\mu \overleftrightarrow{D}_\nu \frac{1}{2}(1-\gamma_5)\lambda)$ by $-\frac{1}{4}M\varepsilon^\nu\bar\lambda \overleftrightarrow{D}_\nu\gamma_5\lambda$. (Note the promised disappearance of a gamma matrix in this manipulation.)

In effect, then, fig. 9 becomes fig. 10, with $2k+1$ external gravitons and one insertion of $-\frac{1}{4}M\varepsilon^\nu\bar\lambda \overleftrightarrow{D}_\nu\gamma_5\lambda$.

As the amplitude of fig. 10 is somewhat complicated, let us proceed in stages. First we carry out the Dirac algebra, after putting the propagators in the usual rationalized form $i(\not{p}+M)/(p^2-M^2)$. The diagram contains exactly one factor of γ_5, at the anomalous vertex labeled Q. Remembering that the polarization vectors have been written $\varepsilon_{\mu\nu}^{(i)} = \varepsilon_\mu^{(i)}\varepsilon_\nu^{(i)}$, so that each non-anomalous vertex contains a factor

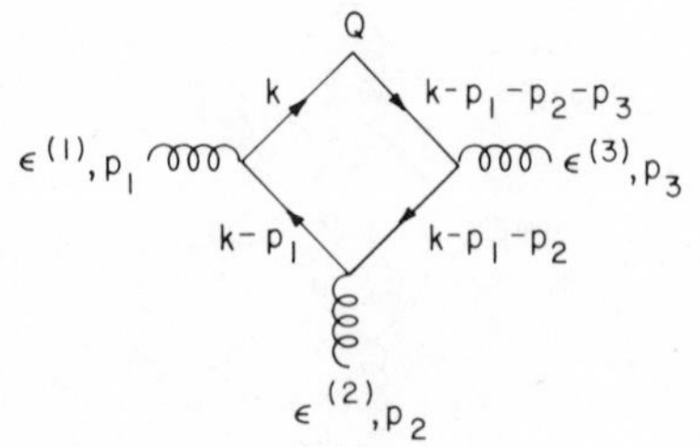

Fig. 10. After manipulations described in the text, the diagram of fig. 10 reduces to a one-loop diagram with one insertion of $Q=-\frac{1}{4}M\varepsilon^{\nu}\bar{\lambda}\tilde{D}_{\nu}\gamma_5\lambda$ and several external gravitons. Fig. 10 is labeled more explicitly than fig. 9, to facilitate comparison with eq. (24).

of $\not\varepsilon^{(i)}$, the Dirac trace that must be performed is

$$A=\mathrm{Tr}\ \gamma_5(\not k+M)\not\varepsilon^{(1)}(\not k-\not p^{(1)}+M)\not\varepsilon^{(2)}(\not k-\not p^{(1)}-\not p^{(2)}+M)$$
$$\times\not\varepsilon^{(3)}\cdots\not\varepsilon^{(2k+1)}(\not k-\not p^{(1)}\cdots-p^{(2k+1)}+M)\,. \qquad (24)$$

The same trace was encountered in ref. [7]. There are at most $4k+3$ gamma matrices multiplying γ_5. A non-zero trace requires $4k+2$ gamma matrices, so we must pick out terms that are precisely linear in M. The various terms obtained by extracting M from different places nearly cancel each other. Bearing in mind that

$$\mathrm{Tr}\ \gamma_5\gamma_{\mu_1}\gamma_{\mu_2}\cdots\gamma_{\mu_{4k+2}}=-2^{2k+1}\varepsilon_{\mu_1\mu\cdots\mu_{4k+2}}\,,$$

the result is $A=2^{2k+1}MR(\varepsilon^{(i)},p^{(j)})$ where

$$R(\varepsilon^{(i)},p^{(j)})=-\varepsilon_{\mu_1\mu_2\cdots\mu_{4k+2}}p^{(1)}_{\mu_1}\varepsilon^{(1)}_{\mu_2}p^{(2)}_{\mu_3}\varepsilon^{(2)}_{\mu_4}\cdots p^{(2k+1)}_{\mu_{4k+1}}\varepsilon^{(2k+1)}_{\mu_{4k+2}}\,. \qquad (25)$$

The important point is that the kinematical factor R depends only on the external momenta and polarization vectors, and not on the loop momentum.

After eliminating the Dirac algebra in this way, the propagators are effectively $i/(p^2-M^2)$ – the propagator of a massive scalar field. At the ith vertex is a factor $-\frac{1}{4}i\varepsilon^{(i)}_{\mu}(p+p')^{\mu}$ where p and p' are the incoming and outgoing momenta of the scalar particles. At the anomalous vertex is a factor $-\frac{1}{4}i\varepsilon_{\mu}(p+p')^{\mu}$ where ε_{μ}, which we will henceforth call $\varepsilon^{(0)}_{\mu}$, is the parameter of an infinitesimal coordinate transformation.

These diagrams will appear formidable. But a little thought shows that the vertex factor $-\frac{1}{4}i\varepsilon^{(i)}_{\mu}(p+p')^{\mu}$ has a simple interpretation. It is the amplitude for the absorption by a charged scalar of charge $\frac{1}{4}$ of a photon of polarization $\varepsilon^{(i)}_{\mu}$ and momentum $(p'-p)^{\mu}$.

A charged scalar also has seagull vertices, of course, where two photons of polarization $\varepsilon^{(1)}$ and $\varepsilon^{(2)}$ are simultaneously absorbed, the amplitude being $2ie^2\varepsilon^{(1)}\varepsilon^{(2)}$. In our problem, however, there are gravitational seagull diagrams, coming from $\mathscr{L}_2$ of eq. (23). It is easy to see that after factoring out the Dirac algebra in the way just described, the gravitational seagulls become electromagnetic seagulls in the analogue problem. Putting the pieces together, the anomalous divergence of the amplitude with $2k+2$ gravitons is

$$I_{1/2}=2^{2k+1}iM^2R(\varepsilon^{(i)},p^{(j)})Z(\varepsilon^{(i)},p^{(j)})\,, \tag{26}$$

where Z is the amplitude for a charged scalar of mass M and charge $\frac{1}{4}$ interacting with $2k+2$ photons of momentum and polarization $p^{(j)}$ and $\varepsilon^{(i)}$, $i, j=0,\ldots,2k+1$.

In general, the evaluation of Z is quite formidable. However, we are interested in the limit as the regulator mass M goes to infinity. Z is gauge invariant and so must be constructed from the electromagnetic field strength $F_{\mu\nu}$ and its derivatives. Since Z is of order $(2k+2)$ in the field strengths, we see by dimensional analysis that in $4k+2$ dimensions, Z vanishes at least as $1/M^2$ for large M. Moreover, terms in Z involving derivatives of field strengths will vanish for large M faster than $1/M^2$ and are negligible. This means that we can regard Z as the amplitude for a scalar propagating in a *constant* electromagnetic field

$$F_{\mu\nu}=-i\sum_{j=0}^{2k+1}(p_\mu^{(j)}\varepsilon_\nu^{(j)}-p_\nu^{(j)}\varepsilon_\mu^{(j)})\,.$$

The amplitude Z for propagation in a constant field is to be evaluated and then expanded in powers of F to extract the term linear in each $p^{(i)}$ and $\varepsilon^{(j)}$.

A diagrammatic evaluation of Z would still be formidable, but the amplitude for propagation in a constant field can be evaluated by a method due to Schwinger [20]. It is convenient to perform the rest of the calculation in euclidean space. The effective action density is

$$\begin{aligned} Z&=\frac{1}{\mathrm{vol}}\ln\det(-D_\mu D^\mu+M^2)\\ &=-\frac{1}{\mathrm{vol}}\int_0^\infty\frac{\mathrm{d}s}{s}\mathrm{Tr}\,\mathrm{e}^{s(D_\mu D^\mu)}\,\mathrm{e}^{-sM^2}\,. \end{aligned} \tag{27}$$

The electromagnetic field can be brought to canonical form

$$F_{\mu\nu}=2\begin{pmatrix} & x_1 & & & & \\ -x_1 & & & & & \\ & & & x_2 & & \\ & & -x_2 & & & \\ & & & \ddots & & x_{2k+1} \\ & & & & -x_{2k+1} & \end{pmatrix}, \tag{28}$$

with "eigenvalues" $2x_1, 2x_2, \ldots, 2x_{2k+1}$. (The factor of 2 is included for later convenience.)

For a particle of charge e interacting in two dimensions with a magnetic field of strength B, it is a classic result that [20]

$$\frac{1}{\mathrm{Vol}}\,\mathrm{Tr}\,\mathrm{e}^{s(D_\mu D^\mu)}=\frac{1}{4\pi}\,\frac{eB}{\sinh(eBs)}\,. \tag{29}$$

With the field brought to the canonical form (28) the "hamiltonian" $H=-D_\mu D^\mu$ in (27) is a sum of $2k+1$ commuting two-dimensional operators. The trace in (27) is therefore a product of two-dimensional traces, so we get

$$Z=-\int_0^\infty \frac{\mathrm{d}s}{s}\prod_{i=1}^{2k+1}\left(\frac{\frac{1}{2}x_i}{4\pi\sinh(x_i s)}\right)\exp(-sM^2)\,. \tag{30}$$

In (30) the factor in brackets is to be expanded to order $2k+2$ in the x_i; terms of lower or higher order are irrelevant. The term of order $(2k+2)$ in x_i is of order s, cancelling the $1/s$ singularity in (30). The s integral is then $\int_0^\infty \mathrm{d}s\,\mathrm{e}^{-sM^2}$. So (30) reduces to

$$Z=-\frac{1}{(4\pi)^{2k+1}}\,\frac{1}{M^2}\prod_{i=1}^{2k+1}\frac{\frac{1}{2}x_i}{\sinh\frac{1}{2}x_i}\,, \tag{31}$$

where it is understood that the right-hand side of (31) is to be expanded in powers of the x_i, with all terms dropped except terms of order $2k+2$.

Combining (26) and (31), we have finally for the anomaly due to a complex spin-$\frac{1}{2}$ Weyl fermion

$$I_{1/2}=-i\frac{1}{(2\pi)^{2k+1}}\,R(\varepsilon^{(i)},p^{(j)})\prod_{i=1}^{2k+1}\frac{\frac{1}{2}x_i}{\sinh\frac{1}{2}x_i}\,. \tag{32}$$

In interpreting this formula, it is to be understood that (32) is regarded as a function of $F_{\mu\nu}$ via eq. (28), and that $F_{\mu\nu}$ is to be expressed in terms of the $\varepsilon_\mu^{(i)}$ and $p_\nu^{(j)}$ by a formula given earlier. The product on the right-hand side of (32) is to be expanded, extracting terms linear in each $\varepsilon^{(i)}$ and $p^{(j)}$.

For $k=0$, (32) agrees with our earlier two-dimensional results (13) and (15). The comparison of (32) with those results is somewhat subtle because (15) was defined to obey Bose symmetry while (32) was computed with a regularization that violates Bose symmetry. The simplest way to make the comparison is to add to (13) suitable contact terms to construct a non-Bose-symmetric functional of the two graviton channels that is conserved in one channel. One then obtains (32) as the anomalous divergence in the other channel.

Eq. (32) is related to the formula for the Dirac index density in $4k+4$ dimensions, in agreement with remarks originally made by M.F. Atiyah (private communication). The linearized Riemann tensor of a graviton is

$$R_{\mu\nu\alpha\beta}=\tfrac{1}{2}(p_\mu p_\alpha\varepsilon_{\nu\beta}+p_\nu p_\beta\varepsilon_{\mu\alpha}-p_\mu p_\beta\varepsilon_{\nu\alpha}-p_\nu p_\alpha\varepsilon_{\mu\beta})\,.$$

If $\varepsilon_{\mu\nu} = \varepsilon_\mu \varepsilon_\nu$ this becomes

$$\tfrac{1}{2}(p_\mu \varepsilon_\nu - p_\nu \varepsilon_\mu)(p_\alpha \varepsilon_\beta - p_\beta \varepsilon_\alpha) .$$

In the mathematical literature one introduces "curvature-two forms"; this amounts to absorbing the $(p_\alpha \varepsilon_\beta - p_\beta \varepsilon_\alpha)$ in our kinematical factor R. The standard mathematical formulae are then written in terms of "eigenvalues" of the "curvature-two forms"; this amounts to working with the eigenvalues x_i of $\frac{1}{2}\sum_j (p^{(j)}\varepsilon_\nu^{(j)} - p_\nu^{(j)}\varepsilon_\mu^{(j)})$, as we have done. The connection of our results with the index theorem in $4k+4$ dimensions will become clearer in sect. 11*.

In sects. 7 and 8 we will calculate the gravitational anomalies due to fields of spin $\frac{3}{2}$ and due to antisymmetric tensor fields. Happily, the tricks we have used in the spin-$\frac{1}{2}$ case will suffice, with a few minor modifications.

7. Gravitational anomalies for fields of spin $\frac{3}{2}$

Now we turn our attention to calculating the gravitational anomaly for fields of other spin. First we will consider the Rarita–Schwinger field $\psi_{\mu\alpha}$; μ is a vector index and α a spinor index. We wish to calculate the anomaly for a field that obeys a Weyl condition $\gamma_5 \psi_\mu = -\psi_\mu$. We will not impose an additional Majorana condition; if this is done our answer must be divided by two.

In the quantization of the Rarita–Schwinger field, it is necessary to introduce several spin-$\frac{1}{2}$ Faddeev–Popov ghosts. Specifically, one needs two ghosts of the same chirality as ψ_μ and one of opposite chirality. Although this counting of ghosts sounds odd at first, it really has a simple explanation. Consider a physical propagating spin-$\frac{3}{2}$ particle of momentum k_μ. The constraints $k_\mu \psi^\mu = 0$ and the gauge invariance under $\psi^\mu \to \psi^\mu + k^\mu \alpha$ (for any spin-$\frac{1}{2}$ field α) remove two spin-$\frac{1}{2}$ degrees of freedom of the same chirality as ψ_μ. The additional constraint $\gamma^\mu \psi_\mu = 0$ removes one spin-$\frac{1}{2}$ degree of freedom of *opposite* chirality to ψ_μ. These conditions leave only a physical spin-$\frac{3}{2}$ particle.

As far as the anomalies are concerned, the effect of the ghosts is very simple. Two ghosts with the same chirality as ψ and one of opposite chirality contribute the same anomaly as one ghost of the same chirality as ψ. Hence we must simply compute the ψ_μ anomaly (using some non-singular lagrangian constructed by gauge fixing) and then subtract the spin-$\frac{1}{2}$ anomaly (already computed in sect. 6) for one ghost field with the same chirality as ψ_μ.

What non-singular lagrangian for ψ_μ may be used? The simplest lagrangian one might hope for would be a simple Dirac-like lagrangian:

$$\mathcal{L} = -\tfrac{1}{2}\bar{\psi}_\mu i\overleftrightarrow{\not{D}}\left(\frac{1-\gamma_5}{2}\right)\psi^\mu . \tag{33}$$

* For gauge theories, such a connection was made by Goldstone (unpublished).

(The minus sign reflects a $(+------\cdots-)$ signature.) Actually, standard gauge fixing in the Rarita–Schwinger lagrangian leads not quite to (33) but to a lagrangian with some extra terms. But now we may make great use of the remarks at the end of sect. 2. The difference between (33) and the standard gauge-fixed Rarita–Schwinger lagrangian does not influence anomalies.

Now we can carry out an analysis similar to the one in sect. 6. Eq. (33) is almost equivalent to a theory of $4k+2$ decoupled spin-$\frac{1}{2}$ fields. The only difference arises because there is an extra term in the covariant derivative of ψ^μ: $D_\nu\psi^\mu = \partial_\nu\psi^\mu + \frac{1}{2}\omega_{\nu ab}\sigma^{ab}\psi^\mu + \Gamma^\mu_{\nu\alpha}\psi^\alpha$. The last term, proportional to the Christoffel symbol $\Gamma^\mu_{\nu\alpha}$, gives rise to additional interaction vertices.

Counting of gamma matrices similar to that which we carried out in sect. 6 shows that, to extract the anomaly in a diagram with $2k+2$ external gravitons, it is adequate to use for $\Gamma^\mu_{\nu\alpha}$ the linearized expression $\Gamma^\mu_{\nu\alpha} = \frac{1}{2}\eta^{\mu\sigma}(\partial_\nu h_{\sigma\alpha} + \partial_\alpha h_{\sigma\nu} - \partial_\sigma h_{\alpha\nu})$. The term proportional to $\partial_\nu h_{\sigma\alpha}$ can be discarded. It cancels upon including the covariant derivatives of ψ and $\bar\psi$ in $\bar\psi i\overleftrightarrow{\not{D}}\psi$ (essentially it cancels because for Majorana fermions, $\bar\psi_\sigma\gamma_\nu\psi_\alpha$ is antisymmetric in σ and α by Fermi statistics). Apart from the obvious generalization of eq. (23),

$$\mathcal{L}_1 = \tfrac{1}{4}ih^{\alpha\beta}\psi_\mu\gamma_\alpha\overleftrightarrow{\partial}_\beta\left(\frac{1-\gamma_5}{2}\right)\psi^\mu\,,$$

$$\mathcal{L}_2 = \tfrac{1}{16}i(h_{\lambda\alpha}\partial_\sigma h_{\tau\alpha})\bar\psi_\mu\Gamma^{\sigma\lambda\tau}\left(\frac{1-\gamma_5}{2}\right)\psi^\mu\,, \tag{34}$$

the new interaction vertex is

$$\mathcal{L}_3 = \tfrac{1}{2}i(\partial_\sigma h_{\alpha\nu} - \partial_\alpha h_{\sigma\nu})\bar\psi^\sigma\gamma^\nu\psi^\alpha\,. \tag{35}$$

Now we carry out an analysis similar to the discussion in sect. 6. For diagrams with $2k+2$ external gravitons coupled to arbitrary vertices of $\mathcal{L}_1$, $\mathcal{L}_2$, or $\mathcal{L}_3$, the Dirac algebra leads to the same kinematical factor $R(\varepsilon_i, p_j)$ defined in sect. 6. After eliminating the Dirac algebra in this way, we are left with a theory of a boson ϕ^μ which in this case has spin one because it inherits the vector index carried by ψ^μ.

As before, in the effective boson theory, the vertices in $\mathcal{L}_1$ and $\mathcal{L}_2$ represent the minimal interaction of ϕ^μ with an electromagnetic field. What about $\mathcal{L}_3$? It has a very simple interpretation – it describes the magnetic moment of ϕ^μ. This arises as follows. For interaction with a graviton of momentum p_μ and polarization $\varepsilon_{\mu\nu} = \varepsilon_\mu\varepsilon_\nu$, (35) becomes $\frac{1}{2}(p_\mu\varepsilon_\nu - p_\nu\varepsilon_\mu)\bar\psi^\mu\not{\varepsilon}\psi^\nu$. The factor of $\not{\varepsilon}$ disappears in doing the Dirac algebra and passing from ψ^μ to an effective boson ϕ^μ. So $\mathcal{L}_3$ reduces to $\frac{1}{2}iF_{\sigma\nu}\phi^{*\sigma}\phi^\nu$ in the effective boson theory.

The effect of this is that the anomaly in the vector-spinor loop is

$$J = 2^{2k+1}iM^2R(\varepsilon^{(i)}, p^{(j)})\tilde{Z}(\varepsilon^{(i)}, p^{(j)})\,, \tag{36}$$

where $\tilde{Z}$ is the effective action for a charged vector meson interacting with the constant electromagnetic field

$$F_{\mu\nu}=-i\sum_{j=0}^{2k+1}(p_\mu^{(j)}\varepsilon_\nu^{(j)}-p_\nu^{(j)}\varepsilon_\mu^{(j)}).$$

The effective hamiltonian for the charged vector meson is (with now a euclidean signature) defined by

$$H\phi_\mu=-(\partial_\sigma+\tfrac{1}{4}iA_\sigma)^2\phi_\mu+\tfrac{1}{2}iF_{\mu\nu}\phi_\nu.$$

As in sect. 6,

$$\tilde{Z}=\mathrm{Tr}\ln H=-\int_0^\infty\frac{\mathrm{d}s}{s}\mathrm{Tr}\,\mathrm{e}^{-sH}. \tag{37}$$

The trace in (37) is simple because $H=H_1+H_2$ where $H_1=-\delta_{\mu\nu}(\partial_\sigma+\frac{1}{4}iA_\sigma)^2$ and $H_2=\frac{1}{2}iF_{\mu\nu}$. For a constant field, H_1 commutes with H_2. Since H_1 acts only on spatial variables and H_2 acts only on the spin, the trace in (37) factorizes as $(\mathrm{Tr}\,\mathrm{e}^{-sH_1})_{\mathrm{space}}\cdot(\mathrm{Tr}\,\mathrm{e}^{-sH_2})_{\mathrm{spin}}$. The spatial trace in that product is the trace discussed in sect. 6. The spin trace is the trace of a constant finite-dimensional matrix; in the notation of eq. (28), this trace is easily seen to be

$$\sum_{j=1}^{2k+1}2\cosh x_j.$$

Evaluating (37) by analogy with the treatment of (27), and remembering to subtract the contribution of a spin-$\frac{1}{2}$ ghost, the anomaly for the spin-$\frac{3}{2}$ field is

$$I_{3/2}=-i\frac{1}{(2\pi)^{2k+1}}R(\varepsilon^{(i)},p^{(j)})$$
$$\times\left(\prod_{i=1}^{2k+1}\frac{\frac{1}{2}x_i}{\sinh(\frac{1}{2}x_i)}\right)\left(\sum_{j=0}^{2k+1}2\cosh x_j-1\right). \tag{38}$$

Again, (38) is to be expanded in a power series in the x_i, only the term of order $2k+2$ in the x_i being relevant.

8. The antisymmetric tensor field

Unlike fermion integrals, which are formal constructions, euclidean space integrals for Bose fields are real, honest integrals. For fermions it is possible to integrate over a field ψ without integrating over its complex conjugate ψ^*. For bosons, instead, one always integrates over both the fields and their complex conjugates. In any theory that has a covariant formulation, the bosonic integration variables form a representation of the O(n) euclidean symmetry group, and it is always a real representation, for the reason just stated. Moreover, for bosons the euclidean action (or at least its real part) must be positive definite, and the kinetic energy is usually a sum of squares $\int dx \sum \theta_i(x)^2$ where the θ_i (linear in derivatives of Bose fields) may have various labels. Such an expression can always be regularized in the manner $\sum \theta_i^2 \to \sum_i ((1 - D_\mu D^\mu / \Lambda^2)\theta_i)^2$ and so leads to an effective action free of anomalies.

The only apparent exception to this reasoning would arise in the case of a boson theory which is Lorentz covariant or generally covariant but does not have a covariant lagrangian. There is no obvious, general way to regularize such a theory or prove the absence of anomalies.

There seems to be only one known case of a covariant Bose field without a covariant lagrangian [17]. This is the case of a theory in $4k+2$ dimensions with a field $A_{\mu_1 \cdots \mu_{2k}}$ that is an antisymmetric tensor with $2k$ indices. The curl of such a field $F_{\mu_1\mu_2\cdots\mu_{2k+1}} = (\partial_{\mu_1} A_{\mu_2\cdots\mu_{2k+1}} + \text{cyclic permutations})$ is an antisymmetric tensor with $2k+1$ indices. $A_{\mu_1\cdots\mu_{2k}}$ is a gauge field; under a gauge transformation it changes as the curl of an object with one less index; but its curl F is gauge invariant. In Minkowski space such $F_{\mu_1\cdots\mu_{2k+1}}$ can be constrained to obey a self-duality condition $F_{\mu_1\cdots\mu_{2k+1}} = (1/(2k+1)!)\varepsilon_{\mu_1\mu_2\cdots\mu_{4k+2}} F^{\mu_{2k+2}\cdots\mu_{4k+2}}$. Together with the Bianchi identity, the self-duality condition serves as a covariant equation of motion. We will consider A and F to be real; if they are complex, the anomaly is twice as big.

In Minkowski space of $4k$ dimensions, the self-duality condition would necessarily contain a factor of i (or $-i$), and CPT would relate the self-dual field to an anti-self-dual field. However, this is not true in $4k+2$ dimension; the self-dual field is self-conjugate under CPT. The self-dual field appears in certain supergravity theories in six and ten dimensions.

It seems that although there are covariant field theories containing the self-dual field, these theories have no covariant lagrangian. This suggests that there might be an anomaly in the coupling of the self-dual field to gravity. Indeed, we can immediately see that this is the case in two dimensions. In two dimensions $k=0$ and the antisymmetric tensor of $2k$ indices is just a scalar field ϕ. In terms of light-cone variables $x^\pm = \sqrt{\tfrac{1}{2}}(x^0 \pm x^1)$, the self-duality condition is just $\partial_-\phi = 0$. Thus, the self-dual field corresponds to "half" of a massless scalar. A massless scalar σ in two dimensions, which obeys $\partial_+\partial_-\sigma = 0$, can be written $\sigma = \sigma_+ + \sigma_-$, where $\partial_+\sigma_- = \partial_-\sigma_+ = 0$; the self-dual field is σ_+.

In two dimensions, by bosonization of fermions [21], a real scalar field is equivalent to a complex spinor field. The real self-dual field ϕ or σ_+ corresponds to the positive chirality complex spinor studied in sect. 5 and has exactly the same anomaly.

Now let us consider the situation in more than two dimensions. Precisely because a covariant lagrangian is not known (and the coupling to gravity has not been worked out even in the light-cone approach of [17]), we will have to be pragmatic and invent suitable Feynman rules.

Our discussion below simplifies slightly if we work in euclidean space. When we use gamma $\gamma_{\mu_1}, \gamma_{\mu_2}, \ldots, \gamma_{\mu_N}$ matrices, they will be real, symmetric $2^{N/2} \times 2^{N/2}$ matrices ($N = 4k+2$) that obey $\gamma_\mu\gamma_\nu + \gamma_\nu\gamma_\mu = 2g_{\mu\nu}$. We define $\gamma_5 = i\gamma_1\gamma_2\cdots\gamma_N$; it is antisymmetric and its square is unity. We also define $\Gamma_{\mu_1\mu_2\cdots\mu_n} =$

$(1/n!)(\gamma_{\mu_1}\gamma_{\mu_2}\cdots\gamma_{\mu_n} \pm \text{permutations})$. Note that the transpose of $\Gamma_{\mu_1\cdots\mu_n}$ is $\Gamma_{\mu_n\cdots\mu_1}$ and that $\operatorname{Tr}\Gamma_{\mu_1\cdots\mu_n}\Gamma_{\nu_n\cdots\nu_1} = 2^{N/2}(g_{\mu_1\nu_1}\cdots g_{\mu_n\nu_n} \pm \text{permutations})$.

The coupling of the graviton to the antisymmetric tensor field $A_{\mu_1\cdots\mu_{2k}}$ is normally $-\frac{1}{2}h^{\mu\nu}T_{\mu\nu}$, where the energy momentum tensor is normally

$$T_{\mu\nu}(F) = \frac{1}{(2k)!}F_{\mu\alpha_1\cdots\alpha_{2k}}F_{\nu\alpha_1\cdots\alpha_{2k}} - \frac{1}{2(2k+1)!}g_{\mu\nu}(F_{\alpha_1\cdots\alpha_k})^2 . \tag{39}$$

Notice that the energy-momentum tensor does not involve $A_{\mu_1\cdots\mu_{2k}}$ explicitly but only the gauge invariant curl $F_{\mu_1\cdots\mu_{2k+1}}$. This means that to construct Feynman diagrams with external gravitons we do not need to construct a propagator for the gauge field A. It is enough to know the gauge invariant free propagator of F:

$$\langle F_{\mu_1\cdots\mu_{2k+1}}(q)F_{\nu_1\cdots\nu_{2k+1}}(-q)\rangle = -\frac{q_{\mu_1}q_{\nu_1}}{q^2}g_{\mu_2\nu_2}\cdots g_{\mu_{2k+1}\nu_{2k+1}}$$

$$\pm \text{permutations of } \mu_i\nu_j . \tag{40}$$

Now, to deal with the self-dual field we will assume that it is correct to use the propagator (40) without modification, while modifying the energy-momentum tensor. We take the interaction with gravity to be $-\frac{1}{2}h^{\mu\nu}T_{\mu\nu}(\frac{1}{2}(F - i\tilde{F}))$, where $\tilde{F}_{\mu_1\cdots\mu_{2k+1}} = 1/(2k+1)!\,\varepsilon_{\mu_1\cdots\mu_{2k+1}\nu_1\cdots\nu_{2k+1}}\cdot F^{\nu_1\cdots\nu_{2k+1}}$. Thus, we permit both self-dual and anti-self-dual fields in propagators, but only self-dual fields can be emitted or absorbed at vertices, since the energy-momentum tensor is constructed from the self-dual part of F only.

In practice, when we actually compute diagrams, we will use the ordinary energy-momentum tensor at every vertex except one, and $T_{\mu\nu}(\frac{1}{2}(F - i\tilde{F}))$ at one vertex only. Since duality is conserved at vertices, in that

$$T_{\mu\nu}(F) = T_{\mu\nu}(\tfrac{1}{2}(F - i\tilde{F})) + T_{\mu\nu}(\tfrac{1}{2}(F + i\tilde{F})) , \tag{41}$$

and since the propagator also conserves duality, it suffices to project out the self-dual part of F at one vertex only.

We now have a well-defined prescription for loop diagrams, and we could proceed, for instance, to compute hexagon diagrams in 10 dimensions (fig. 11). But the algebra would be quite cumbersome.

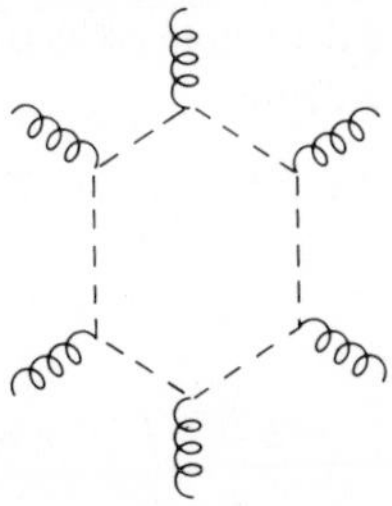

Fig. 11. A hexagon diagram in ten dimensions; the internal lines are antisymmetric tensor fields and the external lines are gravitons.

To achieve some simplification we introduce additional fields that are free of anomalies but permit a covenient reorganization of the algebra. We introduce not just the $2k$ rank tensor field $A_{\mu_1\cdots\mu_{2k}}$ but the whole complement of antisymmetric tensors A, A_μ, $A_{\mu_1\mu_2}$, $A_{\mu_1\mu_2\mu_3}, \ldots, A_{\mu_1\cdots\mu_N}$. (Recall $N=4k+2$ is the dimension of space–time.) And we define the gauge invariant curls

$$\begin{aligned} F &= 0\,, \\ F_\mu &= \partial_\mu A\,, \\ F_{\mu_1\mu_2} &= \partial_{\mu_1} A_{\mu_1\mu_3} + \text{cyclic permutations}\,, \\ &\vdots \\ F_{\mu_1\cdots\mu_N} &= \partial_{\mu_1} A_{\mu_2\cdots\mu_N} + \text{cyclic permutations}\,. \end{aligned} \tag{42}$$

The lagrangian for the antisymmetric fields are simple generalizations of the Maxwell lagrangian:

$$\mathscr{L} = \int \mathrm{d}^N x\,\sqrt{g} - \frac{1}{2\cdot n!}\, g^{\mu_1\nu_1} g^{\mu_2\nu_2} \cdots g^{\mu_n\nu_n} F_{\mu_1\cdots\mu_n} F_{\nu_1\cdots\nu_n}\,. \tag{43}$$

The energy-momentum tensor

$$T^{(n)}_{\mu\nu} = \frac{1}{(n-1)!} F_{\mu\alpha_1\cdots\alpha_{n-1}} F_{\nu\alpha_1\cdots\alpha_{n-1}} - \frac{1}{2\cdot n!}\, g_{\mu\nu} (F_{\alpha_1\cdots\alpha_n})^2\,, \tag{44}$$

and the free propagator

$$\langle F_{\mu_1\cdots\mu_n}(q) F_{\nu_1\cdots\nu_n}(-q)\rangle = -\frac{q_{\mu_1} q_{\nu_1} g_{\mu_2\nu_2}\cdots g_{\mu_n\nu_n}}{q^2} \pm \text{permutations of } \mu_i,\ \nu_j\,, \tag{45}$$

are simple generalizations of previous formulae.

The utility of introducing this host of new fields springs from a simple group theoretical fact. The tensor product of the spinor representation of O(N) with itself is precisely the direct sum of the nth rank tensor representation of O(N) for $0 \leqslant n \leqslant N$. Thus we can describe the whole collection (42) by a single field $\phi_{\alpha\beta}$ with two spinor indices α and β. We define

$$\phi_{\alpha\beta} = 2^{-N/4} \sum_{n=0}^{N} (\Gamma_{\mu_1\cdots\mu_n})_{\alpha\beta} F^{\mu_1\cdots\mu_n}\,. \tag{46}$$

The inverse formula is

$$F_{\mu_1\cdots\mu_n} = 2^{-N/4} (\Gamma_{\mu_n\cdots\mu_1})_{\beta\alpha} \phi_{\alpha\beta}\,, \tag{47}$$

as one may show by simple gamma matrix algebra. The point is that Feynman rules for $\phi_{\alpha\beta}$ are much more manageable than Feynman rules for the $F_{\mu_1\cdots\mu_n}$. Eqs. (45)

are reproduced by a simple equation

$$\langle \phi_{\alpha\beta}(q)\phi_{\gamma\delta}(-q)\rangle = \frac{1}{2q^2}((\gamma_5 \not{q})_{\alpha\gamma}(\gamma_5 \not{q})_{\beta\delta} + q^2 \delta_{\alpha\gamma}\delta_{\beta\delta}) , \tag{48}$$

for the $\phi_{\alpha\beta}$ propagator (recall $\gamma_5 = i\gamma^1\gamma^2 \cdots \gamma^N$). The total energy momentum tensor $T_{\mu\nu} = \sum_{n=0}^{N} T_{\mu\nu}^{(n)}$ (with $T_{\mu\nu}^{(n)}$ in (44)) is

$$T_{\mu\nu} = \tfrac{1}{4}\phi_{\alpha\beta}\phi_{\gamma\delta}((\gamma_\mu\gamma_5)_{\alpha\gamma}(\gamma_\nu\gamma_5)_{\beta\delta}) + (\mu \leftrightarrow \nu) . \tag{49}$$

Eqs. (48) and (49) are much simpler than they may at first look. In (48) the second term lacks a pole at $q^2 = 0$ and can be discarded; it is a non-minimal term that does not affect the anomalies (though it is needed to reproduce (45)). The first term in (48) describes independent propagation of the two spinor indices α and β. The energy momentum tensor also propagates α and β independently. So the combinatories involves a product of two independent Dirac traces and is very simple.

We must generalize the duality operation $F \to \tilde{F}$ to the $\phi_{\alpha\beta}$ field. A suitable generalization, as one may see from (46), is

$$\phi_{\alpha\beta} \to (\gamma_5)_{\alpha\alpha'}\phi_{\alpha'\beta} . \tag{50}$$

For the $(2k+1)$-rank antisymmetric tensor, (50) is the old duality operation. For the other components, (50) is an operation that exchanges n-rank tensors with $(4k+2)$-n-rank tensors. The fact that (50) does not annihilate the n-rank tensors for $n \neq 2k+1$ is unwanted, in the sense that we do not wish to probe for anomalies in those fields, but it is harmless, in the sense that we know that the tensors with $n \neq 2k+1$ have no anomalies.

The Feynman rules (48) and (49) describe a theory with both self-dual and anti-self-dual fields propagating. We wish, instead, to compute the anomaly in a theory with only self-dual fields. As in our previous discussion we accomplish this as follows. At every vertex but one we use the energy-momentum tensor (49), but at one vertex, the anomalous vertex (indicated by a box in fig. 12) we use instead

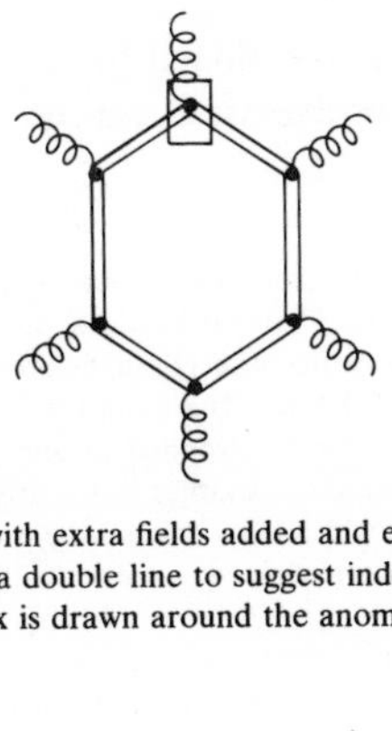

Fig. 12. The diagram of fig. 11, but with extra fields added and expressed in terms of a field $\phi_{\alpha\beta}$ with two spinor indices. $\phi_{\alpha\beta}$ is denoted as a double line to suggest independent propagation of the α and β indices. A box is drawn around the anomalous vertex.

a projected energy-momentum tensor

$$\tilde{T}_{\mu\nu}=\tfrac{1}{4}(\tfrac{1}{2}(1+\gamma_5)_{\alpha\alpha'}\phi_{\alpha'\beta}\cdot\tfrac{1}{2}(1+\gamma_5)_{\gamma\gamma'}\phi_{\gamma'\delta})((\gamma_\mu\gamma_5)_{\alpha\gamma}(\gamma_\nu\gamma_5)_{\beta\delta})+(\mu\leftrightarrow\nu)\,. \quad (51)$$

We now evaluate diagram (12). At the ordinary vertices, the coupling is $-\frac{1}{2}h_{\mu\nu}T^{\mu\nu}$; the ith graviton has momentum $p_\mu^{(i)}$ and polarization $\varepsilon_{\mu\nu}^{(i)}=\varepsilon_\mu^{(i)}\varepsilon_\nu^{(i)}$. At the anomalous vertex, the coupling is $\varepsilon_\nu\partial_\mu\tilde{T}_{\mu\nu}$, where $\varepsilon_\nu=\varepsilon_\nu^{(0)}$ is the parameter of an infinitesimal coordinate transformation.

We regulate diagram (12) by the Pauli–Villars method. The regulator field $\tilde{\phi}_{\alpha\beta}$ has propagator

$$\langle\tilde{\phi}_{\alpha\beta}(q)\tilde{\phi}_{\gamma\delta}(-q)\rangle=\frac{1}{2(q^2+M^2)}((\gamma_5(\not{q}+iM))_{\alpha\gamma}(\gamma_5(\not{q}+iM))_{\beta\delta}+(q^2+M^2)\delta_{\alpha\gamma}\delta_{\beta\delta})\,. \quad (52)$$

Of course, in the regulated diagram, one may use naive manipulations and one need only keep the regulator loop. And the last term in (52) is irrelevant and may be dropped, for reasons mentioned earlier.

The actual evaluation of diagram (12) is rather simple*. Of the two Dirac traces, one contains at the anomalous vertex an insertion of $\not{p}$, p_μ being the graviton momentum. Let us call this trace the α trace and the other one the β trace. The α trace just gives $2^{2k+1}iM^2R(\varepsilon^{(i)},p^{(j)})$ where R is the familiar kinematic factor of sect. 6. The remaining expression is very simple; it is one half the amplitude for a Dirac spinor of charge $\frac{1}{4}$ interacting with operators of momenta $p_\mu^{(i)}$ and polarizations $\varepsilon_\mu^{(i)}$. This should not be too surprising in view of our results of sects. 6 and 7. (Note that after dropping the irrelevant last term and performing the α trace, (52) reduces to $\frac{1}{2}\gamma_5$ times the standard Dirac propagator $(\not{q}+iM)/(q^2+M^2)$. Likewise after performing the α trace (49) becomes a standard Dirac vertex $\not{\varepsilon}$ times $\frac{1}{2}\gamma_5$.)

Our anomaly is hence

$$\tilde{I}=-\tfrac{1}{2}iM^2 2^{2k+1}R(\varepsilon^{(i)},p^{(j)})Z\,, \quad (53)$$

where Z is the amplitude just mentioned for a charged spinor interacting with photons. As in our previous problems, Z can be computed as the amplitude for

* The interested reader should note the following points. One has, of course, a minus sign for the regulator loop. In matrix elements (or Feynman vertices) of the energy-momentum tensor one must include a factor of two from Bose statistics. The γ_5's in the propagators and vertices cancel harmlessly, leaving only γ_5's from projectors $\frac{1}{2}(1+\gamma_5)$. The γ_5 in the α trace goes into making the kinematic factor R, but the F in the β trace can be dropped (it gives terms that vanish as $M\to\infty$). So the $\frac{1}{2}(1+\gamma_5)$ in the β trace gives a factor of $\frac{1}{2}$. Another factor of $\frac{1}{2}$ occurs because the fields are real, but there is a factor of 2 from choosing which Dirac trace is the α trace and which is the β trace. Because of the two factors of $\frac{1}{2}$ and one factor of 2, we get eventually $\frac{1}{2}$ the amplitude for a charged Dirac particle in an external field.

propagation in the constant field $F_{\mu\nu}$ defined in sect, 7. So

$$\begin{aligned} Z &= \mathrm{Tr}\ln(i\not{D}+iM) \\ &= \tfrac{1}{2}\mathrm{Tr}\ln(-\not{D}^2+M^2) \\ &= \tfrac{1}{2}\mathrm{Tr}\ln(-D_\mu D^\mu+M^2+i\Gamma^{\mu\nu}F_{\mu\nu}) \\ &= -\tfrac{1}{2}\int_0^\infty \mathrm{d}s\, \mathrm{e}^{-sM^2}\,\mathrm{Tr}\,\mathrm{e}^{-s(-D_\mu D^\mu+i\Gamma^{\mu\nu}F_{\mu\nu})}\,. \end{aligned} \tag{53}$$

(Here $\Gamma^{\mu\nu}=\frac{1}{2}[\Gamma^\mu,\Gamma^\nu]$.) For a constant field, $-D_\mu D^\mu$ and $i\Gamma^{\mu\nu}F_{\mu\nu}$ commute with each other, so the trace in (53) factorizes as a product of traces. The trace of $\mathrm{e}^{-s(-D_\mu D^\mu)}$ was evaluated in sect. 6, while in the notation of that section

$$\mathrm{Tr}\,\mathrm{e}^{-is\Gamma^{\mu\nu}F_{\mu\nu}}=2^{2k+1}\prod_{i=1}^{2k+1}(\cosh(\tfrac{1}{2}x_i))\,. \tag{54}$$

Evaluating the s integral in (53) as in sect. 6, we find that the anomaly $\tilde{I}$ of a real self-dual antisymmetric tensor field is

$$I=+\tfrac{1}{4}i\frac{1}{(2\pi)^{2k+1}}R(\varepsilon^{(i)},p^{(j)})2^{2k+1}\prod_{i=1}^{2k+1}\frac{\frac{1}{2}x_i}{\sinh(\frac{1}{2}x_i)}\cosh(\tfrac{1}{2}x_i)\,, \tag{55}$$

which, as always, is to be expanded to order $2k+2$ in the x_i. As far as the term of order $2k+2$ is concerned, (55) is equivalent to

$$I=\tfrac{1}{8}i\frac{1}{(2\pi)^{2k+1}}R(\varepsilon^{(i)},p^{(j)})\prod_{i=1}^{2k+1}\frac{x_i}{\tanh x_i}\,. \tag{56}$$

A special case of (56) can easily be tested. In two dimensions, by a bosonization argument mentioned earlier, the anomaly for a real self-dual scalar must equal that of a complex fermion of definite chirality. Hence as one can easily check, for $k=0$ the term of order x^2 in (56) must coincide with the term of order x^2 in eq. (32) for the spin-$\frac{1}{2}$ anomaly.

9. Mixed anomalies

Until now, our object has been to analyze anomalies that arise for theories with matter fields coupled to gravity only.

Higher-dimensional anomalies for theories with matter fields coupled to gauge fields only have been computed previously by several authors [7].

In this section, we will consider a more general problem that encompasses these two cases: we will calculate the anomalies for matter fields coupled to both gravity and gauge fields. Since (as far as is known) antisymmetric tensor fields cannot be consistently coupled to gauge fields, the relevant cases are the fields of spin $\frac{1}{2}$ or spin $\frac{3}{2}$.

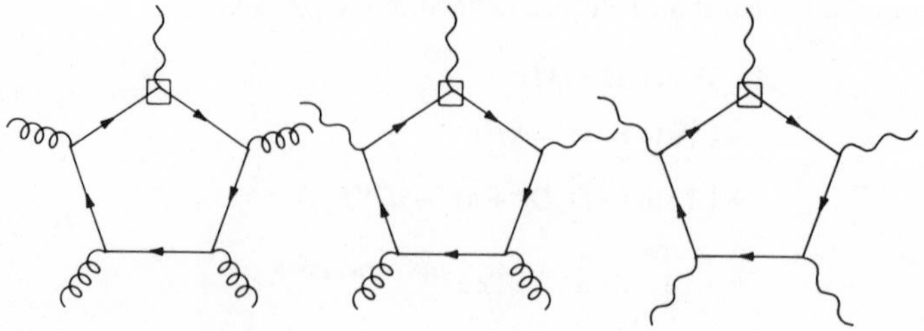

Fig. 13. The anomalous diagrams in eight dimensions. Gluons are wavy lines and gravitons are loopy lines. At one vertex (with a box around it) there is a projection operator $\frac{1}{2}(1-\gamma_5)$.

In any even number of dimensions $2n$, we will study diagrams with $n+1$ external boson lines, which may be gluons or gravitons. We will see that diagrams with r external gluons and $n+1-r$ external gravitons are always anomalous, as long as $n+1-r$ is even. For instance, in eight dimensions, the diagrams with five external gluons, three external gluons and two gravitons, or one external gluon and four gravitons are all potentially anomalous, depending on the gauge group (fig. 13). The cancellation of anomalies in higher-dimensional theories is much more difficult than in four dimensions; implications for some pseudo-realistic models will be described elsewhere [13].

So far, we have only discussed the particular case of diagrams with no external gluons and $n+1$ external gravitons. There is something very special about this case.

The triangle anomaly in four dimensions has two important but fundamentally different interpretations. If one external current is the generator of a global symmetry while the other two are coupled to gauge mesons, the anomaly represents the breakdown of the global symmetry in the presence of gauge fields. As such it can be related to the index theorem [22] for the four-dimensional Dirac operator. The anomalous part of the triangle can be deduced from the index theorem; calculating the triangle is a way to prove the index theorem.

The triangle anomaly has another interpretation, which has been stressed in this paper and is *not* directly related to any four-dimensional index theorem. If all three currents are coupled to gauge mesons, the anomaly represents a breakdown of gauge invariance.

The gravitational anomaly with external gravitons only has the second sort of interpretation. In gauge theories we can distinguish local symmetries from global symmetries; we can distinguish conserved currents coupled to gauge mesons from conserved currents that generate global symmetries. If precisely one of the currents is a global current, an anomalous diagram is related to an index theorem. In gravity we cannot distinguish a "global energy momentum tensor" from a "local energy-momentum tensor"; there is only one energy momentum then, and it couples to gravity. The anomalies evaluated in previous sections represent violation of general covariance. They cannot be interpreted as violations of a global conservation law in the presence of gravity. They cannot be derived from an index theorem or any known mathematical theorem*. This fact was one of the main motivations for the detailed calculations (and exposition) in this paper. By contrast, we now will examine anomalies in diagrams with gravitons *and* currents. By regarding one of the currents as the current of a global symmetry, these anomalies can be interpreted as breakdown of a global conservation law; and they can be deduced from (or used to prove) known index theorems. This was noted for the case of diagrams with external currents only in some of the papers of ref. [7].

We now turn to the detailed evaluation of the relevant diagrams, considering first the case of particles of spin $\frac{1}{2}$. Somewhere in the loop (fig. 13) there is a chirality projection operator $\frac{1}{2}(1-\gamma_5)$. We are really dealing, of course, with the one-loop diagram of a massive regulator field. At a vertex with gluon emission there is a factor

$$-i\gamma^{\mu}T^{a}_{\mathrm{L}}\,, \tag{57}$$

where T^a_{L} is the relevant group generator in the representation furnished by the left-handed fermions. Of course, the group theory factor associated with a given diagram is

$$\mathrm{Tr}\; T^{a_1}_{\mathrm{L}}T^{a_2}_{\mathrm{L}}\cdots T^{a_r}_{\mathrm{L}}\,. \tag{58}$$

After extracting this factor, what appears at a gluon vertex is just $-i\gamma^{\mu}$. At graviton vertices there appears instead

$$-\tfrac{1}{4}i\gamma_{\mu}(p+p')_{\nu}\,. \tag{59}$$

Now, recall that to deal with the graviton vertices our first step was to carry out the gamma matrix algebra. This depends only on the γ_{μ} factor in (59). Since that factor appears also in (57), the gamma matrix algebra goes through in the same way whether the external particles are gluons or gravitons. So all diagrams receive the same kinematic factor $R(\varepsilon^{(i)}, p^{(j)})$ discussed in sect. 5.

After removing this factor and the group theory factor (58), what remains of the graviton vertex is $-\frac{1}{4}i(p+p')^{\mu}$, corresponding to a particle of charge $\frac{1}{4}$ interacting with "photons"; what remains of a former gluon vertex is just $(-i)$, which one can think of as a vertex for absorption of a scalar σ. Of course, our propagators are now $i/(p^2-M^2)$, so we have an effective theory of charged scalars coupled to external photons and scalars (fig. 14).

We now wish to extract the limit as $M^2\to\infty$. Summing over permutations of gluons and gravitons, the one-loop diagrams give an effective action for $F_{\mu\nu}$ and σ. Terms involving derivatives of $F_{\mu\nu}$ or of σ vanish, on dimensional grounds, as

* M.F. Atiyah (private communication) pointed out that a relation between the gravitational anomaly in $4k+2$ dimensions and the index theorem in $4k+4$ dimensions should exist. This remark was based on certain properties of the diffeomorphism group in $4k\times 2$ dimensions. Our formulae for the anomalies are compatible with this idea.

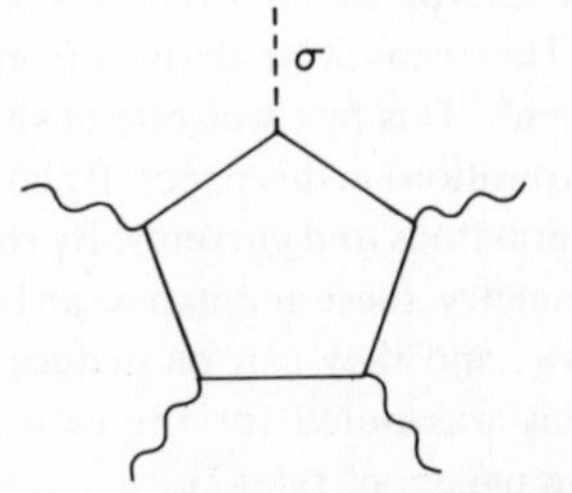

Fig. 14. After performing Dirac algebra, the Dirac propagators reduce to scalar propagators, the graviton vertices reduce to "photon" vertices, and the gluon vertices reduce to vertices for interaction with an effective scalar σ. The first diagram of fig. 13 is redrawn appropriately.

$M\to\infty$, so $F_{\mu\nu}$ can be treated as a constant electromagnetic field and the σ particles can be taken all to have zero momentum.

All dependence on momenta and polarizations of external gluons (which have been reduced to σ particles) is hence contained in $R(\varepsilon^{(i)}, p^{(j)})$. Since this factor is symmetric under permutations of the external lines, the group theory factor (58) must also be symmetrized, yielding what we will call $S\,\mathrm{Tr}\,(T_{\mathrm{L}}^{a_1}\cdots T_{\mathrm{L}}^{a_r})$, the symmetrized trace of the generators $T_{\mathrm{L}}^{a_1}\cdots T_{\mathrm{L}}^{a_r}$.

For emission of a zero-momentum scalar, the vertex $-i$ has a simple and well-known interpretation. It can be interpreted as resulting from differentiation with respect to M^2:

$$\frac{i}{p^2-M^2}(-i)\frac{i}{p^2-M^2}=\frac{\partial}{\partial M^2}\frac{i}{p^2-M^2}\,. \tag{60}$$

So a diagram with external photons and r external σ particles of zero momentum equals the rth derivative with respect to M^2 of a diagram with external photons only. But the amplitude for propagation in an external electromagnetic field only we have already evaluated. We may hence borrow our old results, just replacing (30) by

$$Z'=-S\,\mathrm{Tr}\,T_{\mathrm{L}}^{a_1}\cdots T_{\mathrm{L}}^{a_k}\left(\frac{\partial}{\partial M^2}\right)^r\int_0^\infty\frac{\mathrm{d}s}{s}\prod_{k=1}^{2k+1}\frac{\frac{1}{2}x_i}{4\pi\sinh(\frac{1}{2}x_i s)}\exp(-sM^2)\,, \tag{61}$$

which now is to be expanded to order $\frac{1}{2}(n+2)-k$ in the x_i. The s integral can be done just as before, so

$$Z'=-S\,\mathrm{Tr}\,(T_{\mathrm{L}}^{a_1}\cdots T_{\mathrm{L}}^{a_r})\frac{1}{(4\pi)^{2k+1}}\frac{1}{M^2}\prod_{i=1}^{n/2}\frac{\frac{1}{2}x_i}{\sinh\frac{1}{2}x_i} \tag{62}$$

and the anomaly (since nothing else changes in the derivation of (32)) is

$$I_{1/2}=-S\,\mathrm{Tr}\,(T_{\mathrm{L}}^{a_1}\cdots T_{\mathrm{L}}^{a_r})\frac{i}{(2\pi)^{n/2}}R(\varepsilon^{(i)}, p^{(j)})\prod_{i=1}^{n/2}\frac{\frac{1}{2}x_i}{\sinh\frac{1}{2}x_i}\,. \tag{63}$$

Eq. (63) is to be expanded to order $\frac{1}{2}(n+2)-r$ in the x_i, and expressed in terms of $\varepsilon^{(i)}$ and $p^{(j)}$ by rules explained in sect. 6.

Since (63) is even in each of the x_i, we see there is an anomaly only if $\frac{1}{2}(n+2)-r$ is even. There is potential trouble if the number r of external gluons is $r=\frac{1}{2}n+1$, $\frac{1}{2}n-1, \frac{1}{2}n-3, \cdots$. The condition for cancellation of anomalies (assuming only spin-$\frac{1}{2}$ fields are considered) is

$$S \operatorname{Tr} T_{\mathrm{L}}^{a_1} T_{\mathrm{L}}^{a_2} \cdots T_{\mathrm{L}}^{a_r} = 0\,, \tag{64}$$

for all such r and all $a_1 \cdots a_r$. For large n, (64) is very restrictive, since many values of r must be considered. For $r=\frac{1}{2}n+1$, this result was obtained in ref. [7].

In general, we must also include the contribution of possible charged fields of spin $\frac{3}{2}$. No new features arise in generalizing the conditions of sect. 7, and we get

$$I_{3/2} = -S \operatorname{Tr}(\tilde{T}_{\mathrm{L}}^{a_1} \tilde{T}_{\mathrm{L}}^{a_2} \cdots \tilde{T}_{\mathrm{L}}^{a_r}) \frac{i}{(2\pi)^{n/2}} R(\varepsilon^{(i)}, p^{(j)})$$
$$\times \prod_{i=1}^{n/2} \frac{\frac{1}{2}x_i}{\sinh \frac{1}{2}x_i} \times \left(\sum_{j=1}^{n/2} 2\cosh x_j - 1 \right), \tag{65}$$

where now $\tilde{T}_{\mathrm{L}}^{a}$ are the group generators for left-handed fields of spin $\frac{3}{2}$.

10. Global gravitational anomalies

So far we have considered anomalies that show up in perturbation theory. Such anomalies represent the lack of invariance of the effective action under infinitesimal general coordinate (or gauge) transformations.

Even if perturbative anomalies are absent, we must ask whether the effective action is invariant under coordinate or gauge transformations that cannot be reached continuously from the identity. Actually, in n dimensional euclidean space, we should restrict ourselves to coordinate transformations that approach the identity at infinity. Invariance under such coordinate transformations is needed for the internal consistency of a generally covariant theory, for reasons analogous to similar considerations in gauge theories. Let π be a hypothetical coordinate transformation that approaches the identity at infinity under which the effective action is not invariant. To be specific, suppose (since this is the case we will find) that the fermion integral changes sign under π in some theory. In this case, the theory is inconsistent because the euclidean path integral vanishes. The contribution to the path integral from any metric $g_{\mu\nu}$ would be exactly cancelled by the contribution of the conjugate metric $g_{\mu\nu}{}^{\pi}$ induced from $g_{\mu\nu}$ by the coordinate transformation π. Because π approaches the identity at infinity, one could not exclude $g_{\mu\nu}{}^{\pi}$ by a boundary condition. Moreover, $g_{\mu\nu}{}^{\pi}$ can be reached continuously from $g_{\mu\nu}$ (but not by means of coordinate transformations) by the interpolation $tg_{\mu\nu}+(1-t)g_{\mu\nu}{}^{\pi}$, $0 \leq t \leq 1$, so it is not possible to eliminate $g_{\mu\nu}{}^{\pi}$ by integrating over only "half" of field space.

Let us refer to coordinate transformations that are trivial at infinity and cannot be reached continuously from the identity as disconnected coordinate transformations. Finding such transformations is a difficult problem. It is known that in two or four dimensions there are no disconnected coordinate transformations, but this is unknown in three dimensions. The first example of a disconnected coordinate transformation was given by Milnor, in the six-dimensional case [23] (in constructing so-called exotic seven spheres). Typically, in n dimensions, for large n, there are many disconnected coordinate transformations. For instance, in six dimensions the group of coordinate transformations that are trivial at infinity has 28 components (the identity and 27 disconnected transformations); in eight dimensions this group has 8 components; in ten dimensions it has 992 components; in fourteen dimensions there are 16 256 [24].

There are three situations in which one might envisage global anomalies associated with disconnected general coordinate transformations:

(i) Matter fields in $4k+2$ dimensions which cannot have bare masses because they transform in a complex representation of $O(4k+2)$.

(ii) Theories in $8k$ or $8k+1$ dimensions with a single Majorana Fermi field (or an odd number of them). As discussed at the end of sect. 4, because of Fermi statistics a single Majorana field cannot acquire a mass in $8k$ or $8k+1$ dimensions; only in $8k$ or $8k+1$ dimensions does this phenomena occur.

(iii) Theories in which bare masses are possible.

We will not consider here cases of type (i). In these cases, infinitesimal anomalies appear in perturbation theory, as we have already discussed. The constraints from cancellation of infinitesimal anomalies are much more severe than any additional constraints that would come from global anomalies; however, we do not know if such additional constraints exist.

Regarding theories of type (iii), when bare masses are possible, Pauli–Villars regularization is possible. Since this regularization preserves general covariance, general covariance does not suffer from any local or global anomaly. However, in certain cases Pauli–Villars regularization violates parity, and in those cases there can be a global anomaly leading to breakdown of parity conservation. Although it is outside our main theme we will digress to consider this point.

In an even number of dimensions, Pauli–Villars regularization, if possible at all, always preserves parity. In an odd number of dimensions this is not the case. For instance, a Majorana fermion in $2+1$ dimensions may have a bare mass, but its bare mass violates parity. More generally, in $4k-1$ dimensions the sign of the mass term of a Majorana fermion is odd under parity. An *even* number of Majorana fermions can always receive parity-conserving bare masses (give positive bare masses to half of them and negative bare masses to the other half). With an *odd* number of Majorana fermions in $4k-1$ dimensions, parity conserving bare masses are not possible, and in this case a rather subtle global anomaly can ruin parity conservation.

If a bare mass is possible, the anomaly for any physical field ψ equals the anomaly of a very heavy regulator field χ that may be introduced. Since the whole effective action $\Gamma(\chi)$ of the very massive regulator χ is a local functional, any anomalous behavior of $\Gamma(\chi)$ (under coordinate transformations) can be cancelled by a local functional; therefore, any anomalous behavior of $\Gamma(\psi)$, the physically relevant effective action, can be cancelled by the same local counterterm. However, even if $\Gamma(\psi)$ conserves parity, it may be impossible to choose the local counterterm that cancels its anomalous variation to be parity conserving.

Since there is no trouble with *two* Majorana fermions, $e^{-2\Gamma(\psi)}$ (or $e^{-n\Gamma(\psi)}$ for even n) is generally covariant with no need for parity-violating counterterms. The worrisome possibility is that $e^{-\Gamma(\psi)}$ may change sign under a disconnected coordinate (or gauge) transformation π. If so, the anomalous behavior can be cancelled by adding to $\Gamma(\psi)$ the same local functional that cancels the anomalous behavior of $\Gamma(\chi)$.

$\Gamma(\chi)$ is local, and, like $\Gamma(\psi)$, it is infinitesimally generally covariant. The local functionals of the metric that are infinitesimally but not globally generally covariant are the Chern–Simons secondary characteristic classes Q_α. They are odd under parity and exist in $4k-1$ dimensions (in gauge theories the Chern–Simons classes exist in any odd number of dimensions). Our worry is that the Q_α may appear in $\Gamma(\chi)$. Because the Q_α are multivalued, their coefficients must ordinarily be integers in quantum field theory [25]. However, matter fields can modify this quantization. If $\Gamma(\chi)=\sum c_\alpha Q_\alpha$, with non-integral c_α, then to cancel the anomalous behavior of $\Gamma(\chi)$ – or more pertinently to cancel the equivalent anomaly of $\Gamma(\psi)$ – the coefficients of Q_α in the lagrangian must be $n_\alpha - c_\alpha$, with some integer n_α. In particular, the coefficient of Q_α in the lagrangian cannot vanish, and the quantum theory cannot conserve parity. There are non-trivial examples of this bizarre phenomenon, which is related to discussions of the η invariant in mathematics [26]. For instance, consider an SU(2) gauge theory in $2+1$ dimensions*. Ordinarily the coefficient of the Chern–Simons term (or "topological mass term") must be an integer [25]. If, however, a single Majorana SU(2) doublet is included, one finds (by reasoning similar to that in ref. [16]), that $e^{-\Gamma}$ is odd under a certain disconnected gauge transformation. To compensate for this the Chern–Simons coefficient must be a *half-integer*. It cannot vanish, and the quantum theory cannot conserve parity, contrary to what one would think classically. We do not know under what conditions such phenomena occur in general relativity.

Returning to our main theme, we now consider case (ii) in our previous catalogue of possibilities – the theory of a single Majorana Fermi field in $8k$ or $8k+1$ dimensions. In this case we will actually meet a breakdown of general covariance. We will consider only the case of fermions of spin $\frac{1}{2}$.

As discussed in sects. (2) and (4) (and in the remarks just concluded) for a Dirac field the fermion integral det $i\not{D}$ is perfectly invariant under general coordinate

* This point has been discovered independently and discussed in more detail by N. Redlich, ref. [25].

transformations. The difficulty arises in taking a square root to find the fermion integral $\sqrt{\det i\not{D}}$ of the Majorana field. The sign of the square root is potentially ambiguous.

Is there in $8k$ or $8k+1$ dimensions a disconnected coordinate transformation under which $\sqrt{\det i\not{D}}$ changes sign? Remarkably enough this question has already been answered by Hitchin [27] who proved the existence of such a transformation (this reference was pointed out by M. Atiyah).

Hitchin's interest was actually to prove the existence of riemannian metrics on an arbitrary manifold of $8k$ or $8k+1$ dimensions for which the Dirac operator has a zero eigenvalue. He did this as follows. In $8k$ dimensions (for example) the Dirac operator $\not{D}$ is a real, antisymmetric operator, just like the Dirac operator for SU(2) gauge fields in four dimensions. Its non-zero eigenvalues therefore occur in complex conjugate pairs. Hitchin proved the existence of a disconnected coordinate transformation π with the following property. In interpolating from $g_{\mu\nu}$ to $g_{\mu\nu}{}^{\pi}$ there are, for topological reasons, an odd number of "level crossings" in which an eigenvalue changes place with its complex conjugate (fig. 15). This guarantees the existence of zero eigenvalues somewhere between $g_{\mu\nu}$ and $g_{\mu\nu}{}^{\pi}$. As in ref. [15], it also means that $\sqrt{\det i\not{D}}$ is odd under the transformation π.

We conclude, therefore, that in $8k$ or $8k+1$ dimensions it is inconsistent to couple to gravity an odd number of Majorana fermions of spin $\frac{1}{2}$. We do not know if there is a similar problem for spin-$\frac{3}{2}$ fields, or whether, if so, spin-$\frac{1}{2}$ anomalies can be cancelled by spin-$\frac{3}{2}$ anomalies.

11. An alternative computation of the gravitational anomaly

In sects. 6–9 we have presented a diagrammatic calculation of the gravitational anomaly in $4k+2$ dimensions for Weyl fermions of spin $\frac{1}{2}$ and $\frac{3}{2}$ and for antisymmetric self-dual tensor gauge fields, as well as the combined gravitational and gauge

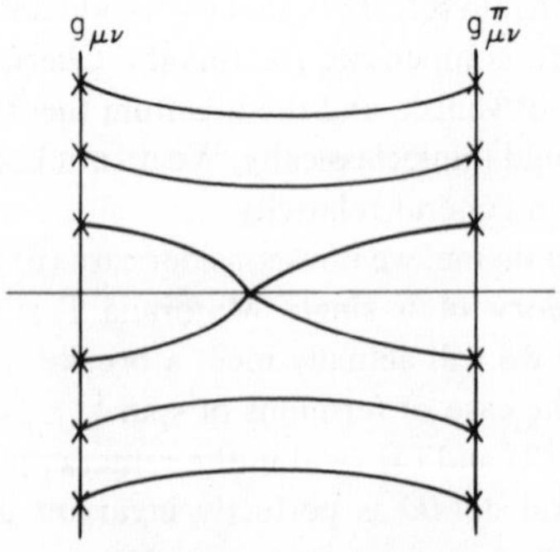

Fig. 15. An odd number of positive eigenvalues become negative in interpolating from $g_{\mu\nu}$ to $g^{\pi}_{\mu\nu}$. This is why $\sqrt{\det i\not{D}}$ is odd under π.

anomalies (figs. 9–14). In the detailed evaluation of the Feynman integrals, we arranged the trace over Dirac matrices and the Lorentz algebra in a convenient way so that the remaining integral could be reduced to the problem of evaluating the propagator of a particle in the presence of a constant external electromagnetic field. The fact that this method works for all the cases considered, suggests that there should be a method of computing the anomaly that exhibits the relation to the propagation of particles in constant external fields from the beginning.

In this section we will present an alternative way of evaluating the gravitational anomalies with or without gauge fields based on a generalization of a procedure first introduced by Fujikawa [28]. The basic idea of Fujikawa's procedure is to notice that the symmetries of the classical action are not necessarily symmetries of the measure in the functional integral which defines the graviton theory, and therefore classical symmetries may cease to be conserved at the quantum level. If one applies this method of computing anomalies to the usual axial anomaly [1], one gets a very clear connection between the anomaly and the Atiyah–Singer index theorem for the Dirac equation [22]. As will be fully explained below, this method proves to be very useful in relating (at least formally) the gravitational anomaly in $4k+2$ dimensions with the index theorem for certain operators in $4k+4$ dimensions, thus supporting the remark by M. Atiyah which was mentioned previously.

We illustrate the method we will use, by first computing the gravitational contribution to the axial anomaly for a fermion coupled to gravity in an arbitrary number of dimensions. Besides purely illustrative purposes, the results of this example will be very useful to us later.

Let $\psi(x)$ be a Dirac fermion defined on a $2n$-dimensional manifold M_{2n} with metric g_{ij}, and euclidean signature. The coupling of $\psi(x)$ to the gravitational field can be read off from (22)

$$\mathscr{L} = e\bar{\psi}(x) i\gamma^{\mu} D_{\mu}\psi \,. \tag{66}$$

(66) is invariant under a global axial U(1) transformation:

$$\psi(x) \to \mathrm{e}^{i\alpha\gamma_5}\psi(x) \,. \tag{67}$$

Under an infinitesimal space–time dependent chiral transformation (67) the action changes by

$$\delta S = \int \mathrm{d}x \, e\alpha(x) D_{\mu}(\bar{\psi}\gamma^{\mu}\gamma_5\psi) \,. \tag{68}$$

In order to check whether the axial current $\bar{\psi}\gamma_{\mu}\gamma_5\psi$ is still conserved at the quantum level, we consider the effective action

$$\mathrm{e}^{-\Gamma(g)} = \int \mathrm{d}\bar{\psi}\, \mathrm{d}\psi \, \exp\{-S(e, \bar{\psi}, \psi)\} \,, \tag{69}$$

and make a change of variables $\psi \to \psi' = \psi + i\alpha(x)\gamma_5\psi(x)$. As was pointed out in [28] the only term which could lead to anomalous contributions to the Ward identities is the measure in (69). The easiest way to define the measure in general, is to expand ψ and $\bar\psi$ in terms of the eigenfunctions of the Dirac equation

$$i\not{D}\psi_n = \lambda_n\psi_n,$$

$$\psi = \sum_n a_n\psi_n, \qquad \bar\psi = \sum_n \psi_n^+(x)\bar b_n,$$

so that the measure becomes $\prod_{n,m} d\bar b_m da_n$. Under the change of variables $\psi \to \psi + i\alpha(x)\gamma_5\psi$, the measure changes by a jacobian factor

$$\prod_{n,m} d\bar b_n da_m \to \left(\prod_{n,m} d\bar b_n da_m\right) \exp\left\{-2i\int dx \sum_n \psi_n^+(x)\alpha(x)\gamma_5\psi_n(x)\right\}. \tag{70}$$

The minus sign on the exponent is due to Fermi statistics. If we let $\alpha(x)$ be a constant for simplicity, we have to evaluate

$$\sum_n \int (dx)\psi_n^\dagger(x)\gamma_5\psi_n(x). \tag{71}$$

This trace is clearly ill-defined. A simple way to define (71) is to use a gaussian cut-off [28]:

$$\sum_n \int \psi_n^\dagger(x)\gamma_5\psi_n(x)e\, dx \equiv \lim_{\beta\to 0}\int (dx)e\psi_n^\dagger(x)\gamma_5\psi_n(x)\, e^{-\beta\lambda_n^2}$$
$$= \lim_{\beta\to 0} \mathrm{Tr}\,\gamma_5\, e^{-\beta(i\not{D})^2}. \tag{72}$$

The anomaly, if any, is given by the β-independent term of the right-hand side of (72). The procedure used in [28] to evaluate (72) becomes very cumbersome when trying to obtain the gravitational contribution to the axial anomaly in an arbitrary number of dimensions. The general philosophy of our procedure is to find a one-dimensional quantum mechanical system defined on the manifold M_{2n} such that its hamiltonian is $(i\not{D})^2$. Thus (72) becomes the partition function for an ensemble with the density matrix $\rho = \gamma_5 \exp[-\beta(i\not{D})^2]$ at temperature β^{-1}; or just $\mathrm{Tr}\,\rho$. Since $\mathrm{Tr}\,\rho$ has a functional integral representation, the evaluation of (72) is equivalent to the high-temperature expansion for the functional integral representation of $\mathrm{Tr}\,\rho$, which as will be shown below is a much simpler problem than the direct evaluation of (72).

The $(0+1)$-dimensional field theory we need is given by:

$$L = \tfrac{1}{2}g_{ij}(x)\frac{dx^i}{d\tau}\frac{dx^j}{d\tau} + \tfrac{1}{2}ig_{ij}(x)\psi^i\left(\frac{d}{d\tau}\psi^j + \Gamma^i{}_{jk}\frac{dx^j}{d\tau}\psi^l\right), \tag{73}$$

where the x^i are the coordinates on M_{2n}, Γ^i_{jk} is the standard Cristoffel symbol constructed in terms of the metric $g_{ij}(x)$, and the $\psi^i(t)$ are one-component real fermionic variables. The form of (73) is suggested by the supersymmetric non-linear σ-model in two dimensions [29]. If we dimensionally reduce the $(1+1)$-dimensional σ-model to $0+1$ dimensions, we obtain

$$L=\tfrac{1}{2}g_{ij}(x)\frac{\mathrm{d}x^i}{\mathrm{d}\tau}\frac{\mathrm{d}x^j}{\mathrm{d}\tau}+\tfrac{1}{2}ig_{ij}(x)\psi^i_\alpha\left(\frac{\mathrm{d}}{\mathrm{d}t}\psi^j_\alpha+\Gamma^i_{jk}\frac{\mathrm{d}x^j}{\mathrm{d}\tau}\psi^k_\alpha\right)+\tfrac{1}{4}R_{ijkl}\psi^i_1\psi^j_1\psi^k_2\psi^l_2\,,$$

$$\alpha=1,2\,, \tag{74}$$

(where R_{ijkl} is the curvature tensor on M_{2n}). If we impose the additional condition that $\psi^i_1=\psi^i_2=\sqrt{\tfrac{1}{2}}\psi^i$, we obtain (73). Before imposing this constraint, (74) is invariant under two supersymmetry transformations generated by two constant anticommuting real numbers ε_1, ε_2. After the constraint is implemented, (73) is still invariant under a single supersymmetry transformation with $\varepsilon_1=-\varepsilon_2=\varepsilon$. The supercharge is given after canonical quantization by $\sqrt{\tfrac{1}{2}}i\not{D}$, so that the hamiltonian of (73) becomes $H=\sqrt{\tfrac{1}{2}}(i\not{D})^2$. If we define a new set of fermion fields $\psi^a=e^a_i(x)\psi^i$, the canonical commutation relations which follow from (73) are:

$$\{\psi^a,\psi^b\}=\delta^{ab}\,, \tag{75}$$

and in terms of ψ^a, ψ^b, the canonical conjugate momentum to x^i is

$$p_i=g_{ij}(x)x^j+\tfrac{1}{4}i\omega_{iab}[\psi^a,\psi^b]\,. \tag{76}$$

Notice that (75) implies that the fermions generate a clifford algebra on M_{2n}, and that $\gamma_5=(-1)^F$; (i.e. $(-1)^F$ is the operator that anti-commutes with all Fermi fields), hence $\mathrm{Tr}\,\rho=\mathrm{Tr}\,(-1)^F\,\mathrm{e}^{-\beta H}$ which is the index for supersymmetric theories introduced in [30]. In fact, $\mathrm{Tr}\,\rho$ in this case is just the index of the Dirac equation [31]. The evaluation of $\mathrm{Tr}\,\rho$ is carried out in terms of its functional integral representation. Standard arguments imply [32]

$$\mathrm{Tr}\,\rho=\int_{\mathrm{PBC}}\mathrm{d}x(\tau)\,\mathrm{d}\psi(\tau)\exp\left\{-\int_0^\beta L_{\mathrm{E}}(\tau)\,\mathrm{d}\tau\right\}, \tag{77}$$

with bosons and fermions integrated over with periodic boundary conditions (PBC) with period β due to the presence of $(-1)^F$ in the trace. $L_{\mathrm{E}}(\tau)$ stands for the euclidean version of (73). In the $\beta\to 0$ limit, (77) is dominated by constant paths $x^i=x^i_0$, $\psi^i=\psi^i_0$, which in this case are zero-action solutions of the classical equations of motion. Hence the leading $\beta\to 0$ behavior of (77) is given by the quadratic term in the expansion of $L_{\mathrm{E}}(\tau)$ around the constant configurations (x^i_0,ψ^i_0). This expansion is greatly simplified if we use normal coordinates on M_{2n} [31, 33]. If $(\xi(\tau),\eta(\tau))$ are respectively the bosonic and fermionic non-constant small fluctuations around (x_0,ψ_0), the small-fluctuation lagrangian is

$$L^{(2)}_{\mathrm{E}}=\tfrac{1}{2}g_{ij}(x_0)\frac{\mathrm{d}\xi^i}{\mathrm{d}\tau}\frac{\mathrm{d}\xi^j}{\mathrm{d}\tau}-\tfrac{1}{4}R_{ijab}\psi^a_0\psi^b_0\xi^i\frac{\mathrm{d}\xi^j}{\mathrm{d}\tau}+\tfrac{1}{2}i\eta^a\frac{\mathrm{d}}{\mathrm{d}\tau}\eta^a\,.$$

Thus, the computation of the trace is reduced to the evaluation of a one-dimensional determinant. Since the fermionic fluctuations do not couple to the curvature to second order, they will cancel with the normalization factor (we normalize the trace with respect to the flat space case). Hence, the normalized trace is:

$$(\mathrm{Tr}\,\rho)_{\mathrm{norm}}=\frac{i^n}{(2\pi)^n}\int(\mathrm{d}x_0)(\mathrm{d}\psi_0)\frac{\det^{-1/2}(-\delta_{ab}\,\mathrm{d}^2/\mathrm{d}\tau^2+R_{ab}\,\mathrm{d}/\mathrm{d}\tau)}{\det^{-1/2}(-\delta_{ab}\,\mathrm{d}^2/\mathrm{d}\tau^2)}, \qquad (79)$$

where $R_{ab}=\frac{1}{2}R_{abcd}\psi_0^c\psi_0^d$.

Despite the dependence of R_{ab} on fermion zero modes, we will treat the $R_{ab}\,\mathrm{d}/\mathrm{d}t$ term as part of the boson kinetic energy. In (79) we have redefined ξ^i to include the vierbein at x_0: $\xi^a=e_i^a(x_0)\xi^i$. Consequently the ξ kinetic term has the standard form, furthermore, the factor of $(2\pi)^{-n}$ comes from the standard Feynman measure for the bosonic degrees of freedom, and the factor of i^n is present because we are integrating over $2n$ real fermionic variables (the ψ_0's). Since R_{ab} is an antisymmetric $2n\times 2n$ matrix, we can skew diagonalize it, and call its eigenvalues x_α $\alpha=1,\ldots,n$. Now (79) becomes

$$\begin{aligned}(\mathrm{Tr}\,\rho)&=\frac{i^n}{(2\pi)^n}\int_{M_{2n}}\mathrm{d\,Vol}\int\prod_\alpha\prod_{n\geq 1}\left(1+\frac{(\frac{1}{2}ix_\alpha)^2}{\pi^2n^2}\right)^{-1}\mathrm{d}\psi_0\\&=\frac{i^n}{(2\pi)^n}\int_{M_{2n}}\mathrm{d\,Vol}\int(\mathrm{d}\psi_0)\prod_\alpha\frac{(\frac{1}{2}ix_\alpha)}{\sinh(\frac{1}{2}ix_\alpha)}.\end{aligned} \qquad (80)$$

Since the x_α are bilinear in the ψ_0's, and the polynomial appearing in the integrand of (80) is even under the interchange of $x_\alpha\to -x_\alpha$ for any number of α's, the number of ψ_0's in each monomial of the Taylor expansion of the integrand of (80) is a multiple of four, and thus the Grassmann integral will vanish unless the manifold has dimension $4k$. Formally, the constant anticommuting numbers ψ_0^a form a realization of the basis of 1-forms on the manifold, and the (ψ_0^a) integral just projects out the term proportional to $\psi_0^1\cdots\psi_0^{2n}$. More geometrically, let $R_{ab}=\frac{1}{2}R_{ab}e^a\wedge e^b$ be the curvature of the manifold referred to orthogonal frames, and let x_α, $\alpha=1,\ldots,n$ be the set of skew eigenvalues of R_{ab}, then (80) can be rewritten as follows

$$(\mathrm{Tr}\,\rho)=\int_{M_{2n}}\left(\prod_\alpha\frac{(x_\alpha/4\pi)}{\sinh(x_\alpha/4\pi)}\right)_{\mathrm{Vol}}. \qquad (81)$$

The subscript "Vol" means that we have to pick out the term in the expansion of the integrand which is proportional to the volume form of M_{2n}. The function of the x_α appearing in (81) is the index density for the Dirac operator, and it is known in the mathematical literature as the Dirac genus, or $\hat{A}$ polynomial [6]. Its expansion in terms of the x_α is

$$A(M_{2n})=1-\tfrac{1}{24}p_1(M)+\tfrac{1}{5760}(7p_1^2-4p_2)+\frac{1}{2^6 15120}(-16p_3+44p_1p_2-31p_1^3)+\cdots; \qquad (82)$$

the $p_i(M)$ are known as Pontryagin classes and are defined by:

$$\det\left(1-\frac{1}{2\pi}R\right)=1+\frac{1}{(2\pi)^2}p_1+\frac{1}{(2\pi)^4}p_2+\frac{1}{(2\pi)^6}p_3+\cdots,$$

$$p_1(M)\equiv\sum_\alpha \omega_\alpha^2=-\tfrac{1}{2}\operatorname{Tr}R^2,$$

$$p_2(M)\equiv\sum_{\alpha<\beta}\omega_\alpha^2\omega_\beta^2=-\tfrac{1}{4}\operatorname{Tr}R^4+\tfrac{1}{8}(\operatorname{Tr}R^2)^2,$$

$$p_3(M)\equiv\sum_{\alpha<\beta<\gamma}\omega_\alpha^2\omega_\beta^2\omega_\gamma^2=-\tfrac{1}{6}\operatorname{Tr}R^6+\tfrac{1}{8}\operatorname{Tr}R^2\operatorname{Tr}R^4-\tfrac{1}{48}(\operatorname{Tr}R^2)^3. \tag{83}$$

Thus, the anomaly in four dimensions is given by $p_1(M)$ [5]:

$$\langle D_\mu J_5^\mu\rangle=-\frac{2i}{24(2\pi)^2}\tfrac{1}{8}R_{abcd}R^{ba}{}_{ef}\varepsilon^{cdef}=-\frac{2i}{384\pi^2}R\tilde{R}. \tag{84}$$

The factor of 2 in the numerator appears because we have been using a Dirac spinor. For a Weyl fermion, the result is half of (84).

The long exercise we have just gone through is not without sense. First it has permitted us to introduce our method and some useful notation, and second, the gravitational anomaly in $4k+2$ dimensions will be shown to be closely related to the axial anomaly in $4k+4$ dimensions.

In applying Fujikawa's method to the gravitational or mixed anomalies, care must be taken in interpreting the results. As it stands, this procedure does not generate the complete solution to the Wess–Zumino consistency conditions. The regulator we will use explicitly violates Bose symmetry because the gravitons have V−A couplings with the matter fields, not the vectorial couplings which appear in $i\not{D}$. However, if we only look for the leading term in the anomaly (the $n+1$ polygon in $2n$ dimensions), i.e. the term with the highest number of external momenta, the results that will be obtained are correct up to a factor of $1/(n+1)$ required to restore Bose symmetry in the leading term as explained thoroughly in sects. 2 and 6. The subleading terms could in principle be obtained using the Wess–Zumino consistency conditions. Conversely, we could start with the complete answer provided by Fujikawa's method which explicitly violates Bose symmetry and then use the Wess–Zumino consistency conditions to obtain the appropriate contact terms which need be subtracted in order to obtain the complete Bose symmetric solution of the consistency conditions. Thus in the final formulae for the anomaly which will follow, the curvature tensor should be understood to stand for $R_{\mu\nu\alpha\beta}=\partial^2_{\mu\alpha}h_{\nu\beta}+\partial^2_{\nu\beta}h_{\mu\alpha}-\partial^2_{\nu\alpha}h_{\mu\beta}-\partial^2_{\mu\beta}h_{\nu\alpha}$, and similarly, the gauge field strength $F^a_{\mu\nu}$ stands for $\partial_\mu A^a_\nu-\partial_\nu A^a_\mu$ as should be in the leading approximation. As will be shown below, this procedure gives the same answer as the diagrammatic calculation and thus provides an independent check on the computation of previous sections.

Let us now turn to the computation of the gravitational anomaly. Following the outline presented above, we first write down the transformation rule for spinors under an infinitesimal general coordinate transformation $x^i \to x^i + \eta^i(x)$. Up to a local Lorentz transformation:

$$\delta_\eta \psi = -\eta^i D_i \psi . \tag{85}$$

Since we are interested in the case when ψ is a Weyl fermion in $2n$ dimensions, the first problem we encounter is that, as explained in sect. 2, the operator $i\not{D}_L = \frac{1}{2} i\not{D}(1-\gamma_5)$ is not self-adjoint. Thus in order to define the measure for the fermionic functional integral, we expand ψ in terms of the eigenfunctions of $(i\not{D}_L)^+(i\not{D}_L)$. We expand $\bar{\psi}$ in terms of the eigenfunctions of $(i\not{D}_L)(i\not{D}_L)^+$ [28]. With this definition of the functional integral and the measure, the jacobian induced by the infinitesimal coordinate transformation (85) is

$$g = \exp\left(\sum_n \int (\mathrm{d}x) e \psi_n^+(x)\eta^i(x) D_i \psi_n(x) - \sum_n \int (\mathrm{d}x) e \phi_n^+(x) \eta^i D_i \phi_n(x)\right). \tag{86}$$

Regularizing the trace as before:

$$\begin{aligned} &\sum_n \int (\mathrm{d}x) e \psi_n^+(x)\eta^i(x) D_i \psi_n(x) - \sum_n \int (\mathrm{d}x) e \phi_n^+ \eta^i D_i \phi_n(x) \\ &\equiv \lim_{\beta\to 0} \int (\mathrm{d}x) e \left(\sum_n \psi_n^+(x) \eta^i D_i\, \mathrm{e}^{-\beta \not{D}_L^+ \not{D}_L} \psi_n - \sum_n \phi_n^+ \eta^i D_i\, \mathrm{e}^{-\beta \not{D}_L \not{D}_L^+} \phi_n\right) \\ &= \lim_{\beta\to 0} \mathrm{Tr}\, \eta^i D_i \gamma_5\, \mathrm{e}^{-\beta(i\not{D})^2}, \end{aligned} \tag{87}$$

which is very similar to (80) and can also be represented as a functional integral associated to (73). In terms of the fields defining the $(0+1)$-dimensional field theory (73), $\eta^i D_i$ can be represented as $i\eta_i(x)\, \mathrm{d}x^i/\mathrm{d}\tau$, and the trace (87) is just

$$\lim_{\beta\to 0} \int_{\mathrm{PBC}} \mathrm{d}x(\tau)\, \mathrm{d}\psi(\tau) \left(-\eta_i(x) \frac{\mathrm{d}x^i}{\mathrm{d}\tau}\right) \exp\left\{-\int_0^\beta L_E(\tau)\, \mathrm{d}\tau\right\}. \tag{88}$$

As before, (88) is dominated in the $\beta \to 0$ limit by the constant configurations so that the only extra bit of information needed to evaluate (88) is the expansion of $\eta_i(x)\, \mathrm{d}x^i/\mathrm{d}\tau$ around a constant configuration (x_0, ψ_0). It is easy to see that to second order the expansion is given by $D_i \eta_j(x_0) \dot{\xi}^i \xi^j$. The computation is further simplified if we exponentiate $D_i \eta_j(x_0) \dot{\xi}^i \xi^j$, and at the end expand the result to first order in η^i. Once we exponentiate the only term which contributes is the antisymmetric part of $D_i \eta_j$, the symmetric part cancelling due to the periodicity of the boundary conditions. Thus, the only change with respect to the previous computation of the axial anomaly is that $R_{ab} = \frac{1}{2} R_{abcd} \psi_0^c \psi_0^d$ is substituted by $R_{ab} + D_a \eta_b - D_b \eta_a \equiv R'_{ab}$. If we denote the skew eigenvalues of R'_{ab} by x_α $\alpha = 1, \ldots, x_n$, and normalize the trace as before. The answer can be literally copied from (80):

$$\frac{i^n}{(2\pi)^n} \int_{M_{2n}} \mathrm{d}\,\mathrm{vol} \int \mathrm{d}\psi_0 \prod_\alpha \frac{\frac{1}{2} i x'_\alpha}{\sinh(\frac{1}{2} i x_\alpha^2)}. \tag{89}$$

Taylor expanding (89), the anomaly will be given by the term which is first order in η^i and contains $2n\psi_0$'s. It is not hard to convince oneself that such a term can only occur if $2n=4k+2$, which provides a nice check on the general arguments presented in sect. 4. To obtain the anomaly, we just expand the integrand in (89) to order $2k+2$ and then extract the term linear in η^i, which is exactly the same thing as writing the gravitational contribution to the axial anomaly in $4k+4$ dimensions with the substitution $R_{ab} \to R_{ab} + D_a\eta_b - D_b\eta_a$, and afterwards extracting the term proportional to η^i, as well as integrating over the constant fermionic variables. It is now a tedious exercise in elementary algebra to insert (82) in (89) and expand to the appropriate order. We will present the result in two ways. First we will give the combination of modified Pontryagin classes p'_i which contain the anomaly, and then we will explicitly display the anomaly in terms of the curvature tensor (by modified Pontryagin classes we mean the polynomials defined in (83) but with the substitution $R_{ab} \to R_{ab} + D_a\eta_b - D_b\eta_a$ understood). In 2 dimensions, the anomaly is contained in $\frac{1}{24}ip'_1$:

$$\int \mathrm{d}^2x\, e\eta^i D^j \langle T_{ij}\rangle = -\frac{i}{48\pi}\int \mathrm{d}^2x\, eD^i\eta^j R_{ijkl}\varepsilon^{kl}\,; \qquad ijkl=1,2\,. \tag{90}$$

For $d=6$, the corresponding coefficient is $-\frac{1}{5760}i(7p_1'^2 - 4p'_2)$:

$$\int \mathrm{d}^6x\, e\eta^i D^j\langle T_{ij}\rangle = -\frac{1}{2880(4\pi)^3}\int D^a\eta^b\varepsilon^{ijklmn} \times (5R_{baij}R^c_{\ jkl}R^d_{\ cmn} + 4R_b{}^c{}_{ij}R_c{}^d{}_{kl}R_{damn})e\,\mathrm{d}^6x\,. \tag{91}$$

Finally, in ten dimensions the anomaly is contained in

$$+\frac{i}{2^6 15120}(-16p'_3 + 44p'_1p'_2 - 31p_1'^3)\,, \tag{92}$$

which yields

$$\begin{aligned}\int \mathrm{d}^{10}x\, e\eta^i D^j\langle T_{ij}\rangle = \frac{i}{(16\pi)^5 5670}\int \mathrm{d}^{10}x\, eD^a\eta^b \\ \times(105R_{abij}R_c{}^d{}_{kl}R_d{}^c{}_{mn}R_e{}^f{}_{pq}R_f{}^e{}_{rs} + 84R_{abij}R_c{}^d{}_{kl}R_d{}^e{}_{mn}R_e{}^f{}_{pq}R_f{}^c{}_{rs} \\ +168R_a{}^c{}_{ij}R_c{}^d{}_{kl}R_{dbmn}R_e{}^f{}_{pq}R_f{}^e{}_{rs} \\ +192R_a{}^c{}_{ij}R_c{}^d{}_{kl}R_d{}^c{}_{mn}R_e{}^f{}_{pq}R_{fbrs})\varepsilon^{ijklmnpqrs}\,.\end{aligned} \tag{93}$$

It is clear that we only need to use the modified Pontryagin classes when looking for anomaly cancellations between fields of different spin. (We return to this subject in the next section.)

As pointed out in sect. 4, we have to calculate the anomaly for spin-$\frac{3}{2}$ Weyl fermions and self-dual antisymmetric tensors of rank $2k+1$ (in $4k+2$ dimensions).

It is quite clear from the above, that we need to extend the $(0+1)$-dimensional theory (73) in order to deal with the spin-$\frac{3}{2}$ case. To do this, let us consider in general a spinor field ψ_A on M_{2n} which contains some tensor indices A. If the index A is a vector index, we have the gravitino field; other type of tensor indices would imply that we are dealing with higher half-integer spin fields. Let $(T^{ab})_{AB}$ be the generators of the corresponding tensor representation of $SO(2n)$, $a, b = 1, 2, \dots, 2n$; $A, B = 1, 2, \dots, \dim T$. In the presence of the external gravitational field, the Dirac equation satisfied by ψ_A is the Dirac equation in the tensor product representation of the spinor representation and the tensor representation generated by T. Thus, if we want to calculate the gravitational anomaly, we have to extend (73) so that the hamiltonian of the new $(0+1)$-dimensional theory is the square of the Dirac operator in the representation of $SO(2n)$ defined by ψ_A. The only new ingredients required to achieve the desired generalization of (73) consist of including a new set of fermionic variables c_A^*, c_B which transform under $SO(2n)$ according to $(T^{ab})_{AB}$. The relevant lagrangian is [31]:

$$L = \tfrac{1}{2} g_{ij}(x) \frac{dx^i}{d\tau}\frac{dx^j}{d\tau} + \tfrac{1}{2} i \delta_{ab} \psi^a \left(\frac{d}{dt} \psi^b + \omega^b_{ic} \frac{dx^i}{d\tau} \psi^c \right)$$
$$+ i c_a^* \left(\frac{d}{dt} c_A + \tfrac{1}{2} i \omega_{iab} (T^{ab})_{AB} \frac{dx^i}{d\tau} c_B \right) + \tfrac{1}{4} i \psi^i \psi^j R_{ijab} c_A^* T^{ab}_{AB} c_B \,. \quad (94)$$

If we were to calculate the axial anomaly, we would have to calculate $\mathrm{Tr}\, \gamma_5 \exp[-\beta(i\not{D})^2]$ in the $\beta \to 0$ limit with a constraint that will be mentioned shortly. The functional integral representation of the trace implies that $(x^i(\tau), \psi^j(\tau))$ are again integrated over with periodic boundary conditions with period β, while c_A, c_A^* are integrated over with antiperiodic boundary conditions.

There is a further constraint to be imposed on the trace, which is that the trace should run only over one-particle states of the c-fermions. This is because we are interested only in the anomaly corresponding to the T representation of $SO(2n)$ and not in any of its tensor products which are carried by the multiparticle states of c-fermions. Though this constraint may be difficult to impose in general, it is rather easy to implement in our case because we are only interested in the leading β behavior. We can briefly check that our procedure and conventions are correct by computing the spin-$\frac{3}{2}$ contribution to the axial anomaly in four dimensions which is known to be minus twenty-one times the spin-$\frac{1}{2}$ contribution [6]. In this case $(T^{ab})_{cd} = -i(\delta^a{}_c \delta^b{}_d - \delta^a{}_d \delta^b{}_c)$. Following the arguments presented for the spin-$\frac{1}{2}$ contribution at the beginning of this section, we have to compute the following trace

$$\mathrm{Tr}'\, \gamma_5\, e^{-\beta(i\not{D})^2 s=3/2} = \int' dx(\tau)\, d\psi(\tau)\, dc^*(\tau)\, dc(\tau) \exp(-S_{3/2}) \,, \quad (95)$$

where the apostrophe means that the computation is restricted to one-particle states of c and c^* and $s_{3/2}$ is the euclidean action associated to (94). In the high-temperature limit we again expand around constant configurations $(x_0, \psi_0, c=0)$. Notice that there are no constant c configurations due to the boundary conditions, so that the lagrangian (94) is automatically second order with respect to the c's. Thus, in the small-β limit the terms involving the c's look like ordinary fermionic oscillators with a curvature dependent mass term, then the trace over one-particle states yields $\mathrm{Tr}\exp(\frac{1}{2}\psi_0^i\psi_0^j R^a_{ijb}(x_0))$, while the integral over $x(\tau)$ and $\psi(\tau)$ gives the same result as for the spin-$\frac{1}{2}$ axial anomaly. Including the ghost contributions discussed in sect. 7, the final result is:

$$\frac{i^n}{(2\pi)^n}\int \mathrm{d\,Vol}\int \mathrm{d}\,\psi_0(\mathrm{Tr}\,\mathrm{e}^R-1)\prod_\alpha \frac{\frac{1}{2}ix_\alpha}{\sinh(\frac{1}{2}ix_\alpha)}, \tag{96}$$

in (30), R represents the matrix $\frac{1}{2}R_{abcd}\psi_0^c\psi_0^d$. If we restrict (30) to a four-dimensional manifold, it follows after some algebra that the spin-$\frac{3}{2}$ contribution to the axial anomaly is twenty-one times the spin-$\frac{1}{2}$ contribution, and of opposite sign. It is now straightforward to compute the spin-$\frac{3}{2}$ contribution to the anomaly in the conservation of the energy momentum tensor. Up to a local Lorentz transformation, the change of ψ_A under an infinitesimal coordinate transformation $x^i \to x^i + \eta^i(x)$ is:

$$-\delta_\eta\psi_A = \eta^i D_i\psi_A + D_a\eta_b(T^{ab})_{AB}\psi_B\,. \tag{97}$$

Following the steps of the spin-$\frac{1}{2}$ case, the jacobian induced by (97) can be rewritten on a trace in term of the theory (94) with $(T^{ab})_{AB}$ being represented by $c^*{}_A(T^{ab})_{AB}c_B$. It then follows that the only change with respect to the computation of the axial anomaly is again the replacement of $\frac{1}{2}\psi_0^i\psi_0^j R_{ijab}$ by $\frac{1}{2}\psi_0^i\psi_0^j R_{ijab}+D_a\eta_b-D_b\eta_a$. Hence (96) with this substitution gives the desired result. The final answer

$$\frac{i^n}{(2\pi)^n}\int \mathrm{d\,Vol}\int \mathrm{d}\psi_0(\mathrm{Tr}\,\mathrm{e}^{R'}-1)\prod_\alpha \frac{\frac{1}{2}ix_\alpha}{\sinh(\frac{1}{2}ix_\alpha)}, \tag{98}$$

and R' is the matrix $R'_{ab}=R_{ab}+D_a\eta_b-D_b\eta_a$. Eq. (98) is again non-vanishing only when the space–time dimension $4k+2$. Writing out the anomaly in terms of the modified Pontryagin classes in two, six and ten dimensions, the result is:

$$A_{3/2}(d=2) = -\frac{23i}{48\pi}\,p_1', \tag{99}$$

in two dimensions

$$A_{3/2}(d=6) = -\frac{i}{(2\pi)^3}\frac{1}{16}\left(\tfrac{55}{72}p_1'^2-\tfrac{49}{18}p_2'\right), \tag{100}$$

in six dimensions, and

$$A_{3/2}(d=10) = -\frac{i}{(2\pi)^5\cdot 2^6}\left(\tfrac{5}{336}p_1'^3-\tfrac{3}{28}p_1'p_2'+\tfrac{11}{21}p_3'\right), \tag{101}$$

if we wanted to obtain the final answer in terms of curvature tensors, we would have to substitute (82)–(83) in (99)–(101) and expand to first order in η^i. Since we will only be interested later on in anomaly cancellations, we will not display the final form of the anomaly in terms of curvature tensors.

With this method of computing the anomaly, we could in principle calculate the contribution to $\langle D^i T_{ij}\rangle$ due to higher half-integer spin fields $\frac{5}{2}, \frac{7}{2}, \ldots$. Since at the moment it is not clear how to quantize spin $\frac{5}{2}$ and higher in the presence of an external gravitational field [34], we won't discuss further this type of field.

Finally, we still have to consider the contribution to the anomaly coming from the self-dual antisymmetric tensor gauge field.

For this type of field, the methods presented so far in this section cannot be applied directly. The reason is that there is no generally covariant lagrangian leading to the equations of motion of the self-dual tensor field, as explained at length in sect. 8. Therefore there is no obvious $(0+1)$-dimensional field theory which could reproduce the anomaly for the self-dual tensor field. We will proceed in a manner similar to sect. 8. There, the anomalous polygon graph was calculated by noticing that the energy-momentum tensor only involves the gauge invariant field strength $F_{\mu_1\cdots\mu_{2k+1}}$, and that it naturally splits between two terms, one containing the self-dual part of F and the other containing the anti-self-dual part. Then the computation of the anomaly for the self-dual fields was accomplished by inserting in all but one vertex the unconstrained energy-momentum tensor, and the projected energy-momentum tensor in the remaining vertex.

After that, we included more tensor fields of lower and higher rank until we transformed the computation of the anomaly for a self-dual tensor field into an equivalent computation in terms of a bispinor field $\phi_{\alpha\beta}$ with positive chirality in each of its spinor indices. This procedure is legitimate because we know that the new fields added to $F_{\mu_1\cdots\mu_{2k+1}}$ in order to obtain $\phi_{\alpha\beta}$ do not contribute to the anomaly.

In order to apply Fujikawa's method, we consider the first-order formalism version of lagrangian (42). This simply means that we are integrating over both $A_{\mu_1\cdots\mu_{2k}}$ and $F_{\mu_1\cdots\mu_{2k+1}}$ as independent fields, and the lagrangian becomes:

$$\mathcal{L}=\frac{2}{n!}g^{\mu_1\nu_1}\cdots g^{\mu_{2k+1}\nu_{2k+1}}F_{\mu_1\cdots\mu_{2k+1}}F_{\nu_1\cdots\nu_{2k+1}}$$
$$-\frac{2}{n!}g^{\mu_1\nu_1}\cdots g^{\mu_{2k+1}\nu_{2k+1}}F_{\mu_1\cdots\mu_{2k+1}}\left(\frac{1}{n!}\partial_{\nu_1}A_{\nu_2\cdots\nu_{2k+1}}\pm\text{permutations}\right).$$

It is easy to check that the equations of motion for $\mathcal{L}$ are the same as those following from (43). From this point of view, the integration measure is $[\mathrm{d}A\,\mathrm{d}F]$. Now we can split F into its self-dual and anti-self-dual parts, so that the integration measure becomes $[\mathrm{d}A\,\mathrm{d}F^+\,\mathrm{d}F^-]$. So far we are still working in Minkowski space, where the self-duality condition is real in $4k+2$ dimensions $(*^2=+1)$. If we perform an infinitesimal coordinate transformation, we will obtain three jacobians from the measure, one for each field A, F^+, F^-. Since $A_{\mu_1\cdots\mu_{2k}}$ is a tensor of rank $2k$, it furnishes a real representation of the Lorentz group and therefore it should not contribute to the anomaly. If we concentrate on the jacobian for the fields F^+, F^-, it is also clear that the anomalies that could arise from each of their jacobians cancel out because F^{+} and F^{-} together generate a real representation of the Lorentz group. However, we argue in analogy with sect. 8, the anomaly for the self-dual will be fully contained in the jacobian generated by the F^{+} field. If we perform an infinitesimal coordinate transformation $x^{i} \to x^{i} + \eta^{i}(x)$, the change in F^{+} is given by

$$-\delta_{\eta} F^{+}_{a_1 \cdots a_{2k+1}} = \eta^{i} D_i F^{+}_{a_1 \cdots a_{2k+1}} + (D_a \eta_b)(T^{+ab})_{a_1 \cdots a_{2k+1}, b_1 \cdots b_{2k+1}} F^{+}_{b_1 \cdots b_{2k+1}}, \tag{102}$$

where $(T^{+ab})_{a_1 \cdots a_{2k+1}, b_1 \cdots b_{2k+1}}$ are the matrices generating the self-dual $(2k+1)$-rank representation of the Lorentz group.

If we were to work in Minkowski space, it is easy to check that the jacobian factor generated by (102) is given by the following trace $\frac{1}{2} \mathrm{Tr}_{(2k+1)} * \delta\eta$, where the subscript $(2k+1)$ means that the trace is taken over $(2k+1)$-rank antisymmetric tensor. Since this trace is very ill-defined in Minkowski space, we can regularize it by going to euclidean space and introducing a gaussian cut-off as we did for the spin-$\frac{1}{2}$ and $\frac{3}{2}$ computations.

Since the duality condition becomes complex in euclidean space ($*^2 = -1$), we have to double the number of fields so that F becomes complex. We can easily take care of this by extending the trace to a complex trace, and dividing by 2 in order to account for the doubling of the number of degrees of freedom. Also, when rotating to euclidean space the duality operation $*$ becomes $i*$. Hence we have to calculate $\mathrm{Tr}_{(2k+1)}\,(i * \delta\eta)/4$. In order to regularize this trace, we notice that the unconstrained field $F_{\mu_1 \cdots \mu_{2k+1}}$ satisfies the equations of motion $\Box F = 0$, where $\Box$ is the laplacian operator acting on antisymmetric tensors. Since we are working on euclidean space, the laplacian is a positive definite operator, and therefore we can compute the trace in the basis which make the laplacian diagonal. Thus we define the trace as the $\beta \to 0$ limit of $\mathrm{Tr}_{(2k+1)}\,(i * \delta\eta\, \mathrm{e}^{-\beta\Box})/4$. Before we proceed, it is worth pointing out that the laplacian commutes with the duality operation i.e. $\Box * = * \Box$, and that the equation $\Box F = 0$ means that F considered as a $2k+1$ form is harmonic, which implies among other things that $F_{\mu_1 \cdots \mu_{2k+1}}$ can be rewritten as $\partial_{\mu_1} A_{\mu_2 \cdots \mu_{2k+1}} \pm$ perm. These two properties of $\Box$ make the choice of regularization very well suited for our computation. In order to simplify the trace we follow arguments of sect. 8. We add tensor fields of lower and higher rank so that the trace runs over tensors of all ranks: $0, 1, 2, \ldots, 2k+1, \ldots, 4k+2$. That this procedure is legitimate was explained in sect. 8 and will not be further discussed here. The final step in the evaluation of the trace, is to find a $(0+1)$-dimensional theory whose hamiltonian is the laplacian on forms of arbitrary rank. Happily, the answer to this question is already known [30]. It is given by the supersymmetric σ-model of eq. (74). In order to make the connection more clear, let us rewrite (74) in terms of complex fermions $\psi^{i} = \sqrt{\frac{1}{2}}(\psi^{i} + i\psi^{i}_{2})$. After simple manipulations we get:

$$L = \tfrac{1}{2} g_{ij}(x) x^{i} x^{j} + \tfrac{1}{2} i g_{ij}(x) \psi^{*i} \frac{\overleftrightarrow{D}}{\mathrm{d}t} \psi^{j} + \tfrac{1}{4} R_{ijkl} \psi^{*i} \psi^{*j} \psi^{k} \psi^{l}, \tag{103}$$

after canonical quantization the fermions obey the usual anticommutation relations $\{\psi^i, \psi^{*j}\} = g^{ij}$, $\{\psi^i, \psi^j\} = 0$. From this point of view, the fermionic vacuum state is given by arbitrary functions over the manifold M_{4k+2} where the theory is defined. The states with one fermion are represented by the action of the fermionic creation operators ψ_i^* on the fermionic vacuum $\psi_i^*|\Omega\rangle$. Since ψ_i^* transforms like a vector under coordinate reparameterization of M_{4k+2}, one-fermion states transform like vectors. Similarly, states with two fermions are represented by a second rank antisymmetric tensor, and so on, until we reach the states with $4k+2$ fermions which correspond to totally antisymmetric tensors on M_{4k+2}. Furthermore, the duality operation which in spinor language is given by $\gamma_5 \otimes 1$ (see sect. 8), is here represented by a discrete symmetry. Q_5, which interchanges creation and annihilation operators [30]. Thus the trace we want to calculate is just $\mathrm{Tr}\, Q_5 \delta_\eta\, e^{-\beta H}$, where H is the hamiltonian associated to (74), (103). In terms of (74), using real fermions ψ_1^i, ψ_2^i, the Q_5 operation is $(\psi_1^i \to -\psi_2^i, \psi_2^i \to \psi_2^i)$, hence the functional integral representation for $\mathrm{Tr}\, Q_5 \delta_\eta\, e^{-\beta H}$ will be as in (77) but with (x^i, ψ_1^i) integrated over with periodic boundary conditions, whereas ψ_2^i has to be integrated over with antiperiodic boundary conditions. Finally, before we compute the trace, we have to identify δ_η in terms of the operators defining (74), (103). This is easily done in analogy with the spin-$\frac{3}{2}$ case. In eq. (96) we consider the infinitesimal coordinate transformation of a spinor field ψ with an extra index (A) valued in an arbitrary representation of the Lorentz group. If we take this extra index to be another spinor index, we get the form of the infinitesimal coordinate transformation of a bispinor, and the c-fermions are now replaced by the ψ_2^i. Thus the evaluation of $\mathrm{Tr}\, Q_5 \delta_\eta\, e^{-\beta H}$ is a rerun of the arguments which led to (98), where the trace in the integrand is carried out in the spinor representation. The final result is now

$$-\frac{1}{4}\frac{i^{2k+1}}{(2\pi)^{2k+1}}\int \mathrm{d}\psi_{10} \prod_\alpha \cosh\left(\tfrac{1}{2}ix'_\alpha\right)\left(\prod_\beta \frac{\frac{1}{2}ix'_\beta}{\sinh\left(\frac{1}{2}ix'_\beta\right)}\right)2^{2k+1}$$

$$= -\frac{i^{2k+1}}{8(2\pi)^{2k+1}}\int \mathrm{d}\psi_{10} \prod_\alpha \frac{ix'_\alpha}{\tanh ix'_\alpha}, \qquad (105)$$

which agrees with eq. (56).

As in previous cases, we will write down the expansion of (103) in terms of the modified Pontryagin classes:

$$A(d=2) = -\frac{ip'_1}{24(2\pi)},$$

$$A(d=6) = \frac{i}{5760(2\pi)^3}(16p_1'^2 - 112p'_2),$$

$$A(d=10) = \frac{i}{967680(2\pi)^5}(-256p_1'^3 + 1664p'_1 p'_2 - 7936p'_3). \qquad (106)$$

Finally, the combined gravitational and gauge anomalies can also be dealt with by the methods of this section in a very simple way. In fact the only non-trivial result necessary to carry out the computation is to generalize (94) so that the connection $\omega_i{}^a{}_b$ and the curvature R_{ijab} appearing in the last two terms of (94) can be substituted by the connection and curvature of an arbitrary gauge field defined on space–time.

For simplicity, we will only consider a Weyl fermion in $2n$ dimensions interacting with external gravitational and gauge fields. The extension of our results to include spin-$\frac{3}{2}$ Weyl fields is straightforward and will not be presented. Let G be an arbitrary gauge group, and assume that the fermion representation is generated by $(T^\alpha)_{AB}$ $\alpha=1,\ldots,\dim \mathrm{G}$, $A, B=1,\ldots,\dim T$. If we use Fujikawa's method [28], we have to find a $(0+1)$-dimensional field theory whose hamiltonian is $(i\not{D})^2$ where D_i is the covariant derivative with respect to the gravitational and gauge fields. In this case, by simple trial and error we can easily find the suitable modification of the lagrangian (94). The only change required is that now the c-fermions couple to the gauge connection $A_i^\alpha(x)$ and the gauge curvature F_{ij}^α and that the generators T^{ab} appearing in (94) are replaced by the gauge group generators $(T^\alpha)_{AB}$:

$$L=\tfrac{1}{2}g_{ij}(x)\frac{\mathrm{d}x^i}{\mathrm{d}\tau}\frac{\mathrm{d}x^j}{\mathrm{d}\tau}+\tfrac{1}{2}i\delta_{ab}\psi^a\left(\frac{\mathrm{d}}{\mathrm{d}\tau}\psi^b+\omega_{ic}^{\ b}\frac{\mathrm{d}x^i}{\mathrm{d}\tau}\psi^c\right)$$

$$+ic_A^*\left(\frac{\mathrm{d}}{\mathrm{d}\tau}c_A+iA_i^\alpha(x)\frac{\mathrm{d}x^i}{\mathrm{d}\tau}(T^\alpha)_{AB}c_B\right)+\tfrac{1}{2}i\psi^i\psi^jF_{ij}^\alpha(x)c_A^*(T^\alpha)_{AB}c_B. \tag{107}$$

Imagine that $\omega_{ib}^{\ a}=0$, $g_{ij}=\delta_{ij}$. In order to check our conventions, we will compute the gauge contribution to the axial anomaly in $2n$ dimensions using our procedure first and then comparing the result with the standard Feynman graph calculation. The computation of the Adler–Bell–Jackiw anomaly [1] beyond 4-dimensions was carried out in [7], and we will borrow some notation from the last reference.

In the evaluation of the axial anomaly using our procedure we have to remember that when we write down the anomaly as a particular partition function for (107), we have to restrict the trace to one-particle states of the c-fermions for exactly the same reason as in the computation of the spin-$\frac{3}{2}$ contribution to the gravitational anomaly. If we rerun now the arguments which led to the computation of the spin-$\frac{3}{2}$ contribution to the axial anomaly with the obvious change that the index A is now an internal rather than a Lorentz index and that we are choosing for simplicity the space–time to be flat, we get for the axial anomaly:

$$+\frac{i^n}{(2\pi)^n}i\int(\mathrm{d}\psi_0)\,\mathrm{Tr}\,(\exp\tfrac{1}{2}i\psi_0^i\psi_0^jF_{ij}^\alpha T^\alpha)$$

$$=\frac{(-1)^n}{(2\pi)^n n!}\varepsilon^{i_1j_1\cdots i_nj_n}F_{i_1j_1}^{\alpha_1}(x_0)\cdots F_{i_nj_n}^{\alpha_n}(x_0)S\,\mathrm{Tr}\,T^{\alpha_1}\cdots T^{\alpha_n}, \tag{108}$$

where

$$S\,\mathrm{Tr}\,T^{\alpha_1}\cdots T^{\alpha_n}=\frac{1}{n!}\sum_{\substack{\text{perm}\\ i_1\cdots i_n}}\mathrm{Tr}\,(T^{\alpha_{i1}}\cdots T^{\alpha_{in}})\,. \tag{109}$$

This result also follows from the index theorem for the Dirac equation [35], because as is well known, the axial anomaly is given by the local density of the Atiyah–Singer index theorem. If we also want to include the gravitational contribution to the anomaly, we just include the $\hat{A}$(M) polynomial (82) in the integrand of (108) as follows from the arguments of the first part of this section. Thus the combined gauge and gravitational contribution to the axial anomaly in $2n$ dimensions is given by

$$\frac{i^n}{(2\pi)^n}\int (\mathrm{d}\psi_0)\,\mathrm{Tr}\exp\left(\tfrac{1}{2}i\psi_0^i\psi_0^j F_{ij}^\alpha T^\alpha\right)\prod_r \frac{\frac{1}{2}ix_r}{\sinh\left(\frac{1}{2}ix_r\right)}\,. \tag{110}$$

We are interested not only in the axial anomaly but also in the gauge invariance of the effective action. When we have both external and gravitational fields, we may expect anomalies in one-loop diagrams with both external gluons and gravitons. These are the anomalies we intend to calculate in the remainder of this section.

If we perform an infinitesimal gauge transformation

$$\delta\psi = i\eta_\alpha T^\alpha\psi\,,$$

and use the (0+1)-dimensional theory (107), the anomaly will be given by:

$$\lim_{\beta\to 0}\mathrm{Tr}\,\gamma_5 i\eta_\alpha(c^*T^\alpha c)\,\mathrm{e}^{-\beta H}\,, \tag{111}$$

where as usual the trace is computed over one-particle states for the c-fermions. If we write (111) in terms of its functional integral representation, and exponentiate the term $\eta_\alpha c^* T^\alpha c$, we obtain

$$\lim_{\beta\to 0}\mathrm{Tr}\,\gamma_5 i\eta_\alpha(c^*T^\alpha c)\,\mathrm{e}^{-\beta H}=\frac{i^n}{(2\pi)^n}\int(\mathrm{d}\psi_0)(\mathrm{Tr}\,\mathrm{e}^{i(F^\alpha+\eta^\alpha)T^\alpha})\hat{A}(M)\,,$$

$$F^\alpha=\tfrac{1}{2}F_{ij}^\alpha\psi_0^i\psi_0^j\,, \tag{112}$$

and the anomaly is extracted from (112) by expanding to first order in η^α. The expansion of (112) in $2n$ dimensions will clearly contain one term with n-gluons, then a term with $n-2$ gluons and 2 gravitons and so on. If instead we wanted to compute the variation of the effective action with respect to an infinitesimal general coordinate transformation, then we just have to set η^α to zero, and replace $\hat{A}$(M) calculated with the standard curvature R_{abcd} by $\hat{A}'$(M) calculated in terms of the modified curvature $R_{ab}+D_a\eta_b-D_b\eta_a$. If one wants to have an effective action that preserves gauge invariance, one would have to require the cancellation of the different symmetrized traces (109) appearing in the anomalous graphs between left- and right-handed representations. For left–right asymmetric theories, this constraint is highly non-trivial beyond four dimensions [7].

Explicit expressions for the combined anomalies can be obtained by combining (112) with the expression for $\hat{A}(M)$ in (82)–(84). In four dimensions we do not obtain any mixed anomaly unless the gauge group contains U(1) factors in which case there is a triangle anomaly with one external U(1) field and two external gravitons. This is the anomaly of refs. [5, 6] discussed at length in sect. 4.

We would like to remind the reader again that formulae (89), (98), (105), and (112) should be understood in terms of the leading term, i.e. $R_{\mu\nu\alpha\beta} \doteq \partial_{\mu\alpha}h_{\nu\beta} + \partial^2_{\nu\alpha}h_{\mu\beta} - \partial^2_{\mu\beta}h_{\nu\alpha} - \partial^2_{\nu\alpha}h_{\mu\beta}$; $F^a_{\mu\nu} = \partial_\mu A^a_\nu - \partial_\nu A_\mu$. Non-leading terms should be obtained through application of the Wess–Zumino consistency conditions.

12. Cancellation of anomalies

We now turn, at last, to "phenomenological" considerations. We have seen that matter fields with chiral couplings to gravity exist only in $4k+2$ dimensions, and that such couplings give rise to anomalies. Of course, in standard phenomenology the individual quark and lepton multiplets have triangle anomalies in their couplings to gauge fields, but the anomalies cancel between multiplets in a non-trivial way. It is natural to ask whether a similar situation can occur with gravity. Although gravitational couplings of chiral fields of various spins give rise to anomalies, is it possible to cancel the anomalies between fields of different spin?

We have seen that in $4k+2$ dimensions, the gravitational anomalies are conveniently written as polynomials of order $2k+2$ in certain quantities x_i, $i = 1 \cdots 2k+1$. The spin-$\frac{1}{2}$, spin-$\frac{3}{2}$, and antisymmetric tensor anomalies are (apart from a factor common to the three cases)*

$$\hat{I}_{1/2} = \prod_{i=1}^{2k+1} \frac{\frac{1}{2}x_i}{\sinh \frac{1}{2}x_i},$$

$$\hat{I}_{3/2} = \left(\prod_{i=1}^{2k+1} \frac{\frac{1}{2}x_i}{\sinh \frac{1}{2}x_i}\right)\left(-1 + \sum_{j=1}^{2k+1} 2\cosh x_j\right),$$

$$\hat{I}_A = -\tfrac{1}{8} \prod_{i=1}^{2k+1} \frac{x_i}{\tanh x_i}. \tag{114}$$

In $4k+2$ dimensions, one is to expand these formulae in powers of the x_i, keeping only the terms of order $2k+2$.

We observe that the formulas in (1) are invariant under permutations of the x_i and under $x_i \leftrightarrow -x_i$ for any i. So each term in their power series expansion has the

* By $I_{1/2}$ and $I_{3/2}$ we mean the anomalies of positive chirality complex Weyl fields. For a real Weyl field (possible in $8k+2$ dimensions) one must divide by two. I_A is the anomaly for a *self-dual* real antisymmetric tensor field. By a self-dual (rather than anti-self-dual) tensor we mean the representation that arises in combining two positive chirality spinors. For a complex self-dual tensor one must multiply I_A by two. The minus sign in I_A arises from the Bose (rather than Fermi) regulator loop; the factor of $\frac{1}{8}$ has a more complicated origin, which was explained in sects. 9 and 10.

same symmetries. To count the independent tensor structures appearing in the anomalies, it is enough to count the homogeneous polynomials of relevant degree and symmetry. In two dimensions ($k=0$), there is one x and one homogeneous second order polynomial x^2. In six dimensions there are three x_i and two linearly independent homogeneous fourth-order polynomials with the right symmetry, namely $\sum x_i^4$ and $(\sum x_i^3)^2$. In ten dimensions there are five x_i and three relevant polynomials, namely $\sum x_i^6$, $(\sum x_i^4)\cdot(\sum x_j^2)$, and $(\sum x_i^2)^3$. Cancellation of gravitational anomalies requires cancelling the coefficient of each dangerous operator. Beyond ten dimensions, the number of dangerous polynomials whose coefficients must cancel to avoid anomalies rapidly increases.

We could expand the anomalies as functions of $A_n=\sum_i x_i^{2n}$. Instead, following mathematical usage, we recall the origin of the x_i as eigenvalues of a matrix

$$R=\begin{pmatrix} 0 & x_1 & & & & & & \\ -x_1 & 0 & & & & & & \\ & & 0 & x_2 & & & & \\ & & -x_2 & 0 & & & & \\ & & & & \ddots & & & \\ & & & & & & 0 & x_{2k+1} \\ & & & & & & -x_{2k+1} & 0 \end{pmatrix} \tag{115}$$

and we define polynomials p_i as follows. We note that $\det(1-R/2\pi)=\prod_i(1+x_i^2/(2\pi)^2)$. We write a power series expansion of the determinant:

$$\det(1-R/2\pi)=\sum_{n=0}^{\infty}\frac{p_n}{(2\pi)^{2n}}, \tag{116}$$

where p_n is a polynomial of order $2n$. Thus, $p_0=1$, $p_1=\sum x_i^2$, $p_2=\sum_{i<j}x_i^2x_j^2$, $p_3=\sum_{i<j<k}x_i^2x_j^2x_k^2$, etc. Every even symmetric polynomial of order $2n$ can be expanded in the p_m of $m\leq n$. We will express our results in this way*.

The first case that arises is two dimensions. Of course, this case is of mathematical interest only. In two dimensions

$$\hat{I}_{1/2}=-\tfrac{1}{24}p_1\,,\qquad \hat{I}_{3/2}=\tfrac{23}{24}p_1\,,\qquad \hat{I}_A=-\tfrac{1}{24}p_1\,. \tag{117}$$

Evidently, the anomaly can be cancelled in various ways. The fact that $\hat{I}_{1/2}=\hat{I}_A$ in two dimensions reflects the fact that a positive chirality fermion is equivalent to a right-moving scalar.

* We are grateful to P. Ginsparg for correcting a variety of numerical errors in an earlier version of this section.

For Kaluza–Klein theory our interest is in six or more dimensions, $k \geq 1$. In six dimensions power series expansion of eq. (114) yields

$$\hat{I}_{1/2} = \tfrac{1}{5760}(7p_1^2 - 4p_2)\,,$$
$$\hat{I}_{3/2} = \tfrac{1}{5760}(275p_1^2 - 980p_2)\,,$$
$$\hat{I}_A = \tfrac{1}{5760}(16p_1^2 - 112p_2)\,. \qquad (118)$$

It may be seen that any two of these expressions are linearly independent, so that anomaly cancellation is possible only if all three spins are present. However, since there are three fields and only two independent anomalies (p_1^2 and p_2), there inevitably is a linear combination of these expressions that vanishes. The simplest non-trivial solution is $21\hat{I}_{1/2} - \hat{I}_{3/2} + 8\hat{I}_A = 0$. Thus a six-dimensional theory with 21 positive chirality spin-$\frac{1}{2}$ fields, one negative chirality gravitino, and eight self-dual antisymmetric tensor fields is free of anomalies. Although these numbers might seem clumsy, six-dimensional supergravity theories with this field content (modulo anomaly-free fields) do exist and might be of interest. It is a very favorable fact that the minimal solution only requires one gravitino; while there can be any number of spin-$\frac{1}{2}$ or antisymmetric tensor fields in six-dimensional supergravity, the number of gravitinos is necessarily ≤ 4.

Turning now to ten dimensions, we find by power series expansion of (114)

$$\hat{I}_{1/2} = \tfrac{1}{967\,680}(-31p_1^3 + 44p_1p_2 - 16p_3)\,,$$
$$\hat{I}_{3/2} = \tfrac{1}{967\,680}(225p_1^3 - 1620p_1p_2 + 7920p_3)\,,$$
$$\hat{I}_A = \tfrac{1}{967\,680}(-256p_1^3 + 1664p_1p_2 - 7936p_3)\,. \qquad (119)$$

Since there are three fields and three linearly independent anomalies, one would not *a priori* expect non-trivial calculation of gravitational anomalies to be possible in ten dimensions. But now we meet a real surprise, which is by far the most striking result of this paper. The expressions for $\hat{I}_{1/2}$, $\hat{I}_{3/2}$, and $\hat{I}_A$ in (119) are linearly dependent. In addition, the minimal solution is remarkably simple: $-\hat{I}_{1/2} + \hat{I}_{3/2} + \hat{I}_A = 0$. Thus, a ten-dimensional theory with one (complex) negative chirality spin-$\frac{1}{2}$ field, one (complex) positive chirality spin-$\frac{3}{2}$ field, and one (real) self-dual antisymmetric tensor is free of anomalies. What is more, modulo fields that do not contribute anomalies, this is precisely the field content of the chiral $n=2$ supergravity theory in ten dimensions [11], which is the naive low-energy limit of one of the ten-dimensional supersymmetric string theories [12]. Since this theory cannot be coupled to supersymmetric matter multiplets, and cannot be extended to a theory with $n>2$ supersymmetry (Nahm, ref. [11]), it appears to be the unique theory in ten dimensions with non-trivial cancellation of gravitational anomalies.

Are non-trivial anomaly cancellations possible beyond ten dimensions? In fourteen dimensions,

$$\hat{I}_{1/2}=\frac{1}{464\,486\,400}[381p_1^4-904p_1^2p_2+512p_1p_3+208p_2^2-192p_4]\,,$$

$$\hat{I}_{3/2}=\frac{1}{464\,486\,400}[6393p_1^4-42\,472p_1^2p_2-70\,144p_1p_3+102\,544p_2^2-94\,656p_4]\,,$$

$$\hat{I}_{A}=\frac{1}{464\,486\,400}[12\,288p_1^4-90\,112p_1^2p_2+290\,816p_1p_3+77\,824p_2^2 -1\,560\,576p_4]\,. \tag{120}$$

These expressions are linearly independent, so non-trivial anomaly cancellations are impossible in fourteen dimensions. (The situation becomes even worse if one considers that independent of anomalies consistent theories with massless spin-$\frac{3}{2}$ fields presumably do not exist in fourteen dimensions.)

A simple argument, similar to one in sect. 5, now shows that because non-trivial anomaly cancellation is impossible in fourteen dimensions, it is impossible in $14+4n$ dimensions for any $n\geqslant 1$. Consider a $(14+4n)$-dimensional manifold of topology $M^{14}\times B$, M^{14} being fourteen-dimensional Minkowski space and B a compact manifold of dimension $4n$ on which the Dirac, Rarita–Schwinger, and antisymmetric tensor equations have non-zero index. (For instance, B may be a product of n copies of K3.) An arbitrary chiral theory in $14+4n$ dimensions will reduce on $M^{14}\times B$ to a fourteen-dimensional chiral theory. Since this fourteen-dimensional chiral theory is necessarily anomalous, it follows that any chiral theory that might have been considered in $14+4n$ dimensions also has anomalies.

In conclusion, in any number of dimensions non-trivial cancellation of gravitational anomalies requires massless spin-$\frac{3}{2}$ fields and hence supergravity. In six dimensions there are various theories with non-trivial cancellation of gravitational anomalies. They require a fairly elaborate field content, but may be of interest. In ten dimensions the unique theory with such non-trivial cancellation is the chiral $n=2$ supergravity theory which is the low-energy limit of one of the superstring theories. Beyond ten dimensions non-trivial cancellations of gravitational anomalies does not occur. We will not explore here the phenomenological consequences of mixed gauge-gravitational anomalies.

We wish to thank M.F. Atiyah for valuable comments that helped motivate this work, and P. Ginsparg for useful discussions and for writing a computer program to generate the complicated power series expansions in sects. 11 and 12. We wish to thank J. Bagger, P. Frampton, D.J. Gross, J. Schwarz, and A. Zee for valuable discussions. We also thank R. Jackiw for discussions of global anomalies, and for drawing our attention to ref. [36], where an anomalous commutator closely related to our discussion in sect. 5 is evaluated. (See footnote on p. 295.)

References

[1] S. Adler, Phys. Rev. 177 (1969) 2426; Lectures on elementary particles and quantum field theory, eds. S. Deser et al. (MIT, 1970);
J. Bell and R. Jackiw, Nuovo Cim. 60A (1969) 47;
R. Jackiw, in Lectures on current algebra and its applications (Princeton, 1972);
S.L. Adler and W. Bardeen, Phys. Rev. 182 (1969) 1517;
W.A. Bardeen, Phys. Rev. 184 (1969) 1848;
R.W. Brown, C.-C. Shi and B.-L. Young, Phys. Rev. 186 (1969) 1491;
J. Wess and B. Zumino, Phys. Lett. 37B (1971) 95;
A. Zee, Phys. Rev. Lett. 29 (1972) 1198

[2] J. Steinberger, Phys. Rev. 76 (1949) 1180;
J. Schwinger, Phys. Rev. 82 (1951) 664;
L. Rosenberg, Phys. Rev. 129 (1963) 2786;
R. Jackiw and K. Johnson, Phys. Rev. 182 (1969) 1459;
S. Adler and D.G. Boulware, Phys. Rev. 184 (1969) 1740;
S.L. Adler, B.W. Lee, S.B. Treiman and A. Zee, Phys. Rev. D4 (1971) 3497;
R. Aviv and A. Zee, Phys. Rev. D5 (1972) 2372;
M.V. Terentiev, JETP Lett. 14 (1971) 140;
A.M. Belavin, A.M. Polyakov, A.S. Schwarz and Yu.S. Tyupkin, Phys. Lett. 59B (1975) 85;
G. 't Hooft, Phys. Rev. Lett. 37 (1976) 8; Phys. Rev. D14 (1976) 3432;
C. Callan, R. Dashen and D.J. Gross, Phys. Lett. 63B (1976) 334;
R. Jackiw and C. Rebbi, Phys. Rev. Lett. 37 (1976) 172

[3] D.J. Gross and R. Jackiw, Phys. Rev. D6 (1972) 477;
C. Bouchiat, J. Iliopoulos and Ph. Meyer, Phys. Lett. 38B (1972) 519;
H. Georgi and S. Glashow, Phys. Rev. D6 (1972) 429

[4] G. 't Hooft, in Recent developments in gauge theories, G. 't Hooft et al., eds. (Plenum, New York, 1980);
A.A. Ansel'm, JEPT Lett. 32 (1980) 138;
A. Zee, Phys. Lett. 95B (1980) 290;
Y. Frishman, A. Schwimmer, T. Banks and S. Yankielowicz, Nucl. Phys. B177 (1981) 157;
S. Coleman and B. Grossman, Nucl. Phys. B203 (1982) 205;
G.R. Farrar, Phys. Lett. 96B (1980) 273;
S. Weinberg, Phys. Lett. 102B (1981) 401;
C.H. Albright, Phys. Rev. D24 (1981) 1969;
I. Bars, Phys. Lett. 109B (1982) 73;
T. Banks, S. Yankielowicz and A. Schwimmer, Phys. Lett. 96B (1980) 67;
A. Schwimmer, Nucl. Phys. B198 (1982) 269

[5] R. Delbourgo and A. Salam, Phys. Lett. 40B (1972) 381;
T. Eguchi and P. Freund, Phys. Rev. Lett. 37 (1976) 1251

[6] S.W. Hawking and C.N. Pope, Nucl. Phys. B146 (1978) 381;
M.J. Perry, Nucl. Phys. B143 (1978) 114;
N.K. Nielsen, H. Romer and B. Schroer, Nucl. Phys. B136 (1978) 475;
N.K. Nielsen, M.T. Grisaru, R. Romer and P. van Nieuwenhuizen, Nucl. Phys. B140 (1978) 477;
A.J. Hanson and H. Romer, Phys. Lett. 80B (1978) 58;
R. Critchley, Phys. Lett. 78B (1978) 410;
S.M. Christensen and M.J. Duff, Phys. Lett. 76B (1978) 571; Nucl. Phys. B154 (1979) 301;
T. Eguchi, P.B. Gilkey and A.J. Hanson, Phys. Reports 66 (1980) 213

[7] P.H. Frampton and T.W. Kephart, Phys. Rev. Lett. 50 (1983) 1343, 1347;
P.K. Townsend and G. Sierra, Nucl. Phys. B222 (1983) 493;
B. Zumino, W. Yong-Shi and A. Zee (Univ. Washington preprint, 1983);
Y. Matsuki, and Y. Matsuki and A. Hill, OSU preprints, 1983

[8] J.H. Schwarz, Phys. Reports 89 (1982) 223, and references therein;
C.H. Ragadiakos and J.G. Taylor, Phys. Lett. 124B (1983) 201

[9] S.M. Christensen and S.A. Fulling, Phys. Rev. D15 (1977) 2088;
T.S. Bunch and P.C.W. Davies, Proc. Roy. Soc. A356 (1977) 569;

L.S. Brown, Phys. Rev. D15 (1977) 1469;
H. S. Tsao, Phys. Lett. 69B (1977) 79;
J.S. Dowker and R. Critchley, Phys. Rev. D16 (1977) 3390

[10] E. Cremmer, B. Julia and J. Scherk, Phys. Lett. 76B (1978) 409;
E. Cremmer and B. Julia, Nucl. Phys. B159 (1979) 141;
A. Chamseddine, Nucl. Phys. B185 (1981) 403; Phys. Rev. D24 (1981) 3065

[11] W. Nahm, Nucl. Phys. B135 (1978) 149;
M.B. Green and J.H. Schwarz, Phys. Lett. 109B (1982) 444; 122B (1983) 143;
J. Schwarz and P. West, Phys. Lett. 126B (1983) 301;
P.S. Howe and P. West, King's College preprint (1983)

[12] M.B. Green and J.H. Schwarz, Nucl. Phys. B181 (1981) 502; B198 (1982) 252, 441; Phys. Lett. 109B (1982) 444;
J.H. Schwarz, Phys. Reports 89 (1982) 223

[13] E. Witten, Fermion quantum numbers in Kaluza–Klein theory, Princeton preprint, to appear

[14] B. de Witt, Phys. Reports 19 (1975) 295

[15] A. Salam and P.T. Matthews, Phys. Rev. 90 (1953) 690;
J. Schwinger, Phys. Rev. 93 (1954) 615

[16] E. Witten, Phys. Lett. 117B (1982) 324

[17] N. Marcus and J. Schwarz, Phys. Lett. 115B (1982) 111

[18] L. Brink, J.H. Schwarz and J. Scherk, Nucl. Phys. B121 (1977) 77;
N.S. Manton, Nucl. Phys. B193 (1981) 391;
G. Chapline and N. Manton, Nucl. Phys. B184 (1981) 391;
G. Chapline and R. Slansky, Nucl. Phys. B209 (1982) 461;
R. Coquereaux, Phys. Lett. 115B (1982) 389;
T. Kugo and P. Townsend, Nucl. Phys. B221 (1983) 357;
C. Wetterich, Phys. Lett. 110B (1982) 379; 113B (1982) 377; Nucl. Phys. B211 (1983) 117; B222 (1983) 20

[19] G. 't Hooft, Nucl. Phys. B75 (1974) 461

[20] J.S. Schwinger, Phys. Rev. 82 (1951) 664;
C. Itzykson and J.-B. Zuber, Quantum field theory (McGraw-Hill, 1980)

[21] S. Coleman, Phys. Rev. D11 (1975) 2088;
S. Mandelstam, Phys. Rev. D11 (1975) 3026

[22] M.F. Atiyah and I.M. Singer, Ann. of Math. 87 (1968) 485, 546; 93 (1971) 1, 119, 139;
M.F. Atiyah and G.B. Segal, Ann. of Math. 87 (1968) 531

[23] J. Milnor, Ann. of Math. (2) 64 (1956) 399

[24] S. Smale, Ann. of Math. (2) 74 (1961) 392

[25] S. Deser, R. Jackiw and S. Templeton, Phys. Rev. Lett. 48 (1982) 975; Ann. of Phys. 140 (1982) 372;
N. Redlich, MIT preprint, to appear

[26] M.F. Atiyah, V.K. Parodi and I.M. Singer, Proc. Camb. Philos. Soc. 77 (1975) 43; 78 (1975) 405; 79 (1976) 71

[27] N. Hitchin, Adv. Math. 14 (1974) 1

[28] K. Fujikawa, Phys. Rev. Lett. 42 (1979) 1195; 44 (1980) 1733; Phys. Rev. D21 (1980) 2848; D22 (1980) 1499 (E); D23 (1981) 2262;
M.B. Einhorn and D.R.T. Jones, U.M.-T.H. 83-3 preprint

[29] P. di Vecchia and S. Ferrara, Nucl. Phys. B130 (1977) 93;
E. Witten, Phys. Rev. D16 (1977) 2991;
D.Z. Freedman and P.K. Townsend, Nucl. Phys. B177 (1981) 282

[30] E. Witten, Nucl. Phys. B202 (1982) 253

[31] L. Alvarez-Gaumé, Harvard preprints HUTP-82/A029 and A035

[32] S. Cecotti and L. Girardello, Phys. Lett. 110B (1982) 39

[33] L. Alvarez-Gaumé, D.Z. Freedman and S. Mukhi, Ann. of Phys. 134 (1981) 85

[34] P.V. Nieuwenhuisen, Phys. Reports 68 (1981) 189

[35] R. Jackiw and C. Rebbi, Phys. Rev. D14 (1976) 517;
N.K. Nielsen, H. Römer and B. Schroer, Phys. Lett. 70B (1977) 445

[36] S. Fubini, R. Jackiw and A. Hanson, Phys. Rev. D7 (1973) 1732

CURRENT ALGEBRA, BARYONS, AND QUARK CONFINEMENT[†] [*]

Edward Witten
Joseph Henry Laboratories
Princeton University
Princeton, New Jersey 08544, USA

[†]Reprinted with permission from *Nuclear Physics* **B223** (1983) 433-444.
[*]This work was supported in part by National Science Foundation under Grant No. PHY80 - 19754.

It is shown that ordinary baryons can be understood as solitons in current algebra effective lagrangians. The formation of color flux tubes can also be seen in current algebra, under certain conditions.

The idea that in some sense the ordinary proton and neutron might be solitons in a non-linear sigma model has a long history. The first suggestion was made by Skyrme more than twenty years ago [1]. Finkelstein and Rubinstein showed that such objects could in principle be fermions [2], in a paper that probably represented the first use of what would now be called θ vacua in quantum field theory. A gauge invariant version was attempted by Faddeev [3]. Some relevant miracles are known to occur in two space-time dimensions [4]; there also exists a different mechanism by which solitons can be fermions [4].

It is known that in the large-N limit of quantum chromodynamics [5] meson interactions are governed by the tree approximation to an effective local field theory of mesons. Several years ago, it was pointed out [6] that baryons behave as if they were solitons in the effective large-N meson field theory. However, it was not clear in exactly what sense the baryons actually *are* solitons.

The first relevant papers mainly motivated by attempts to understand implications of QCD current algebra were recent papers by Balachandran et al. [7] and by Boguta [8].

We will always denote the number of colors as N and the number of light flavors as n. For definiteness we first consider the usual case $n = 3$. Nothing changes for $n > 3$. Some modifications for $n < 3$ are pointed out later. Except where stated otherwise, we discuss standard current algebra with global $SU(n) \times SU(n)$ spontaneously broken to diagonal $SU(n)$, presumably as a result of an underlying $SU(N)$ gauge interaction.

Standard current algebra can be described by a field $U(x)$ which (for each space-time point x) is a point in the SU(3) manifold. Ignoring quark bare masses, this field is governed by an effective action of the form

$$I = -\tfrac{1}{16}F_\pi^2 \int d^4x \,\mathrm{Tr}\, \partial_\mu U \partial_\mu U^{-1} + N\Gamma + \text{higher order terms}. \tag{1}$$

Here Γ is the Wess-Zumino term [9] which cannot be written as the integral of a manifestly SU(3) × SU(3) invariant density, and $F_\pi \simeq 190$ Mev. In quantum field theory the coefficient of Γ must a priori be an integer [10], and indeed we will see that the quantization of the soliton excitations of (1) is inconsistent (they obey neither bose nor fermi statistics) unless N is an integer.

Any finite energy configuration $U(x, y, z)$ must approach a constant at spatial infinity. This being so, any such configuration represents an element in the third homotopy group $\pi_3(\mathrm{SU}(3))$. Since $\pi_3(\mathrm{SU}(3)) \simeq Z$, there are soliton excitations, and they obey an additive conservation law. Actually, higher-order terms in (1) are needed to stabilize the soliton solutions and prevent them from shrinking to zero size. We will see that such higher-order terms (which could be measured in principle by studying meson processes) must be present in the large-N limit of QCD and are related to the bag radius. Our remarks will not depend on the details of the higher-order terms.

A technical remark is in order. To study solitons, it is convenient to work with a euclidean space-time M of topology $S^3 \times S^1$. Here S^3 represents the spatial variables, and S^1 is a compactified euclidean time coordinate. A given non-linear sigma model field $U(x)$ defines a mapping of M into SU(3). We may think of M as the boundary of a five-dimensional manifold Q with topology $S^3 \times D$, D being a two-dimensional disc. Using the fact that $\pi_1(\mathrm{SU}(3)) = \pi_4(\mathrm{SU}(3)) = 0$, it can be shown that the mapping of M into SU(3) defined by $U(x)$ can be extended to a mapping from Q into SU(3). Then as in ref. [10] the functional Γ is defined by $\Gamma = \int_Q \omega$, where ω is the fifth-rank antisymmetric tensor on the SU(3) manifold defined in ref. [10]. By analogy with the discussion in ref. [10], Γ is well-defined modulo 2π. (It is essential here that because $\pi_2(\mathrm{SU}(3)) = 0$, the five-dimensional homology classes in $H_5(\mathrm{SU}(3))$ that can be represented by cycles with topology $S^3 \times S^2$ are precisely those that can be represented by cycles with topology S^5. There are closed five-surfaces S in SU(3) such that $\int_S \omega$ is an *odd* multiple of π, but they do not arise if space-time has topology $S^3 \times S^1$ and Q is taken to be $S^3 \times D$.)

Now let us discuss the quantum numbers of the current algebra soliton. First, let us calculate its baryon number (which was first demonstrated to be non-zero in ref. [7], where, however, different assumptions were made from those we will follow). In previous work [10] it was shown that the baryon-number current has an anomalous piece, related to the $N\Gamma$ term in eq. (1). If the baryon number of a quark is $1/N$, so that an ordinary baryon made from N quarks has baryon number 1, then the anomalous piece in the baryon number current B_μ was shown to be

$$B_\mu = \frac{\varepsilon_{\mu\nu\alpha\beta}}{24\pi^2} \mathrm{Tr}\left(U^{-1}\partial_\nu U\right)\left(U^{-1}\partial_\alpha U\right)\left(U^{-1}\partial_\beta U\right). \tag{2}$$

So the baryon number of a configuration is

$$B = \int \mathrm{d}^3x\, B_0 = \frac{1}{24\pi^2}\int \mathrm{d}^3x\, \varepsilon^{ijk}\mathrm{Tr}\left(U^{-1}\partial_i U\right)\left(U^{-1}\partial_j U\right)\left(U^{-1}\partial_k U\right). \tag{3}$$

The right-hand side of eq. (24) can be recognized as the properly normalized integral expression for the winding number in π_3(SU(3)). In a soliton field the right-hand side of (3) equals one, so the soliton has baryon number one; it is a baryon. (In ref. [7] the baryon number of the soliton was computed using methods of Goldstone and Wilczek [11]. The result that the soliton has baryon number one would emerge in this framework if the elementary fermions are taken to be quarks.)

Now let us determine whether the soliton is a boson or a fermion. To this end, we compare the amplitude for two processes. In one process, a soliton sits at rest for a long time T. The amplitude is $\exp(-iMT)$, M being the soliton energy. In the second process, the soliton is adiabatically rotated through a 2π angle in the course of a long time T. The usual term in the lagrangian $L_0 = \frac{1}{16}F_\pi^2 \mathrm{Tr}\,\partial_\mu U\,\partial_\mu U^{-1}$ does not distinguish between the two processes, because the only piece in L_0 that contains time derivatives is quadratic in time derivatives, and the integral $\int \mathrm{d}t\,\mathrm{Tr}(\partial U/\partial t)(\partial U^{-1}/\partial t)$ vanishes in the limit of an adiabatic process. However, the anomalous term Γ is linear in time derivatives, and distinguishes between a soliton that sits at rest and a soliton that is adiabatically rotated. For a soliton at rest, $\Gamma = 0$. For a soliton that is adiabatically rotated through a 2π angle, a slightly laborious calculation explained at the end of this paper shows that $\Gamma = \pi$. So for a soliton that is adiabatically rotated by a 2π angle, the amplitude is not $\exp(-iMT)$ but $\exp(-iMT)\exp(iN\pi) = (-1)^N\exp(-iMT)$.

The factor $(-1)^N$ means that for odd N the soliton is a fermion; for even N it is a boson. This is uncannily reminiscent of the fact that an ordinary baryon contains N quarks and is a boson or a fermion depending on whether N is even or odd.

These results are unchanged if there are more than three light flavors of quarks. How do they hold up if there are only two light flavors? The field $U(x)$ is then an element of SU(2). Because π_3(SU(2)) = Z, there are still solitons. The baryon-number current has the same anomalous piece, and the soliton still has baryon number one. But in SU(2) current algebra, there is no Γ term, so how can we see that the soliton can be a fermion?

The answer was given long ago [2]. Although π_4(SU(3)) = 0, π_4(SU(2)) = Z_2. With suitably compactified space-time, there are thus two topological classes of maps from space-time to SU(2). In the SU(2) non-linear sigma model, there are hence two

"θ-vacua": fields that represent the non-trivial class in $\pi_4(\mathrm{SU}(2))$ may be weighted with a sign $+1$ or -1. An explicit field $U(x, y, z, t)$ which goes to 1 at space-time infinity and represents the non-trivial class in $\pi_4(\mathrm{SU}(2))$ can (fig. 1) be described as follows (a variant of this description figures in recent work by Goldstone [12]). Start at $t \to -\infty$ with a constant, $U = 1$; moving forward in time, gradually create a soliton-anti-soliton pair and separate them; rotate the soliton through a 2π angle without touching the anti-soliton; bring together the soliton and anti-soliton and annihilate them. Weighting this field with a factor of -1, while a configuration without the 2π rotation of the soliton is homotopically trivial and gets a factor $+1$, corresponds to quantizing the soliton as a fermion. Thus, internally to $\mathrm{SU}(2)\times\mathrm{SU}(2)$ current algebra, one sees that the soliton can be a fermion. In $\mathrm{SU}(3)\times\mathrm{SU}(3)$ current algebra one finds the stronger result that the soliton *must* be a fermion if and only if N is odd.

Our results so far are consistent with the idea that quantization of the current algebra soliton describes ordinary nucleons. However, we have not established this. Perhaps there are ordinary baryons and exotic, topologically excited solitonic baryons. However, certain results will now be described which seem to directly show that the ordinary nucleons are the ground state of the soliton.

For simplicity, we will focus now on the case of only two flavors. Soliton states can be labeled by their spin and isospin quantum numbers, which we will call J and I, respectively. We will determine semiclassically what values of I and J are expected for solitons. A semiclassical description of current algebra solitons will be accurate quantitatively only in the limit of large N. (Since F_π^2 is proportional to N, N enters the effective lagrangian (1) as an overall multiplicative factor. Hence, N plays the role usually played by $1/\hbar$.) So we will check the results we find for solitons by comparing to the expected quantum numbers of baryons in the large-N limit.

Let us first determine the expected baryon quantum numbers. We make the usual assumption that the multi-quark wave function is symmetric in space and antisymmetric in color, and hence must have complete symmetry in spin and isospin. The spin-isospin group is $\mathrm{SU}(2)\times\mathrm{SU}(2) \sim \mathrm{O}(4)$. A quark transforms as $(\frac{1}{2}, \frac{1}{2})$; this is the

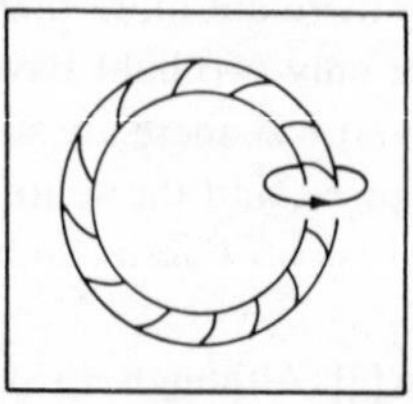

Fig. 1. A soliton-antisoliton pair is created from the vacuum; the soliton is rotated by a 2π angle; the pair is then annihilated. This represents the non-trivial homotopy class in $\pi_4(\mathrm{SU}(2))$.

vector representation of O(4). We may represent a quark as ϕ_i, where $i = 1 \ldots 4$ is a combined spin-isospin index labeling the O(4) four-vector.

We must form symmetric combinations of N vectors ϕ_i. As is well known, there is a quadratic invariant $\phi^2 = \Sigma_{i=1}^{4}\phi_i^2$. One can also form symmetric traceless tensors of any rank $A^{(p)}_{i_1\ldots i_p} = (\phi_{i_1}\phi_{i_2}\ldots\phi_{i_p} - \text{trace terms})$; this transforms as $(\frac{1}{2}p, \frac{1}{2}p)$ under $\mathrm{SU}(2)\times\mathrm{SU}(2)$. The general symmetric expression that we can make from N quarks is $(\phi^2)^k A^{(N-2k)}_{i_1\ldots i_{N-2k}}$, where $0 \leqslant k \leqslant \frac{1}{2}N$. So the values of I and J that are possible are the following:

$$N \text{ even}, \qquad I = J = 0, 1, 2, 3, \ldots,$$

$$N \text{ odd}, \qquad I = J = \tfrac{1}{2}, \tfrac{3}{2}, \tfrac{5}{2}, \tfrac{7}{2}, \ldots. \tag{4}$$

For instance, in nature we have $N = 3$. The first two terms in the sequence indicated above are the nucleon, of $I = J = \frac{1}{2}$, and the delta, of $I = J = \frac{3}{2}$. If the number of colors were five or more, we would expect to see more terms in this series. Moreover, simple considerations involving color magnetic forces suggest that, as for $N = 3$, the mass of the baryons in this sequence is always an increasing function of I or J.

Now let us compare to what is expected in the soliton picture. (This question has been treated previously in ref. [7].) We do not know the effective action of which the soliton is a minimum, because we do not know what non-minimal terms must be added to eq. (1). We will make the simple assumption that the soliton field has the maximum possible symmetry. The soliton field cannot be invariant under I or J (or any component thereof), but it can be invariant under a diagonal subgroup $I + J$. This corresponds to an ansatz $U(x) = \exp[iF(r)]\boldsymbol{T}\cdot\boldsymbol{x}$, where $F(r) = 0$ at $r = 0$ and $F(r) \to \pi$ as $r \to \infty$.

Quantization of such a soliton is very similar to quantization of an isotropic rigid rotor. The hamiltonian of an isotropic rotor is invariant under an $\mathrm{SU}(2)\times\mathrm{SU}(2)$ group consisting of the rotations of body fixed and space fixed coordinates, respectively. We will refer to these symmetries as I and J, respectively. A given configuration of the rotor is invariant under a diagonal subgroup of $\mathrm{SU}(2)\times\mathrm{SU}(2)$. This is just analogous to our solitons, assuming the classical soliton solution is invariant under $I + J$.

The quantization of the isotropic rigid rotor is well known. If the rotor is quantized as a boson, it has $I = J = 0, 1, 2, \ldots$. If it is quantized as a fermion, it has $I = J = \frac{1}{2}, \frac{3}{2}, \frac{5}{2}, \ldots$. The agreement of these results with eq. (4) is hardly likely to be fortuitous.

In the case of three or more flavors, it may still be shown that the quantization of collective coordinates gives the expected flavor quantum numbers of baryons. The analysis is more complicated; the Wess-Zumino interaction plays a crucial role.

So far, we have assumed that the color gauge group is SU(N). Now let us discuss what would happen if the color group were O(N) or Sp(N). (By Sp(N) we will mean the group of $N \times N$ unitary matrices of quaternions; thus $\mathrm{Sp}(1) \simeq \mathrm{SU}(2)$.) We will see that also for these gauge groups, the topological properties of the current algebra theory correctly reproduce properties of the underlying gauge theory.

In an $\mathrm{O}(N)$ gauge theory, we assume that we have n multiplets of left-handed (Weyl) spinors in the fundamental N-dimensional representation of $\mathrm{O}(N)$. There is no distinction between quarks and antiquarks, because this representation is real. (If n is even, the theory is equivalent to a theory of $\frac{1}{2}n$ Dirac multiplets.) The anomaly free flavor symmetry group is $\mathrm{SU}(n)$. Simple considerations based on the most attractive channel idea suggest that the flavor symmetry will be spontaneously broken down to $\mathrm{O}(n)$, which is the maximal subgroup of $\mathrm{SU}(n)$ that permits all fermions to acquire mass. In this case the current algebra theory is based on a field that takes values in the quotient space $\mathrm{SU}(n)/\mathrm{O}(n)$.

In an $\mathrm{Sp}(N)$ gauge theory, we assume the fermion multiplets to be in the fundamental $2N$-dimensional representation of $\mathrm{Sp}(N)$. Since this representation is pseudoreal, there is again no distinction between quarks and antiquarks. In this theory the number of fermion multiplets must be even; otherwise, the $\mathrm{Sp}(N)$ gauge theory is inconsistent because of a non-perturbative anomaly [2] involving $\pi_4(\mathrm{Sp}(N))$. If there are $2n$ multiplets, the flavor symmetry is $\mathrm{SU}(2n)$. Simple arguments suggest that the $\mathrm{SU}(2n)$ flavor group is spontaneously broken to $\mathrm{Sp}(n)$, so that the current algebra theory is based on the quotient space $\mathrm{Su}(2n)/\mathrm{Sp}(n)$. This corresponds to symmetry breaking in the most attractive channel; $\mathrm{Sp}(n)$ is the largest unbroken symmetry that lets all quarks get mass.

In $\mathrm{O}(N)$, since there is no distinction between quarks and antiquarks, there is also no distinction between baryons and anti-baryons. A baryon can be formed from an antisymmetric combination of N quarks; $B = \varepsilon_{i_1 i_2 \ldots i_N} q^{i_1} q^{i_2} \ldots q^{i_N}$. But in $\mathrm{O}(N)$, a product of two epsilon symbols can be rewritten as a sum of products of N Kronecker deltas:

$$\varepsilon_{i_1 i_2 \ldots i_N} \varepsilon_{j_1 j_2 \ldots j_N} = \left(\delta_{i_1 j_1} \delta_{i_2 j_2} \ldots \delta_{i_N j_N} \pm \text{permutations} \right).$$

This means that in an $\mathrm{O}(N)$ gauge theory, two baryons can annihilate into N mesons.

On the other hand, in an $\mathrm{Sp}(N)$ gauge theory there are no baryons at all. The group $\mathrm{Sp}(N)$ can be defined as the subgroup of $\mathrm{SU}(2N)$ that leaves fixed an antisymmetric second rank tensor γ_{ij}. A meson made from two quarks of the same chirality can be described by the two quark operator $\gamma_{ij} q^i q^j$. In $\mathrm{Sp}(N)$ the epsilon symbol can be written as a sum of products of N γ's:

$$\varepsilon_{i_1 i_2 \ldots i_{2N}} = \left(\gamma_{i_1 i_2} \gamma_{i_3 i_4} \ldots \gamma_{i_{2N-1} i_{2N}} \pm \text{permutations} \right).$$

So in an $\mathrm{Sp}(N)$ gauge theory, a single would-be baryon can decay to N mesons.

Now let us discuss the physical phenomena that are related to the topological properties of our current algebra spaces SU(n)/O(n) and SU(n)/Sp(n). We recall from ref. (10) that the existence in QCD current algebra with at least three flavors of the Wess-Zumino interaction, with its a priori quantization law, is closely related to the fact that $\pi_5(\mathrm{SU}(n)) = \mathrm{Z}$, $n \geqslant 3$. The analogue is that

$$\pi_5(\mathrm{SU}(n)/\mathrm{O}(n)) = \mathrm{Z}, \qquad n \geqslant 3,$$
$$\pi_5(\mathrm{SU}(2n)/\mathrm{Sp}(n)) = \mathrm{Z}, \qquad n \geqslant 2. \tag{5}$$

So also the O(N) and Sp(N) gauge theories possess at the current algebra level an interaction like the Wess-Zumino term, provided the number of flavors is large enough. Built into the current algebra theories is the fact that in the underlying theory there is a parameter (the number of colors) which a priori must be an integer.

Now we come to the question of the existence of solitons. These are classified by the third homotopy group of the configuration space, and we have

$$\pi_3(\mathrm{SU}(n)/\mathrm{O}(n)) = \mathrm{Z}_2, \quad n \geqslant 4,$$
$$\pi_3(\mathrm{SU}(2n)/\mathrm{Sp}(n)) = 0, \qquad \text{any } n. \tag{6}$$

Thus, in the case of an O(N) gauge theory with at least four flavors, the current algebra theory admits solitons, but the number of solitons is conserved only modulo two. This agrees with the fact that in the O(N) gauge theory there are baryons which can annihilate in pairs. In current algebra corresponding to Sp(N) gauge theory there are no solitons, just as the Sp(N) gauge theory has no baryons.

For O(N) gauge theories with less than four light flavors we have

$$\pi_3(\mathrm{SU}(3)/\mathrm{O}(3)) = \mathrm{Z}_4,$$
$$\pi_3(\mathrm{SU}(2)/\mathrm{O}(2)) = \mathrm{Z}. \tag{7}$$

Thus, the spectrum of current algebra solitons seems richer than the expected spectrum of baryons in the underlying gauge theory. The following remark seems

TABLE 1
Some homotopy groups of certain homogeneous spaces

	SU(n)	SU(n)/O(n)	SU(2n)/Sp(n)
π_2	0	Z_2, $n \geqslant 3$	0
π_3	Z, all n	Z_2, $n \geqslant 4$	0
π_5	Z, $n \geqslant 3$	Z, $n \geqslant 3$	Z, $n \geqslant 3$

appropriate in this connection. It is only in the multi-color, large-N limit that a semiclassical description of current algebra solitons becomes accurate. Actually, large-N gauge theories are described by weakly interacting theories of mesons, but it is not only Goldstone bosons that enter; one has an infinite meson spectrum. Corresponding to the rich meson spectrum is an unknown and perhaps topologically complicated configuration space P of the large-N theory. Plausibly, baryons can always be realized as solitons in the large-N theory, even if all or almost all quark flavors are heavy. Perhaps $\pi_3(\mathrm{P})$ is always Z, Z_2, or O for SU(N), O(N), and Sp(N) gauge theories. The Goldstone boson space is only a small subspace of P and would not necessarily reflect the topology of P properly. Our results suggest that as the number of flavors increases, the Goldstone boson space becomes an increasingly good *topological* approximation to P. In this view, the extra solitons suggested by eq. (7) for O(N) gauge theories with two or three flavors become unstable when SU(2)/O(2) or SU(3)/O(3) is embedded in P.

One further physical question will be addressed here. Is color confinement implicit in current algebra?

Do current algebra theories in which the field U labels a point in SU(n), SU(n)/O(n), or SU($2n$)/SP(n) admit flux tubes? By a flux tube we mean a configuration $U(x, y, z)$ which is independent of z and possesses a non-trivial topology in the x-y plane. To ensure that the energy per unit length is finite, U must approach a constant as $x, y \to \infty$. The proper topological classification involves therefore the *second* homotopy group of the space in which U takes its values. In fact, we have

$$\pi_2(\mathrm{SU}(n)) = 0,$$

$$\pi_2(\mathrm{SU}(n)/\mathrm{O}(n)) = Z_2, \qquad n \geq 3,$$

$$\pi_2(\mathrm{SU}(2n)/\mathrm{Sp}(n)) = 0. \tag{8}$$

Thus, current algebra theories corresponding to underlying SU(N) and Sp(N) gauge theories do not admit flux tubes. The theories based on underlying O(N) gauge groups do admit flux tubes, but two such flux tubes can annihilate.

These facts have the following natural interpretation. Our current algebra theories correspond to underlying gauge theories with quarks in the fundamental representation of SU(N), O(N), or Sp(N). SU(N) or Sp(N) gauge theories with dynamical quarks cannot support flux tubes because arbitrary external sources can be screened by sources in the fundamental representation of the group. For O(N) gauge theories it is different. An external source in the spinor representation of O(N) cannot be screened by charges in the fundamental representation. But two spinors make a tensor, which can be screened. So the O(N) gauge theory with dynamical quarks supports only one type of color flux tube: the response to an external source in the spinor representation of $O(N)$. It is very plausible that this color flux tube should be identified with the excitation that appears in current algebra because $\pi_2(SU(n)/O(n)) = Z_2$.

The following fact supports this identification. The interaction between two sources in the spinor representation of $O(N)$ is, in perturbation theory, N times as big as the interaction between two quarks. Defining the large-N limit in such a way that the interaction between two quarks is of order one, the interaction between two spinor charges is therefore of order N. This strongly suggests that the energy per unit length in the flux tube connecting two spinor charges is of order N. This is consistent with our current algebra identification; the whole current algebra effective lagrangian is of order N (since $F_\pi^2 \sim N$), so the energy per unit length of a current algebra flux tube is certainly of order N.

In conclusion, it still remains for us to establish the contention made earlier that the value of the Wess-Zumino functional Γ for a process consisting of a 2π rotation of a soliton is $\Gamma = \pi$.

First of all, the soliton field can be chosen to be of the form

$$V(x_i) = \left(\begin{array}{c|c} W(x_i) & 0 \\ \hline 0 & 1 \end{array}\right), \tag{9}$$

where the SU(2) matrix W is chosen to be invariant under a combined isospin rotation plus rotation of the spatial coordinate x_i. This being so, a 2π rotation of V in space is equivalent to a 2π rotation of V in isospin. Introducing a periodic time coordinate t which runs from 0 to 2π, the desired field in which a soliton is rotated by a 2π angle can be chosen to be

$$U(x_i, t) = \begin{pmatrix} e^{it/2} & & \\ & e^{-it/2} & \\ & & 1 \end{pmatrix} V(x_i) \begin{pmatrix} e^{-it/2} & & \\ & e^{it/2} & \\ & & 1 \end{pmatrix}. \tag{10}$$

Note that $U(x_i, t)$ is periodic in t with period 2π even though the individual exponentials $\exp(\pm\frac{1}{2}it)$ have period 4π. Because of the special form of V, we can equivalently write U in the much more convenient form

$$U(x_i, t) = \begin{pmatrix} 1 & & \\ & e^{-it} & \\ & & e^{it} \end{pmatrix} V(x_i) \begin{pmatrix} 1 & & \\ & e^{it} & \\ & & e^{-it} \end{pmatrix}. \tag{11}$$

This field $U(x_i, t)$ describes a soliton that is rotated by a 2π angle as t ranges from 0 to 2π. We wish to evaluate $\Gamma(U)$.

To this end we introduce a fifth parameter ρ $(0 \leqslant \rho \leqslant 1)$ so as to form a five-manifold of which space-time is the boundary; this five-manifold will have the topology of three-space times a disc. A convenient choice is to write

$$\tilde{U}(x_i, t, \rho) = A^{-1}(t, \rho) U(x_i, t) A(t, \rho), \tag{12}$$

where

$$A(t, \rho) = \begin{pmatrix} \mathrm{i} & 0 & 0 \\ 0 & \rho e^{it} & \sqrt{1-\rho^2} \\ 0 & -\sqrt{1-\rho^2} & \rho e^{-it} \end{pmatrix}. \tag{13}$$

Note that at $\rho = 0$, $A(t, \rho)$ is independent of t. So we can think of ρ and t as polar coordinates for the plane, ρ being the radius and t the usual angular variable. Also $\tilde{U}(x_i, t, 1) = U(x_i, t)$ so the product of three space with the unit circle in the ρ-t plane can be identified with the original space-time. According to eq. (14) of ref. (10), what we must calculate is

$$\Gamma(U) = -\frac{i}{240\pi^2} \int_0^1 \mathrm{d}\rho \int_0^{2\pi} \mathrm{d}t \int \mathrm{d}^3x \, \varepsilon^{ijklm}$$

$$\times \left[\mathrm{Tr}\, \tilde{U}^{-1} \partial_i \tilde{U} \tilde{U}^{-1} \partial_j \tilde{U} \tilde{U}^{-1} \partial_k \tilde{U} \tilde{U}^{-1} \partial_l \tilde{U} \tilde{U}^{-1} \partial_m \tilde{U} \right], \tag{14}$$

where i, j, k, l, and m may be ρ, t, x_1, x_2, or x_3. The integral can be done without undue difficulty (the fact that W is invariant under spatial rotations plus isospin is very useful), and one finds $\Gamma(U) = \pi$.

This calculation can also be used to fill in a gap in the discussion of ref. (10). In that paper, the following remark was made. Let $A(x, y, z, t)$ be a mapping from space-time into SU(2) that is in the non-trivial homotopy class in $\pi_4(\mathrm{SU}(2))$. Embed A in SU(3) in the trivial form

$$\hat{A} = \left(\begin{array}{cc|c} \multicolumn{2}{c|}{\multirow{2}{*}{A}} & 0 \\ & & 0 \\ \hline 0 & 0 & 1 \end{array} \right).$$

Then $\Gamma(\hat{A}) = \pi$. In fact, as we have noted above, the non-trivial homotopy class in $\pi_4(\mathrm{SU}(2))$ differs from the trivial class by a 2π rotation of a soliton (which may be one member of a soliton-antisolition pair). The fact that $\Gamma = \pi$ for a 2π rotation of soliton means that $\Gamma = \pi$ for the non-trivial homotopy class in $\pi_4(\mathrm{SU}(2))$.

The following important fact deserves to be demonstrated explicitly. As before, let A be a mapping of space-time into SU(2) and let $\hat{A}$ be its embedding in SU(3). Then $\Gamma(\hat{A})$ depends only on the homotopy class of A in $\pi_4(\mathrm{SU}(2))$. In fact, suppose $\hat{A}$ is

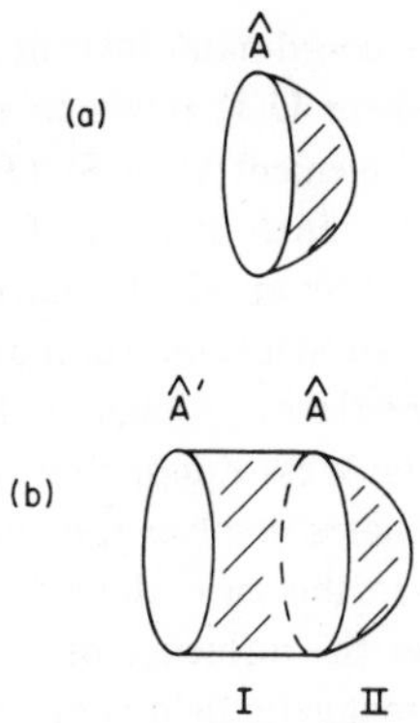

Fig. 2. A demonstration that Γ is a homotopy invariant for SU(2) mappings.

homotopic to $\hat{A}'$: Let us prove that $\Gamma(\hat{A}) = \Gamma(\hat{A}')$. To compute $\Gamma(\hat{A})$ we realize space-time as the boundary of a disc, extend $\hat{A}$ to be defined over that disc, and evaluate an appropriate integral (fig. 2a). To evaluate $\Gamma(\hat{A}')$ we again must extend $\hat{A}'$ to a disc. This can be done in a very convenient way (fig. 2b). Since $\hat{A}$ is homotopic to $\hat{A}'$, we first deform $\hat{A}'$ into $\hat{A}$ via matrices of the form $\left(\begin{array}{c|c} X & 0 \\ \hline 0 & 1 \end{array}\right)$ (matrices that are really SU(2) matrices embedded in SU(3)) and then we extend $\hat{A}$ over a disc as before. The integral contribution to $\Gamma(\hat{A}')$ from part I of fig. 2b vanishes because the fifth rank antisymmetric tensor that enters in defining Γ vanishes when restricted to any SU(2) subgroup of SU(3). The integral in part II of fig. 2b is the same as the integral in fig. 2a, so $\Gamma(\hat{A}) = \Gamma(\hat{A}')$.

The fact that Γ is a homotopy invariant for SU(2) mappings also means that Γ can be used to prove that $\pi_4(\mathrm{SU}(2))$ is non-trivial. Since Γ obviously is 0 for the trivial homotopy class in $\pi_4(\mathrm{SU}(2))$, while $\Gamma = \pi$ for a process containing a 2π rotation of a soliton, the latter process must represent a non-trivial element in $\pi_4(\mathrm{SU}(2))$. What cannot be proved so easily is that this is the only non-trivial element.

I would like to thank A.P. Balachandran and V.P. Nair for interesting me in current algebra solitons.

Note added in proof

Many physicists have asked how the soliton quantum numbers can be calculated if there are three flavors. Following is a sketch of how this question can be answered.

We assume that for SU(3) × SU(3) current algebra, the soliton solution is simply an SU(2) solution embedded in SU(3). Such a solution is invariant under combined spin-isospin transformations; and it is also invariant under hypercharge rotations.

There are now seven collective coordinates instead of three. They parametrize the coset space X = SU(3)/U(1), where U(1) refers to right multiplication by hypercharge. Thus a point in X is an element U of SU(3) defined up to multiplication on the right by a hypercharge transformation. The space X has flavor SU(3) symmetry (left multiplication of U by an SU(3) matrix) and rotation SU(2) symmetry (right multiplication of U by an SU(3) matrix that commutes with hypercharge).

The crucial novelty of the three-flavor problem is that even when restricted to the space of collective coordinates, the Wess-Zumino term does not vanish. As usual, the quantization of collective coordinates involves the quantum mechanics of a particle moving on the manifold X, but in this case, the effect of the Wess-Zumino term is that the particle is moving under the influence of a simulated "magnetic field" on the X manifold. Moreover, this magnetic field is of the Dirac monopole type; it has string singularities which are unobservable if the Wess-Zumino coupling is properly quantized.

The wave functions of the collective coordinates are "monopole harmonics" on the X manifold with quantum numbers that depend on the "magnetic charge." For charge three (three colors) the lowest monopole harmonic is an SU(3) octet of spin $\frac{1}{2}$, and the next one is an SU(3) decuplet of spin $\frac{3}{2}$.

References

[1] T.H.R. Skyrme, Proc. Roy. Soc. A260 (1961) 127
[2] D. Finkelstein and J. Rubinstein, J. Math. Phys. 9 (1968) 1762
[3] L.D. Faddeev, Lett. Math. Phys. 1 (1976) 289
[4] S. Coleman, Phys. Rev. D11 (1975) 2088;
R. Jackiw and C. Rebbi, Phys. Rev. Lett. 36 (1976) 1116;
P. Hasenfratz and G. 't Hooft, Phys. Rev. Lett. 36 (1976) 1119
[5] G. 't Hooft, Nucl. Phys. B72 (1974) 461; B75 (1974) 461
[6] E. Witten, Nucl. Phys. B160 (1979) 57
[7] A.P. Balachandran, V.P. Nair, S.G. Rajeev and A. Stern, Phys. Rev. Lett. 49 (1982) 1124; Syracuse University preprint (1982)
[8] J. Boguta, Phys. Rev. Lett. 50 (1983) 148
[9] J. Wess and B. Zumino, Phys. Lett. 37B (1971) 95
[10] E. Witten, Nucl. Phys. B223 (1983) 422
[11] J. Goldstone and F. Wilczek, Phys. Rev. Lett. 47 (1981) 986
[12] J. Goldstone, private communication

SKYRMIONS AND QCD[†]

Edward Witten
Joseph Henry Laboratories
Princeton University
Princeton, New Jersey 08544, USA

[†]This contribution first appeared in *Solitons in Nuclear and Elementary Particle Physics*, eds., A. Chodos, E. Hadjimichael and C. Tze (World Scientific, Singapore, 1984).

Skyrme's remarkable idea [1] that baryons are solitons in a meson theory has, in recent years, received partial confirmation as a result of developments in our understanding of QCD. In these notes, I will sketch the connection of Skyrmions with fundamental QCD theory and comment briefly on recent and possible future developments. Some of my comments overlap with those of Balachandran [2].

In the extremely low energy limit (energies much less than Λ_{QCD}), QCD is equivalent to a theory of Goldstone bosons (pions). This is, however, not the limit appropriate for Skyrmions or baryons, since the energy of a quark bound in a baryon is of order Λ_{QCD}. Departing from the low energy limit, in what sense is QCD equivalent to a local field theory of mesons (and glueballs)? Such an assumption is often made as a phenomenological ansatz, but if we want to learn something solid and lasting about QCD and Skyrmions, QCD must be equivalent in some well-defined way — and not just in the limit of low energies — to a meson theory.

Such a possibility was first established by 't Hooft [3], who considered QCD generalized from SU(3) to an SU(N) gauge group. One way to state 't Hooft's conclusion is as follows. If one assumes confinement, then QCD is equivalent to a theory of mesons (and glueballs) in which the (quartic) meson coupling constant is $1/N$. As I have discussed these matters at length elsewhere [4] I will not go into detail here. I would like, however, to stress two points. First, QCD is equivalent for any N, including $N = 3$, to a meson theory in which $1/N$ is the coupling constant. Large N is special only in that the meson coupling is weak, and the tree approximation to the meson theory is valid. For $N = 3$, the meson coupling is only moderately weak. Second, although QCD is equivalent to a meson theory, this is not a simple theory such as the non-linear sigma model or Skyrme model. It is an extremely complex and unknown theory with an infinite number of elementary fields (the latter point follows from asymptotic freedom predictions, as discussed for instance in [4]).*

*Incidentally, although the equivalence of QCD to a meson theory with coupling $1/N$ is valid for any N, including $N = 3$, it is virtually impossible to discover or prove this equivalence if one is willing to think only about the case $N = 3$. The equivalence of QCD to a meson theory emerges naturally [3,4] when one thinks about the whole class of QCD-like theories of arbitrary N.

If the meson theory that is equivalent to QCD is complicated and unknown, of what use is it to know that such a theory exists? One answer [3,4] is that in the weak coupling (large N) limit of QCD, some general predictions can be made which do not depend on details of the unknown equivalent meson theory. Typical of such predictions are Zweig's rule and the existence of an extremely large number of narrow resonances. These properties are fairly well obeyed in nature and in fact are significant aspects of the strong interactions as we observe them. There is thus experimental support for the idea that the strong interactions are equivalent to a complicated meson theory with a moderately weak coupling and precisely the selection rules that emerge from 't Hooft's large N diagrammatics.

How do baryons fit into this picture? At first sight, this is puzzling, since baryons seem to be described by hopelessly complicated Feynman diagrams. However, in [4], I showed that for large N, baryon masses are of order N, while the baryon size and the baryon-baryon and baryon-meson scattering cross sections are both of order one. This is very different from the behavior of mesons (the meson mass is of order one, while the meson-meson scattering amplitude is of order $1/N$). On the other hand, the scaling properties of baryons for $N \to \infty$ are reminiscent of the behavior of solitons. (For weak coupling, a soliton mass grows as the inverse of the coupling, while the soliton size and cross sections are of order unity.) It is therefore natural to suspect that baryons are solitons in the meson theory that is equivalent to QCD. This should, presumably, be true for any N, but if N is large, the meson theory is weakly coupled and the solitons can be treated semiclassically.

The development of this idea was hampered by puzzles connected with the baryon quantum numbers. To begin with, why would a soliton in a meson theory have non-zero baryon number? Even more perplexing, baryon quantum numbers depend on N in a peculiar way. A baryon is made of N quarks, so for even N, baryons are bosons; for odd N, they are fermions. In the two flavor case, the ground state baryon has $I = J = 0$ (I and J are isospin and spin, respectively) if N is even and $I = J = 1/2$ if N is odd. (At least, this is the prediction of simple quark models.) For three or more flavors, the spin and flavor quantum numbers depend on N in an even more intricate way. These results are peculiar because in the meson theory $1/N$ is a coupling constant, and the quantum numbers of particles do not ordinarily depend on coupling constants. How can it be that the baryon/soliton is a boson if $1/N = 1/10^6$ and a fermion if $1/N = 1/(10^6 + 1)$? Moreover, there seems to be a logical paradox. In the meson theory $1/N$ is a coupling constant, and coupling constants can usually be continually adjusted. If the solition is a boson for $1/N = 1/10^6$ and a fermion for $1/N = 1/(10^6 + 1)$, what happens for $1/N = 1/(10^6 + .5)$?

These puzzles all turn out to have definite and conclusive answers. Taking the questions in reverse order, the paradox just alluded to is avoided because [5] the meson action is multivalued, being defined only modulo $2\pi N$; as a result the quantum mechanics of the meson theory makes sense only if N is an integer. The proper spin and flavor quantum numbers of baryons emerge if the solution of the meson theory has the right collective coordinates. For instance, in the two flavor case, the right quantum number emerge [6] if the soliton solution of the meson theory is invariant not under I or J but only under $I + J$. Such is the case, for instance, in the Skyrme model. To obtain the right spin and flavor quantum numbers in the three-flavor case depends on the collective coordinates and on the proper treatment [7] of the Wess-Zumino interaction [8]. As for the statistics of solitons, it was shown long ago [9] that with two flavors, in models like the Skyrme model, the soliton can be quantized as a fermion. The same is true [10] for three or more flavors if (and only if) the Wess-Zumino term is included; and in that case, the soliton *must* be quantized as a fermion precisely if N is odd. Finally, we come to a sine qua non of Skyrmion physics: in models such as the Skyrme model, the solitons do indeed have baryon number because [11] of the anomaly-induced coupling of the baryon current to Goldstone bosons.

These considerations remove all of the obstacles to interpreting baryons as the solitons of meson physics, with the semiclassical soliton expansion being equivalent to the $1/N$ expansion in QCD. I believe that the arguments of [4] and [10] are quite convincing in this regard. I will not repeat those arguments here, but I cannot resist mentioning one example of the connection of Skyrmions with microphysics.

For Skyrme solitons, with two flavors, there is a rotational term $\mathbf{J}^2/2I$ in the energy. How is this related to the quark model? In the quark model of ground state baryons, there are N quarks in a common spatial wave function. Between any pair of quarks there is a spin-dependent interaction $\lambda\, \sigma_i \cdot \sigma_j$, λ being some constant. The total spin dependent interaction is thus $\lambda \Sigma\, \sigma_i \cdot \sigma_j = \lambda \mathbf{J}^2/2 - 3\lambda N$ where $\mathbf{J} = \Sigma\, \sigma_i$. Thus, the Skyrme and quark model parameters are related by $I = 1/\lambda$.

Granted that baryons can correctly be viewed as solitons in meson physics, to what extent can this viewpoint lead to a basis for calculations?

The large N effective action of QCD is complicated and unknown. This being so, it is logical to test the reasonableness of the "baryon as soliton" idea with calculations in simple models that incorporate the minimum necessary features. Such calculations (which, incidentally, tend to be rather lengthy) have been carried out in the Skyrme model [12] and in a model with π and ω fields [13]. The static properties of nucleons computed in these models in the semiclassical approximation are generally within about 30% of experiment. Similar accuracy emerges in studies of excited baryon resonances in the Skyrme model [14, 15, 16].

There obviously is considerable interest in trying to reduce these 30% errors. I also believe that the following question is of fundamental importance. To what extent does the 30% error of the Skyrme and $\pi\omega$ models reflect use of the semiclassical approximation, and to what extent does it reflect the fact that the Skyrme and $\pi\omega$ models are crude approximations to meson physics? I suspect the error mainly has the latter origin. If the proper meson theory were known, the semiclassical approximation would be equivalent to the $1/N$ expansion, and I personally suspect the error would be much less than 30%.

How can this question be answered? We have little prospect of determining the proper meson theory (it would be necessary to sum the planar diagrams of QCD). My own viewpoint is as follows. The $1/N$ expansion per se, without our being able to sum planar diagrams, has no predictive power for mesons, except for some selection rules such as Zweig's rule (and these should be incorporated in any model). However, the $1/N$ expansion makes the remarkable prediction that baryon physics can be deduced from meson physics via solitons. To test this claim, since we cannot calculate meson physics, we should take meson physics from experiment.

Ideally, one should construct a realistic model of the first twenty or thirty meson resonances, with quantum numbers, masses, and couplings taken from experiment. Looking for baryons as solitons in such a meson theory and quantizing them semiclassically would give a rigorous test of the $1/N$ expansion.

There should be no embarrasment in replacing the Skyrme model of mesons with a model with 10^3 or 10^6 mesonic parameters as long as the parameters are taken from experiment. After all, in this project, no predictive power for mesons is claimed; the goal is to see whether baryon properties can be deduced from meson properties via semiclassical quantization of solitons.

This project is not practical in the form just stated because we do not have detailed experimental knowledge of meson scattering amplitudes. However, a less ambitious project is probably feasible. We know in nuclear physics that a realistic description of nucleons requires π, ρ, and ω fields at the least. We know experimentally their masses and the basic couplings $F_\pi, g_{\rho\pi\pi}$, and $g_{\omega\rho\pi}$. A chirally invariant theory of π, ρ, and ω with observed masses and observed values of F_π, $g_{\rho\pi\pi}$, and $g_{\omega\rho\pi}$ would be far more realistic than the Skyrme model. If the general philosophy under discussion here is corect, semiclassical quantization of the soliton solutions of such a model should significantly reduce the 30% errors of the Skyrme model.*

Development of such a model is an important way to test the ideas considered here. An equally important way to test these ideas is to develop what might be called model independent tests of the $1/N$ expansion.

By this I mean relations among observables that hold in semiclassical quantization of any soliton model of baryons. For instance, let μ_p, μ_n, and $\mu_{N\Delta}$ be the proton and neutron magnetic moments and the delta-nucleon transition magnetic moments. Then in [12] it was shown that $\mu_{N\Delta} = 1/\sqrt{2}\,(\mu_p - \mu_n)$ in any semiclassical soliton model (so that, in QCD, this relation follows from the $1/N$ expansion). This relation and the analogous one [12] $g_{\pi N\Delta} = \frac{3}{2} g_{\pi NN}$ are valid to within a few percent. Similar relations involving strange baryons have been derived by several authors [17] but there is much more work to be done along these lines.

I suspect that it will be found that model independent relations among non-strange baryons will be more successful than relations that include strange baryons. The reason for this is that with three flavors, the spin and flavor quantum numbers of baryons depend on N in an awkward way, so the extrapolation from large N to the real world is clumsy. For two flavors the ground state baryon has $I = J = 1/2$ for any odd N and the extrapolation is more smooth.

I have been discussing ways to test what I consider the fundamental ideas in the connection between Skyrmions and QCD. At a more pragmatic level, one may ask whether there are problems for which soliton models of baryons are actually useful in clarifying yet unknown physics or are competitive with other models. Here are a few possibilities.

1. The H

Quark models suggest [18] that although the short range interaction between two nucleons is repulsive, the short range interaction between two Λ^0's is attractive. As a result, a bound state of two Λ^0's, with a binding energy of perhaps 100-200 MeV, is predicted; it has been called the H.

It was quite exciting that a similar prediction appears in the three flavor Skyrme model [19]. The Skyrme and quark models make the same prediction

*Although there are some differences in approach, D. Caldi (see reference [2]) independently advocated inclusion of ρ and ω in soliton models.

concerning the quantum numbers of the H and its first excitation. Hopefully, this agreement will spur experimental searches for the H — and perhaps computer simulations.

2. The Callan-Rubakov Effect

Callan and Rubakov predicted [20] that magnetic monopoles can catalyze the violation of baryon number. For a realistic theory of this effect, one must take the asymptotic states to be nucleons and mesons rather than free quarks. Standard baryon models cannot be conveniently coupled to magnetic monopoles, but this is quite feasible for the Skyrme model. The Callan-Rubakov effect emerges naturally when the Skyrme model is coupled to monopoles [21]; I believe this is the one example to date where the soliton model of baryons gives the most realistic approach to a physical phenomenon.

3. Baryon Resonances

The crude Skyrme model gives a surprisingly realistic account of baryon resonances [14, 15, 16] and baryon-meson scattering [15, 16]. This encourages one to hope that a more realistic soliton model, perhaps the $\pi\rho\omega$ model suggested earlier, would be competitive with other approaches to these phenomena.

In conclusion, I want to stress that whether or not we might wish otherwise, QCD with the SU(3) gauge group is not logically unique but has an SU(N) generalization. Moreover, although we might wish the Skyrmion theory to be a special miracle uniquely suited to reproducing the properties of SU(3) baryons only, there is no evidence that this is so. There is every evidence that Skyrmion physics with varying couplings reproduces QCD baryons of varying N. Basic facts of life of Skyrmion physics, like the occurrence for weak coupling of a band of states of $I = J = 1/2, 3/2, 5/2, \ldots$ (extending up to large values not realized in nature but present for large N) show that weak coupling in Skyrmion theory corresponds to large N in QCD. Innumerable facts about Skyrmions only make sense if one bears in mind that the classical Skyrme equations, which do *not* depend on N (N is a parameter that scales out of those equations) simple do not know if the physicist studying then is interested in SU(3) or in SU($10^6 + 3$). This emerges only upon quantization. The $1/N$ expansion of QCD, as understood from Feynman diagrams [3,4], is the road map which makes the success of Skyrmion physics rationally comprehensible.

References

[1] T. H. R. Skyrme, Proc. Roy. Soc. A260 (1961) 127.

[2] A. P. Balachandran in: Solitons in Nuclear and Elementary Particle Physics, ed. A. Chodos et al. (World Scientific, Singapore 1984).

[3] G. 't Hooft, Nucl. Phys. B72 (1974) 461, B75 (1974) 461.

[4] E. Witten, Nucl. Phys. B160 (1979) 57.

[5] E. Witten, Nucl. Phys. B223 (1983) 422.

[6] T. H. R. Skyrme, [1]; N. K. Pak and H. C. Tze, Ann. Phys. 117 (1979) 164; A. P. Balachandran, V. P. Nair, S. G. Rajeev, and A. Stern, Phys. Rev. D27 (1983) 1153; E. Witten, Nucl. Phys. B223 (1983) 433.

[7] E. Witten, Nucl. Phys. B223 (1983) 433; E. Guadagnini, Princeton preprint (1983).

[8] J. Wess and B. Zumino, Phys. Lett. 37B (1971) 95.

[9] D. Finkelstein and J. Rubinstein, J. Math. Phys. 9 (1968) 1762.

[10] E. Witten, Nucl. Phys. B223 (1983) 433.

[11] A. P. Balachandran, V. P. Nair, S. G. Rajeev, and A. Stern, Phys. Rev. Lett. 49 (1982) 1124, Phys. Rev. D27 (1983) 1153.

[12] G. Adkins, C. Nappi, and E. Witten, Nucl. Phys. B228 (1983) 552; M. Rho, A. S. Goldhaber, and G. E. Brown, Phys. Rev. Lett. 51 (1983) 747; A. D. Jackson and M. Rho, Phys. Rev. Lett. 51 (1983) 751; G. Adkins and C. Nappi, Nucl. Phys. B223 (1984) 109.

[13] G. Adkins and C. Nappi, Phys. Lett. 137B (1984) 251.

[14] C. Hajduk and B. Schwesinger, Stony Brook preprint (1984); J. Dey and J. Le Tourneux, Montreal preprint (1984); A. Hayashi and G. Holzwarth, Siegen University preprint (1984); L. Biedenharn, Y. Dothan, and M. Tarlini, University of Texas preprint (1984).

[15] J. Breit and C. Nappi, IAS preprint (1984).

[16] H. Walliser and G. Eckart, Siegen preprint (1984); A. Hayashi, G. Eckart, G. Holzwarth, and H. Walliser, Siegen preprint (1984).

[17] E. Guadagnini, Princeton preprint (1983); J. Bijnens, H. Sonoda, and M. B. Wise, Cal. Tech preprint 68-1096 (1984); G. Adkins and C. Nappi, Princeton preprint, to appear.

[18] R. L. Jaffe, Phys. Lett. 38 (1977) 195.

[19] A. P. Balachandran, F. Lizzi, V. G. J. Rodgers, and A. Stern, Syracuse University preprint SU-4222-277 (1983); F. Lizzi in: Solitons in Nuclear and Elementary Particle Physics, ed. A. Chodos et al. (World Scientific, Singapore 1984).

[20] V. Rubakov, Nucl. Phys. 203 (1982) 311, JETP Lett. 33 (1981) 644; C. G. Callan, Jr., Phys. Rev. D25 (1982) 2141, Phys. Rev. D26 (1982) 2058, Nucl. Phys. B212 (1983) 391, and in Problems in Unification and Supergravity, ed. G. Farrar and F. Henyey (AIP, 1984).

[21] C. G. Callan, Jr., and E. Witten, Nucl. Phys. B239 (1984) 161.